European Studies in Philosophy of Science

Volume 10

This new series results from the synergy of EPSA - European Philosophy of Science Association - and PSE - Philosophy of Science in a European Perspective: ESF Networking Programme (2008–2013). It continues the aims of the Springer series "The Philosophy of Science in a European Perspective" and is meant to give a new impetus to European research in the philosophy of science. The main purpose of the series is to provide a publication platform to young researchers working in Europe, who will thus be encouraged to publish in English and make their work internationally known and available. In addition, the series will host the EPSA conference proceedings, selected papers coming from workshops, edited volumes on specific issues in the philosophy of science, monographs and outstanding Ph.D. dissertations. There will be a special emphasis on philosophy of science originating from Europe. In all cases there will be a commitment to high standards of quality. The Editors will be assisted by an Editorial Board of renowned scholars, who will advise on the selection of manuscripts to be considered for publication.

More information about this series at http://www.springer.com/series/13909

Florian J. Boge

Quantum Mechanics Between Ontology and Epistemology

Springer

Florian J. Boge
Interdisciplinary Centre for
Science and Technology Studies (IZWT)
Bergische Universität Wuppertal
Wuppertal, Germany

ISSN 2365-4228 ISSN 2365-4236 (electronic)
European Studies in Philosophy of Science
ISBN 978-3-030-07093-9 ISBN 978-3-319-95765-4 (eBook)
https://doi.org/10.1007/978-3-319-95765-4

Softcover re-print of the Hardcover 1st edition 2018

This Springer imprint is published by the registered company Springer Nature Switzerland AG
The registered company address is: Gewerbestrasse 11, 6330 Cham, Switzerland

For Nikki and Junis

Acknowledgments

I would first of all like to thank Andreas Hüttemann for the constant support and encouragement, for many helpful comments, for interesting discussions, and for valuable advice on practical and philosophical matters and Claus Kiefer for an open mind and a supportive attitude and for teaching me some deep insights into the foundations of quantum mechanics.

Special thanks are due to my wife for hanging in there with me and to her and my parents for the emotional and financial support throughout the years.

And otherwise, I would like to thank (in "almost" alphabetical order by last name): Andreas Bartels for helpful commentary; Harvey Brown for exchange and support; George Darby and David Glick for several discussions; Christian Feldbacher for his support in EPSA 15; Florian Fischer for continuing exchange and practical support; Simon Friederich for a number of discussions; Alexander Gebharter for demanding my expertise on occasion and introducing me to others in the field; Sheldon Goldstein for encouraging feedback; Siegfried Jaag, Marie I. Kaiser, and Christian Loew for many helpful comments and discussions over the years; Rochus Klesse for sharp and helpful commentary on issues of the many worlds interpretation; Anna Marmodoro for outstanding support in preparation of and during my stay in Oxford; Paul Näger for some thought-provoking discussion in SOPhiA 15; Nina Retzlaff for involving me in her enjoyable workshop on causality and quantum mechanics; Markus Schrenk for stimulating intellectual exchange, constant encouragement, and financial support and for being a great employer; Gerhard Schurz for much of my philosophical background knowledge; Chris Timpson for a dense series of enlightening advice; Paul Weingartner for interesting impulses…, and of course everyone else who contributed in some way or other and who I may have forgotten.

A significant number of revisions on the manuscript for this book was made during my employment with the research unit *The Epistemology of the Large Hadron Collider*, funded by the German Research Foundation (DFG; Grant FOR2063). I hence acknowledge the funding.

Contents

List of Figures

List of Tables

List of Acronyms

ADT	—	Armstrong, Dretske, and Tooley
AQFT	—	Algebraic quantum field theory
AUT	—	Autonomy, assumption in Bell's theorem
BCS	—	Bardeen, Cooper, and Schrieffer
BM	—	Bohmian mechanics
BM_H^N	—	Humean-nomological Bohmian mechanics
BM_O^N	—	Ontological-nomological Bohmian mechanics
BSA	—	Best system analysis
CEL	—	Causal Einstein locality
CHSH	—	Clauser, Horne, Shimony, and Holt (inequality)
CI	—	Collapse interpretations
CI_H^N	—	Humean-nomological collapse interpretations
CI_O^N	—	Ontological-nomological collapse interpretations
CMC	—	Causal Markov condition in causal graph theory
CSL	—	Continuous spontaneous localization model(s)
CSLm	—	Mass-density ontology for CSL
DAG	—	Directed acyclic graph
Δ	—	Overlap lemma in the PBR theorem
DE	—	Dirac equation
DEC	—	Desideratum of epistemological clarity
DOC	—	Desideratum of ontological clarity
EE-link	—	Eigenvalue-eigenstate link
EPR	—	Einstein, Podolsky, and Rosen
EPRB	—	Einstein, Podolsky, Rosen, and Bohm
EQM	—	Epistemological interpretation of QM
F	—	Faithfulness condition in causal graph theory
FACT	—	(Local) Factorization, assumption in Bell's theorem
FUNC	—	Functional composition rule, assumption in the KS theorem
FUNC*	—	Contextualized functional composition rule
GNS	—	Gelfand, Naimark, and Segal
GR	—	General relativity

GRW	—	Ghirardi, Rhimini, and Weber (model)
GRWf	—	Flash ontology for GRW
GRWm	—	Mass-density ontology for GRW
INVAR	—	Invariance, assumption in Hardy's theorem
KB	—	Knowledge balance principle, assumption in Spekkens' toy model
KGE	—	Klein-Gordon equation
KS	—	Kochen and Specker
LC	—	Local causality, assumption in Bell's theorem
LOCC	—	Local operations and classical communication
MAC	—	Minimal adequacy criterion
MWI	—	many worlds interpretation
NQMC	—	Non-quantum magnitude claim
OI	—	Outcome independence, assumption in Bell's theorem
OM	—	Ontological model
ONB	—	Orthonormal basis
OP	—	Outcome problem
Outc.	—	Assumption of well-defined outcome probabilities in the PBR theorem
PBR	—	Pusey, Barrett, and Rudolph
PC	—	Possibilistic completeness, assumption in Hardy's theorem
PCC	—	Principle of the common cause
PCC*	—	Weak principle of the common cause
P.-Indep.	—	Preparation independence assumption in the PBR theorem
PI	—	Parameter independence, assumption in Bell's theorem
Pos.	—	Positivity assumption in a variant of the PBR theorem
POVM	—	Positive operator valued measure
PP	—	Projection postulate
PQM	—	Healey's pragmatist interpretation of QM
Prod. 1	—	Assumption of factorizing quantum states in the PBR theorem
Prod. 2	—	Assumption of factorizing probabilities in the PBR theorem
PVM	—	Projector-valued measure
QED	—	Quantum electrodynamics
QFT	—	Quantum field theory
QM	—	Quantum mechanics
QSA	—	Quantum state assignment
QSA$^+$	—	QSA plus all available information in non-quantum terms
ROD	—	Restricted ontic indifference, assumption in Hardy's theorem
SE	—	Schrödinger equation
SEP	—	Einstein separability
Sep.	—	Separability assumption in the PBR theorem
Sep.*	—	Weaker separability assumption in a variant of the PBR theorem
SQUID	—	Superconducting quantum interference device
SR	—	Special relativity

SSE	—	Stationary Schrödinger equation
TDSE	—	Time-dependent Schrödinger equation
TOT	—	Totality assumption in Hardy's theorem
VD	—	Value definiteness, assumption in the KS theorem
VLOC	—	Value locality, assumption for a hidden variables model
VSEP	—	Value separability, assumption for a hidden variables model
vNE	—	von Neumann equation

Chapter 1
Introduction

Physics does not consist only of atomic research, science does not consist only of physics, and life does not consist only of science.

—Schrödinger in a letter to W. Wien, 1926 (cf. Moore 2015, p. 226)

1.1 Why Philosophize About Quantum Mechanics?

Sometimes upon watching a fly bump its head against a window over and over, we might feel unpleasantly reminded of our human attempts to understand what we call 'the natural world'. In a more benevolent analogy, our bafflement about many phenomena that modern science has predicted, or even brought into existence, may remind us of the bafflement of astronaut Dr. David Bowman in Kubrick's classic *2001: A Space Odyssey* upon gazing into the black monolith. A remarkably successful scientific theory which reflects the analogy more clearly than any other, as it has notoriously managed to escape our intuitive grasp, is *quantum mechanics* (QM). Its predictive and technical-implementational success in the twentieth century, witnessed in the following quotes, has inspired talk of a 'scientific revolution':

> Quantum mechanics is the most accurate theory in all of science. An extreme test is the calculation of the 'gyromagnetic ratio of the electron' with a precision of one part in a trillion. [...] In fact, one-third of our economy involves products based on quantum mechanics. (Rosenblum and Kuttner 2011, p. 116)

> The spectacular advances in chemistry, biology, and medicine—and in essentially every other science—could not have occurred without the tools that quantum mechanics made possible. Without quantum mechanics there would be no global economy to speak of, because the electronics revolution that brought us the computer age is a child of quantum mechanics. So is the photonics revolution that brought us the Information Age. The creation

F. J. Boge, *Quantum Mechanics Between Ontology and Epistemology*, European Studies in Philosophy of Science 10,
https://doi.org/10.1007/978-3-319-95765-4_1

of quantum physics has transformed our world, bringing with it all the benefits—and the risks—of a scientific revolution. (Kleppner and Jackiw 2000, p. 893)

And yet there is still no consensus as to how to interpret QM. Feynman (1965, p. 129) once famously said: "I think I can safely say that nobody understands quantum mechanics." It is as if we had learned how to *use* the monolith, or to *fly around* the window, even though we do not understand how any of the two works.

Sean Carroll (2013) has indeed gone so far as to call the lack of consensus, reflecting the general lack of understanding, an *embarrassment*, alongside the fact that the interpretation which *is* seemingly favored by the most physicists, as reflected e.g. in polls by Tegmark (1998) and Schlosshauer et al. (2013), is what is usually called the 'orthodox' or 'Copenhagen' interpretation.

What is wrong with this interpretation? John Bell, quite in the spirit of the distinction between use and understanding that we have appealed to, once claimed that "'The Copenhagen interpretation' is a very ambiguous term. Some people use it just to mean the sort of practical quantum mechanics that you can do—like you can ride a bicycle without really knowing what you're doing." (Mann and Crease 1988, p. 86) In other words: Most physicists do not *care* about how to understand QM, and they do not have to in order to apply the theory successfully.

A second aspect to Bell's quote is that it is notoriously difficult to pin down what exactly should count as the 'Copenhagen' or 'orthodox' interpretation. And in fact, we shall not even use the two notions synonymously in what follows. As regards the 'Copenhagen interpretation', elements from the somewhat differing views of some of the founding fathers of QM are usually summarized under this label, which views may, upon closer inspection, not even all be compatible with one another (cf. Gomatam 2006; Howard 2002; Strapp 1972). But what is worse is that some of these views may even be disputed to *count* as 'interpretations' at all, and may rather be viewed as a kind of 'instruction manual' for "the sort of practical quantum mechanics" that Bell was talking about.

In his 2005 *A Philosopher Looks at Quantum Mechanics (Again)*, Hilary Putnam, for instance, recalls an undisclosed physicist colleague of his[1] beginning a lecture with the words "*There is no* Copenhagen interpretation of quantum mechanics" (my emphasis—FB), after a period of fourteen years of continued discussion on the subject. In a similar spirit, physicist David Mermin (1989) once summarized his understanding of the Copenhagen view by one simple commandment: "Shut up and calculate!" But just as physicists like Bell and Mermin did (or do, in the latter case), many philosophers feel at unease with this attitude and prefer to "rather celebrate the strangeness of quantum theory than deny it [. . .]." (Mermin 1989, p. 9)

This unease has inspired what Ney and Albert (2013, p. xi) call a "joint project" between physics and philosophy. This book is intended to provide a contribution to that project. To this end, we will contrast attempts of furnishing QM with an *ontology* that can accommodate everything that seems 'strange' or 'weird' about it

[1]Putnam's intellectual autobiography reveals that it was M. Gell-Mann (cf. Putnam 2015, p. 68).

with attempts of harvesting QM's *epistemological* implications instead. These two general approaches may be seen, in a sense, as rival programs. But of course, in both approaches there are also subtle *internal* differences. There is, in particular, an important subdivide among the interpretations that broadly count as 'epistemic', as we shall see in Chap. 4, as some of these try to preserve our common sense intuitions to the extent that this is possible while others radically challenge those very intuitions.

More precisely, it has become fashionable again in recent years to attempt to interpret QM epistemically by introducing new formal tools, including what is typically known as *hidden variables*, 'hidden' (true) states of micro-systems not described by any element of the QM formalism. If successful, such an interpretation would allow to view QM as reflecting a lack of knowledge about an otherwise rather 'classical' underlying reality, i.e., one that can easily be accounted for in terms of the configurations of systems described by those hidden variable-descriptions and allows to explain quite directly why we see houses, chairs, tables, and so forth.

On the other hand, there have been more recent advances in making sense of QM in terms of knowledge (or even *belief*) without explicitly assuming additional such variables or states. Approaches of this kind, however, come with the implication that there may be *irreducible* limitations as to what can be known or even *said* about any given object of investigation, or about the world 'as a whole' and our relation to it. This, of course, reminds us of certain strains in traditional epistemology, so it will be interesting too see how these approaches fit with those strains.

Opposed to this, we have the well-established schemes for interpreting QM by providing some ontology of the 'quantum world', i.e. explaining what the world is (or may be) like *according to* QM. But none of these schemes has surfaced as a clear winner so far (Carroll's embarrassment), and one should ask *why* exactly this is. This will be a task we have to face as well, in the course of this book.

Our road to all of these interpretations will be an 'unorthodox' one, as we will basically proceed from intuitive to less and less intuitive by some standards. This means that certain well-known results will not be discussed right away, but only in the appropriate context, to dispel the appeal of specific kinds of interpretation. We start off, in the next chapter, with a gentle introduction to the general physics and the general philosophical concerns raised by the theory and by experiment. This should give philosophers who are only loosely familiar with the subject matter a chance to tag along. Readers well-acquainted with the subject may, of course, feel free to skip large parts, though it may be useful to see which conventions are being applied.

Additionally, in a mathematical appendix (A) we provide the mathematics indispensable for an understanding of the discussion, and a little more beyond that for the interested reader. It may be advisable to read (part of) that appendix first, and then scroll back to it whenever the concepts are needed in the context of discussion.

Finally, we note a bunch of philosophical introductions to the subject that also provide short (and in some cases maybe more accessible) introductions to the mathematics, in varying detail and depth. Particularly simple and accessible ones are Albert (1992, chapter 2), Ney (2013), and Maudlin (2011, p. 260 ff.), a more comprehensive one is Hughes (1989), and a very compact one is given by Redhead (1987). A non-philosophical introduction which may be recommended to absolute beginners is Susskind and Friedman (2014); and of course other textbooks such as those of Shankar (1994) or Sakurai (1994) provide detailed and 'pedagogically suitable' expositions.[2]

1.2 A Few Words on Interpretation

Considering different interpretations of QM and their philosophical implications inevitably raises the question of *what constitutes* an interpretation. The explication of this concept is, however, a notoriously difficult task, as has been pointed out e.g. by Jammer (1974, p. 9). Moreover, Jaeger (2009, p. 96) notes "a tendency [...] to inject philosophical biases or concerns into the very conception of interpretation," referencing the differing views of Bub (1974, p. 143), Teller (1995, p. 5), and Mittelstaedt (1998, p. 1).

It is nonetheless desirable, and should be possible, to give a sufficiently general characterization of what is meant by 'interpretation', without thereby siding with either of the philosophically more involved conceptions or injecting own philosophical biases. To do so, let us begin by contrasting two characterizations from two comprehensive works on the philosophy of QM, both from different eras of the philosophy of science, which both purport to present the standard or most widely accepted account at the time.

The first one is Jammer's account as explained in his 1974 classic *The Philosophy of Quantum Mechanics*. Jammer, obviously under the influence of logical empiricism, thinks that "a physical theory is a partially interpreted formal system." (p. 10) This he explains by dividing a given theory T into an abstract *formalism* F, "the logical skeleton of the theory, [...] a deductive, usually axiomatized calculus devoid of any empirical meaning" and a set R of *correspondence rules*, which connect F to experience. F not only contains constants and mathematical expressions, but also nonlogical terms, like 'particle' and 'state function' (cf. ibid.). The 'partial interpretation' now consists in connecting elements of F to experience *via* R, which hence leads to the partially interpreted F_R. A different set R' of correspondence rules would lead to a different partially interpreted theory $F_{R'}$ accordingly. But F also includes non-primitive, defined terms, which are the *theoretical terms* not directly connected to experience by R.

This approach of Jammer's is an expression of the *syntactic* view of theories, rooted especially in the work of Carnap (cf. 1956; 1958). As we can see, the theory is treated here as a predominantly *linguistic* entity, and a lot of weight is put on the direct connection to experience. Nice and tidy as this approach may seem, it raises

[2]German readers may also profit from Nortmann (2008) and Friebe et al. (2015), both of which are philosophical introductions to the topic.

various questions about axiomatizability, the observational-theoretical dichotomy, the appropriate tools of formalization... which is why it was abandoned by most philosophers of science from the late sixties on, most notably by Hempel, one of its chief proponents up to that point (cf. Suppe 2000, pp. 102–103, for some historical details).

A prominent alternative is the so-called *semantic* view of theories, rooted in the works of Suppes (e.g. his 2002, for a detailed exposition), and endorsed more recently by Ruetsche in her 2011 *Interpreting Quantum Theories*. Ruetsche paraphrases the task of interpretation according to this view as follows:

> [T]o interpret a physical theory is to characterize the worlds possible according to that theory. Two phases of this characterization can be distinguished. One phase identifies the theory's structures: its states, observables, and dynamics. The other characterizes the physical situations that count as models of the theory so structured. Interpretation is an exercise in nomic articulation: a theory's laws guide the characterization of its possible worlds; the interpretation of a theory is at the same time an explication of the notion of nomological possibility allied with the theory. (Ruetsche 2011, p. 9)

Ruetsche's version of the semantic approach exhibits a strong emphasis on modality, predominant in the philosophy of science at least since the writings of Lewis and Kripke in the 1970s (cf. Soames 2014, p. 139 ff.). The semantic view roughly construes a theory as a family of *models*, i.e. *not* in the first place a *linguistic* entity. This does not mean that theories generally cannot be given a formulation in some language or be axiomatized in favorable cases (e.g. the discussion in da Costa and French 2003, p. 27 on 'Suppes predicates'); it is just that the formulation *is* not the theory but the family of models is. The latter is systematically prior to the former.

The semantic view is still quite popular today, but it equally raises a bunch of questions. For instance: it raises disagreement about how the theory, the family of models, precisely relates to its formulation (cf. Chakravartty 2001, p. 326), what the *ontological status* of the models is, or how they *represent* their target systems (cf. Frigg 2006, p. 50); and Ruetsche's own formulation obviously raises further questions about the ontology and semantics of *modal* statements.

Helping ourselves to "uncomplicated and appealing (if vague) ideas", as does Ruetsche (2011, p. 6) in her acceptance of the semantic view, we here choose a suitably neutral position in the light of the problems of syntactic *and* semantic approaches. This neutral position constitutes a *pragmatic view* of theories if you will, according to which "a theory is a more or less amorphous entity consisting of sentences, models, problems, standards, skills, practices, etc." (Mormann 2007, p. 137)

Besides problems and discrepancies, there are also *commonalities* between Jammer's and Ruetsche's approaches though, that should probably figure in any account of theory-interpretation in physics. Both recognize that there is a part where the abstract formal or structural ingredients of the theory are identified, including its states and dynamics. And there is a part of connecting the abstract with the concrete, and with empirical content. In both approaches, there is also a residue of the theory which is not directly connected to experience; certainly not all elements of the nomological structure implied by a family of models are directly accessible to sense experience, and neither, of course, are the entities presumed to be the referents of theoretical terms in the syntactic approach.

We can hence distinguish two 'levels' of interpretation. In the spirit of Redhead (1987, p. 44), and to avoid a confusion with the specific use of 'partiality' in the syntactic approach, we will call a *minimal interpretation* of QM any set of rules or postulates that establishes what the states and the dynamics of the theory are, and how these connect to experience—basically the *instruction manual* mentioned above. An interpretation that additionally tells us "how [...] the world [could] possibly be the way this theory says it is", and thereby answers van Fraassen's (1991, p. 193) "foundational question *par excellence*," will be called a *non-minimal interpretation*. With this amount of clarity on general philosophy of science-issues, we should be able to make some first sense in a concrete investigation of QM's foundations.

Chapter 2
Some Quantum Mechanics, Its Problems, and How Not to Think About Them

...if one is not shocked about the quantum theory at first, one cannot possibly have understood it.

—Attributed to N. Bohr by Heisenberg (cf. 1969, p. 241)

2.1 Non-relativistic Quantum Mechanics: Off to a Gentle Start

2.1.1 A Tale of Waves and Particles?

QM is notoriously associated with a certain 'strangeness' or 'weirdness' (e.g. Rosenblum and Kuttner 2011, p. 4; Davies 2004, p. 11) which stems, in the first place, from the divergence of the phenomena that it describes and predicts from our pre-quantum expectations. By 'phenomenon' we here mean, for practical reasons, something along the lines of Bogen and Woodward (1988, pp. 305–306), according to whom the phenomenon is rather what the theory predicts, which may not even be observable, whereas the *data* are the observables that serve as evidence for phenomena. A nice summary of the distinguishing features between the two terms in Bogen and Woodward's understanding is given by da Costa and French (2003, p. 69; references omitted):

> Data can be straightforwardly observed, are idiosyncratic to particular experimental contexts, are the result of a "highly complex and unusual coincidence of circumstances", and are "relatively easy to identify, classify, measure, aggregate, and analyze in ways that are reliable and reproducible by others". Phenomena, on the other hand, are not observable, are not idiosyncratic, have "stable, repeatable characteristics which will be detectable by means of a variety of different procedures" involving different kinds of data, and in general are constant and stable across different experimental contexts.

F. J. Boge, *Quantum Mechanics Between Ontology and Epistemology*, European Studies in Philosophy of Science 10,
https://doi.org/10.1007/978-3-319-95765-4_2

Thus in the present context our use obviously diverges from the original Greek φαινόμενον, which denotes something apparent to the senses (cf. Perschbacher 1990, p. 425), and with it from certain philosophical traditions. This is at least somewhat in line with Pauli's (1961, pp. 93–95) thinking on the matter, who suggested that in principle any kind of conscious content such as perceptions, thoughts, and ideas could count as phenomena, but that *physical* phenomena are complex and require a theoretical-interpretational background for their description, involving especially previous experience with certain apparatuses. Note, however, that we are not thereby taking sides on epistemological issues here; the terminological choice is a mere *convenience* that allows loose talk of 'phenomena' as it occurs in the literature, without further explanation.

In essence the strangeness of QM thus lies, on the one hand, in the fact that we may come to think of something as 'fishy' while looking at a computer screen attached to some specific apparatus, say, or comparing lists of numbers which we regard as 'measurement outcomes' and hence as data indicative of some phenomenon, and that we might then feel the need to contemplate how to fit these phenomena into the preoccupations we endorse based on the remainder of our total experience. As Fine (1989, p. 130) puts it: "What surprises or puzzles is relative to context, which includes at least psychological set and background beliefs." And on the other hand, the strangeness stems from the fact that the theory we have come to use (with remarkable success) to accommodate all of these unusual phenomena predicts many further ones.

QM was originally constructed[1] as a theory of radiation and the fundamental constituents of matter, i.e. atoms, electrons, and further, later discovered subatomic particles. For now we restrict out attention to non-relativistic 'single particle' QM, since the central problems thus established pervade also the more advanced forms. Talk of 'single particle QM' certainly raises the question of what, precisely, is meant by 'particle'. Intuitively one may think of particles as tiny little 'dot-like' entities, tiny carriers of properties, the *fundamental* ones of which are thought to be no further divisible. These intuitions are borrowed from the concept of a 'mass point' endorsed in pre-quantum physics. In the light of what is established in this chapter, however, we will already find the need for revision, and a more cautious treatment will be provided in Sect. 2.1.3.

The first kind of 'fishyness' can be established very well from this intuitive particle-notion and the fact that the very same entities that we can think of, in this sense, as particles in *some* contexts seem to behave like *waves* in others. An example to this effect can be set up by appeal to *polarization*, a property originally familiar from *electromagnetic waves*.

From his equations of electromagnetism (in the vacuum), Maxwell derived a wave equation, $\Delta \boldsymbol{E} = \frac{1}{c^2}\frac{\partial^2}{\partial t^2}\boldsymbol{E}$, concluding that electromagnetic disturbances,

[1]Caveat: Parts of our exposition of the historical details will be quite 'whig' in the sense of Butterfield (1931), meaning that one "studies the past with reference to the present[...]." (ibid., section 2)

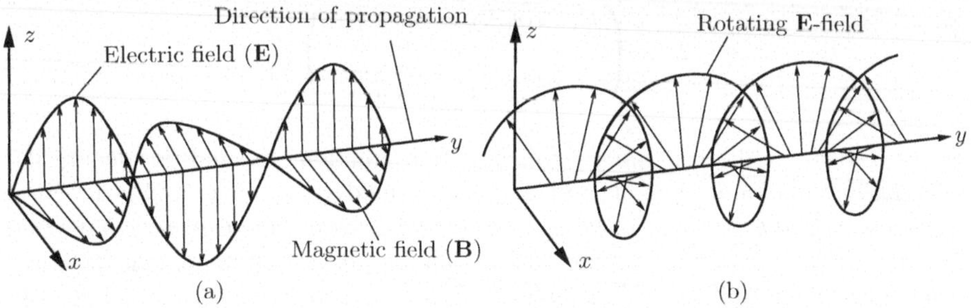

Fig. 2.1 (**a**) A simple electromagnetic wave with linear polarization. (**b**) Electric field of a circularly or elliptically polarized electromagnetic wave

fluctuating fields (or 'waves'), should, under suitable conditions, propagate freely in space. Here, Δ is the *Laplacian operator* (that we will meet with again shortly), $\boldsymbol{E}$ the electric field, and c the speed of light. An analogous equation holds for the magnetic field ($\boldsymbol{B}$). Due to experimentation by Hertz, this consequence of the Maxwellian theory could be confirmed in the late nineteenth century (cf. Harman 2003; Newton 2007, p. 141). And according to electromagnetic theory, *visible light* is nothing but electromagnetic radiation, electromagnetic fields traveling as waves, differing from other sorts of electromagnetic radiation only in frequency, i.e. in the number of oscillations per second (e.g. Griffiths 1995, p. 306; Rae 2004, p. 4).

An electromagnetic wave can be visualized by a combination of two space and time dependent vectors ('pointers'), hence each constituting a *vector-field*, and depicting the oscillating amplitudes of the electric and magnetic fields $\boldsymbol{E}$ and $\boldsymbol{B}$ respectively (Fig. 2.1). These inevitably come together in an electromagnetic wave phenomenon, according to Maxwell's equations, and can eventually be identified as two aspects of the same phenomenon in special relativity (cf. Jackson 1990, p. 558). We can attribute a *polarization* to the electromagnetic wave as the direction along which the electric field-vector oscillates.

In the simplest case, the electromagnetic wave is polarized *linearly*, as depicted in Fig. 2.1a. This means that the wave has one $\boldsymbol{E}$-vector oscillating in one plane (here: the x-z-plane), instead of many superimposed $\boldsymbol{E}$-vectors oscillating in all kinds of directions, in which case the wave would be *un*polarized. Moreover, $\boldsymbol{E}$ does not rotate around the axis of propagation in this simplest case; otherwise it would be called either *circularly* or *elliptically* polarized, depending on the 'phase' of its components (cf. Fig. 2.1b).

We can make things precise by representing the $\boldsymbol{E}$-vector of polarized light by a *column*[2]

[2]For the wave depicted in Fig. 2.1a the x-component is actually zero, as its $\boldsymbol{E}$-field only oscillates in z-direction, whereas the $\boldsymbol{E}$-field of a more general wave as described by Eq. (2.1) would oscillate in some direction in the x-z-plane for $\varphi_0 = 0$ and $E_x \neq 0$.

$$\boldsymbol{E} \doteq \begin{pmatrix} E_x \cdot \cos(ky - \omega t - \varphi_0) \\ 0 \\ E_z \cdot \cos(ky - \omega t) \end{pmatrix}, \tag{2.1}$$

where ω denotes the (*angular*) *frequency* of oscillation, E_x and E_z denote the *amplitudes* ('maximum heights') of the electric field vector in y- and z-direction respectively, and $k = \frac{2\pi}{\lambda}$ denotes the *wave-number*, with λ the *wavelength*, i.e. the distance between two neighboring peaks of the wave. It then holds that the wave is polarized linearly if $\varphi_0 = 0$, circularly if $\varphi_0 = \frac{\pi}{2}$ and $E_x = E_z$, and eliptically else (cf. Aharonov and Rohrlich 2005, p. 29; Demtröder 2009, pp. 196–199). The arguments of the cosines are what we have called the *phase* above, and φ_0 is called a *relative* phase between both components.

Additionally, whether the wave rotates clockwise or counter-clockwise in the direction of propagation depends on the signs of the components displaced by φ_0 (here a minus-sign in front of, say, the z-component would mean that is rotates counter-clockwise). One uses trigonometric functions (here: the cosine) in the description of the waves since, for one, their graphs actually *look like* (smooth, harmonic) waves, and, more importantly, since they solve the corresponding (differential) wave-equation that can be found on purely theoretical grounds. Any arbitrarily 'ugly'-looking wave can then be 'synthesized' out of these trigonometric functions by means of Fourier methods (cf. later), i.e. essentially by imagining waves of different period (of repetition) and amplitude to be put on top of each other, the possibility of which is ensured by the *linearity* of the wave equation.

Let us now consider only the simple case of linearly polarized light. One can produce such light by sending it through a *polarizer*, implemented e.g. by means of a crystal or a grid made out of a reflective materials (cf. Demtröder 2009, p. 253 ff.; Walker et al. 2012, p. 985 ff.). The latter kind of such polarizers will only transmit light with a particular polarization, whence they are occasionally referred to as '*yes-no polarizes*', whereas crystals can be used to split a beam of incident light into two differently polarized beams. The latter fact justifies to generally think of polarizers as resolving a beam of light into two components, "one parallel to [the polarizer's direction—FB], the other perpendicular", and a yes-no polarizer, we can take it, simply "absorbs the perpendicular component[...]." (Maudlin 2011, p. 9)

Assume that we have a beam of light emanent from a yes-no polarizer. This should count as a *preparation* of the light in a polarized state. A second polarizer, then called an *analyzer* (cf. Fig. 2.2), can be used to check for what was prepared: If *no* light passes the analyzer, the polarization of the light cannot be in line with the analyzer's direction of polarization; it should be perpendicular, according to the above considerations. If the beam of light is maximally bright, then its polarization must be exactly in line with the direction of the analyzer, as (almost) all of the light will apparently pass. For any orientation in between, the intensity will in fact go down depending on the degree of misalignment between the two directions of polarization, meaning that *not all of* the light can have passed the analyzer.

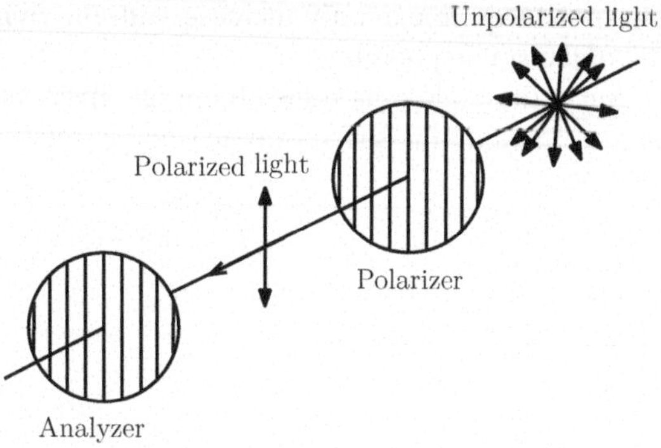

Fig. 2.2 The incident unpolarized light hits the polarizer and the fraction which passes it can only pass the analyzer as well if its polarization is in line with the analyzer's direction (indicated by the vertical lines)

But light, famously, seems to have a strange 'double nature'. In 1905, Einstein used Planck's idea of a *discretization* of energy to explain the so called *photoelectric effect*, the observation that if light is directed onto a metal plate, electrons are emitted from the plate in consequence and in a fashion rather unexpected at that time. For light of a particular frequency, the kinetic energy each electron acquires in the process is the same; it can be given by $E_{\text{kin}} = \hbar\omega - W_e$ (cf. Einstein 1905, p. 146; Demtröder 2010, p. 34), with ω defined as above, $\hbar$ *Planck's reduced constant*, derived from his original quantization constant h $(= 2\pi\hbar)$ which he used to explain black body radiation in terms of quantized energy emission (e.g. Gearhart 2009, pp. 39–40), and W_e the work that needs to be performed on the electrons so that they can exit the binding potential of the metal plate. Thus the energy the electrons acquire is dependent on the *frequency* of the incident light, the number of oscillations per second.

This dependence on frequency was in need of explanation, as from the point of view of the wave picture established above, one should expect that the energy of the electrons varies with the *intensity* I of the incident light-wave instead, that is, with the average energy over time and area transmitted by the wave, which in turn depends on its *amplitude* (on how strongly it 'wiggles').[3] But with greater intensity the *number* of electrons emitted over time increases in the photoelectric effect, not the energy each electron acquires. Einstein explained this by considering the beam to consist of discrete packets of energy $E = \hbar\omega = h\nu$ which he named "Lichtquanten" (Einstein 1905, p. 144), i.e. *light quanta*, and which only later acquired the now common name '*photons*' (cf. Hentschel 2009a, pp. 339–344; Rae 2004, pp. 4–7). The more intense the beam of light, the more photons arrive and the more likely it becomes that electrons are emitted from ('kicked out of') the plate. But since the energy each photon of a particular frequency carries is identical, the energy the

[3]More specifically, the classical expression for intensity can be calculated from the average magnitude of the *Poynting-vector* $\boldsymbol{S} = \frac{1}{\mu_0}\boldsymbol{E} \times \boldsymbol{B}$ as $\langle S \rangle = \frac{1}{\mu_0}\langle EB \rangle = \frac{1}{c\mu_0}\langle E^2 \rangle$, with μ_0 the vacuum permeability, c the velocity of light, and E and B the magnitudes of the electric and magnetic field vectors (e.g. Walker et al. 2012, p. 981).

electrons acquire can only increase with the frequency of the light, i.e. the energy $\hbar\omega$ of the single photons.

Since these photons transmit energy, they can be associated with a momentum $p = \frac{E}{c} = mc$ (c the speed of light), and a (relativistic) mass m. Thus from

$$E = \hbar\omega \quad (\Leftrightarrow \omega = \frac{E}{\hbar}) \tag{2.2}$$

we get,

$$p = \frac{\hbar\omega}{c}. \tag{2.3}$$

A general fact about waves is that the product of wavelength and frequency defines a velocity,[4] i.e., for electromagnetic waves $c = \lambda\nu = \frac{\lambda\omega}{2\pi}$. But inserted into Eq. (2.3) this gives us

$$p = \frac{2\pi\hbar}{\lambda} = \frac{h}{\lambda}. \tag{2.4}$$

$\lambda = \frac{h}{p}$ is called a *de Broglie-wave length*, after Louis de Broglie, who was the first to also speculate about the wave nature of *matter*, in virtue of the fact that light waves could exhibit properties of material particles (cf. de Broglie 1925, p. 92; Landau and Lifshitz 1965, p. 52). The wave number k was defined by $k = \frac{2\pi}{\lambda}$ above, so from (2.4) we obtain

$$p = \hbar k \Leftrightarrow k = \frac{p}{\hbar}. \tag{2.5}$$

This means that, from the relation between energy and frequency, one finds that the wave number k is proportional to a momentum. This seems like a weird admixture of concepts describing waves, i.e. spread out fluctuating entities, and concepts describing particles, i.e. tiny concentrated (typically massive) objects that are intuitively thought to be impenetrable and incapable of spreading out.

[4]The frequency has the dimension 1/time, the wavelength is a length, so the product has the dimension length/time, which is that of a velocity. Strictly speaking, what we appeal to above is the so called *phase velocity* which may in fact exceed the speed of light, c, when multiple waves travel together as a 'packet', a narrow lump of oscillations. So it is basically the 'false kind of velocity'. The velocity of interest is the *group velocity* $\frac{d\omega}{dk}$, which, however, in the present case of a single wave coincides with the phase velocity, so that no harm comes from the simplification. Note that no energy or information can be transported with phase velocity: pictorially it describes how fast the 'ripples' in the packet propagate, but the ripples 'diminish' while approaching the boundary of the packet, and so no energy or information is transmitted with a speed $> c$ (e.g. Griffiths 1995, pp. 47–48; Griffiths 1999, p. 399 for details, examples, and illustrations). This is, of course, important for consistency with relativity.

Moreover, if we take the photoelectric effect to be suggestive of the existence of 'light particles' in a straightforward sense, what becomes of the polarization? Is it associated with every single such 'photon'? In fact there are reasons to think so. De-exciting atoms provide a source of light that can be used in such a way that typically only one photon (packet of energy) is emitted at a time, i.e., only 'single dots', minimal energy transmissions from the atom to a suitable measuring device, will be measured at any instant of time (cf. Grangier et al. 1986).

Sending a beam of unpolarized light through a polarizer one obtains a comparatively less 'bright' beam (its intensity goes down), and if we construe the beam as being made up of photons, the intensity can be construed as having gone down because photons with the 'wrong' polarization have been 'sorted out' by the polarizer. For yes-no polarizers one hence either measures, at each instant of time, a photon which passes the polarizer or one does not. But the case can be made even stronger when one uses a calcite crystal that *splits* a beam of unpolarized light into two beams with mutually perpendicular polarizations. Such a crystal can hence be used to construct a device which gives off a signal after registering a photon in either channel and thus (indirectly) determines its polarization (e.g. Rae 2004, pp. 19–23). Photons themselves seem to possess a property that is sensitive to the adjustments of polarizers.

What, however, still compels us to believe that *waves* play any role here at all? A trademark of waves are so-called *interference phenomena*, and a crucial experiment to reveal such phenomena is the (infamous) *double slit experiment*. Consider that when light from a suitable source is emitted onto a metal plate with a narrow slit carved into it, passes the slit and hits some screen behind the slit, it leaves a bright pattern, mostly in the center of the screen but also somewhat distributed off-center in a characteristic way (cf. Fig. 2.3a). This is so far still compatible with a particle interpretation as suggested by the observations from the photoelectric effect. Parts of a total beam of photons might be scattered at the slit in such a way that most of them land in the central area but some also in the de-central bright places.

But a more compelling explanation can be given in terms of waves: Consider a plane wave being emitted from the source and hitting the slit. In accordance with Huygens' principle (cf. Walker et al. 2012, p. 1072), a new set of (spherical) waves forms due to the interaction with the slit. Depending on path difference, these waves will *interfere* with each other either *constructively* or *destructively*, i.e. in some places the peaks of two wave-crests will coincide, thus giving rise to an 'even more peaked' wave, and in some places a peak and a trough will coincide, thus erasing the wavieness altogether. And in a more thorough treatment, the existence of the off-central part of the distribution is thus straightforwardly *predicted* by such a wave-approach, as a part of an *interference pattern*.

Using a metal plate with *two* tiny slits instead, a *double slit*, one can add credence to the wave approach, since in the double slit case, an even more obvious interference pattern appears (cf. Fig. 2.3b). This pattern is also familiar from similar experiments with water- or sound waves. But as we have noted above, the energy of a beam of light is transmitted in a dot like fashion, i.e. related to the 'number of

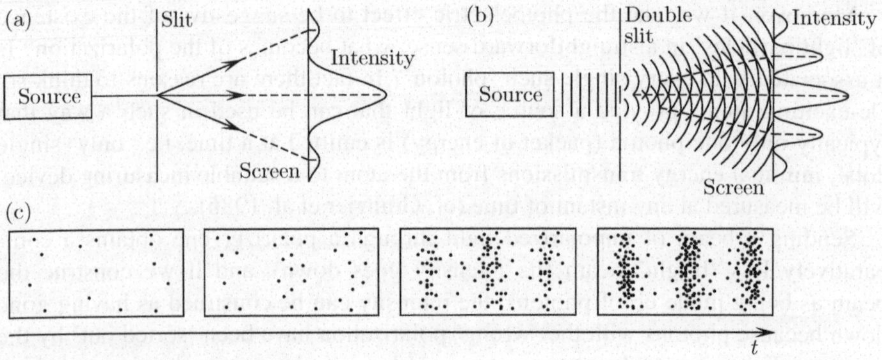

Fig. 2.3 (**a**) The light emitted from the source passes the slit and gives rise to a pattern which can be interpreted in terms of waves as well as particles. (**b**) Using a double slit, a pattern of intensity characteristic of wave interference is observed. (**c**) An interference pattern successively builds up as observed e.g. in an experiment with electrons by Tonomura et al. (1989) (three central maxima are illustrated here)

photons' it consists of; so one is confronted with the difficulty of incorporating the particle-like aspects into this behavior.

At this point one might be tempted to think that it must be the whole *collection* of particles that travels in a wavelike fashion, when together in a beam, and hence accounts for the interference pattern. This is not a viable interpretation of the situation though, since one can adjust the incident beam to an appropriately low intensity so that only single dots appear on the screen, one after the other. But one will *still* observe that, after sufficient time, the distribution of these dots looks exactly like the original interference pattern (cf. Fig. 2.3c). It seems that somehow each single photon "*interferes only with itself*" (Dirac 1930, p. 9; my emphasis—FB), which also avoids difficulties arising from energy-conservation and the apparent *annihilation* of photons that would be required to otherwise account for the interference pattern (cf. ibid.). One hence usually encounters talk of *single particle interference* in this connection (e.g. Thaller 2005, p. 189).

But what does all of this mean? What kind of an entity is the photon, that it somehow seems to be a particle that can 'interfere with itself'? Is each photon capable of spreading out in a wavelike fashion at the slit, only to collapse down to a narrow point again in proximity of the screen? When one tries to sort these things out experimentally, another remarkable thing happens. Placing detectors behind each of the two slits to measure which path the photons 'actually take' will always reveal a detection in *one path only*, not both. But at the same time, the interference pattern *vanishes* in this kind of experiment: the distribution of transmitted energy in setups with detectors behind the slit looks like one which would arise from 'ordinary' particles passing through either of the two slits and accumulating in two proximate lumps in succession of each of the two slits.

Do the photons *alter* their behavior from wave-like to particle-like when their position is determined, either by detectors behind the slit or by the screen itself, upon incidence? Roger Penrose (2004, p. 517) actually claims something quite similar:

> the impression could be gained that the particle-like aspects of a wave/particle are what show up in a measurement, whereas it is the wavelike ones that show up between measurements. This is not so far from the truth of what quantum mechanics tells us [...].

Still, all of this constitutes a kind of puzzle, a 'mystery' if you will, and Feynman (2010 [1965a], p. 1–2; emphasis in original) even went so far as to call it "the *only* mystery." We shall eventually see that this may in fact have been a little premature, as Feynman much later somewhat acknowledged himself (cf. Feynman 1982, p. 485). When it comes to considerations of joint states of multiple systems, QM holds even deeper mysteries than this one.

Now we mentioned before that de Broglie speculated about the wave nature of *matter* as well, and experiments of the sort described above have indeed been performed with matter particles such as electrons (Davisson and Germer 1927; Möllenstedt and Jönsson 1959; Tonomura et al. 1989), neutrons (Gähler and Zeilinger 1991), as well as larger (C_{60}, C_{70}) *molecules* (Arndt et al. 1999; Arndt et al. 2001). For the larger molecules, the interference pattern is slightly more washed out and the 'degree of washing out', if you will, depends on the *isolation* from an environment of air molecules in such experiments; Arndt et al. (1999), for instance, used a vacuum chamber with a pressure of about one fifty-billionth of normal atmospheric pressure. This is in itself a very important point: interference phenomena (usually) require isolation from the environment. We will come back to the role of the environment in Chaps. 6 and 7, when we confront *quantum decoherence*.

What to make of the double-slit experiments discussed above? Are light and matter indeed both waves in some sense, only exhibiting particle-like behavior under suitable conditions, as basically suggested in the above quote by Penrose? This is, of course, the old question of *wave-particle duality* that has accompanied and maybe blurred discussions about QM for a long time. Let us formulate a first stab at an interpretation based on the experiments and considerations investigated so far, which we shall call *the naïve view*:

Conjecture 0 (The naïve view) QM is a theory about little things which can behave in a spread-out, wavelike manner in some circumstances, and can equally behave in a condensed, particle-like manner in other circumstances.

In a sense we are here putting up a straw man just to put it down again, because there are numerous good reasons (that will soon become clear) why the naïve view cannot possibly be true as it stands. And Penrose (2004, p. 517) equally continues his considerations on wave-particle duality by the comment that "the two wave/particle aspects are by no means so simply delineated[...]." However, as mentioned earlier, comments suggestive of something along these lines can still be found in textbooks, as evidenced e.g. in the following (beautifully straightforward) introductory passage from a modern textbook on quantum *field* theory (QFT):

> The advent of quantum mechanics convinced people that things that had previously been thought of as particles were in fact waves. For example, it was found that electrons and neutrons were not simply little rigid bits of matter but they obey a wave equation known as the Schrödinger equation. [...] It was also realized that things that had been previously thought of as waves were in fact particles. (Lancaster and Blundell 2014, p. 19)

Such an attitude to the matter is certainly entertained by numerous practitioners disinterested in interpretational subtleties. A similar observation is made by Falkenburg (2010), who writes:

> Philosophers of science are inclined to think that wave-particle duality is an obsolete concept, because according to quantum mechanics there are neither waves nor particles in a classical sense. But in physical practice, wave-particle duality is alive. (p. 31; emphasis omitted—FB)

Falkenburg's reference is, in particular, to a talk of Nobel prize winner Wolfgang Ketterle, in which he claimed that "after several years of physical practice one gets used to preparing waves and detecting particles." (as cited in Falkenburg 2010, p. 34) It hence seems instructive to give some careful thought to how far the analogy between a *quantum description* of certain phenomena and either a *wave- or particle description* can be taken.

As for the scope of this analogy one should note that it is not only interference phenomena in the double slit experiment that are suggestive of 'waves being involved'. Greenstein and Zajonc (2006), for instance, discuss the resemblance of *modulation* in atomic decay to the modulation of *sound waves* as another example in their introductory-level book on experimental aspects of QM. To wit, when an atom decays from one excited level into its ground state *via* a large series of closely spaced intermediate levels, it will emit light with exponentially decaying intensity.[5] However, if the upper level from which the atom decays is split up (by means of a weak magnetic field, say) into two levels E and E', the measured intensity of light from the decay will be modulated, much like the fluctuating sound that results from two tuning forks of different frequency being struck at once (cf. Greenstein and Zajonc 2006, pp. 102–105). In this sense the 'frequencies' $\omega = E/\hbar$ and $\omega' = E'/\hbar$ associated with the two close-by energy levels behave like the frequencies of two sound waves, and atomic exponential decay exhibits similarities to phenomena traditionally explicitly understood in terms of waves.[6]

Thus taking the 'wavieness' quite seriously, a possible analysis of the double slit experiment in terms of waves that also exhibit particle-like behavior under suitable

---

[5]E.g. Basdevant and Dalibard (2002, p. 343) for the details.

[6]The actual experiment described by Greenstein and Zajonc (2006, pp. 103) in fact involves *multiple* atoms. Therein detectors are used, however, that accept only single photons at a time, so that Greenstein and Zanjonc deem the arrangement "very close to an ideal experiment, in which we work with one atom and one photon at a time." (ibid.)

circumstances could run as follows. When a photon hits the photographic screen behind the double slit and is 'detected as a particle', "the wave function has changed its shape. What used to be a broad wave packet representing the photon extending across the film has collapsed to a single sharp peak centered on the atom that registered the event." (Greenstein and Zajonc 2006, p. 217; emphasis omitted) This is the intuition behind a *collapse* or *reduction of the wave packet* in an experimental situation, and it is usually traced back to Heisenberg (1927, p. 186), even though Heisenberg—for convincing reasons—did not think of the quantum mechanical wave packet in the sense of Conjecture 0. For now, however, let us assume that something of this sort could explain the transition from wave-like to particle-like.

How do the wavy aspects appear in the *formalism* of QM? To elaborate, let us first note the identity

$$e^{i\theta} = \cos\theta + i\sin\theta, \tag{2.6}$$

the so called *Euler formula* which can easily be worked out in terms of the power series expansions of the three functions. Here $i = \sqrt{-1}$ is the *imaginary unit*. Using (2.6) one can give a very general and useful description of some arbitrary wave by writing

$$\psi(\theta) = Ae^{i\theta}, \tag{2.7}$$

with A the wave's amplitude (which we here take to be real-valued). In pre-quantum physics (e.g. classical electromagnetism) the use of this complex representation is merely a computational convenience; one usually appeals to the complex description throughout the calculation, and then considers only the real or imaginary part (the cosine or the sine) at the end, to describe the 'real', 'physical' wave. In QM, however, things do not turn out this way.

We can replace θ in (2.7) by $ky - \omega t$ to obtain a description of a wave traveling forward in y-direction. To adjust this expression such as to provide a simple model for the double-slit experiment, we should appeal to spherical rather than plane waves, originating at the two slits in the metal plate respectively, which we do by using rather the *radial distances* r_j to a respective slit $j \in \{1, 2\}$ in the description of the waves at a point $\boldsymbol{x}_0$ on the screen. Here $r_j = |\boldsymbol{x}_0 - \boldsymbol{x}_j|$, with $\boldsymbol{x}_j$ the position of a respective slit, and $|\cdot|$ refers to the Euclidean norm (cf. Appendix A). Let us also drop the time dependence for now, which is ultimately possible because $e^{i(ky-\omega t)} = e^{iky}e^{-i\omega t}$ (i.e. the temporal part 'factors out').

For each of the slits we now obtain a description $\psi_1(r_1) = A_1 e^{ikr_1}$ and $\psi_2(r_2) = A_2 e^{ikr_2}$, and since, in the absence of detectors, our wave may pass through any of the two slits, we should add the two functions together to obtain the total wave function

$$\psi_{1+2}(r_1, r_2) = \psi_1(r_1) + \psi_2(r_2) = A_1 e^{ikr_1} + A_2 e^{ikr_2}.$$

For 'ordinary' waves, one can now predict the *intensity distribution* w.r.t. points $\boldsymbol{x}_0$ on the screen by computing the squared modulus of the wave function:

$$\begin{aligned}|\psi_{1+2}(\boldsymbol{x}_0)|^2 &= \psi^*_{1+2}(\boldsymbol{x}_0)\psi_{1+2}(\boldsymbol{x}_0)\\ &= (A_1e^{-ikr_1} + A_2e^{-ikr_2}) \cdot (A_1e^{ikr_1} + A_2e^{ikr_2})\\ &= A_1^2 + A_2^2 + 2A_1A_2\cos(k(r_2 - r_1)) \qquad (2.8)\end{aligned}$$

(e.g. Greenstein and Zajonc 2006, pp. 15–17; Shankar 2010, p. 13).

$2A_1A_2\cos(k(r_2 - r_1))$ is called an *interference term*, which gives rise to the characteristic pattern, whereas the first two terms, A_1^2 and A_2^2, can be interpreted to each represent what would have been obtained in case one of the slits would have been blocked or how particles, each traveling through one of the slits only, would have distributed in such an experiment. And in case detectors are used to determine through which of the slits something passes, one obtains the result $A_1^2 + A_2^2$, as mentioned above.

But interpreting the squared modulus as an intensity distribution of a wave is not really adequate since we also have to incorporate the dot-like properties of the pattern into our description. Thus in QM (2.8) is reinterpreted as a *probability density*, so that $|\psi_{1+2}(\boldsymbol{x}_0)|^2\mathrm{d}^3\boldsymbol{x}$ defines the probability of finding a dot on the screen in a 'small volume' $\mathrm{d}^3\boldsymbol{x}$ around $\boldsymbol{x}_0$. According to Conjecture 0 and our subsequent 'collapse-considerations', this could be interpreted as the probability of the wave *collapsing* to the tiny volume $[x_0, x_0 + \mathrm{d}x] \times [y_0, y_0 + \mathrm{d}y] \times [z_0, z_0 + \mathrm{d}z]$ (for $\boldsymbol{x}_0 = \begin{pmatrix} x_0 \\ y_0 \\ z_0 \end{pmatrix}$).

Before ((2.4) and (2.5)) we noted that $k = p/\hbar$ and that $\omega = E/\hbar$, and we can use these relations to define a wave function $\Psi(x, t) = Ae^{\frac{i(px-Et)}{\hbar}}$. With $\boldsymbol{k} = \begin{pmatrix} k_x \\ k_y \\ k_z \end{pmatrix}$ representing a wave *vector*, so that $|\boldsymbol{k}| = \frac{2\pi}{\lambda}$, we can equally write $\Psi(\boldsymbol{x}, t) = Ae^{\frac{i(\boldsymbol{px}-Et)}{\hbar}}$ for a '3D(imensional) wave'.[7] And these waves somehow 'encode' the momentum and energy associated with the 'particles' measured upon collapse.

Since our 'waves' could also be 'electron waves', we have used the connections given by the de Broglie wavelength in a generalized fashion here. Historically, when Schrödinger first showed interest in de Broglie's work on matter waves, he was confronted with serious opposition, as was de Broglie himself for that matter. One point of contention (raised by Debye; cf. Mehra and Rechenberg 1987, p. 421 ff.) was the lack of a *wave equation*, such as that for electromagnetic waves mentioned before. As the story goes, Schrödinger then went on to find such an equation, which is now famously known as the *Schrödinger equation* (SE). Using what we have established so far, we can give a simple heuristic 'derivation' of it.

We start by realizing that $\Psi(\boldsymbol{x}, t) = Ae^{\frac{i(\boldsymbol{px}-Et)}{\hbar}}$ is a function of definite energy and momentum, since it is only variable in x and t, and if we differentiate it w.r.t. space, we will get back the same function, multiplied by something proportional

[7] Recall that $\boldsymbol{p} \cdot \boldsymbol{x}$ is an inner product which computes $p_x x + p_y y + p_z z$, which is why $E \cdot t$ can be meaningfully subtracted from it.

to $\boldsymbol{p}$. More precisely, one can define what is known as the *momentum operator* (in *position representation*) by $\hat{\boldsymbol{p}} = -i\hbar\nabla$, with ∇ the partial derivative or *Nabla* operator that has the simple form $\nabla \doteq \begin{pmatrix} \frac{\partial}{\partial x} \\ \frac{\partial}{\partial y} \\ \frac{\partial}{\partial z} \end{pmatrix}$ in Cartesian coordinates. Applying this to our wave function, we get

$$\hat{\boldsymbol{p}}\Psi(\boldsymbol{x},t) = -i\hbar\nabla A e^{\frac{i(\boldsymbol{px}-Et)}{\hbar}} \doteq -i\hbar A \begin{pmatrix} \frac{\partial}{\partial x} e^{\frac{i(p_x x + p_y y + p_z z - Et)}{\hbar}} \\ \frac{\partial}{\partial y} e^{\frac{i(p_x x + p_y y + p_z z - Et)}{\hbar}} \\ \frac{\partial}{\partial z} e^{\frac{i(p_x x + p_y y + p_z z - Et)}{\hbar}} \end{pmatrix} = \begin{pmatrix} p_x \\ p_y \\ p_z \end{pmatrix} A e^{\frac{i(\boldsymbol{px}-Et)}{\hbar}} \doteq \boldsymbol{p}\Psi(\boldsymbol{x},t).$$

Equally, taking the time derivative of $\Psi(\boldsymbol{x},t)$, multiplied by $i \cdot \hbar$, will simply give back $\Psi(\boldsymbol{x},t)$ multiplied by E. Form classical (non-relativisitc) physics one has the total energy of a particle given as $E = \frac{p^2}{2m} + V$, with V a (possibly time- and place-dependent) potential energy function. Multiplying both sides by our wave function Ψ and making the (heuristic) substitutions $E \rightsquigarrow i\hbar\frac{\partial}{\partial t}$ and $\boldsymbol{p} \rightsquigarrow -i\hbar\nabla$, we thus get the *time dependent* Schrödinger equation,

$$\left(-\frac{\hbar^2}{2m}\Delta + V\right)\Psi(\boldsymbol{x},t) = i\hbar\frac{\partial}{\partial t}\Psi(\boldsymbol{x},t) \tag{TDSE}$$

with $\Delta = \nabla^2 = \frac{\partial^2}{\partial x^2} + \frac{\partial^2}{\partial y^2} + \frac{\partial^2}{\partial z^2}$ the *Laplacian operator* we have already met in the context of electromagnetic waves.[8] A time-independent or 'stationary' form of the Schrödinger Equation can be obtained by factoring out (and ignoring) the time component of the wave function, provided that the potential energy is also time-independent:

$$\left(-\frac{\hbar^2}{2m}\Delta + V\right)\psi(\boldsymbol{x}) = E\psi(\boldsymbol{x}). \tag{SSE}$$

$\hat{H} := -\frac{\hbar^2}{2m}\Delta + V$ is called the *Hamiltonian (operator)*, the quantum analogue of the *Hamilton function* of classical mechanics.

In the time *dependent* form, the SE is indeed somewhat reminiscent of a wave equation. Not only that, but by appeal to concepts from classical (Hamiltonian) mechanics, one can similarly give a heuristic derivation of the SSE as an actual *instance* of an equation of the form $\Delta\Psi = \frac{1}{u^2}\frac{\partial^2\Psi}{\partial t^2}$, with $u = \frac{E}{\sqrt{2m(E-V)}}$ a 'wave-velocity', whose form is motivated by the classical considerations (cf. Capri 2002, pp. 43–45). So far this fits rather well with our naïve view, since now we have waves with a wave equation and with the connection to particle-like aspects spelled out in terms of a 'wave collapse'.

[8]Note that $\frac{(\hat{p})^2}{2m} = \frac{(-i\hbar\nabla)^2}{2m} = -\frac{\hbar^2}{2m}\Delta$.

The SE is certainly the most important equation of QM, but it comes with certain restrictions. First of all, it is *non-relativistic*; and in using notions like "3D-space" etc. we have been talking so far as if space and time were simply two entirely separate and absolute categories. But the special and general theories of relativity have taught us otherwise, and this should be reflected in the dynamical equations of QM. Furthermore, the SE is not always exactly solvable. For 'many-particle' systems, one is forced to work with approximations instead. We will return to both these points below, many-particle systems and relativity, in confronting some aspects of quantum *field* theory (QFT).

The crucial role of the TDSE in QM is to describe the temporal evolution of a given system, i.e. its *dynamics*,[9] but it also *selects* the functions that can be used to describe a system under given circumstances (with a given potential V, suitable boundary conditions, etc.). Above, we have appealed to a function $\Psi(\boldsymbol{x}, t) = \Psi_{\boldsymbol{p}}(\boldsymbol{x}, t)$ of definite momentum $\boldsymbol{p}$ as a solution of the TDSE, but now assume that for a given V it is equally solved by another function $\Phi_{\boldsymbol{p}'}(\boldsymbol{x}, t)$ of different momentum $\boldsymbol{p}'$ (and energy E' accordingly). Crucially, the TDSE is *linear*, which means that $(i\hbar\frac{\partial}{\partial t} - \hat{H})(\alpha\Psi + \beta\Phi) = \alpha(i\hbar\frac{\partial}{\partial t} - \hat{H})\Psi + \beta(i\hbar\frac{\partial}{\partial t} - \hat{H})\Phi$ $(\alpha, \beta \in \mathbb{C})$, and since $(i\hbar\frac{\partial}{\partial t} - \hat{H})\Psi = (i\hbar\frac{\partial}{\partial t} - \hat{H})\Phi = 0$, $\Upsilon = \alpha\Psi + \beta\Phi$ defines another possible solution. This leads us to a so called *dynamical superposition principle*:

Principle of Superposition I (Dynamical) Any two solutions of the TDSE, Ψ and Φ, can be superposed in the form $\alpha\Psi + \beta\Phi$ $(\alpha, \beta \in \mathbb{C})$ to form a new solution.[10]

This superposition principle, implied by the linearity of the TDSE (and the SSE, for that matter), actually makes much sense on account of our naïve view, and it is of course also present in other, classical wave equations. Note, first of all, that summing up can also be understood in terms of integration. This is the case when the system under consideration is *free*, i.e. not subject to any potential. Thus setting $V = 0$ in the TDSE, we obtain the general solution

$$\Psi(\boldsymbol{x}, t) = \int_{\mathbb{R}^3} A(\boldsymbol{p}) e^{\frac{i}{\hbar}\boldsymbol{p}\boldsymbol{x} - Et} \mathrm{d}^3\boldsymbol{p}. \tag{2.9}$$

This wave function, or rather its real and imaginary parts taken separately, mathematically describe(s) a so called *wave packet*. Recall how superposition of waves, as discussed in the double slit experiment, leads to constructive and destructive interference. Remembering also that $\boldsymbol{p} \propto \boldsymbol{k}$ (read: '$\boldsymbol{p}$ is proportional

---

[9]Terminology may be confusing here, since the Greek δύναμις, from which the term derives, actually means 'power, ability' (cf. Perschbacher 1990, p. 108). The modern use of the term can be connected to Newtonian physics, where considerations of *forces* give rise to the differential equations describing the time evolution of systems. The introduction of term into physics is typically traced back to the dynamism of Leibniz (cf. Bernstein 1981, p. 97).

[10]Cf. Joos et al. (2003, p. 7).

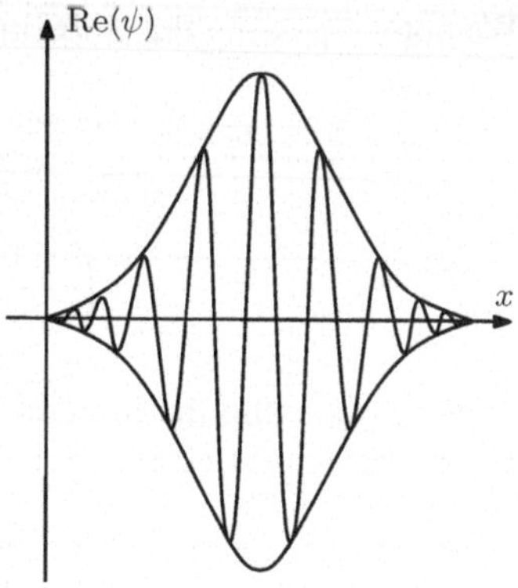

Fig. 2.4 Example of a narrow wave packet (the real part is illustrated). The black outer 'envelope' describes how the group of superimposed waves travels as a whole

to $\boldsymbol{k}$') and $|\boldsymbol{k}| = \frac{2\pi}{\lambda}$, the above integral can be viewed as 'putting (plane) waves of different *wavelengths* on top of one another', whence these will have peaks and troughs in different places and *interfere* such as to give rise to a particularly shaped resulting wave.

$A(\boldsymbol{p})$ here functions as a kind of *weighting*, so that the waves will mostly stem from a certain range of values of $\boldsymbol{p}$ (and λ respectively). The resulting picture is that of Fig. 2.4, a narrowly lumped up but still wavy entity. Thus interpreted in terms of waves, the above superposition principle makes perfect sense, and the nice little wave packet raises hopes for 'solving the riddle' in terms of waves that can behave like (or 'collapse into') particles after all. But we must stress that this is a false impression, and that *quantum* superposition, later understood as a *kinematic* feature, will eventually turn out to be quite different from classical superposition.

From the treatment of wave packets, one also arrives at a first, intuitive version of the infamous *uncertainty relations*, discovered by Heisenberg (1927). Today these are given as $\Delta x \Delta p_x \geq \frac{\hbar}{2}$ and $\Delta E \Delta t \geq \frac{\hbar}{2}$, where p_x is the linear momentum in x-direction, E the energy, and x and t are position and time respectively. Δ here represents a 'spread' in value, not the Laplacian. Let us see why the content of the first one, that there is a lower limit as to the joint 'uncertainties' (or rather: well-definednesses) of $\boldsymbol{x}$ and $\boldsymbol{p}$, arises from wave packets. First note the so called *Fourier transformation*, which makes it possible to convert some function of position $\boldsymbol{x}$ into a function of $\boldsymbol{k}$ (and back again), and equally so for ω and t. Suppressing time components once more, the two transformations for $\boldsymbol{x}$ and $\boldsymbol{k}$ are given as

$$f(\boldsymbol{x}') = \frac{1}{\sqrt{(2\pi)^3}} \int_{\mathbb{R}^3} e^{i\boldsymbol{k}\boldsymbol{x}'} \tilde{f}(\boldsymbol{k}) \mathrm{d}^3\boldsymbol{k} \tag{2.10}$$

and

$$\tilde{f}(\boldsymbol{k}) = \frac{1}{\sqrt{(2\pi)^3}} \int_{\mathbb{R}^3} e^{-i\boldsymbol{k}\boldsymbol{x}} f(\boldsymbol{x}) \mathrm{d}^3\boldsymbol{x}. \tag{2.11}$$

We note in passing that substituting (2.11) for $\tilde{f}(\boldsymbol{k})$ into (2.10), we obtain

$$f(\boldsymbol{x}') = \frac{1}{\sqrt{(2\pi)^3}} \int_{\mathbb{R}^3} e^{i\boldsymbol{k}\boldsymbol{x}'} \left(\frac{1}{\sqrt{2\pi}} \int_{\mathbb{R}^3} e^{-i\boldsymbol{k}\boldsymbol{x}} f(\boldsymbol{x}) \mathrm{d}^3\boldsymbol{x} \right) \mathrm{d}^3\boldsymbol{k}$$
$$= \int_{\mathbb{R}^3} \left(\frac{1}{(2\pi)^3} \int_{\mathbb{R}^3} e^{i\boldsymbol{k}(\boldsymbol{x}'-\boldsymbol{x})} \mathrm{d}^3\boldsymbol{k} \right) f(\boldsymbol{x}) \mathrm{d}^3\boldsymbol{x}$$

which means that the integral in the bracketed expression simply maps $f(\boldsymbol{x})$ onto value $f(\boldsymbol{x}')$, so we can conclude that

$$\frac{1}{(2\pi)^3} \int_{\mathbb{R}^3} e^{i\mathbf{k}(\boldsymbol{x}'-\boldsymbol{x})} \mathrm{d}^3\boldsymbol{k} = \delta^3(\boldsymbol{x}' - \boldsymbol{x}), \tag{2.12}$$

where the latter expression is the *Dirac-δ* discussed in Appendix A.[11]

Since $\boldsymbol{k} = \frac{\boldsymbol{p}}{\hbar}$ we have $\frac{\mathrm{d}k}{\mathrm{d}p_j} = \frac{1}{\hbar}$ for each component ($j \in \{x, y, z\}$), and we can (symbolically) say that $\mathrm{d}^3\boldsymbol{k} = \frac{1}{\hbar^3}\mathrm{d}^3\boldsymbol{p}$. Thus, substituting $\boldsymbol{k} \rightsquigarrow \boldsymbol{p}$ in expression (2.10) the integral becomes

$$f(\boldsymbol{x}') = \frac{1}{\sqrt{(2\pi\hbar)^3}} \int_{\mathbb{R}^3} \tilde{f}(\boldsymbol{p}) e^{\frac{i}{\hbar}\boldsymbol{p}\boldsymbol{x}'} \mathrm{d}^3\boldsymbol{p}. \tag{2.13}$$

Our wave packet from above is obviously of this general form, for the choice $A(\boldsymbol{p}) = \frac{1}{\sqrt{(2\pi\hbar)^3}} \tilde{f}(\boldsymbol{p})$ and including a time-component. Since a wave packet is a somewhat *localized* entity, it will exhibit a small spread in width (directly related to Δx). But one has to use many different wavelengths to obtain such a narrow packet and hence a large range of different momenta, so that the spread in the value of momentum (directly related to Δp_x) will be comparatively large for a fairly localized packet. Fourier transforming a Gaußian bell curve in position space for instance, i.e. allowing for the amplitude $A(\boldsymbol{p})$ to be proportional to $e^{-\frac{(p-p_0)^2}{2(\sigma\hbar)^2}}$ (for specifiable parameters σ, $\boldsymbol{p}_0$), leads to a *broader* bell curve in the momentum space. And transforming the function $f(\boldsymbol{x}) = 1$ which is uniform over *all* positions, *everywhere*, or equally an unrestricted plane wave $e^{\frac{i}{\hbar}\boldsymbol{p}\boldsymbol{x}}$, leads to a Dirac-δ, i.e. a *maximally concentrated* wave function in momentum space.

Notably, these relations hold *regardless* of whether we use $\boldsymbol{p}$ or $\boldsymbol{k}$ and regardless of whether we choose a complex representation or not; an 'uncertainty relation' between the width of a wave packet and the spread in its wave number is already present in *classical* wave optics (cf. Demtröder 2010, p. 101).

[11]Equation (2.12) can also be straightforwardly proven, for instance by appeal to a convergence generating factor.

The *interpretation* of $\boldsymbol{k}$ *as a momentum* (up to a scaling factor), however, leads to interpretational difficulties as regards the wave- and particle aspects. It is not the association of momenta to waves *in general* that creates a problem: in classical electrodynamics an electromagnetic field, traveling as a wave, carries a *field momentum* that can be computed by integrating its Poynting-vector (divided by c^2) over some volume (e.g. Jackson 1990, p. 261). But the mathematical relations here are different, and there is no obvious way to reinterpret the momentum in QM as a sort of field momentum. How, then, *is* the momentum $\boldsymbol{p}$, carried by the 'particle' when the wave 'collapses', related to the wave itself? What does it *mean* that momentum is 'encoded' in the wave function, as we have claimed above?

2.1.2 Abstractions, Examples, and Further Peculiarities

The questions raised at the end of the last section surely introduce some first unease about the naïve view. A second unease should be raised by the existence of polarization and the fact that have not really incorporated it into our description so far. How is the overall polarization of a wave connected to the polarization associated with each photon?

We noted that in virtue of their polarization, single photons will do one of two things when incident on a polarizer, go up or down, pass or be absorbed. This view of polarization makes it a '*two-state*' quantity, paradigmatic for a larger range of examples. In fact, we can get access to almost all of the 'strangeness' associated with QM by telling a completely different story from that of Sect. 2.1.1 in terms of such two-state quantities.

Let us begin by considering a particular such quantity called *spin*. This spin is a quantity with a sense of 'directionality', whence one would classically represent it as a (Euclidean) vector $\boldsymbol{s}$, a pointer that indicates the direction and strength of the magnitude in question on a particular system. Spin has to be attributed to quantum mechanical 'systems'[12] such as protons or electrons to account for the fact that they exhibit a change in behavior under the influence of magnetic fields. This makes it reasonable to attribute *magnetic* properties to them, and the spin ultimately codifies these.

It was already known in nineteenth century physics that rotating electric fields are inevitably accompanied by changing magnetic fields, as reflected in *Faraday's law of induction*, $\nabla \times \boldsymbol{E} = -\frac{\partial \boldsymbol{B}}{\partial t}$ (e.g. Walker et al. 2012, p. 949). Initially it was hence theorized, first by Kronig and later by Uhlenbeck and Goudsmit, and on the basis of such phenomena as the Zeeman-effect and multiplicity in spectral lines

[12]We here take the widely used notion of a 'system', which Bell (1990a, p. 34) complained, should be purged from a physically precise theory altogether, to be simply ontologically as non-committal as the term 'entity' in philosophy, and hence without any general implication of an involvement of, say, parts and wholes.

(cf. Tomonaga 1974, p. 33), that charged particles such as the electron possess an *intrinsic angular momentum*, i.e. revolve around a symmetry axis as the earth does on a daily basis, giving rise to a magnetic field. This conception of spin as a rotation of the electron, however, proved untenable early on in the development of QM as it lead, among other things, to the prediction of unreasonably high energies and rotational velocities much greater than the speed of light (cf. Basdevant and Dalibard 2002, p. 232; Demtröder 2010, p. 187).

Nowadays, spin if often simply referred to as a "decidedly nonclassical concept" (Hentschel 2009b, p. 726) which "has no classical counterpart" (Shankar 1994, p. 373). Moreover, it is sometimes even disputed in what sense the spin of a *single* electron can actually be 'observed' (cf. Morrison 2007), although there are experiments that are usually taken to do just that. There is a general problem here, associated with the contrivedness of the experiments in question, and hence a general problem of *theory-ladenness* in physical experimentation and 'observation'. This will become more important in the later discussion.

We will here get access to the concept of spin by appeal to a series of 'rather straightforward' experiments performed on silver atoms, and hence avoid deeper discussion of the aforementioned issues with the spin of elementary particles for now. Note that we will also restrict our attention here to one of "the usual textbook 'caricatures'" (Busch et al. 1995, p. 7) and ignore the connection to spatial degrees of freedom, wave packet spreading, the influence of the environment... and so forth, all to be thematized later. Our 'caricature' will fully suffice, at this point, to introduce what is crucial.

Thinking of the aforementioned atoms as 'little magnets' for now, we may associate a 'magnetic moment' $\boldsymbol{\mu}$ to them, proportional to their spin $\boldsymbol{s}$[13] and characterizing the magnitude and direction of the associated magnetic properties. The expression $\boldsymbol{\mu} \cdot \boldsymbol{B}$ can then be used to compute the energy of the particle in a magnetic field $\boldsymbol{B}$, and $-\nabla(\boldsymbol{\mu}\boldsymbol{B})$ provides and expression for the force it experiences, which does not vanish in case $\boldsymbol{B}$ is *inhomogeneous* (so that the gradient is non-zero).

The 'directedness' of spin can be understood, in the (faulty) image (providing, however, a suitable 'mental crutch' for now), by recognizing that an intrinsic angular momentum can go either clockwise or counterclockwise, giving rise to differently oriented magnetic moments. One could hence identify the spin-vector, viewed as a little pointer parallel to the direction of the magnetic moment, with the thumb on a right handed 'thumbs-up', while the curled up fingers would represent the direction of the electron's rotational motion.

Depending on the number of electrons in an atom, spins can pair up in such a way that any two of them 'point in opposite directions' and their respective magnetic moments cancel. Silver atoms, however, have an uneven number of electrons (47 to

---

[13]The proportionality can be given in the form $\boldsymbol{\mu} = g_s \frac{\mu_s}{\hbar} \boldsymbol{s}$ where g_s varies with the particle sort, and $\mu_s = \frac{q\hbar}{2m_0 c}$ is called a *magneton*, with $\hbar$ Planck's reduced constant, q the particle's charge, c the speed of light, and m_0 its rest mass (cf. Haken and Wolf 1996, p. 188; Mayer-Kuckuk 2002, p. 58).

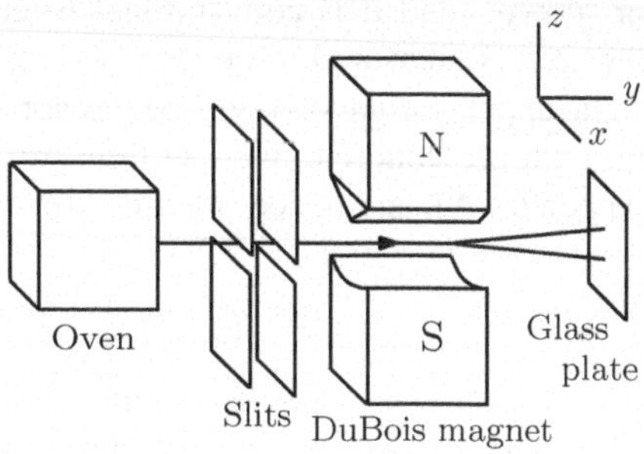

Fig. 2.5 A beam of silver atoms passes the magnetic field and is then detected on the glass plate

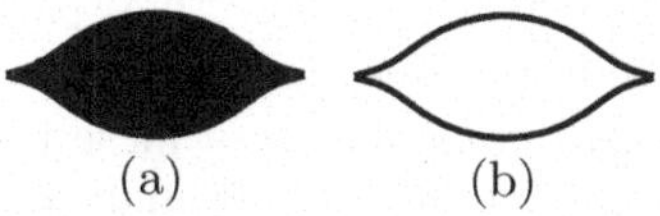

Fig. 2.6 A pattern similar to (**a**) would be expected to appear on the glass plate if spins were just like little magnets; a pattern as in (**b**) is actually observed

be exact) and hence one spin is left over without a cancellation partner. This gives rise to a total spin or net-magnetic moment of the atom (cf. Hughes 1989, p. 4; Sakurai 1994, p. 2).

In 1922 Otto Stern and Walther Gerlach performed an experiment where they heated up silver atoms in an oven, and the silver atoms thus prepared would escape through a small opening and then be collimated into a narrow beam. This beam would subsequently pass an inhomogeneous magnetic field produced by a specific kind of magnet (sometimes called a *DuBois magnet*; cf. Hughes 1989, p. 2), and then hit a glass plate (cf. Fig. 2.5).

Subjected to the force exerted by the inhomogeneous field, one would expect that the orientation of the atoms' magnetic moments (and hence, spins) should be changed more or less strongly in accordance with their previous, random orientation. Accordingly, a continuous blob on the screen should be expected where the silver atoms hit it (Fig. 2.6a). But this is not what was observed, and not what QM tells us either. Instead, only two lines, marking off the boundary of the 'classically' expected area appeared in the experiment, similar to the pattern in Fig. 2.6b.

The now-standard interpretation of this outcome is that the component of the spin of the atoms (and hence supposedly also of the electrons) in horizontal direction can only take on one of two values: it can either point up or down, and some of the atoms have their spin up, some have it down (e.g. Hughes 1989, p. 3; Sakurai 1994, p. 4).

We can set up a simple QM description of this experiment by appeal to the formalism discussed in Appendix A, and represent an atom's state with spin up along this direction by $|\uparrow\rangle$, and the corresponding spin down-state by $|\downarrow\rangle$. If we only consider the spin degrees of freedom, we can call this a *two state system*, since there are only two possible states w.r.t. the direction of interest that can be recognized in a spin measurement of the sort described above. One of course has to introduce a *reference frame* (and a coordinate system) to make sense of notions such as 'up' or 'down', and it is conventional to choose the horizontal direction along which the spin is measured to be the z-axis (as indicated by the coordinate system in the right upper corner of Fig. 2.5). Accordingly, we write $|\uparrow_z\rangle$ and $|\downarrow_z\rangle$ respectively. To be more concrete, we can think of this as a two-dimensional vector space (cf. Appendix A), and hence 'identify' (choosing a representation and basis)

$$|\uparrow_z\rangle \doteq \begin{pmatrix} 1 \\ 0 \end{pmatrix}, \qquad |\downarrow_z\rangle \doteq \begin{pmatrix} 0 \\ 1 \end{pmatrix},$$

so that the two states are viewed as column-vectors forming an *orthonormal basis* (ONB) of the (complex) *2D Hilbert space* $\mathbb{C}^2$, the space of all columns with two complex entries.[14] A state with spin along an arbitrary direction w.r.t. the z-axis can now be represented by a *linear combination* of these two vectors. For example, a (normalized) spin vector pointing to the right from the direction in which the atoms are moving in Fig. 2.5 (out of the paper and in the positive x-direction), can be expressed as

$$|\rightarrow_z\rangle = \tfrac{1}{\sqrt{2}}(|\uparrow_z\rangle + |\downarrow_z\rangle) = |\uparrow_x\rangle\,, \tag{2.14}$$

with column-representation

$$|\uparrow_x\rangle \doteq \begin{pmatrix} \frac{1}{\sqrt{2}} \\ \frac{1}{\sqrt{2}} \end{pmatrix}.$$

Equally, if we want to represent a spin pointing down along the x-axis, we can simply represent it by

$$|\downarrow_x\rangle = \tfrac{1}{\sqrt{2}}(|\uparrow_z\rangle - |\downarrow_z\rangle).$$

These two are easily seen to correspond to another ONB of $\mathbb{C}^2$, since

$$\sqrt{\langle\uparrow_x|\uparrow_x\rangle} \doteq \sqrt{\left(\tfrac{1}{\sqrt{2}}, \tfrac{1}{\sqrt{2}}\right)\begin{pmatrix} \frac{1}{\sqrt{2}} \\ \frac{1}{\sqrt{2}} \end{pmatrix}} = \sqrt{\left(\tfrac{1}{\sqrt{2}}\right)^2 + \left(\tfrac{1}{\sqrt{2}}\right)^2} = \sqrt{1} = 1,$$

$$\sqrt{\langle\uparrow_x|\downarrow_x\rangle} \doteq \sqrt{\left(\tfrac{1}{\sqrt{2}}, \tfrac{1}{\sqrt{2}}\right)\begin{pmatrix} \frac{1}{\sqrt{2}} \\ -\frac{1}{\sqrt{2}} \end{pmatrix}} = \sqrt{\left(\tfrac{1}{\sqrt{2}}\right)^2 - \left(\tfrac{1}{\sqrt{2}}\right)^2} = 0 \quad \text{(and so forth)}.$$

[14] If terminology is starting to sound unfamiliar, please at any rate consult Appendix A.

We can also put a number on the measurable value of the two possible spins along the chosen axis, namely $\frac{\hbar}{2}$ for spin up and $-\frac{\hbar}{2}$ for spin down.[15] Not all quantum systems have this kind of spin (if any); they can rather be grouped into two general classes: *fermions*, whose spin is a half-integer multiple of $\hbar$, and *bosons*, whose spin is an integer multiple of $\hbar$, and both of these classes obey different statistical rules (cf. Griffiths 1995, p. 179; Sakurai 1994, p. 362).[16]

Suppose that we do the same kind of experiment as described above, but rotate our Stern-Gerlach apparatus by 90°, so that the north pole of the magnet lies in positive x-direction. What should we expect to observe? Supposedly not much should change, except that the spin can now be up or down along the x- instead of the z-axis; and in fact, this expectation is met in experiment. What happens if we measure atoms along the z-axis again which have emerged, say, upwards in the z-direction from a previous magnet? Here we will supposedly expect that not much changes about these atoms, and they will, in fact, all still have their spins up in that direction (cf. Fig. 2.7a).

But funny things start to happen if we measure along two different directions in succession. Imagine, for instance, a rotated DuBois magnet behind the first one, as in Fig. 2.7b, and the atoms that emerge in z-down direction blocked, so that only the z-up ones are measured for x. What should we expect now? As a matter of fact, the outcome for the x-spin is then just as it was for the z-spin in the first, simpler experiment; sometimes the result is $\frac{\hbar}{2}$, sometimes it is $-\frac{\hbar}{2}$ (e.g. Hughes 1989, p. 3; Nortmann 2008, p. 146). It seems the measurement of the z-spin as 'up' has no bearing on the value of the x-spin.

Now let us make things a little more complicated by putting yet another DuBois magnet in the row, this time aligned along the z-axis again, just as in Fig. 2.7c. We will call atoms which have passed the first magnet '*prepared* in state $|\uparrow_z\rangle$' when the lower beam is blocked. Intuitively, we may expect that all of these atoms will have their spins up when they emerge out of the third magnet, because that is what we had prepared, and the x-measurement has seemingly no bearing on the value of the z-spin. But they do not! The measurement along the x axis completely randomizes the system w.r.t. z-spin again, and hence destroys the preparation effected by the first magnet.

But can we be sure that the atoms entering the second magnet actually *did* all have spin up along z? Maybe the property gets lost on the way and the atom simply starts 'spinning' in a random direction again. This possibility is excluded by our previous observation that the z-spin preparation is 'faithful' in a sense (Fig. 2.7a),

---

[15] In terms of units or physical dimensions we can thus see how the spin is still 'reminiscent' of an angular momentum; the dimension of $\hbar$ is energy $\times$ time which is equal to $\frac{\text{mass}\times\text{lenght}^2}{\text{time}}$, the dimension of an (intrinsic) angular momentum $I\omega$, with I the moment of inertia.

[16] Of course there is a debate on the existence of so called *paraparticles* which obey a third kind of statistics (cf. Massimi 2005, p. 154 ff.). The connection between spin and statistics, moreover, can only be thoroughly established in the formalism of QFT, and it here appears merely as an inductive generalization. But we will not pursue either of these issues any further in this book.

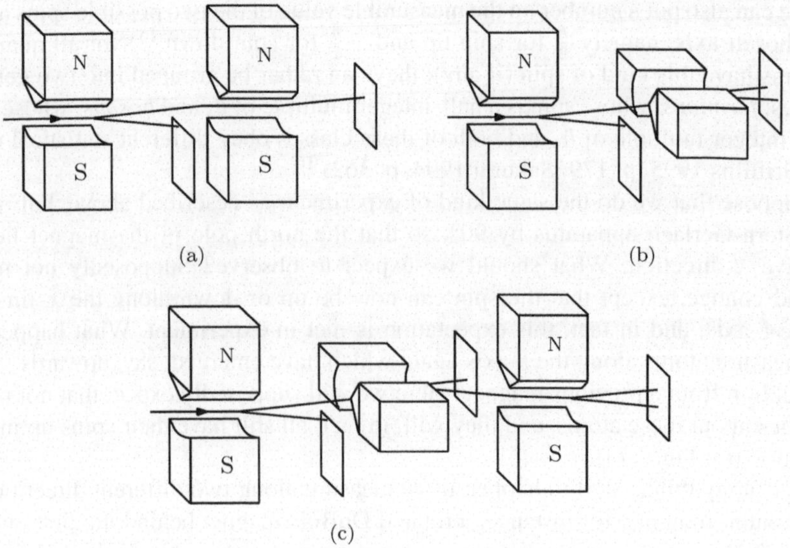

Fig. 2.7 (**a**) If we prepare the system as $|\uparrow_z\rangle$ and then measure z-spin shortly after, we will measure $|\uparrow_z\rangle$ again. (**b**) If we prepare the system as in (**a**), then measure spin along the x-axis, the outcome will be random. (**c**) If we take the beam of atoms prepared as $|\uparrow_x\rangle$ from the previous setting, and measure the spin along the z-axis again, the z-result is completely randomized by the intermediate x-measurement

and treating the system as free in between the magnets, this is also what QM predicts.

So we may come to wonder what kind of a spin the system actually has before we measure it *at all*. But for all we know so far, we cannot really say. All that we can do is give a *probability* for the outcome of a measurement along a given axis, conditional on our knowledge of how the system was *prepared*. If we know nothing about the state of the system w.r.t. the property in question, i.e. if we have not prepared it in a state in which it appears to assume some definite value, we should choose a uniform distribution of probabilities over the possible results of the measurement. If we know that the spin is up along the z-axis, then based on what we have observed so far, we can also only predict that there is a 50–50 chance (using 'chance' deliberately loosely here) of the spin being up along the x-axis as well. And in fact, QM does just this, and it does so using the *squared modulus* of the inner product of $|\uparrow_z\rangle$ and $|\uparrow_x\rangle$, i.e.:

$$\Pr_{s_x}^{|\uparrow_z\rangle}(\uparrow_x) = |\langle\uparrow_x|\uparrow_z\rangle|^2 = \frac{1}{2}, \tag{2.15}$$

where $\Pr_{s_x}^{|\uparrow_z\rangle}(\uparrow_x)$[17] denotes the probability of measuring spin up along the x-axis given that the system was prepared in the state $|\uparrow_z\rangle$ and that spin along the x-axis is measured for (s_x).

It should be obvious[18] that we are dealing with a *conditional* probability here. More generally, we could thus use an expression like $p(\underline{s_x} = +\frac{\hbar}{2}|\underline{M} = s_x, \underline{s_z} = +\frac{\hbar}{2})$ for an arbitrary probability function p.[19] The symbols appearing on the left hand side of the equality signs should be thought of as *(random) variables*, whereas the symbols on the right may be thought of as their values. 'Random variable' is meant here not necessarily in the narrow, measure-theoretic sense of Appendix A, but in the general sense as used e.g. by Pearl (2009, p. 8):

> By a *variable* we will mean an attribute, measurement or inquiry that may take on one of several possible outcomes, or *values*, from a specified domain. If we have beliefs (i.e., probabilities) attached to the possible values that a variable may attain, we will call that variable a *random variable*." (emphasis in original)

The notation may still be a little confusing though, since variables are certainly not identical to their values. We have here tacitly assumed that the random variable pertains to something. Following e.g. Schurz and Gebharter (2016, p. 1076), we can understand random variables $\underline{X}$ (which we will occasionally write with an underline for distinction) as functions $D \xrightarrow{X} V_{\underline{X}}$ from a domain D of individuals into a set $V_{\underline{X}}$ of values the variable can take on. Thus we should actually write $\underline{s_x}(S) = +\frac{\hbar}{2}$, which describes the event that $\underline{s_x}$, i.e. 'spin along the x-axis' takes on the value 'up' for a system $S \in D$. Similarly $\underline{M}(S) = s_x$ describes the event that the variable $\underline{M}$ ('physical measurement') takes on the value s_x for system S. Of course in principle specifications of space-time points (S takes on value y for variable $\underline{Y}$ *here, now*) could be added to make things more precise, but whether this is even meaningful depends on the interpretation of the probability function (i.e. single case or not).

Events (or rather event *types*[20]) and properties (such as 'spin up along x') can be understood as values of random variables in this sense. The probability function p (or Pr, which symbol we reserve for the quantum probabilities in what follows) then maps from an algebra (in the sense of Appendix A, Definition A.4) over the value-set of a random variable in question, or from the Cartesian product of the value-sets of multiple random variables in the case of a joint probability $p(\underline{X}_1 = x, \underline{X}_2 =$

[17]This is the notation also used by Redhead (1987, p. 8).

[18]Note that we are effectively avoiding such things as the 'big vs. many'-debate by simply stipulating that what is conditioned on *has* probabilities (cf. Wroński 2014, p. 45ff.). For convenience, we will do so throughout this book.

[19]'Function' is meant here in the neutral, set-theoretic sense, not in the specific sense of calculus. If you prefer this, you can replace it in thought by the more neutral 'map', which also covers measures (cf. Appendix A for details).

[20]Again, depending on the interpretation of p, it must measure the probability that variable $\underline{X}$ takes on a certain value for *some* individual $S \in D$, not a specific one.

$y, \ldots$),[21] into the interval [0, 1] (cf. Schurz and Gebharter 2016, p. 1076). Since the function maps an algebra of *sets* over the value space of some $\underline{X}$ into [0, 1], one should actually write expressions of the form $p(\{x\})$, but in line with common practice, we will usually directly write values as the argument of the probability function.

Let us now check why (2.15) corresponds to a 50%-chance. We know that

$$\langle\uparrow_x|\uparrow_z\rangle \doteq \left(\frac{1}{\sqrt{2}}\ \frac{1}{\sqrt{2}}\right)\begin{pmatrix}1\\0\end{pmatrix} = \frac{1}{\sqrt{2}}\cdot 1 + \frac{1}{\sqrt{2}}\cdot 0 = \frac{1}{\sqrt{2}} \tag{2.16}$$

so obviously

$$|\langle\uparrow_x|\uparrow_z\rangle|^2 = \left(\frac{1}{\sqrt{2}}\right)^2 = \frac{1}{2}, \tag{2.17}$$

as expected.[22]

Note that from (2.14) we can see that $\frac{1}{\sqrt{2}}$ is the *expansion coefficient* for both $|\uparrow_z\rangle$ and $|\downarrow_z\rangle$ in the expansion of $|\uparrow_x\rangle$ in the basis defined by $|\uparrow_z\rangle$ and $|\downarrow_z\rangle$. It is not hard to figure out that we could also use $|\uparrow_x\rangle$ and $|\downarrow_x\rangle$ as a basis of the same vector space, and expand

$$|\uparrow_z\rangle = \frac{1}{\sqrt{2}}(|\uparrow_x\rangle + |\downarrow_x\rangle) \doteq \frac{1}{\sqrt{2}}\left(\begin{pmatrix}\frac{1}{\sqrt{2}}\\\frac{1}{\sqrt{2}}\end{pmatrix} + \begin{pmatrix}\frac{1}{\sqrt{2}}\\-\frac{1}{\sqrt{2}}\end{pmatrix}\right) = \begin{pmatrix}1\\0\end{pmatrix}.$$

Hence $\frac{1}{\sqrt{2}}$ is equally the expansion coefficient for $|\uparrow_z\rangle$ in the spin-x-basis. According to (2.16) $\langle\uparrow_x|\uparrow_z\rangle$ is equal to $\frac{1}{\sqrt{2}}$, and so is $\langle\downarrow_x|\uparrow_z\rangle$, which we can use to write

$$|\uparrow_z\rangle = \langle\uparrow_x|\uparrow_z\rangle\,|\uparrow_x\rangle + \langle\downarrow_x|\uparrow_z\rangle\,|\downarrow_x\rangle\,. \tag{2.18}$$

The state of definite spin w.r.t. the z-axis is hence represented as an equal sum of possible states w.r.t. the x-axis, and expressions such as $\langle\uparrow_x|\uparrow_z\rangle$ intuitively describe the 'overlap' of states $|\uparrow_x\rangle$ and $|\uparrow_z\rangle$ or 'how much they have in common'. A state description like (2.18) is called a *quantum superposition*, and we have hence stumbled upon a very interesting and central issue in the philosophy of QM; namely the general *aptness* of such descriptions. The assumption of this 'aptness' constitutes a (if not *the*) central *postulate* of the theory:

---

[21] As should be clear by now, joint probabilities are a delicate matter in QM; for what, say, is the joint probability $p(s_{\underline{x}}(S) = +\frac{\hbar}{2}, s_{\underline{z}}(S) = +\frac{\hbar}{2})$, understood as 'equal time'? There does not seem to be an answer; both measurement procedures mutually exclude each other. That the matter is 'even more delicate' than mere limitations of joint measurability will become obvious in the following.

[22] Taking the squared modulus $|\langle\cdot|\cdot\rangle|^2$ not only ensures real values, but also that the function Pr satisfies the first Kolmogorov axiom (cf. Appendix A) $\Pr(a) \geq 0$ for all a in the domain of Pr.

Principle of Superposition II (Kinematical) Any two physical states, $|1\rangle$ and $|2\rangle$, whatever their meaning, can be superposed in the form $\alpha_1 |1\rangle + \alpha_2 |2\rangle$ ($\alpha_1, \alpha_2 \in \mathbb{C}$) to form a new physical state.[23]

The content of this principle is actually very different from that of the *dynamical* superposition principle (although from a certain standpoint, both are intimately connected in QM, as shall become clear later). We are here not talking about *waves* overlapping to give new waves, or about sums of *solutions to a differential equation* forming a new solution; we are talking about *physical states* being 'added up' to give a *new state*. This feature is quite peculiar to QM, and Zeh points out that "while the physical meaning of classical superpositions is usually obvious, that of quantum mechanical superpositions has to be somehow determined." (Joos et al. 2003, p. 8)

So in principle any kind of vector is admissible to represent a state, and in some basis this will be a 'superposition state', due to the vector-space structure of the state-space. But there are certain circumstances where it is indicated to impose physically motivated rules that disallow certain kinds of superposition, e.g. of states of different charges, spin-numbers, or masses (cf. Joos et al. 2003, p. 12). These are called *superselection rules*. Two vectors $|\psi\rangle$, $|\phi\rangle$ are said to be separated by a *selection rule* if $\langle\psi|\hat{H}|\phi\rangle = 0$ ($\hat{H}$ the Hamiltonian), i.e. if transitions from one state to the other are inhibited for the dynamical evolution as given by $\hat{H}$. They are said to be separated by a *super*selection rule if for *all physically realizable observables* A (with operators $\hat{A}$) it holds that $\langle\psi|\hat{A}|\phi\rangle = 0$.[24] *A fortiori*, the operators for physically realizable observables must form a proper subset of all self-adjoint ones, since e.g. $\hat{O} = |\psi\rangle\langle\phi| + |\phi\rangle\langle\psi|$ is self-adjoint, but $\langle\psi|\hat{O}|\phi\rangle \neq 0$ (cf. Giulini 2009, p. 772).

Note that since $|1\rangle$ or $|2\rangle$ in the kinematical superposition principle may already be superposition states, this implies arbitrarily large (countable) sums. But we can equally generalize the principle to allow for integrals (i.e. uncountable superpositions of states), as will become clear later. We have called this (again, possibly confusingly) the *kinematical* version, as it is formulated in terms of *states*, and the kinematics is generally understood as the part of a physical theory concerned with what *counts* as a state, and how to determine these.[25]

---

[23] Cf. Joos et al. (2003, p. 7).

[24] That this implies that no 'genuine' superpositions of the form $|\xi\rangle = \alpha |\psi\rangle + \beta |\phi\rangle$ can exist becomes clear by appeal to the *density operator*, thoroughly introduced later. With $\langle\psi|\hat{A}|\phi\rangle = 0$, one has $\langle\xi|\hat{A}|\xi\rangle = |\alpha|^2 \langle\psi|\hat{A}|\psi\rangle + |\beta|^2 \langle\phi|\hat{A}|\phi\rangle$ which would equally result from $\mathrm{Tr}\left(\hat{\rho}\hat{A}\right)$ with $\hat{\rho} = |\alpha|^2 |\psi\rangle\langle\psi| + |\beta|^2 |\phi\rangle\langle\phi|$ the density operator of a proper mixture (cf. 2.1.5), so that a coherent superposition of $|\psi\rangle$ and $|\phi\rangle$ is indistinguishable from a proper mixture (cf. Giulini 2009, p. 773).

[25] κίνεσις is (ancient) Greek for 'motion' (cf. Perschbacher 1990, p. 240), so the term 'kinematics' is again related to Newtonian physics and the fact that therein states are obtained by solving an equation of motion. It (or rather the french cinématique) was first suggested by Ampère (1838) in his *Essai sur la philosophie des sciences*, "for a field of mechanics that would be concerned with motion independent of its causes." (Koetsier 1994, p. 994)

The superposition principle, construed kinematically, has been called "a hallmark of all quantum theories" (Teller 1995, p. 7), and in what follows we will see how far reaching its consequences are. Of no less important stature is the rule we have appealed to in (2.15), called *Born's rule* (after Max Born 1926, p. 805 ff. and especially 1927, p. 241). It provides the pivotal algorithm for computing what outcome frequencies to expect in a measurement, according to QM. Born's rule obviously yields the same result as in (2.17) for the probability of $|\downarrow_x\rangle$ given $|\uparrow_z\rangle$, and it seems that we have two properties of which we can have no definite knowledge at the same time. We cannot design an experiment which definitely tells us the x- *and* the z- or the x- *and* the y-component of a system's spin $\boldsymbol{s} = \begin{pmatrix} s_x \\ s_y \\ s_z \end{pmatrix}$ simultaneously.

Quantum mechanics implements this feature by a replacement of s_x, s_y, and s_z with *self-adjoint operators* (cf. Appendix A, (A.21)) $\hat{s}_x$, $\hat{s}_y$, and $\hat{s}_z$ respectively, and of the vector $\boldsymbol{s}$ by the *vector operator* $\hat{\boldsymbol{s}}$. Dividing the component operators by $\frac{\hbar}{2}$, we obtain the (very useful) *Pauli matrices*

$$\hat{\sigma}_x \doteq \begin{pmatrix} 0 & 1 \\ 1 & 0 \end{pmatrix}, \quad \hat{\sigma}_y \doteq \begin{pmatrix} 0 & -i \\ i & 0 \end{pmatrix}, \quad \hat{\sigma}_z \doteq \begin{pmatrix} 1 & 0 \\ 0 & -1 \end{pmatrix},$$

due to Wolfgang Pauli (1927, p. 608). They have eigenvectors $|\uparrow_j\rangle$, $|\downarrow_j\rangle$ ($j \in \{x, y, z\}$) with eigenvalues $+1$ and -1 respectively, and we can decompose them as

$$\hat{\sigma}_j = +1\,|\uparrow_j\rangle\langle\uparrow_j| + (-1)\,|\downarrow_j\rangle\langle\downarrow_j|\,.$$

As mentioned before, the vector space we have appealed to is simply $\mathbb{C}^2$ (or any one isomorphic to it), i.e. the space of 2-entry columns of complex numbers, endowed with an appropriate sum operation, scalar multiplication, and a scalar product, and we can use $\hat{s}_z$'s eigenvectors $\mathcal{E}(\hat{s}_z) = \{|\uparrow_z\rangle, |\downarrow_z\rangle\}$ as a basis of $\mathbb{C}^2$. The third pair, $|\uparrow_y\rangle$ and $|\downarrow_y\rangle$, is then given by

$$|\uparrow_y\rangle = \tfrac{1}{\sqrt{2}}(|\uparrow_z\rangle + i\,|\downarrow_z\rangle) \doteq \begin{pmatrix} \frac{1}{\sqrt{2}} \\ \frac{i}{\sqrt{2}} \end{pmatrix}, \qquad |\downarrow_y\rangle = \tfrac{1}{\sqrt{2}}(|\uparrow_z\rangle - i\,|\downarrow_z\rangle) \doteq \begin{pmatrix} \frac{1}{\sqrt{2}} \\ -\frac{i}{\sqrt{2}} \end{pmatrix}$$

(i the imaginary unit). If we want to measure the spin along some arbitrary axis in space, given that we have chosen a certain coordinate system, the matrix corresponding to this observable can be obtained by treating the vector operator $\hat{\boldsymbol{\sigma}}$ like an ordinary Euclidean vector and taking its (Euclidean) inner product with a unit vector pointing in the desired direction. Call this vector $\boldsymbol{n}_{\theta\varphi}$ and express it in spherical coordinates as

$$\boldsymbol{n}_{\theta\varphi} = \begin{pmatrix} \sin\theta\cos\varphi \\ \sin\theta\sin\varphi \\ \cos\theta \end{pmatrix},$$

which can be motivated by geometric inspection (e.g. Wong 2013, p. 57). Then the scalar product yields

$$\begin{aligned}\hat{\boldsymbol{\sigma}} \cdot \boldsymbol{n}_{\theta\varphi} &= \sin\theta\cos\varphi \cdot \hat{\sigma}_x + \sin\theta\sin\varphi \cdot \hat{\sigma}_y + \cos\theta \cdot \hat{\sigma}_z = \\ &= \begin{pmatrix} \cos\theta & \sin\theta\cos\varphi - i\sin\theta\sin\varphi \\ \sin\theta\cos\varphi + i\sin\theta\sin\varphi & -\cos\theta \end{pmatrix} = \\ &= \begin{pmatrix} \cos\theta & \sin\theta\cdot e^{-i\varphi} \\ \sin\theta\cdot e^{i\varphi} & -\cos\theta \end{pmatrix},\end{aligned} \tag{2.19}$$

and this will be very useful for later reference.

We have seen above that the *order* of measuring the different spin directions makes a difference. If we measure the z-spin first and then measure it again, $|\uparrow_z\rangle$ stays $|\uparrow_z\rangle$. If we do an intermediate x-spin measurement, this is not so. The means to express this in QM is by appeal to the *commutator* (cf. Appendix A, (A.25)) of the operators, which is non-zero for two such magnitudes for which the order of measurements makes a difference. For the present case we have:

$$[\hat{\sigma}_j, \hat{\sigma}_k] \neq 0, \ \forall j, k \in \{x, y, z\} \text{ such that } j \neq k.$$[26]

As was mentioned in passing, physically measurable magnitudes such as the spin, represented by self-adjoint operators, are called *observables* in QM, which is intended to mean that their value can be determined by means of an experiment. $\boldsymbol{\sigma}$, however, is *not* an observable ($\hat{\boldsymbol{\sigma}}$ is not self-adjoint), so QM tells us that we cannot construct an experiment to determine the definite spin of an electron (or atom) w.r.t. all spatial directions. But $\boldsymbol{\sigma}^2$ *is* an observable and $\hat{\boldsymbol{\sigma}}^2$ commutes with $\hat{\sigma}_z$, say, so that we can know the total *magnitude* of the spin *and* its z-component simultaneously, but *never* its *direction*.

We may object at this point that experiments like the one described above can hardly count as *observations* in any narrower sense of the word. Multiple abductive inferences go into the interpretation of this process as a determination of the values in question, whence concerns about theory-ladenness once more come to mind. And in fact Popper (1967, p. 41) also critically remarked that "all 'observables' are calculated and inferred on theoretical grounds, rather than observed or directly measured. Thus what is 'observable' always depends upon the theory we use." This very point, that the theory determines what counts as observable, was also made

[26] In fact, the commutator will here give back another Pauli matrix, up to a multiple of i; the commutation relations can be summarized in the form $[\hat{\sigma}_k, \hat{\sigma}_\ell] = i2\sum_{m=1}^{3} \epsilon_{k\ell m}\hat{\sigma}_m$ with $\{x, y, z\}$ replaced by $\{1, 2, 3\}$, and where ϵ is the so called *Levi-Civita symbol* which gives back 1 if k, ℓ, and m are cyclical permutations of 1, 2, 3, -1 for anti-cyclical ones, and 0 if two of the numbers are identical. In virtue of this property, the Pauli matrices form an abstract commutator algebra (a *Lie algebra*), as do other types of angular momentum in QM (e.g. Schwindt 2013, pp. 197 ff. and 256 ff.).

by Einstein, as Heisenberg recalls in his autobiographic and philosophical writings (cf. Heisenberg 1969, p. 80, 2011, p. 31). What Einstein actually meant by this is an intricate matter that will make a reprise in Chap. 7, and it will have some significance for the interpretation advanced therein—although we will *not* follow *his* particular path.

Popper, at any rate, nevertheless conceded that most of us seem to understand what is *meant* by 'observable' in this context, and even the empiricist Reichenbach (1944, pp. 20–21) thought that there is always a kind of physical experiment with a quite *directly* observable outcome that we can regard as an indicator of the 'observable' in question taking on some value. So we may also avail ourselves of standard nomenclature.[27]

We have seen how QM has something to say about outcome statistics of measurements, and a second important statistical aspect of QM is that, given the Born rule, we can define a formula for the (theoretical) *average* or *expectation value* $\langle O \rangle$ of an observable O. The *arithmetic mean* over a series of measurements is obtained by summing up all measured values, multiplied by the relative frequency of their occurrence (their 'sample probability', if you will). The (theoretical) average is accordingly given (independently of QM) by summing up (integrating) the *possible* values multiplied by their respective probabilities according to some theoretically fabricated probability distribution. Restricting ourselves to the discrete case, we can express this quantum mechanically as

$$\langle O \rangle_\psi = \sum_j \mathrm{Pr}_O^{|\psi\rangle}(o_j) \cdot o_j = \sum_j |\langle o_j|\psi\rangle|^2 o_j = \sum_j \langle \psi|o_j\rangle\langle o_j|\psi\rangle o_j,$$

where the index $_\psi$ indicates that this average is relative to the quantum state $|\psi\rangle$ of the measured system. Realizing that $o_j|o_j\rangle = \hat{O}|o_j\rangle$, we can rewrite this expression to yield

$$\langle O \rangle_\psi = \sum_j \langle \psi|\hat{O}|o_j\rangle\langle o_j|\psi\rangle = \langle \psi|\hat{O}|\psi\rangle, \tag{2.20}$$

where we have made use of the resolution of the identity operator $\mathbb{1} = \sum_j |o_j\rangle\langle o_j|$ in deriving the last line, and of the fact that $\langle \psi|$ and $\hat{O}$ are *linear* (cf. Appendix A on both points).

Form the average one can also compute the *standard deviation* of a physical quantity, which is a measure for how much any measurable value of a quantity in question is expected to differ (on average) from the quantity's average. In an actual sample of data-points one would compute a *sample* standard deviation quite

---

[27]Still, it is clear that this point is actually of greater concern and has the potential to raise controversy. Moreover, note that *not every self-adjoint operator* can correspond to an observable (cf. Footnote 24) whereas the converse might just be the case (e.g. d'Espagnat 1995, p. 98).

intuitively by taking the square root of the sum over the squared differences[28] from the arithmetic mean $\bar{Q}$ over all the values, divided by the number of sample points, i.e.:

$$\Delta Q_{\text{samp.}} = \sqrt{\frac{\sum_{j=1}^{n}(\bar{Q} - q_j)^2}{n}}.$$

This corresponds to taking the square root of the mean of the squared deviations from the mean value. For the theoretically predicted quantity, one can hence use an elegant, general expression which reads (with $\bar{Q}$ replaced by $\langle Q \rangle$)

$$\Delta Q = \sqrt{\langle (Q - \langle Q \rangle)^2 \rangle}. \tag{2.21}$$

The sample standard deviation may be used as an 'estimator' for the theoretical one (cf. Fornasini 2008, p. 52 ff.), and the theoretical one constitutes a *predictor* for the sample deviation (which is the 'more empirical' viewpoint). In QM contexts this is also referred to as the *uncertainty* of the observable in question. The expression under the square root, identical to $\langle Q^2 \rangle - \langle Q \rangle^2$, is also sometimes called the *dispersion* of Q (e.g. Jaeger 2009, p. 8).

But there are some subtleties with the terminology here. Griffiths (1995, p. 112), for instance, complains about the use of the word 'uncertainty' instead of 'standard deviation' (because that is what the expression really provides), and uses σ instead of Δ (the latter symbol is often also rather used for the standard error of the mean). The more profound reason for criticism, however, is that 'uncertainty' suggests something *merely epistemic*. But *does* the 'uncertainty' associated with a QM observable, due to QM's probabilistic nature, *merely* reflect that we can (or typically do) not *know* its exact value? This is not really clear from the 'bare formalism', and the associated questions will be a major concern later.[29]

With these definitions, one can derive (cf. Griffiths 1995, p. 108 ff.; Shankar 1994, p. 128) what is sometimes called the *generalized uncertainty relation*,

$$\Delta Q \Delta P \geq \frac{1}{2} |\langle [\hat{Q}, \hat{P}] \rangle|, \tag{2.22}$$

for two operators $\hat{Q}$ and $\hat{P}$. Plugging in the commutator for $\hat{\sigma}_z$ and $\hat{\sigma}_x$, say, we obtain $\Delta\sigma_z \Delta\sigma_x \geq \frac{1}{2}|\langle 2i\hat{\sigma}_y \rangle| = \langle \hat{\sigma}_y \rangle$, a value which does not generally vanish. Two observables which satisfy such an uncertainty relation are also called *incompatible*. Of all possible observables, one can always only single out a subset

[28] The square ensures that all the deviations are counted positively.

[29] Again we stick to the standard QM-terminology here nevertheless, as was the case with 'observable' or 'system'.

of simultaneously measurable observables, then usually called a *complete*[30] *set of compatible observables* (cf. Dirac 1930, p. 57; Shankar 1994, p. 133).

Recall the two relations first found by Heisenberg (1927), $\Delta x \Delta p_x \geq \frac{\hbar}{2}$ and $\Delta E \Delta t \geq \frac{\hbar}{2}$. The former is a straightforward instance of (2.22), in virtue of the commutator $[\hat{x}, \hat{p}_x] = i\hbar\mathbb{1}$.[31] The latter, however, can only be derived by appeal to some time dependent observable, due to the nonexistence of a *time-operator* in QM.

Regardless of this fact, one can hence, according to QM, never 'know to arbitrary accuracy' how fast a system is moving and where it is at any given time, and one can also never gain such knowledge w.r.t. the energy a system has at a certain point in time. This is certainly remarkable, but what, precisely, does it *mean*? What is this limitation of knowledge *due to*? Is it the case that the two properties cannot simultaneously *exist* on a given system? Or do we, as finite human beings, simply face limitations in *access* to the *true* state of the system? Again, this is a question that will require considerable attention below; but for now we should merely keep it in the back of our (itching) heads.

Our treatment of spin is straightforwardly comparable to a possible treatment of polarizations. More precisely, it is not just comparable, but in circularly polarized photons the direction (clockwise or counter-clockwise) of the field vector rotating ('*spinning*') around the axis of propagation of the photon, also called its *helicity*, also gives rise to a kind of spin with two possible values, namely $\pm\hbar$—photons are *bosons*. Hence one could define a basis $\mathcal{B} = \{|\circlearrowright\rangle, |\circlearrowleft\rangle\}$ of positive and negative helicity states and set up a mathematical story exactly like that of the electron spin above. Alternatively, linear polarization states $|\leftrightarrow\rangle$ and $|\updownarrow\rangle$ (denoting horizontal and vertical polarization respectively) could be used as a basis, where the helicity states would appear as superpositions $|\pm\rangle = \frac{1}{\sqrt{2}}(|\leftrightarrow\rangle \pm i\,|\updownarrow\rangle)$ (the plus sign is the $|\circlearrowright\rangle$ state). This treatment corresponds closely to that of spin, and one can also use Pauli matrices for the description of clockwise or counterclockwise polarization observables respectively.

How is this *at all* related to our discussion of waves in Sect. 2.1.1? First note that, using the means of Fourier transformation, we can not only bring wave functions but also the SE (in both versions) into a momentum-form. More generally speaking, both representations may be considered as special instances of the two abstract ('basis invariant') equations

---

[30]For asserting 'completeness' it is here assumed to be sufficient that for N-tuples of eigenvalues $(\lambda_1, \ldots \lambda_N)$ of the operators in a commuting set $\left\{\hat{A}_k\right\}_{k=1}^{N}$ there are simultaneous eigenvectors $\left|\psi_{\lambda_1,\ldots\lambda_N}\right\rangle$, such that the set of these simultaneous eigenvectors span the space $\mathcal{H}$ (cf. Ruetsche 2011, p. 200).

[31]Here $\hat{p}_x$ may be represented as $-i\hbar\frac{\partial}{\partial x}$ on a space of functions of position, $\psi(x)$, and $\hat{x}$ merely multiplies the latter by x. Then one sees immediately that $x\left(-i\hbar\frac{\partial\psi}{\partial x}\right) - \left(-i\hbar\frac{\partial}{\partial x}(x\psi)\right) = -i\hbar x\frac{\partial\psi}{\partial x} + i\hbar x\frac{\partial\psi}{\partial x} + i\hbar\psi\frac{\partial x}{\partial x} = i\hbar\psi$. Note that in a space of momentum-dependent functions, $\tilde{\psi}(p)$, $\hat{x}$ would be a derivative w.r.t. p_x, and $\hat{p}_x$ a multiplication-operator.

$$i\hbar \frac{\partial}{\partial t} |\psi(t)\rangle = \hat{H} |\psi(t)\rangle \tag{2.23}$$

$$\hat{H} |\psi\rangle = E |\psi\rangle . \tag{2.24}$$

In fact the SSE, $\hat{H}\psi = E\psi$, is now recognizable an *eigenvalue-equation* like the ones we had considered in the spin examples above. $\hat{H}$ is a self-adjoint operator, representing an observable (energy). Similarly, the momentum operator we had introduced before is the representation of a momentum observable.

But how do wave functions and -packets relate to bras and kets? To understand the connections between the formalisms, consider that an expression for the vector that solves the SE in its abstract form can be given in a manner similar to that of the spin vector in (2.18). This expansion in terms of basis vectors of some suitable (separable or finite; cf. Appendix A) Hilbert space $\mathcal{H}$ with basis $\{|\varphi_j\rangle\}_{j\in J}$ (J a set of indices) can then be written as

$$|\psi\rangle = \mathbb{1}\, |\psi\rangle = \sum_j |\varphi_j\rangle\langle\varphi_j|\, |\psi\rangle = \sum_j \langle\varphi_j|\psi\rangle\, |\varphi_j\rangle . \tag{2.25}$$

For the more general case of a non-separable (rigged Hilbert) space (again, cf. Appendix A) with 'basis' $\{|\boldsymbol{p}\rangle\}_{\boldsymbol{p}\in Q}$ (Q some suitable indexing set for this case) we can write

$$|\psi\rangle = \mathbb{1}\, |\psi\rangle = \int_{\mathbb{R}^3} |\boldsymbol{p}\rangle\langle\boldsymbol{p}|\, |\psi\rangle\, \mathrm{d}^3\boldsymbol{p} = \int_{\mathbb{R}^3} \langle\boldsymbol{p}|\psi\rangle\, |\boldsymbol{p}\rangle\, \mathrm{d}^3\boldsymbol{p} , \tag{2.26}$$

and multiplying from the left with $\langle\boldsymbol{x}|$, we obtain

$$\langle\boldsymbol{x}|\psi\rangle = \int_{\mathbb{R}^3} \langle\boldsymbol{p}|\psi\rangle\, \langle\boldsymbol{x}|\boldsymbol{p}\rangle\, \mathrm{d}^3\boldsymbol{p} . \tag{2.27}$$

Realizing that $\hat{\boldsymbol{p}}\,\langle\boldsymbol{x}|\boldsymbol{p}\rangle = \langle\boldsymbol{x}|\hat{\boldsymbol{p}}|\boldsymbol{p}\rangle = \boldsymbol{p}\,\langle\boldsymbol{x}|\boldsymbol{p}\rangle$ and imposing a 'normalization' requirement (see below), the expression $\langle\boldsymbol{x}|\boldsymbol{p}\rangle$ can be regarded as yielding $\langle\boldsymbol{x}|\boldsymbol{p}\rangle = (2\pi\hbar)^{-3/2} e^{\frac{i}{\hbar}\boldsymbol{p}\boldsymbol{x}}$, so that we retain the familiar form

$$\psi(\boldsymbol{x}) = \frac{1}{\sqrt{(2\pi\hbar)^3}} \int_{\mathbb{R}^3} e^{\frac{i}{\hbar}\boldsymbol{p}\boldsymbol{x}}\, \tilde{\psi}(\boldsymbol{p})\mathrm{d}^3\boldsymbol{p} \tag{2.28}$$

of a wave packet. The functions $\langle\boldsymbol{x}|\boldsymbol{p}\rangle = (2\pi\hbar)^{-3/2} e^{\frac{i}{\hbar}\boldsymbol{p}\boldsymbol{x}}$ may, in fact, be thought of as 'normalized' *eigenfunctions* of the momentum operator $\hat{\boldsymbol{p}}$ in position space, and as forming an 'orthonormal' basis of a rigged Hilbert space or a 'pseudo basis' of the Hilbert space $L^2(\mathbb{R}^3)$ of square integrable wave functions (e.g. Schwindt 2013, p. 80 ff.). This means that for two given momenta $\boldsymbol{p}_0$ and $\boldsymbol{p}_1$ we have

$$\langle \boldsymbol{p}_1|\boldsymbol{p}_0\rangle = \int_{\mathbb{R}^3} \langle \boldsymbol{p}_1|\boldsymbol{x}\rangle\langle \boldsymbol{x}|\boldsymbol{p}_0\rangle \, \mathrm{d}^3\boldsymbol{x} = \int_{\mathbb{R}^3} \left((2\pi\hbar)^{-3/2}\right)^2 e^{-i\frac{\boldsymbol{p}_1}{\hbar}\boldsymbol{x}} e^{i\frac{\boldsymbol{p}_0}{\hbar}\boldsymbol{x}} \mathrm{d}^3\boldsymbol{x}$$
$$= \frac{1}{(2\pi\hbar)^3} \int_{\mathbb{R}^3} e^{i\frac{\boldsymbol{p}_0-\boldsymbol{p}_1}{\hbar}\boldsymbol{x}} \mathrm{d}^3\boldsymbol{x} = \delta^3(\boldsymbol{p}_0 - \boldsymbol{p}_1)$$

(which incidentally explains the factor $(2\pi\hbar)^{-3/2}$ as a 'normalization', in this sense). Their Fourier transformed versions are Dirac-δs, and the Dirac-δs in position space which constitute the eigen-'functions' of $\hat{\boldsymbol{x}}$ in that space have transforms $\langle \boldsymbol{p}|\boldsymbol{x}\rangle = (2\pi\hbar)^{-3/2} e^{-\frac{i}{\hbar}\boldsymbol{p}\boldsymbol{x}}$. The Fourier transformation roughly acts as a *change of basis* between the two representations. We can hence see how the functions in question also constitute a formal *vector space*.

Above, in the discussion of the generalized uncertainty relation, we also mentioned a *position operator* $\hat{x}$, which can be generalized to a vector operator $\hat{\boldsymbol{x}}$ with 'eigenvectors' $|\boldsymbol{x}_0\rangle$ of eigenvalue $\boldsymbol{x}_0$.[32] Since we demand $\hat{\boldsymbol{x}}\,|\boldsymbol{x}_0\rangle \stackrel{!}{=} \boldsymbol{x}_0\,|\boldsymbol{x}_0\rangle$, we can define such an operator either simply as 'multiplication by $\boldsymbol{x}_0$', or we can give it a concrete look by appeal to the identity operator $\mathbb{1}$ in the continuous space of $|\boldsymbol{x}\rangle$s,[33]

$$\hat{\boldsymbol{x}} = \int_{\mathbb{R}^3} \boldsymbol{x}\,|\boldsymbol{x}\rangle\langle\boldsymbol{x}|\,\mathrm{d}^3\boldsymbol{x} \tag{2.29}$$

(cf. Ballentine 2000, p. 23; Manoukian 2007, p. 38) so that we obtain

$$\hat{\boldsymbol{x}}\,|\boldsymbol{x}_0\rangle = \int_{\mathbb{R}^3} \boldsymbol{x}\,|\boldsymbol{x}\rangle\langle\boldsymbol{x}|\boldsymbol{x}_0\rangle\,\mathrm{d}^3\boldsymbol{x} = \int_{\mathbb{R}^3} \boldsymbol{x}\,|\boldsymbol{x}\rangle\,\delta^3(\boldsymbol{x}_0 - \boldsymbol{x})\,\mathrm{d}^3\boldsymbol{x} = \boldsymbol{x}_0\,|\boldsymbol{x}_0\rangle$$

as desired.

Obviously, all three spatial components are predicted to be simultaneously measurable, which is a desirable consequence. The use of continuous kets $|\boldsymbol{x}\rangle$, however, constitutes a gross (if useful) idealization. One could think of them as 'functions' $\phi_{\boldsymbol{x}_0}(\boldsymbol{x})$ of space, in the sense that $\phi_{\boldsymbol{x}_0}(\boldsymbol{x}) = \langle\boldsymbol{x}|\boldsymbol{x}_0\rangle = \delta^3(\boldsymbol{x}_0 - \boldsymbol{x})$. This means that their value at position $\boldsymbol{x}$ is either zero or *infinity*, the latter being incidentally the *norm* ('length') of the corresponding ket-vectors. They may be viewed (and are typically used) as the mathematical description of *perfectly* localized 'point particles'.

But this of course suggest that not *all* of the QM formalism can be indicative of something physical. The formalism needs some careful sorting-out. Teller (1995, p. 48), for instance, calls the Dirac-δ "a most effective tool in spite of constituting a physical fiction", and Dirac himself also speculated that "the infinite length of the ket vectors corresponding to these eigenstates is connected with their unrealizability,

[32] With an eye on Appendix A and the brief discussion of rigged Hilbert spaces therein, this means, strictly speaking, that $\langle\varphi\hat{\boldsymbol{x}}|\boldsymbol{x}_0\rangle = \boldsymbol{x}_0\,\langle\varphi|\boldsymbol{x}_0\rangle$ for $\varphi \in \Phi$ (cf. de la Madrid 2005, p. 302).

[33] A more thorough definition of the position operator is possible in terms of operator valued measures (cf. Appendix A and Heinosaari and Ziman 2012, pp. 128–131).

and that all realizable states correspond to ket vectors that can be normalized and that form a Hilbert space." (Dirac 1930, p. 48) It is clear that 'point particles' would not only be immeasurable, but that the very concept raises a whole bunch of further conceptual worries—if by 'pointlike' one truly means *extensionless*. As computational tools Dirac-δs and position kets are perfectly fine, and they simplify a range of applications a whole lot.

Consider e.g. the generalization of (2.20) to the continuous case. Using position-kets, we can write such things as

$$\langle Q\rangle_{\psi} = \langle\psi|\hat{Q}|\psi\rangle = \int \langle\psi|\boldsymbol{x}\rangle\langle\boldsymbol{x}|\,\hat{Q}\,|\psi\rangle\,\mathrm{d}^3\boldsymbol{x} = \int \psi^*(\boldsymbol{x})\hat{Q}\psi(\boldsymbol{x})\mathrm{d}^3\boldsymbol{x}. \tag{2.30}$$

Now surprisingly, our *wave function* here appears merely as a 'statistical tool', something akin to a probability density function; and in Sect. 2.1.1, we did use the wave function to also define a probability density $|\psi(\boldsymbol{x})|^2 = \psi^*(\boldsymbol{x})\psi(\boldsymbol{x}) = \langle\psi|\boldsymbol{x}\rangle\langle\boldsymbol{x}|\psi\rangle$, which we can use in expressions like $\Pr^{\psi}_{\boldsymbol{x}}(\boldsymbol{x} \in \Delta) = \int_{\Delta} |\psi(\boldsymbol{x})|^2\,\mathrm{d}^3\boldsymbol{x}$ to compute the probability of finding a particle in some volume Δ. It may be tempting to suppose that this is *precisely* and *exclusively* the role of the wave function: a statistical tool. But it needs to be spelled out very carefully what that means, as the subsequent discussion will show. Importantly, we here speak of the probability of '*finding* a particle in some volume', which is a standard way of putting things—something that *Bell* (1990a, p. 39) found particularly objectionable—and we will see in Chap. 4 whence the caution against 'being' instead of 'finding'.

We have now abstracted very much from the suggestive use of wave functions as representatives of waves. And we can see that the calculations can be perfectly done without ever really talking about waves. In fact, historically an *abstract* calculus for providing quantum mechanical predictions about atomic events came *first*, namely Heisenberg's (1925) *matrix mechanics*, extended and elaborated on subsequently in (partly) joint work with Born and Jordan (1925) and Born et al. (1926).[34] Schrödinger (1926) very soon noticed that the two formalisms could be connected to one another. But it was not until the proof of the so called *Stone-von Neumann theorem* that a rigorous sort of *equivalence* between Heisenberg's matrices and Schrödinger's wave functions could be demonstrated.[35]

The talk here is of *unitary equivalence*. It made both original formalisms (wave functions and matrices) identifiable as special instances of the general Hilbert space structure which is most simply expressed in terms of kets. Two Hilbert spaces $\mathcal{H}$ and $\mathcal{H}'$, equipped with families of operators $\left\{\hat{O}_j\right\}_{j\in J}$ and $\left\{\hat{O}'_j\right\}_{j\in J}$, are called unitarily equivalent in case there is a unitary (i.e. bijective, linear, norm-preserving) map U which maps the vectors of $\mathcal{H}$ onto those of $\mathcal{H}'$ and connects the operators

34 Weinberg (2013, pp. 14–21) gives a nice overview of some matrix mechanics; so the interested reader may be referred there.

35 For general discussion cf. Ruetsche (2011, chapters 2 and 3); for a statement of the theorem cf. p. 41 therein; and for proofs cf. the references therein.

$\left\{\hat{O}_j\right\}_{j\in J}$ defined on $\mathcal{H}$ to that defined on $\mathcal{H}'$ *via* $\hat{O}_j = U^{-1}\hat{O}'_j U$ (cf. Ruetsche 2011, p. 26). But unitary equivalence is a somewhat subtle matter: simple examples of *quantized theories*—i.e., loosely speaking, theories wherein *operators* occur in places where corresponding classical theories would have *functions*; cf. Ruetsche (2011, chapter 2) for a by far more rigorous treatment—with a free parameter can be constructed (using e.g. a particle confined to a ring), which are not unitarily equivalent to one another for different values of that parameter. More generally speaking this phenomenon occurs whenever the phase space of the underlying classical theory is not $\mathbb{R}^{2n}$ (cf. Ruetsche 2011, p. 57 ff.). For many practical purposes these subtleties do not matter, and ordinary QM calculations can be handled by blunt appeal to kets in virtue of the Stone-von Neumann theorem. We will touch on some further implications of unitary *in*equivalence, and in how far it must be taken to suggest *physical* inequivalence, in the context of QFT though (Sect. 2.2).

Another gross idealization, like the use of Dirac-δs, should be noted here: in constructing our wave packet above, we have considered the example of a *free* system which may itself constitute a kind of physical fiction. For when can a system ever 'truly' count as free? As far as we know, there will always be an abundance of other systems in the universe that it can interact with; even Neutrinos interact with other matter, if ever so weakly. A similar remark is made by Auyang (1995, p. 37), who calls the free particle in QFT "an approximation or idealization, for particles form an interacting system." But she also notes that this idealization is of course put to good use in the form of initial and final states of scattering problems.

The same thing can be said about free particles in non-relativistic QM, in scattering problems or, say, in solid state physics, where the free electron gas proves to be a very useful (approximate) model for many purposes (cf. Ashcroft and Mermin 1976, p. 29 ff.). But to investigate a certain system's behavior, we inevitably have to subject it to a *measurement procedure* and hence confine it to the extensions of some laboratory equipment, thereby imposing a (not generally negligible) potential upon it. Besides the fact that a discussion of the notion of 'measurement' is indicated (which will follow in Sects. 2.1.4 and 2.1.5), it will hence be instructive to investigate a simple but somewhat more realistic example *with* a potential.

In general the restrictions imposed upon the wave function (normalizability in some interval, direction of travel, periodicity, smoothness at boundary points,...) by a problem with given potential often lead to interesting consequences. In *atomic* physics, for instance, one of QM's first major successes, spherical coordinates (r, ϑ, φ) instead of Cartesian ones (x, y, z) are preferable for the description of the atomic electrons, due to the spherical symmetry of the problem. The Laplacian then assumes the form $\Delta = \frac{1}{r^2}\frac{\partial}{\partial r}\left(r^2\frac{\partial}{\partial r}\right) + \frac{1}{r^2 \sin\vartheta}\frac{\partial}{\partial\vartheta}\left(\sin\vartheta\frac{\partial}{\partial\vartheta}\right) + \frac{1}{r^2\sin^2\vartheta}\frac{\partial^2}{\partial\varphi^2}$ in virtue of the coordinate transformation, and since one finds that the merely angular part is proportional to the square of the orbital angular momentum operator $\hat{\boldsymbol{L}} = \boldsymbol{r} \times \hat{\boldsymbol{p}}$, this leads to a theory of discretized angular momentum for the electrons in question. The eigenfunctions $Y_{\ell,m}(\vartheta, \varphi)$ of $\hat{\boldsymbol{L}}^2$ are parametrized by an integer parameter ℓ (e.g.

Foot 2005, p. 23 ff) which, according to the naïve view, could be thought of (turning a blind eye on some subtleties for the moment) as indicating how 'electron waves', localized in the proximity of the nucleus, can take on various shapes for different values of ℓ, the familiar *orbitals* often depicted in textbooks (e.g. Basdevant and Dalibard 2002, p. 199).[36]

While we have largely abandoned it already, atomic physics hence gives a certain *prima facie* plausibility to our Conjecture 0; well known phenomena such as discrete energy spectra could be understood somewhat intuitively in terms of reconfigurations of 'electron-waves' and the associated emission of energy in the form of photons. Still, it is not clear how the naïve view squares with everything else that has been said, and we will shortly provide good reasons to abandon it completely.

Let us first, however, study a simple exemplary system with a potential in a little more detail,[37] which will serve to introduce a few further (important) details, and incidentally constitute a toy example for more involved systems of greater philosophical relevance to be discussed later. Our example is a model of the ammonia (NH_3) molecule for low energies, in which only a finite set of energy eigenvectors $|E_j\rangle$ turns out to be relevant. In a simple model of the molecule, the three hydrogen atoms (H_3) together may be thought of as defining a triangle with the nitrogen atom (N) located on either side ('left' or 'right') of it. The potential energy function that can be used to describe the 'confinement' of the N-atom on either side of the H_3-triangle is a so-called *double well potential* (cf. Fig. 2.8a). In the present case it may be approximated by an even simpler 'two-box'-potential with 'infinite outer walls', where the latter represent the neglect of the possibility that the N-atom could exit the system altogether (cf. Fig. 2.8b).

In this approximation and for low energies, the SSE can be solved comparatively easily with the aid of certain restrictive considerations (continuity in all domains, vanishing at the boundaries...). And when one does, two solutions are found that fit the constraints equally well. One of these is called *symmetric* (ψ_s) and the other one *antisymmetric* (ψ_a), reflecting the mirror symmetry of their graphs w.r.t. the origin (cf. Fig. 2.8b). What is crucial is that these two solutions correspond to different energy levels (the two lowest ones) that can be measured with the aid of spectroscopy, and are offset by some value $E_a - E_s = \Delta E$.

Since we are here only concerned with *two* (relevant) states, the NH_3-molecule can be thought of as another example of a two-state system, at least when sufficiently shielded so that it cannot be excited to higher energy states. The two (normalized) eigenvectors $|E_s\rangle$ and $|E_a\rangle$ of the (low-energy) Hamiltonian, corresponding to ψ_s

[36] Note, however, that atomic physics already constitutes an example where fully *analytic* treatments without approximations are rare; strictly speaking this is only possible for the single-electron problem, i.e. for the hydrogen atom.

[37] We here closely follow Basdevant and Dalibard (2002, p. 80 ff. and p. 120 ff.).

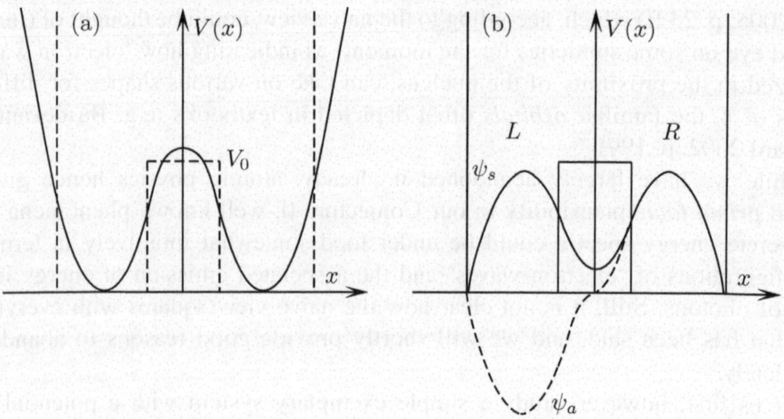

Fig. 2.8 (**a**) The typical double well potential (solid curve) can be approximated by a potential that is infinite at the boundaries and has a finite value (V_0) in the middle (dashed lines), so as to allow for a simple treatment. (**b**) The two possible wave functions which solve the SE for low energies. (Cf. also Basdevant and Dalibard 2002, pp. 79–80)

and ψ_a respectively, form an ONB for this model, and the Hamiltonian can then be put in a simple diagonal matrix form

$$\hat{H} \doteq \begin{pmatrix} E_s & 0 \\ 0 & E_a \end{pmatrix}$$

where

$$|E_s\rangle \doteq \begin{pmatrix} 1 \\ 0 \end{pmatrix}, \quad |E_a\rangle \doteq \begin{pmatrix} 0 \\ 1 \end{pmatrix}, \qquad \hat{H}\,|E_s\rangle = E_s\,|E_s\rangle\,, \quad \text{and} \quad \hat{H}\,|E_a\rangle = E_a\,|E_a\rangle\,.$$

According to the dynamical superposition principle, any linear combination $|\phi\rangle = \mu\,|E_a\rangle + \nu\,|E_s\rangle$ is also a solution, with $|\mu|^2 + |\nu|^2 = 1$ in order for μ and ν to determine the probabilities of finding E_s or E_a on the system. To describe the time evolution of the (unmeasured, isolated) system, we may consider each of the aforementioned vectors $|E_a\rangle$ and $|E_s\rangle$ as functions of time in an initial state at an initial time $t_0 = 0$. A so called *time evolution operator* that 'encodes' the Schrödinger dynamics[38] can then be developed as

[38]The more general case of $t_0 \neq 0$ would require $\hat{U}(t_0; t) = e^{-\frac{i}{\hbar}\hat{H}(t-t_0)}$. If we then let $t - t_0 = \epsilon$, we can write $|\psi(t_0 + \epsilon)\rangle = e^{-\frac{i}{\hbar}\hat{H}\epsilon}\,|\psi(t_0)\rangle = (1 - \frac{i}{\hbar}\hat{H}\epsilon + \mathcal{O}(\epsilon^2))\,|\psi(t_0)\rangle \Leftrightarrow i\hbar\frac{|\psi(t_0+\epsilon)\rangle - |\psi(t)\rangle}{\epsilon} = \hat{H}\,|\psi(t_0)\rangle + \mathcal{O}(\epsilon)\,|\psi(t_0)\rangle$, where $\mathcal{O}(\epsilon^k)$ means 'terms of order ϵ^k' (i.e. wherein ϵ occurs with powers $\geq k$). The last equation obviously gives the TDSE for $\epsilon \to 0$.

$$\hat{U}(t) = e^{-\frac{i}{\hbar}\hat{H}t} = \sum_{n=0}^{\infty} \frac{(-it/\hbar)^n}{n!}\hat{H}^n \doteq \sum_{n=0}^{\infty} \frac{(-it/\hbar)^n}{n!} \begin{pmatrix} E_s^n & 0 \\ 0 & E_a^n \end{pmatrix} =$$

$$= \begin{pmatrix} \sum_{n=0}^{\infty} \frac{(-iE_st/\hbar)^n}{n!} & 0 \\ 0 & \sum_{n=0}^{\infty} \frac{(-iE_at/\hbar)^n}{n!} \end{pmatrix} = \begin{pmatrix} e^{-iE_st/\hbar} & 0 \\ 0 & e^{-iE_at/\hbar} \end{pmatrix},$$

where $\hat{H}^n$ means n successive applications of $\hat{H}$, here represented by an n-fold matrix product, and where we have appealed to the power series expansion $e^{\hat{A}} = \sum_{n=0}^{\infty} \frac{\hat{A}^n}{n!}$, which defines the exponential map on operators. Using this time evolution operator on, say, $|E_s\rangle$ gives $\begin{pmatrix} e^{-iE_st/\hbar} \\ 0 \end{pmatrix}$, or equally $\langle x|\hat{U}(t)|E_s\rangle = \psi_s(x) \cdot e^{-iE_st/\hbar} =: \Psi_s(x,t)$, if we revert back to thinking of $|E_s\rangle$ in terms of a function of position.

These solutions will represent the particle as 'spread out',[39] much like the momentum eigenstates we encountered before; the energy eigenstates do not assign 'sharp', definite positions either, and typically energy eigenstates are also eigenstates of momentum.[40] Defining an approximate (coarse grained) position operator $\hat{X}$ with eigenvectors $|L\rangle$ (hydrogen atom approximately left of the H_3 triangle) and $|R\rangle$ (hydrogen atom approximately right of the H_3 triangle), it turns out that this operator (matrix) is not diagonal in the basis $\{|E_a\rangle, |E_s\rangle\}$, so that $|R\rangle = \frac{1}{\sqrt{2}}(|E_a\rangle + |E_s\rangle)$ and $|L\rangle = \frac{1}{\sqrt{2}}(|E_a\rangle - |E_s\rangle)$.[41] Thus when the molecule is measured to have a certain energy, the N-atom cannot be assigned a definite position, and when the N-atom is approximately localized, the system has none of the definite (measurable) energy values!

Assuming that we start out at $t = 0$ with a state approximately localized in the right domain (R in Fig. 2.8b), we have $|\phi(0)\rangle = |R\rangle = \frac{1}{\sqrt{2}}(|E_a\rangle + |E_s\rangle)$, and the time-evolved states is given by $|\phi(t)\rangle = \frac{1}{\sqrt{2}}(|E_a\rangle\, e^{-iE_at/\hbar} + |E_s\rangle\, e^{-iE_st/\hbar}) = \frac{e^{-iE_st/\hbar}}{\sqrt{2}}(|E_a\rangle\, e^{-i\Delta Et/\hbar} + |E_s\rangle)$. As will be demonstrated later, an overall factor like $e^{-iE_st/\hbar}$ has no observable consequences, and so using the discretization of the two energy states, $\Delta E = \hbar\omega$, we can think of the system as *oscillating* back and forth between the two localized states with frequency $\omega = \Delta E/\hbar$ (cf. Basdevant and Dalibard 2002, p. 83). This is so because (keeping in mind the Euler formula, (2.6))

---

[39]One sometimes reads the term 'dislocalization' in this connection, but it does usually more harm than good. In solid state physics, 'dislocalization' has the more specialized meaning of electron wave functions in partially filled bands of the solid having strongly overlapping supports, whence none is really 'localized' at a particular nucleus. This phenomenon is in fact connected to such familiar properties as *conductivity* (cf. Gross and Marx 2012, p. 134 ff.). This is to be sharply distinguished from '*quantum nonlocality*' though, as we shall see in Chap. 4 (cf. also Zeh 2012, p. 84 on this potential confusion).

[40]Subtleties arise e.g. for *Bloch waves* in crystals (cf. Gross and Marx 2012, p. 341).

[41]For the given choices $\hat{X} = \frac{1}{2}\begin{pmatrix} R+L & R-L \\ R-L & R+L \end{pmatrix}$ or, with $1 = L = -R$, $\hat{X} = \begin{pmatrix} 0 & 1 \\ 1 & 0 \end{pmatrix}$ (cf. Basdevant and Dalibard 2002, p. 122).

for times $t = (2\ell + 1)\pi/\omega$ ($\ell \in \mathbb{Z}$), the state will be proportional to $|L\rangle$, and for times $t = 2\ell\pi/\omega$, the state will be proportional to $|R\rangle$ again. This insight is not 'merely academic' but has empirical consequences: the existence of this frequency is exploited in technological implementations such as the stimulated emission used in a *maser* (read: 'microwave-laser'; cf. Basdevant and Dalibard 2002, p. 124 ff.).

But we here also encounter another interesting quantum phenomenon not discussed so far. Since there is a potential wall in the middle which the system is capable of crossing, we have an instance of *quantum tunneling* on our hands, which features in the explanation of well-known process such as the α-decay of nuclids or scanning-tunneling microscopy (e.g. Bleck-Neuhaus 2013, pp. 189–190). There is a non-zero probability of the N-atom being in the *middle* (i.e. inside the potential wall) which is a situation unfamiliar from classical physics. This tunneling-feature is another 'decidedly quantum'-phenomenon. Again closing an eye on many details, according to our present paradigm (Conjecture 0) this 'tunneling' need not appear as so much of a shock to our intuition, since the tunneling—as we have depicted the situation in Fig. 2.8b—could then be understood as the wave's ability to cross the barrier set up by the potential. And that a wavelike entity could penetrated something like a 'wall' is certainly less difficult to imagine than a 'solid ball' suddenly appearing on the opposite side of this wall (think e.g. of a radio sounding through concrete). This is basically the (faulty) image suggested by many a textbook treatment.

From our example we can, however, also extract some quite general features of the theory, regardless of interpretation. For all systems whose Hamiltonian is not time-dependent, the time evolution is described by an operator of the general form $e^{-\frac{i}{\hbar}\hat{H}t}$ (although not always with an actual matrix representation). For systems with time-dependent Hamiltonian, the method of compiling the time-evolution operator becomes sightly more involved and requires integration and time-ordering (cf. Schwabl 2007, p. 293 ff.). This does not, however, change the fact that time evolution hence defined is always *unitary*, which means that $\hat{U}^\dagger\hat{U}(t) = \hat{U}\hat{U}^\dagger(t) = \mathbb{1}$ ($\forall t$). This, in turn, implies that inner products, *viz. the probabilistic relations*, between state vectors are preserved over time.

This unitary evolution hence implies a surprising sort of *determinism*: the wave function at one point in time determines the wave function at all other points in time in a way that preserves the magnitudes relevant for probabilistic predictions; a feature that Hughes (1989, p. 116) refers to as *statistical determinism*, and to which Born (1926, p. 804) similarly remarked that, while particles described by the wave function behave probabilistically, *probability itself* was governed by a 'causal law'.[42]

Despite the intuitability of the pictorial descriptions given above, our naïve view has certainly already suffered many 'hits' to its plausibility from the previous discussion. How are all these *abstract* features of QM to be interpreted in terms

[42]It is obvious that Born had a deterministic notion of causality in mind, which we know is not necessarily always apt (e.g. Paul and Hall 2013, p. 63 ff., for some discussion).

of a wavy, field-like entity that can collapse onto narrow spots? Moreover, it seems that the TDSE cannot be all there is to the evolution of the system on account of Conjecture 0, since nothing in it predicts or even allows for the kind of 'collapse of the wave-packet' required by that conjecture. This may merely suggest *modifications* to the dynamics, but we noted in several cases that one has to 'close an eye on the details' to make sense of things in terms of waves in the first place. So while we had above given some reasons why many physicists still like to talk in terms of waves and particles, we should now go into the decisive reasons for abandoning our naïve view altogether, i.e. the reasons why philosophers of science typically "think that wave-particle duality is an obsolete concept", as Falkenburg (2007, p. 31) had it.

2.1.3 Quantum 'Waves' Are No Ordinary Waves... Nor Are There Ordinary Particles

The first thing to emphasize is that we have never really left the *complex* domain in our discussion of quantum 'waves'—a fact that Schrödinger (1926a, p. 139) was well aware of in his original development of wave mechanics, and which he considered to constitute a 'certain difficulty' at the time.[43] Above, we have *only* dealt with real-valued quantities when we were concerned with probabilities or probability densities. So one could be tempted to think that the probability density is actually the truly physical magnitude after all, which then collapses upon measurement; and this temptation is also reflected, for instance, in the fact that the aforementioned 'wavelike' graphs associated with atomic orbitals *are* actually depictions of the *probability densities* (this was one subtlety that we closed our eyes on). Yet they are often thought of as depicting the electron's 'real situation' by practitioners. In fact, Schrödinger (1926a, p. 134 ff.) himself originally attempted to interpret $|\psi|^2$ as a physical magnitude instead of ψ; but straightforward reasons can be given why this can at best be a pragmatic approximation under favorable circumstances.

A first reason is that in computing the probability density for a double slit experiment, say, we had to appeal to the *wave functions* first, and *not* to the probability densities directly, in order to derive the correct predictions ($|\psi_1|^2 + |\psi_2|^2 \neq |\psi_{1+2}|^2$). This already strongly suggests that the interpretation of $|\psi|^2$ as describing the distribution of something wave-like is not at all viable. But not even in the case of a single free particle, traveling undisturbed in space, does resorting to

[43] German: "Eine gewisse härte liegt ohne Zweifel zurzeit noch in der Verwendung einer *komplexen* Wellenfunktion." (emphasis in original)

probability densities provide a suitable physical picture, as Heisenberg discovered already in 1927 (cf. p. 187 ff.). To see this, consider a free wave packet[44]

$$\psi(x,t) = \frac{1}{\sqrt{2\pi\hbar}} \int \mathrm{d}p\, A(p) e^{i(px-Et)/\hbar}.$$

Assume now that the weighting $A(p)$ is a Gaußian, centered around some value p_0, i.e., $A(p) = \pi^{-\frac{1}{4}}(\sigma\hbar)^{-\frac{1}{2}} e^{-\frac{(p-p_0)^2}{2(\hbar\sigma)^2}}$. We can then execute the integral to obtain

$$\psi(x,t) = \frac{1}{\sqrt{2\sigma\pi^{\frac{3}{4}}\hbar}} \int \mathrm{d}p\, e^{-\frac{(p-p_0)^2}{2(\sigma\hbar)^2}} e^{\frac{i}{\hbar}(px-Et)} = \frac{1}{\sqrt{2\sigma\pi^{\frac{3}{4}}\hbar}} \int \mathrm{d}p\, e^{\frac{ipx}{\hbar} - \frac{(p-p_0)^2}{2(\sigma\hbar)^2} - \frac{ip^2t}{2m\hbar}} =$$

$$= \frac{1}{\sqrt{2\sigma\pi^{\frac{3}{4}}\hbar}} e^{-\frac{p_0^2}{2(\sigma\hbar)^2}} \int \mathrm{d}p\, e^{-\frac{1}{2}\left(\frac{1}{(\sigma\hbar)^2} + \frac{it}{m\hbar}\right)p^2 + \left(\frac{p_0}{(\sigma\hbar)^2} + \frac{ix}{\hbar}\right)p},$$

and defining $a := \frac{1}{(\sigma\hbar)^2} + \frac{it}{m\hbar}$ and $b := \frac{p_0}{(\sigma\hbar)^2} + \frac{ix}{\hbar}$ this computes

$$\psi(x,t) = \frac{1}{\sqrt{2\sigma\pi^{\frac{3}{4}}\hbar}} e^{-\frac{p_0^2}{2(\sigma\hbar)^2}} \int \mathrm{d}p\, e^{-\frac{a}{2}p^2 + bp} = \frac{1}{\sqrt{a\sigma\pi^{\frac{1}{4}}\hbar}} e^{-\frac{p_0^2}{2(\sigma\hbar)^2}} e^{\frac{b^2}{2a}} =$$

$$\overset{\text{inserting } a,b}{=} \frac{1}{\pi^{\frac{1}{4}}\sqrt{\frac{1}{\sigma} + \frac{i\hbar\sigma}{m}t}} \exp\left(\frac{\left(\frac{p_0}{(\sigma\hbar)^2} + \frac{ix}{\hbar}\right)^2}{2\left(\frac{1}{(\sigma\hbar)^2} + \frac{it}{m\hbar}\right)}\right) e^{-\frac{p_0^2}{2(\sigma\hbar)^2}}.$$

It is a straightforward (if tedious) exercise to confirm that at $t = 0$, our wave packet satisfies the minimum of the uncertainty relation, $\Delta x_0 \Delta p_0 = \frac{\hbar}{2}$, where $\Delta x_0 = \frac{1}{\sigma\sqrt{2}}$, $\Delta p_0 = \frac{\sigma\hbar}{\sqrt{2}}$, so we can replace $\sigma = \frac{1}{\Delta x_0\sqrt{2}}$. In another (equally tedious) calculation, one obtains the probability density

$$|\psi(x,t)|^2 = \frac{1}{\sqrt{2\pi}\,\Delta x(t)} e^{-\frac{(x-v_0t)^2}{2(\Delta x(t))^2}} \tag{2.31}$$

with $\Delta x(t) = \sqrt{(\Delta x_0)^2 + \left(\frac{\Delta p_0}{m}t\right)^2}$, and $v_0 = \frac{p_0}{m}$ the velocity of the center of the wave packet.

At $t = 0$ this function hence describes a narrow wave packet, a nicely tied up lump which could very well be considered as describing something capable of accounting for particle-like behavior. But the characteristic *width* of this wave

[44] Since the generalization to $3D$ is straightforward, we are here limiting our attention to one dimension, and we also omit the reference to the domain of integration for simplicity. We shall avail ourselves of both these simplifications more often in what follows.

packet, $\Delta x(t)$, is now *time dependent*, and the packet will *spread* very rapidly.[45] Schlosshauer (2007, p. 117), for instance, describes the behavior of an electron as predicted by the evolution of the wave packet above, using the experimental value for the electron's size as an initial spread. Initially, the electron would thus be a well-localized entity, but in just an amount of one second the probability density would be spread out as far as 1000 km, so that the initially localized electron should extend from, say, Cologne to Rome, with about an equal probability of being found as a dot-like entity anywhere in between. This certainly constitutes a massive blow to our zeroth conjecture.

We had argued above, however, that the 'free electron' may in a sense be a fictitious concept anyways, and we will later see that there are resources available in modern QM that overcome at least this difficulty of the spreading wave-packet when the *environment* is taken into account.

But even though the spreading can be overcome in principle, the interpretation of state vectors or wave functions ψ as representing actual waves is still not straightforwardly viable, and the most compelling reason to reject this picture of 'real waves' traveling in $3D$ space is that in the majority of cases the wave function ψ cannot be considered as a function of merely one set of $3D$ coordinates ($\psi(\boldsymbol{x})$), but instead depends, in an *inseparable* fashion,[46] on many (N) such coordinates ($\psi(\boldsymbol{x}_1, \boldsymbol{x}_2, \ldots, \boldsymbol{x}_N)$). These cases can arise when N 'particles' or systems have to be considered at once, and here ψ is not defined over a $3D$ space anymore, but only over an abstract *configuration space*, specifying the joint configuration of all the N particles in question (e.g. Ballentine 2000, p. 99; Ney 2013, p. 15 ff.). This insight incidentally served as a key motivation for Born (1926, p. 240) to develop a thoroughly 'statistical' interpretation,[47]—the likes of which we will find wanting however, by and large in virtue of exactly the same phenomenon which precludes the naïve wave-view (non-separability/entanglement).

Schrödinger (1926b, p. 526) was of course well aware of this problem as well, and he originally suggested to deal with it by integrating out the remaining coordinates in the $|\psi|^2$ of a many-particle system so as to obtain a description of one particle alone. *Via* multiplication by the electric charge e, he thought, one could

[45]Schrödinger was, in fact, not fond of the idea of *collapses* at all, but sought for a theory *purely* in terms of waves. Besides the conflict with (almost) point-like measurements, he also wrestled with this difficulty of the spreading wave packet. In doing so, he discovered an outstanding example, the coherent states of the harmonic oscillator potential, which most closely mimic classical behavior. But the generalization was not straightforwardly possible, and the example remained a solitary one (cf. Bitbol 1996, p. 46; Schlosshauer 2007, p. 117).

[46]'Inseparability' can have multiple levels here: for one it can mean that a separation-ansatz for the TDSE does not work, in the sense that one cannot factor a solution of the TDSE for the different coordinates of *one* system. But the meaning of interest here concerns multiple systems and will be made precise only in Sect. 2.1.5.

[47]"[D]ie Schrödingerschen Wellen laufen ja gar nicht im gewöhnlichen Raume, sondern im 'Konfigurationsraume', der soviele Dimensionen hat, als die Anzahl der Freiheitsgrade des betrachteten Systems beträgt ($3N$-Dimensionen für N Partikel)." (Born 1926, p. 240)

then understand it as a *charge density* (cf. Schrödinger 1926a, p. 134). Sometimes such a treatment is still pragmatically applied, for instance in *nuclear physics* (e.g. Blatt and Weisskopf 1979, p. 24), where comparatively large masses are involved and a 'semi-classical' treatment often becomes possible. But Schrödinger viewed this only as a first approximation for various good reasons (cf. Bacciagaluppi 2010, pp. 15–16), and it *cannot* yield a general solution to the dependence on multiple coordinates, as will become clear from our later discussion of *entanglement*.

One particular way in which the dependence on multiple coordinates can arise is the following. Consider two quantum systems, e.g. two electrons, which are not distinguishable by any of their measurable, purportedly intrinsic properties (spin, charge, mass). These could be the electrons of some atom or we could equally think of particles in a scattering experiment. Depending on the circumstances, we can attribute an individual quantum state to each of them, but since they are not distinguishable in the aforementioned sense, we should recognize that it cannot make any observable difference *which* of the electrons is in *which* of the two possible states. This can be expressed by requiring that an *exchange* of the two particles in the overall wave function $\Psi(\boldsymbol{x}_1, \boldsymbol{x}_2)$ that describes the total two-electron system should result in a wave function which makes the *exact same predictions*.

In our discussion of polarization, we mentioned a 'phase argument' in the sinusoidal functions, which would shift the components of the $\boldsymbol{E}$-vector relative to one another. Since our 'waves' in QM are represented by complex exponentials, we can introduce a phase by multiplying the wave function by $e^{i\theta}$, with θ the phase argument. But $e^{i\theta}$ is equally just a complex number of unit modulus that essentially leaves the state vector unchanged. For consider the kets $|\chi\rangle$ and $|\chi'\rangle = e^{i\theta}\,|\chi\rangle$. Then the average of some observable O will be

$$\langle\chi'|\hat{O}|\chi'\rangle = \langle\chi|e^{-i\theta}\,\hat{O}e^{i\theta}|\chi\rangle = \langle\chi|e^{0}\hat{O}|\chi\rangle = \langle\chi|\hat{O}|\chi\rangle,$$

and equally

$$|\langle\phi|\chi'\rangle|^2 = \langle\chi'|\phi\rangle\langle\phi|\chi'\rangle = e^{-i\theta}\,\langle\chi|\phi\rangle\,\langle\phi|\chi\rangle\,e^{i\theta} = \langle\chi|\phi\rangle\,\langle\phi|\chi\rangle = |\langle\phi|\chi\rangle|^2,$$

so all statistical predictions remain unchanged. Above, we have assumed that θ is a *global* phase, but multiplication by a *local* one, i.e. one where $\theta = \theta(\boldsymbol{x}, t)$, is an entirely different matter. This requires a modification of the Schrödinger equation, in particular the introduction of additional *gauge field* terms, such as the electromagnetic potential (e.g. Aharonov and Rohrlich 2005, p. 45 ff.; Dick 2012, p. 258). And a *relative* phase between two vectors in a superposition state *does* matter statistically (global *or* local), as we will see in detail later.

Since the overall global phase does *not* change the predictions, however, we can rephrase the requirement of observational invariance of the total wave function $\Psi(\boldsymbol{x}_1, \boldsymbol{x}_2)$ under particle exchange by defining a *permutation operator* $\mathbb{P}_{12}$ which simply swaps the coordinates of the two systems ($\mathbb{P}_{12}\Psi(\boldsymbol{x}_1, \boldsymbol{x}_2) = \Psi(\boldsymbol{x}_2, \boldsymbol{x}_1)$), and by further requiring that

$$\mathbb{P}_{12}\Psi(\boldsymbol{x}_1, \boldsymbol{x}_2) = e^{i\theta}\Psi(\boldsymbol{x}_1, \boldsymbol{x}_2).$$

But it should also be the case, that two switches in position bring back *exactly* the old state, i.e.

$$\mathbb{P}_{12}(\mathbb{P}_{12}\Psi(\boldsymbol{x}_1, \boldsymbol{x}_2)) = \mathbb{P}_{12}^2\Psi(\boldsymbol{x}_1, \boldsymbol{x}_2) = \Psi(\boldsymbol{x}_1, \boldsymbol{x}_2).$$

This can only hold if the eigenvalues of $\mathbb{P}_{12}$ are in fact $+1$ and -1 (i.e. $\theta = n\pi, n \in \mathbb{Z}$). Hence, there are only two possible sorts of wave functions which satisfy the requirement, those with eigenvalue $+1$ for the permutation operator (called 'symmetric') and those with -1 (called 'antisymmetric').

But to cover *both* of these cases, it cannot generally hold that $\Psi(\boldsymbol{x}_1, \boldsymbol{x}_2) = \psi_a(\boldsymbol{x}_1)\psi_b(\boldsymbol{x}_2)$, because then $\mathbb{P}_{12}\Psi(\boldsymbol{x}_1, \boldsymbol{x}_2) = \Psi(\boldsymbol{x}_2, \boldsymbol{x}_1) = \psi_a(\boldsymbol{x}_2)\psi_b(\boldsymbol{x}_1) \neq -\psi_a(\boldsymbol{x}_1)\psi_b(\boldsymbol{x}_2)$ (for $a = b$, the $+1$ case is obviously possible). Functions of the form

$$\Psi(\boldsymbol{x}_1, \boldsymbol{x}_2) = \frac{1}{\sqrt{2}}(\psi_a(\boldsymbol{x}_1)\psi_b(\boldsymbol{x}_2) \pm \psi_a(\boldsymbol{x}_2)\psi_b(\boldsymbol{x}_1))$$

however satisfy the symmetry requirement. Including also spin vectors will lead to further possibilities; one then obtains a so called *Weyl-* or *Pauli-spinor*[48] which can be represented as a column vector of wave functions and is an element (for the special case of two fermionic particles) of the tensor product space $L^2(\mathbb{R}^6) \otimes \mathbb{C}^4$.[49] This does not change the general theme.

Experimental evidence suggests that there are indeed two general classes of quantum systems, and these are the *bosons* and *fermions* we mentioned in the earlier discussion of spin (cf. Sect. 2.1.2), where the former are described by an overall *symmetric* wave function (+) and the latter by an overall *antisymmetric* one (−). In case of multiple particles, one of course has to include more terms and the appropriate states can be expanded in terms of determinants and permanents (cf. Lancaster and Blundell 2014, p. 40). For fermions, a symmetric wave function has to be combined with an anti-symmetric spin vector, whereas the anti-symmetric one can be combined with any symmetric spin vector, and for bosons the situation is opposite. Hence for a fermionic two particle system one obtains the (well-known) possibilities:

[48]It is called a 'Weyl spinor' in virtue of Hermann Weyl's (1950) extensive investigation of the mathematics of spin, and a 'Pauli spinor' since it satisfies a modified version of the Schrödinger equation with spin-terms which is due to Wolfgang Pauli (1927, p. 618; cf. also Schwabl 2007, p. 192).

[49]More generally, the wave function of any single particle of spin s $(= 0, \frac{1}{2}, 1, \frac{3}{2}, \ldots)$ will be an element of the space $L^2(\mathbb{R}^3) \otimes \mathbb{C}^{2s+1}$, or equally $L^2(\mathbb{R}^3; \mathbb{C}^{2s+1})$. And for N indistinguishable particles, this will be $L^2(\mathbb{R}^{3N}; \mathbb{C}^{(2s+1)N})$ (cf. Gustafson and Sigal 2011, pp. 22 and 35).

$$\text{Anti-symmetric: } \frac{1}{\sqrt{2}}(|\uparrow\rangle\,|\downarrow\rangle - |\downarrow\rangle\,|\uparrow\rangle) \quad (\text{'singlet'})$$

$$\text{Symmetric: } \begin{cases} |\uparrow\rangle|\uparrow\rangle \\ |\downarrow\rangle|\downarrow\rangle \\ \frac{1}{\sqrt{2}}(|\uparrow\rangle\,|\downarrow\rangle + |\downarrow\rangle\,|\uparrow\rangle) \end{cases} \quad (\text{'triplet'})$$

From purely algebraic considerations (cf. Ballentine 2000, p. 162; Griffiths 1995, p. 148) one can work out that the square $\hat{\boldsymbol{j}}^2$ of some angular momentum operator $\hat{\boldsymbol{j}}$ has eigenvalues $j(j+1)\hbar^2$, where j is called the *quantum number* of $\hat{\boldsymbol{j}}$; and for $\hat{\boldsymbol{s}}^2$ this yields $s(s+1)\hbar^2$, which gives $\frac{3}{4}\hbar^2$ for the case $s = \frac{1}{2}$. To generalize this to the joint states above, one can instead define single particle operators $\hat{\boldsymbol{s}} \otimes \mathbb{1}$, $\mathbb{1} \otimes \hat{\boldsymbol{s}}$ and a total spin operator $\hat{\boldsymbol{S}} = \hat{\boldsymbol{s}} \otimes \mathbb{1} + \mathbb{1} \otimes \hat{\boldsymbol{s}}$, for whose square one obtains the value 0 on the singlet and $2\hbar^2$ on all triplet states, so that the *total* spin quantum number S must be 0 for the singlet and 1 for the triplet.

Combining, say, the singlet with the symmetric position wave function yields

$$\boldsymbol{\Psi}(\boldsymbol{x}_1, \boldsymbol{x}_2) = \langle \boldsymbol{x}_1 | \boldsymbol{x}_2 \rangle \left(\frac{1}{\sqrt{2}}(|\psi_a\rangle\,|\psi_b\rangle + |\psi_b\rangle\,|\psi_a\rangle) \otimes \frac{1}{\sqrt{2}}(|\uparrow\rangle\,|\downarrow\rangle - |\downarrow\rangle\,|\uparrow\rangle) \right),$$

which can be represented more concretely as

$$\begin{aligned} \boldsymbol{\Psi}(\boldsymbol{x}_1, \boldsymbol{x}_2) &= \frac{1}{2}(\psi_a(\boldsymbol{x}_1)\psi_b(\boldsymbol{x}_2) + \psi_a(\boldsymbol{x}_2)\psi_b(\boldsymbol{x}_1)) \cdot \begin{pmatrix} 0 \\ 1 \\ -1 \\ 0 \end{pmatrix} \\ &= \begin{pmatrix} 0 \\ \frac{\psi_a(\boldsymbol{x}_1)\psi_b(\boldsymbol{x}_2) + \psi_a(\boldsymbol{x}_2)\psi_b(\boldsymbol{x}_1)}{2} \\ -\frac{\psi_a(\boldsymbol{x}_1)\psi_b(\boldsymbol{x}_2) + \psi_a(\boldsymbol{x}_2)\psi_b(\boldsymbol{x}_1)}{2} \\ 0 \end{pmatrix}, \end{aligned}$$

where we use a boldface $\boldsymbol{\Psi}$ to indicate that the object in question has the form of a column-vector(-field). To ensure antisymmetry for such a spinor, permutations must be taken to also affect the ordering in the spin states though, so it is better to think of permutations as exchanging the *kets*, rather than spatial coordinates:

$$\begin{aligned} \mathbb{P}_{12}\boldsymbol{\Psi}(\boldsymbol{x}_1, \boldsymbol{x}_2) =& \langle \boldsymbol{x}_1 | \boldsymbol{x}_2 \rangle\, \mathbb{P}_{12} \left(\frac{1}{\sqrt{2}}(|\psi_a\rangle\,|\psi_b\rangle + |\psi_b\rangle\,|\psi_a\rangle) \otimes \frac{1}{\sqrt{2}}(|\uparrow\rangle\,|\downarrow\rangle - |\downarrow\rangle\,|\uparrow\rangle) \right) = \\ &= \langle \boldsymbol{x}_1 | \boldsymbol{x}_2 \rangle \left(\frac{1}{\sqrt{2}}(|\psi_b\rangle\,|\psi_a\rangle + |\psi_a\rangle\,|\psi_b\rangle) \otimes \frac{1}{\sqrt{2}}(|\downarrow\rangle\,|\uparrow\rangle - |\uparrow\rangle\,|\downarrow\rangle) \right) = -\boldsymbol{\Psi}(\boldsymbol{x}_1, \boldsymbol{x}_2). \end{aligned}$$

It is certainly puzzling that the state for a system of multiple indistinguishable particles should be given as such a complex 'entangled' superposition state. For what actually *is* the state of any of the involved particles, in an *intuitable* sense of the word?[50] However, the empirical predictions derived by appeal to these kinds of states are supported by numerous impressive confirmations such as multiplet structures in atomic spectroscopy (e.g. Demtröder 2010, p. 198) or the significant differences between cross sections in scattering experiments with distinguishable and indistinguishable particles (e.g. Bleck-Neuhaus 2013, pp. 145–152).

Any state which is *inevitably* described as a superposition of products of single particle states, i.e. *cannot* be written as a product state $|\psi\rangle\,|\phi\rangle$ in *any* basis, is called *entangled*. Moreover, it is a theorem that any vector $|\Psi\rangle$ from a tensor product space $\mathcal{H}_1 \otimes \mathcal{H}_2$ can be written in the form $|\Psi\rangle = \sum_{j=1}^{d} \alpha_j\,|\phi_j\rangle\,|\psi_j\rangle$, with $\{|\phi\rangle_j\}_{j=1}^{d_1}, \{|\psi\rangle_j\}_{j=1}^{d_2}$ *orthogonal* (or orthonormal) bases of $\mathcal{H}_1$ and $\mathcal{H}_2$ respectively, and $d = \min\{d_1, d_2\}$. This is called the *Schmidt-* or *biorthogonal decomposition*.[51] A vector may then equivalently be called entangled *iff* there are *at least two* coefficients α_j which are non-zero in the Schmidt decomposition (cf. Heinosaari and Ziman 2012, p. 263). How deep the philosophical implications of the occurrence of such states in QM run will become clear in the subsequent discussion.

The symmetry constraints discussed above incidentally motivate the infamous *Pauli Principle*: No $\otimes$-factor in a term of a permutation-symmetric pure state occurs more than once.[52] The reason should be obvious for the two-fermion wave function $\Psi(\boldsymbol{x}_1, \boldsymbol{x}_2) = \frac{1}{\sqrt{2}}(\psi_a(\boldsymbol{x}_1)\psi_b(\boldsymbol{x}_2) - \psi_a(\boldsymbol{x}_2)\psi_b(\boldsymbol{x}_1))$, as $a = b$ would here simply give $\Psi = 0$. Indistinguishability considerations and the requirement of invariance under permutations is but one source of entanglement, and other, equally important examples will be discussed later.

With the configuration space we have yet another high dimensional space on our hands, besides the complex, high dimensional Hilbert spaces $\mathcal{H}$ of which the state functions or kets are elements. If ψ is a function describing the positions of N quantum systems in $3D$ space, the number of dimensions of the configuration space will be $3N$. So *only* for a single, isolated quantum system will the configuration space coincide with the 'physical' $3D$ space, which can be seen as a major source of confusion about the status of 'wave'-functions in QM.

---

[50] The complex of problems arising from this is incidentally one of the most thoroughly discussed ones in the philosophy of QM (traditionally in the context of Leibniz' principle of the identity of indiscernibles). A nice historical and systematic overview can be found in Muller and Saunders (2008, pp. 505–508). See also Hawley (2009), and French's and Krause's (2006) comprehensive exposition of the topic for more details on the philosophical debate.

[51] For separable, countably infinite-dimensional spaces, the same can be established in terms of Fock space representations (cf. Horodecki et al. 2009, p. 918).

[52] We here appeal to Muller's (2014, p. 426) version which is less ambiguous than the standard (textbook) one. The textbook version has it that no two fermions can be in the same state. But it has been emphasized (e.g. Muller and Saunders 2008, p. 511; Muller 2014, p. 422) that each fermion in a compound of multiple fermions of the same type may too occupy the 'same state' as given by its reduced density operator (see later).

A wave-interpretation in the sense of Conjecture 0 appears to be straightforwardly impossible, or at the least *very* hard to establish.[53] The role of the configuration space clearly requires clarification in this context. But what does QM talk about then, if not waves? Could it simply be strangely behaving particles instead? Duane (1923) in fact suggested a pure particle interpretation of certain diffraction processes in crystals and Landé (1965a) based an attempted to provide new foundations for QM on Duane's work. These attempts, however, were at first largely ignored (compare Landé's complaints in his (1965b) article *Why Do Quantum Theorists Ignore the Quantum Theory?*), and later harshly criticized on various grounds (cf. Mehra and Rechenberg 1987, p. 1202 ff.). We will later turn to *Bohmian Mechanics* though, an example of a particle-interpretation that is still vividly discussed today; but we will then also see that Bohmian Mechanics requires us to revise many of our 'classical' convictions about particles and measurements.

Some general clarification of the particle concept seems indicated in any case, since the word 'particle' is used quite frequently in modern physics. After all, a whole, vivid branch of modern physics carries the name 'particle physics'; ironically one of the most important applications of quantum *field* theory. What are these 'particles' according to QM or QFT? So far we have entertained a rather intuitive ('classical') notion of 'particles' as tiny little objects which occupy some point (or rather: small region) in space at any given time. But from all that has been said so far, this image has become very unlikely. This has lead Muller (2014, p. 424), among others, to confront the following dilemma:

> must we, as philosophers of science and of physics, (i) charge modern physicists with conceptual confusion and ontological delusion because of their persistent talk and detection of particles whereas they have no particle concept and thus are babbling incoherently when they utter the word "particle", or (ii) conclude that there are particles in QM and QFT, as in CM [classical mechanics—FB], just as there are bears in America, Asia and Europe, but that the particles in QM, QFT and CM differ in kind, just as the cinnamon bears of Colorado, the Tibetan blue bears and the brown bears of the Pyrenees differ in subspecies.

The case for taking horn (ii) of the dilemma has been made, among others, by Falkenburg (2007). Falkenburg has compiled lists of requirements that anything has to fulfill in order to either count as a classical or a quantum mechanical particle, a light quantum, field quantum, virtual particle, or quasi particle. This is followed, in each case, by an extensive discussion of the meanings and interrelations of these concepts. Subsequently Muller (2014, p. 424) has condensed Falkenburg's lists of criteria for classical particles, QM particles, field quanta, and light quanta into one short list of what he simply calls *particles*:

---

[53] A potential counterexample is Cramer's (1986) *transactional* interpretation which *does* interpret QM in terms of waves in spacetime—but on the cost of also accepting waves that can travel *backwards in time*. Most importantly, Cramer's interpretation is riddled with difficulties, whence we deliberately choose not to bother with it any further here (for details, the reader is referred to Maudlin 2011, p. 180 ff.).

(PROP)	they have some intrinsic properties,
(INDEP)	are independent of each other,
(POINT)	point-like in interactions,
(CONS)	obey conservation laws,
(LOCD)	localizable by a detector,
(DISC)	discontinuous, i.e. they come in quanta (of matter or radiation).

Debatable candidates[54] for *intrinsic* properties of light quanta (photons), say, are spin and energy; for the equally massless gluons (the 'force carriers' of the strong interaction), one could add the color charge; for other classes of particles, mass, charge etc. come to mind.

Importantly, a definite *localization* at all times is *not* presupposed here, but merely such features as discontinuity (i.e. only discrete spots or single clicks are observed in experiments), localizability by detectors (either this detector in a lab over there clicks, or the other one over here), and a weak form of independence, in the sense of the very *possibility* of uncoupled states and uncorrelated initial conditions (cf. Falkenburg 2007, p. 212). This is why it is even meaningful to talk about particles at all. Such a restricted concept is certainly indicated, whence, when we use the word 'particle' bluntly in the following, we will usually mean nothing more than something which satisfies Muller's condensed list of criteria. In fact, since this is such a radical departure from the classical concept, we will occasionally refer to particles in this sense as 'partĩcles' instead, when highlighting the contrast to the classical (or classical-like) concept is indicated.

So we have established a cautious particle-concept, with which we can proceed. Note, however, that this does not mean that there truly is a distinguished class of *fundamental* entities which satisfies *even these* quantum mechanically informed criteria. Particles (or rather: the *impression* of such) could be created from something more fundamental, something entirely different. Weinberg (1995, p. 1), for instance, thinks that: "The underlying theory might not be a theory of fields *or* particles, but perhaps of something quite different, like strings." (emphasis in original) And Ruetsche (2011) has even gone so far as to suggest that the particle notions *used* by physicists in QFT are not even "fundamental in the '*physicist's*' sense." (p. 248; my emphasis—FB) This "physicist's sense" is distinguished from the metaphysician's sense in that former is (roughly) concerned with magnitudes in terms of which all other magnitudes can be determined, whereas for the latter this need not be so. She contrasts such a fundamental notion with a *phenomenological* one, that "makes sense of explanatory and experimental practices [. . .]." (p. 248) The Muller-Falkenburg list arguably provides a phenomenological particle concept.

[54] To each of these supposedly intrinsic properties one could obviously object that their meaning is only defined w.r.t. the interaction of the system with some other system, which one could flesh out to yield a thorough relationalism.

Since *probability* is such a central issue in QM, it seems desirable to also assign a more precise meaning to the word 'probability' as well, in order to really make sense of what is *meant* by statements such as 'the probability of finding a particle in this or that region is p'. An exhaustive and satisfying treatment is obviously impossible here, but we will give a quick overview of different positions in a first philosophical interlude (I), and address some further issues in Chap. 7. For now an important aim of ours should be to identify possible interpretational *directions* in which to proceed, given that a naïve treatment in terms of waves *or* particles (not particles) is untenable. As a very first step towards this, we should discuss what has come to be known as the 'orthodox' interpretation of QM, and with it the so called *outcome problem*, the central problem associated with the issues raised in this chapter.

2.1.4 The Quantum Postulates, the Outcome Problem, and the Orthodox Interpretation

We are now in a good position to sum up the very foundational ideas underlying QM in the form of a few postulates. These postulates are sometimes also referred to as 'axioms', but as van Fraassen (1980, p. 65) has pointed out, they do not seem to constitute axioms in the narrower sense of the word, as used in logic or geometry. The postulates are given in various forms in the literature; our exposition rests on the formulations of Held (2012, p. 75), Schwindt (2013, pp. 15–16), and Shankar (1994, p. 116).

(I) A quantum system S is associated with a Hilbert space $\mathcal{H}$ and its state at time t is represented by a vector $|\psi(t)\rangle$ in $\mathcal{H}$.
(II) If $|a\rangle, |b\rangle \in \mathcal{H}$ represent states, then so does any linear combination $|\psi\rangle = \lambda |a\rangle + \mu |b\rangle$, $\lambda, \mu \in \mathbb{C}$ (unless prohibited by a superselection rule).
(III) A physical observable A is represented by a (self-adjoint) operator $\hat{A}$ on $\mathcal{H}$ and the values of A for S are represented by numbers in the spectrum of $\hat{A}$.
(IV) The temporal evolution of the vector $|\psi(t)\rangle$ associated with S is governed by a unitary time evolution $|\psi(t)\rangle = \hat{U}(t_0; t)\,|\psi(t_0)\rangle$, where $\hat{U}(t_0; t)$ is an exponential in the Hamilton operator $\hat{H}$, representing the total energy of S.
(V) Observable A has value a on S *iff* the state of S is given by $|a\rangle$ (with $\hat{A}\,|a\rangle = a\,|a\rangle$ and $\hat{A}$ representing A).
(VI) If S is in a state represented by the (normalized) state vector $|\psi(t)\rangle$, A is an observable for S with some value a, and $|a\rangle$ is a state such that $\hat{A}|a\rangle = a|a\rangle$, then $|\langle a|\psi(t)\rangle|^2$ gives the probability of finding value a for A on S in some measurement procedure for A.

So far this is not really news, but in essence a summary of what was established with the aid of examples above. The sixth postulate is simply Born's rule, which may of course be read with 'probability density' instead of 'probability' in case $\hat{A}$ has a continuous spectrum. Postulate (II) is the (kinematical) superposition principle, which is often times omitted, presumably because many (textbook) expositions take to be implied by (I) or (IV). But we already noted the difference between a kinematical and a dynamical view of superposition, and nowhere in (I) it is stated that *any* $|\psi\rangle \in \mathcal{H}$ (not prohibited by a superselection rule) qualifies as the representation of a state; it could just be the elements of some preferred orthonormal basis, neither of which would then be a superposition of the others.

Postulate (V) is usually called the *eigenvalue-eigenstate link* (e.g. Bub 1997, p. 29; short: *EE-link*), which was explicitly stated by Dirac (1930, p. 35) and assumed equally by von Neumann (1932, e.g. p. 216). The second part of the condition, that observable A *only* has value a on S if S is in state $|a\rangle$, is certainly non-trivial and raises part of the controversy in QM, as we will see later.

A bunch of questions offer themselves when we take a critical look on these postulates. First of all, how is it that we come to 'measure' definite eigenvalues of position (or coarse-grainings thereof), say, when the solutions to the TDSE hardly represent well localized states? What happens in the measurement process? Are we deluded about what *really* goes on when we believe to find definite values in certain measurement processes? Or is QM simply an incomplete assessment of the actual physical situation, since its unitary evolution cannot tell the whole story?

The questions arising from this tension are often subsumed under the term *measurement problem*. But it is not easy to point out what exactly 'the' measurement problem is; Maudlin (1995) alone distinguishes three separate measurement problems, as does Schlosshauer (2007, p. 50), but not (all) the same ones. We should hence think of 'the measurement problem' as short for 'the measurement problem-*complex*', and we will touch on some further elements of this complex later. But possibly the most important and 'most drastic' problem, invoked by both, Schlosshauer and Maudlin, is what is often called the *outcome problem* (OP), and—due to its importance—sometimes even identified as *the* measurement problem (e.g. Esfeld 2012, p. 88; Jaeger 2009, p. 77; Bub 1997, p. 2):

> The following three claims are mutually inconsistent.
>
> A The wave-function of a system is *complete*, i.e. the wave-function specifies (directly or indirectly) all of the physical properties of a system.
> B The wave-function always evolves in accord with a linear dynamical equation (e.g. the Schrödinger equation).
> C Measurements of, e.g., the spin of an electron always (or at least usually) have determinate outcomes, i.e., at the end of the measurement the measuring device is either in a state which indicates spin up (and not down) or spin down (and not up).
>
> (Maudlin 1995, p. 7; emphasis in original)

The truly problematic point about the postulates, stated without further qualification, is hence that they imply a contradiction. If the unitary time evolution is taken to be a complete description of the behavior of a system, then it should, for instance, never (or hardly ever; cf. the oscillations between coarse-position eigenstates in the toy example of the ammonia molecule in Sect. 2.1.1) be found in an eigenstate of position. Historically, the problem was 'solved' by Dirac (1930, p. 7) and von Neumann (1932, p. 217) by adding a seventh postulate, usually called the *projection postulate* (PP):

(VII) When observable A is measured on system S in state $|\psi(t)\rangle$, the state of S undergoes a sudden change $|\psi\rangle \mapsto \frac{\hat{P}_a|\psi\rangle}{\|\hat{P}_a|\psi\rangle\|}$ with probability $|\langle a|\psi(t)\rangle|^2$.

$\hat{P}_a = |a\rangle\langle a|$ is the projection operator onto the subspace spanned by $|a\rangle$, and so upon conclusion of this process, the sate of the system is the (normalized) eigenstate $|a\rangle$ of A. This final set of postulates we refer to (essentially following Bub (1997) and Stapp (2009)) as the *orthodox interpretation* of QM. But this set of postulates certainly only constitutes a *minimal* interpretation in the sense of Sect. 1.2; it *only* provides an algorithm (or rather: heuristic) for how to connect (parts of) the formalism to experience. Thus beyond ensuring physical *practice*, it leaves us with a bunch of open questions as to the underlying *reality*, i.e., as to the processes and structures in virtue of which this algorithm can be used.

One obvious such question is what the sudden change according to postulate (VII) actually amounts to. We have ruled out the naïvely plausible interpretation in terms of collapsing waves (Conjecture 0) on the basis of wave functions being generally defined only over an abstract $3N$-dimensional configuration space, and similar obstacles. It is thus still unclear what it *is* that is being reduced in the measurement process (besides the obscure 'quantum state'), and *how* this reduction takes place: Over which time scales? In virtue of what kind of dynamics?

The view that an endorsement of the EE-link and the PP should even *count* as an 'orthodox interpretation' has been challenged by Wallace (2016), based on observations that seem to indicate that these postulates are not really being put to use in practice after all. Among his reasons are that the PP does not appear in higher level courses on QM, and that there are quarrels about the "black hole information loss paradox", which has at its heart a violation of unitarity that should not worry anybody who embraces the—non-unitary—dynamics of the PP (cf. his pp. 4–5). But these objections seem to misfire; the "black hole information loss paradox" does not represent a measurement(like) situation, and orthodoxy only proclaims the 'sudden change' for these. And regardless of whether the PP or the EE-link are being *discussed explicitly* in higher level textbooks, they seem to be somewhat presupposed (at least as limiting cases; cf. the technical details on measurement theory in Sect. 2.1.5) by actual experimental *practice*. How else could, say, Vaziri et al. (2002, p. 1) write that they "were able to demonstrate that an individual photon can be prepared in eigenstates of external angular momentum", if not by assuming that, when subjected to the preparation procedure, the photons are 'forced' into that particular eigenstate, and that their measurement techniques revealed that to be the case? And photons *are*, strictly speaking, a subject matter theoretically treated by *quantum electrodynamics*, a higher level topic. The orthodox view seems to be implicit in physical practice at all levels.[55]

This does not, however, make it *satisfying* at all, and it is still not very clear what actually qualifies as a 'measurement'; i.e., the very *notion* of measurement cries out for explanation in a context as subtle as this. We will give a few more details about how QM formally deals with measurements in the next section, and thereby also touch on questions of the notion of a *property*. But for now, let us first make the following intuitive remarks: *Prima facie*, a measurement consists of some*one* (the 'observer') *using* some*thing* (the 'equipment') to determine some property of something *third* (the 'measured system'). For instance, you could be using a ruler to measure the length of a pencil by comparing the pencil to the scale imprinted on the ruler and reading off a value of that scale. Strictly speaking, we may hence discern three 'stages' of this entire process (cf. also Boge 2016b, p. 7): (i) interaction of the equipment with the measured system, (ii) interaction of the observer with the equipment, and (iii) registration of a value (measurement result) by the observer. A lot to do with the OP depends on the interpretation of (i)–(iii). We should first confront the question of whether *QM itself* is the appropriate framework for a physical analysis of stages (i) and (ii). The traditional answer to this question was 'no', and QM was taken to apply to a special class of measured systems only, whereas the entirety of physical equipment used to investigate was considered as describable only in 'classical terms'. This attitude is evident, for instance, in Landau and Lifshitz's classic, where they write:

> The possibility of a quantitative description of the motion of an electron requires the presence also of physical objects which obey classical mechanics to a sufficient degree of accuracy. If an electron interacts with such a 'classical object', the state of the latter is, generally speaking, altered. The nature and magnitude of this change depend on the state of the electron, and therefore may serve to characterise it quantitatively. In this connection the 'classical object' is usually called *apparatus*, and its interaction with the electron is spoken of as *measurement*. (Landau and Lifshitz 1965, p. 2; emphasis in original.)

The writings of Landau and Lifshitz are obviously heavily inspired by Bohr's views on the subject, who equally (multiply) emphasized the "distinction between the *objects* under investigation and the *measuring instruments* which serve to define, in classical terms, the conditions under which the phenomena appear." (Bohr 1949, p. 30; emphasis in original) What Bohr *exactly* meant by 'classical terms' and to what extent these are the terms of classical *physics* is subject to considerable debate (e.g. Howard 1994 vs. Bokulich and Bokulich 2005). There seems to be at least some agreement that Bohr's 'classicality' mostly subsumed "everyday concepts, eventually refined by the terminology of classical physics[...]." (Bohr 1938, p. 269)

[55]Note that Wallace's (2016, p. 19) cosmological considerations hardly impair this point; Fuchs and Peres (2000) liken the required *selective* applications of QM in cosmology to "a few collective degrees of freedom" to applications of QM in SQUIDs (cf. later) and see "no difference in principle". All evidence gathered about the universe requires 'definite outcomes' in *some* basis, so the orthodox interpretation or something closely related seems to be implicitly at play in the evaluation of the data.

Much in contrast to Landau, Lifshitz and Bohr, Wigner is sometimes quoted to have asked: "But *why* must I describe it [the measurement—FB] in classical terms? What will happen to me if I don't?" (Wigner 1974, as quoted in Jaynes 1980, p. 40; emphasis in original) And many physicists must have felt the same way, whence despite the insistence of Bohr and others on classicality about the measuring apparatus, today there is a flourishing field of measurement theory in QM.

The first thing to note here is that this theory of measurement in QM "has two branches, one dealing with the changes experienced by the measured system, the other one considering measurements as physical processes." (Busch et al. 1995, p. 34) As for the first branch, we will outline some formal details of the QM treatment of state changes in the next section, but we should here emphasize the 'least' reason for this branch's existence.

A central assumption of classical physics is that measurements could at least *in principle* be as subtle and precise as desired, and would not *necessarily* affect the measured system to any considerable extent. But consider yourself looking for a football, say, which is known to lie somewhere in a room that is entirely dark before the search begins. This search for the football may count as a 'measurement' of its position. In order to see the football, one has to turn on the lights and thus hit the football with a tremendous number of photons. Since photons carry energy, they will definitely alter the football's state, but the change will be so subtle that from the point of view of the searcher, the ball may be considered effectively unaltered by the turning on of the light. Nevertheless, for much, much smaller systems, i.e. atoms, electrons, or muons, the changes effected by most measurement processes are not so subtle. In comparison to them, almost anything that can be used by a human being to gain information about them carries a considerable amount of energy, and hence their states must (at least intuitively) be altered to a much larger degree if information is to be gained about them at all.

Heisenberg (1930, pp. 21–22) famously concerned himself with the determination of an electron's position through a microscope as a physical example for the unavoidability of the uncertainty relations, a thought experiment which has become known as the *Heisenberg microscope*. In 1958 (pp. 48–49) he replaced it by an imaginary γ*-ray* microscope because the short wavelengths of the γ-rays would result in an even higher accuracy in position-determination. From the resolving power of the microscope he then deduced the uncertainty in position of the electron. For the very narrow position determination of the γ-ray microscope, there would be a large 'kick' to the electron, giving rise to a high momentum uncertainty, since the electron receives a recoil that can be quantified (*via* Compton scattering) as a function of h/λ (cf. Heisenberg 1930, pp. 21). Thus, as Heisenberg puts it in his 1958 *Physics and Philosophy* (p. 49),

> in the act of observation at least one light quantum of the γ-ray must have passed the microscope and must first have been deflected by the electron. Therefore, the electron has been pushed by the light quantum, it has changed its momentum and its velocity, and one can show that the uncertainty of this change is just big enough to guarantee the validity of the uncertainty relations.

These considerations suggest that the uncertainty relations, in particular, should be understood in terms of a *disturbance* of the system by the act of measurement, and that a lot of the strangeness of QM hence derives from an impossibility to properly *access* a certain domain of physical investigation ('the microcosm' or 'the quantum realm'). In 1930 Heisenberg in fact spoke of a "destruction of the *knowledge* of a particle's momentum by an apparatus determining its position[...]." (p. 21; my emphasis—FB) But one can give a range of arguments to the effect that the disturbance interpretation of the uncertainty relations is incoherent (cf. Redhead 1987, p. 68), and neither did Heisenberg ultimately retain this interpretation nor did it ever do full justice to Bohr's more subtle convictions. In virtue of the arguments given in Chap. 4, we will find ourselves in good company with them.

Crucially, any kind of treatment to the above effect requires the Bohrian distinction between investigated object and apparatus as quantum and classical. These features, however, are not present in what we have called 'the orthodox interpretation' above, and at least von Neumann (1932, e.g. pp. 4 and 6) *explicitly* advocated that QM should in principle be a universally applicable physical theory.

But in fact, the von Neumannian universalist stance towards QM takes the OP to a whole new level, as can be explained by appeal to a quantum mechanical treatment of the measurement process. Turning thus to the latter branch of quantum measurement theory (measurements as physical processes), we note that for any process to truly count as a precise measurement of the value of the observable property in question (on a given system), there is a sensible requirement for part (i) of our analysis (interaction of system and equipment), called the *calibration condition*. This requirement is that "whenever the system is in an eigenstate, the apparatus should indicate the corresponding eigenvalue unambiguously after the interaction has ceased." (Busch and Lahti 2009, p. 374) The requirement can be weakened to include so called *unsharp* measurements which only give good estimates of the value in question (cf. the next section). The *minimal* requirement related to this kind of measurement, called the *probability reproducibility condition*, "stipulates that the probability measure [for the measured system] is 'transcribed' into a probability measure for the pointer observable in the apparatus state reached after a suitable measurement coupling." (Busch et al. 1996, p. 25) Notably, the use of "pointer observable" here already somewhat suggests that apparatus and system are basically on the same footing, i.e. that the measuring device *can* be treated *quantum mechanically*.

Carrying through with this assumption, we can hence model the situation as follows. Prior to the actual observation, system S and measuring apparatus M will interact, exchange energy, and through this common evolution, M should evolve into a state which is indicative of that of S. This process, as we will see,[56] can be described in terms of a unitary operator, and is nowadays often referred to as

[56]In what follows, we provide a modified and partly extended version of the analysis given in Mittelstaedt (1998, p. 29) and similarly in Joos et al. (2003, p. 48 ff.), which are both themselves adaptations of von Neumann's (1932, p. 422) original treatment.

a *premeasurement* interaction (e.g. Busch and Lahti 2009, p. 375). Von Neumann (1932, p. 351) instead called the changes effected by the unitary evolution a *second* kind of intervention, and thought of these 'interventions' as "automatic changes which occur with passage of time". The process associated with the act of measurement he called an intervention of the *first* kind, and analyzed it in terms of projection operators (cf. also Joos et al. 2003, p. 20; Schlosshauer 2007, p. 58).

The premeasurement interaction ensures that the calibration condition is satisfied. The simplest case, on which we will focus here, is a measurement which hardly disturbs the state of S, sometimes called an *ideal* or *quantum nondemolition measurement* (cf. Schlosshauer 2007, p. 52). But a generalization of the story given in what follows is also available for the non-ideal cases (see comments below). Let us say that S is in some definite state $|A_j\rangle$ w.r.t. some observable A, and M is in some 'ready state' $|Z_0\rangle$. The premeasurement should thus shift the state of the combined system consisting of S and M from $|A_j\rangle\,|Z_0\rangle$ to $|A_j\rangle\,|Z_j\rangle$, that is, from a state where M is ready to measure ($|Z_0\rangle$) to a state in which M is indicative of S's state ($|Z_j\rangle$).

We noted before that energy and time satisfy an uncertainty relation just as much as position and momentum do. And we also noted that a unitary operator of the form $\hat{U}(t) = e^{-\frac{i}{\hbar}\hat{H}t}$ will effect a shift in time in a state $|\psi(0)\rangle$, i.e. $\hat{U}(t)\,|\psi(0)\rangle = |\psi(t)\rangle$. But the Hamiltonian is also the operator whose eigenvalues are energies, so a shift in time is essentially effected by the operator whose corresponding observable satisfies an uncertainty relation with time. Indeed, this theme is more general in QM and a shift in position, say, will be effected by an operator of the form $\hat{U}(x) = e^{-\frac{i}{\hbar}x\hat{p}}$. This can easily be seen, keeping in mind the definition of the exponential of an operator, in the application to a position state:

$$\begin{aligned}
\hat{U}(x)\,|\tilde{x}\rangle &= \hat{U}(x)\int \mathrm{d}\tilde{p}\,|\tilde{p}\rangle\langle\tilde{p}||\tilde{x}\rangle = \int \mathrm{d}\tilde{p}\,\hat{U}(x)\,|\tilde{p}\rangle\, e^{-\frac{i}{\hbar}\tilde{p}\tilde{x}} = \\
&= \int \mathrm{d}\tilde{p}\, e^{-\frac{i}{\hbar}\tilde{p}x}\,|\tilde{p}\rangle\, e^{-\frac{i}{\hbar}\tilde{p}\tilde{x}} = \int \mathrm{d}\tilde{p}\, e^{-\frac{i}{\hbar}\tilde{p}(\tilde{x}+x)}\,|\tilde{p}\rangle \\
&= \int \mathrm{d}\tilde{p}\,|\tilde{p}\rangle\langle\tilde{p}|\tilde{x}+x\rangle = |\tilde{x}+x\rangle
\end{aligned} \tag{2.32}$$

(cf. Binney and Skinner 2014, p. 68; cf. also Nakahara 2003, p. 14–15 for a more rigorous proof). Generalization to $3D$ is immediate ($\hat{U}(\boldsymbol{x})\,|\tilde{\boldsymbol{x}}\rangle = |\tilde{\boldsymbol{x}}+\boldsymbol{x}\rangle$), and the adjoint, acting on bras, will effect a positive shift in a given spatial wave function: $\langle \boldsymbol{x}|\hat{U}(\tilde{\boldsymbol{x}})^{\dagger}|\psi\rangle = \langle \boldsymbol{x}+\tilde{\boldsymbol{x}}|\psi\rangle = \psi(\boldsymbol{x}+\tilde{\boldsymbol{x}})$.

This fact can be used to develop a unitary operator which represents the transformation of the system SM[57] due to the premeasurement interaction, the calibration between the system to be measured and the measuring apparatus.

[57] Here we write SM to denote the otherwise unspecified *composition* of systems S and M; although cf. Greaves and Wallace (2013) for some insights on how systems compose.

Consider the observable A and a *pointer observable* Z, which we could think of here, for illustrative purposes, as the position of an actual pointer, but which may of course also refer to something much more general. Assume also that there is a quantity P (the 'pointer momentum') whose associated operator satisfies the same commutation relation with the pointer observable's operator $\hat{Z}$ as do $\hat{x}$ and $\hat{p}$, i.e. $[\hat{Z}, \hat{P}] = i\hbar$. The unitary operator we are presently searching for should satisfy three requirements: It should (a) bring about the time development during the interaction, and (b) effect a shift from $|Z_0\rangle$ to $|Z_j\rangle$, which (c) should indicate the fact that A has value A_j on S. An operator which does this job is given by $\hat{U}_A = e^{-\frac{i}{\hbar}\hat{H}_{\text{int}}\Delta t} = e^{-i\lambda(\hat{A}\otimes\hat{P})}$, where $\hat{H}_{\text{int}} = \frac{\lambda\hbar}{\Delta t}(\hat{A}\otimes\hat{P})$ is the Hamiltonian of the interaction,[58] and λ is a parameter which represents duration (Δt) and strength of this interaction (cf. Mittelstaedt 1998, p. 29).

When applied to the state $|A_j\rangle\;|Z_0\rangle$, this operator will give

$$\hat{U}_A\,|A_j\rangle\;|Z_0\rangle = e^{-i\lambda(\hat{A}\otimes\hat{P})}\,|A_j\rangle\;|Z_0\rangle = \sum_{n=0}^{\infty}\frac{1}{n!}(-i\lambda)^n(\hat{A}\otimes\hat{P})^n\,|A_j\rangle\;|Z_0\rangle =$$

$$= \sum_{n=0}^{\infty}\frac{1}{n!}(-i\lambda)^n(\hat{A})^n\,|A_j\rangle\,(\hat{P})^n\,|Z_0\rangle = \sum_{n=0}^{\infty}\frac{1}{n!}(-i\lambda)^n(A_j)^n\,|A_j\rangle\,(\hat{P})^n\,|Z_0\rangle$$

$$= |A_j\rangle\,e^{-i\lambda(A_j\hat{P})}\,|Z_0\rangle\,.$$

We can now use the similarity of the resulting operator $e^{-i\lambda(A_j\hat{P})}$ and the position-shift operator $\hat{U}(x)$ from above to understand what the former does to $|Z_0\rangle$: it will effect a shift of the value Z_0 to $(Z_0 + \lambda A_j)$, a value which in turn depends on the value measured (by M) for S's observable A, and on the strength and duration of the interaction, i.e., $|Z_0\rangle \xmapsto{e^{-i\lambda(A_j\hat{P})}} |Z_0 + \lambda A_j\rangle$.

This almost establishes how the interaction shifts the pointer to the indicative position of value Z_j, but we still need to motivate why $(Z_0 + \lambda A_j)$ should be this value. In general, the value to be read off of the measurement device will be a *function* f of the pointer observable's value as, for instance, "a measurement of the particle's potential energy is equivalent to a position measurement (up to degeneracy) if the function $V(\boldsymbol{r})$ is given." (Joos et al. 2003, p. 18; emphasis omitted) In our special case we can think of the values we read off of an actual *scale* with an actual pointer, which will then be the number pointed to. The particular number will generally depend on the distance the pointer has moved from its initial position, which in turn depends on the interaction strength. Thus the choice $f(Z_j) := \frac{Z_j - Z_0}{\lambda} = A_j$, i.e. letting the shift in Z relative to Z_0, scaled by the interaction strength λ, correspond to the value of A, establishes a

[58] We stress that not every interaction can be described by a Hamiltonian of this kind, which has *factorizing* eigenstates. For a discussion see Joos et al. (2003, p. 48).

suitable correspondence between readings on M's scale and actual values of A on S. Then $Z_0 + \lambda A_j = Z_j$, and hence, $|Z_0 + \lambda A_j\rangle = |Z_j\rangle$. All in all, we thus have $|A_j\rangle\,|Z_0\rangle \overset{\hat{U}_A}{\longmapsto} |A_j\rangle\,|Z_j\rangle$.

But the same operator brings about the most peculiar features of the measurement process (when analyzed in this fashion), and maybe the most worrisome part of the OP, or the most confusing implication of the kinematical superposition principle. Namely, consider a situation in which the state of the system is a superposition state $\sum_j \alpha_j\,|A_j\rangle$. Then we will obtain[59]

$$\hat{U}_A \sum_j \alpha_j\,|A_j\rangle\,|Z_0\rangle = \sum_j \alpha_j e^{-i\lambda(\hat{A}\otimes\hat{P})}\,|A_j\rangle\,|Z_0\rangle =$$

$$= \sum_j \alpha_j \sum_{n=0}^{\infty} \frac{1}{n!}(-i\lambda)^n (\hat{A}\otimes\hat{P})^n\,|A_j\rangle\,|Z_0\rangle = \sum_j \alpha_j\,|A_j\rangle\,e^{-i\lambda(A_j\hat{P})}\,|Z_0\rangle$$

$$= \sum_j \alpha_j\,|A_j\rangle\,|Z_j\rangle\,. \tag{2.33}$$

The final state of this premeasurement interaction is an *entangled state*, like that of the indistinguishable electrons discussed at the end of Sect. 2.1.3. But where exactly is our pointer pointing now? The unitary time evolution has put the *measuring device* into a superposition state of different *outcomes*, intimately connected to the superposition state of the measured system. This is, on the face of it, an absurdity, and we obviously need some further element in the theory which leads to just *one definite value* for the pointer. After all, that is what we observe in experiments, so this prediction of QM is, *prima facie*, *empirically inadequate*. Even worse: what, in fact, should it even *mean* that the pointer is in a 'superposition of states'?

Surely, here is where the projection postulate has to 'kick in'. Applying a (joint) projection operator $|A_j\rangle\langle A_j|\otimes|Z_j\rangle\langle Z_j|$ to such an entangled state would reduce the sum to a state with definite eigenvalues A_j and Z_j for the operators $\hat{A}\otimes\mathbb{1}$ and $\mathbb{1}\otimes\hat{Z}$. So if there is indeed a dynamical process as indicated by the projection postulate, this would immediately solve the OP (although the dynamics would still require specification). But *where* exactly is the 'here', where the projection postulate comes into play? Somewhere in the larger environment of the system? In the proximity of the measuring apparatus? In the brain of the observer? In the observer's *mind*? This

[59] In the non-ideal case where the system is demolished by the measurement, the evolution may usually be assumed to proceed in a similar fashion as $\sum_j \alpha_j\,|A_j\rangle\,|Z_0\rangle \overset{\hat{U}}{\mapsto} \sum_{j,k} \alpha_{jk}\,|A_j\rangle\,|Z_k\rangle = \sum_k \tilde{\alpha}_k\,|\tilde{A}_k\rangle\,|Z_k\rangle$, with $|\tilde{\alpha}_k|^2 \approx |\alpha_k|^2$ and $|\tilde{A}_k\rangle \approx |A_k\rangle$ so that all that differs is a minor change in the coefficients and the state of the system (e.g. Bub 1997, p. 150). In the most drastic case though, the system gets *destroyed* (i.e., decomposed and/or absorbed into the apparatus). One then merely requires the apparatus states to carry, after the interaction, information about the state of S *before* its destruction (e.g. Wallace 2003, p. 420).

is the question of the so called *Heisenberg cut*, named after Heisenberg's (1934) discussion of the relation between the two evolutions (cf. also Joos et al. 2003, p. 28). *Somewhere* between stages (i)–(iii) in our above analysis, the reduction has to take place, either during the interaction of equipment and system, or during the interaction of observer and equipment, or during the recognition by the observer.

Von Neumann (1932, p. 420) viewed the situation as follows:

> [T]he measurement or the related process of the subjective perception is a new entity relative to the physical environment and is not reducible to the latter. [...] We wish to measure a temperature. If we want, we can pursue this process numerically until we have the temperature of the environment of the mercury container of the thermometer, and then say: this temperature is measured by the thermometer. [W]e can carry the calculation further, [b]ut in any case, no matter how far we calculate [...] at some time we must say: and this is perceived by the observer. That is, we must always divide the world into two parts, the one being the observed system, the other the observer.

To ensure a "psycho-physical parallelism" however (his p. 419), he was satisfied with placing the cut somewhere *outside* the (conscious) observer. Schrödinger, in contrast, was much less at ease with the measurement process described in von Neumann's fashion, as he noticed it to yield all kinds of bizarre consequences:

> A cat is penned up in a steel chamber, along with the following diabolical device (which must be secured against direct interference by the cat): in a Geiger counter there is a tiny bit of radioactive substance, so small, that perhaps in the course of one hour one of the atoms decays, but also, with equal probability, perhaps none; if it happens, the counter tube discharges and through a relay releases a hammer which shatters a small flask of hydrocyanic acid. If one has left this entire system to itself for an hour, one would say that the cat still lives if meanwhile no atom has decayed. The first atomic decay would have poisoned it. The ψ-function of the entire system would express this by having in it the living and the dead cat (pardon the expression) mixed or smeared out in equal parts. (Schrödinger 1935b, p. 157)

Do we really have to accept that cats can be in mixed up states of being dead and alive at the same time? Or should the cat be considered capable of observing its own state, therefore collapsing the wave function of the atom-Geiger counter system? Wigner (1961, p. 172), in fact, considered a similar scenario (anticipated before him by Everett III (1973, pp. 4–6) in his at first unpublished thesis from the 1950s), which basically lead him to this exact conclusion. He, however, first put a human being (and a less violent 'measurement') in place of the cat:

> What is the wave function if my *friend* looked at the place where the flash [on a photographic plate] might show at time t? [...] One could attribute a wave function to the joint system: friend plus object, and this joint system would have a wave function [...] after the interaction [...]. I can then enter into interaction with this joint system by asking my friend whether he saw a flash. If his answer gives me the impression that he did, the joint wave function of friend + object will change into one in which [...] the total wave function is a product [...] and the wave function of the object is ψ_1. If he says no, the wave function of the object is ψ_2 [...]. However, even in this case, in which the observation was carried out by someone else, the typical change in the wave function occurred only when some information (the yes or no of my friend) entered *my consciousness*. (my emphasis—FB)

To avoid the conclusion that the friend could end up "in a state of suspended animation", dependent on Wigner's own choice to 'measure' him, Wigner thus attributed to "*consciousness* [...] a different role in quantum mechanics than [to] the inanimate measuring device [...]." (p. 873; my emphasis—FB) Wigner here gives a special significance to part (iii) of the measurement process, as we have analyzed it above, and effectively believes that conscious observation could *directly* alter physical reality.

What about large inanimate objects like chairs, tables, houses, or planets then? Are they sitting around in funny 'quantum states', waiting to be observed just to assume their to-us familiar form upon this observation? Indeed, this was the content of John Bell's criticism of the orthodox account of QM with the projection postulate in place: "Was the wavefunction of the world waiting to jump for thousands of millions of years until a single-celled living creature appeared? Or did it have to wait a little longer, for some better qualified system... with a PhD?" (Bell 1990a, p. 34) A similarly piercing remark must have been made by Einstein in a personal discussion with Hilary Putnam, who paraphrases him as follows: "Look, I don't believe that when I am not in my bedroom my bed spreads out all over the room, and whenever I open the door and come in it jumps into the corner." (Einstein 1953, as quoted in Putnam 2005, p. 624)

A question that naturally offers itself is whether QM is even relevant or *applicable* at the *scales* of measuring devices, cats and the like. Why should there be a quantum state of the cat or a 'pointer momentum' in the sense of QM, when QM was originally developed as a theory of tiny *atoms*, and when even nuclei allow at least for a *semi*-classical treatment in many respects? Maybe the Bohrian insistence on the treatment of apparatuses as classical (in some sense), and of QM as applicable only to a limited, microscopic domain was apt after all. And maybe this suggests a way of dealing with the OP: The cat or the measuring device cannot be in a quantum superposition, because they are *classical* objects simply in virtue of their *size*. In short: Maybe the Heisenberg cut is just a matter of size.

Indeed, there exists a well-known class of results which seem to suggest something alike. Loosely speaking, they suggest that *on average* a quantum system will behave in such a way as to mimic classical behavior. These results go by the name *Ehrenfest's theorem*, and they can be derived by evaluating the change of the average of an observable in time (cf. Ballentine 2000, p. 390; Schwabl 2007, p. 29). One instance of Ehrenfest's theorem is the relation $\frac{\mathrm{d}\langle \boldsymbol{p}\rangle}{\mathrm{d}t} = -\langle \nabla V(\boldsymbol{x})\rangle$. In classical physics, the (negative) gradient of a potential energy is a *force*, and so is the time derivative of a momentum, whence this relation is indeed reminiscent of classical mechanics, where the formula holds without the averages. With $\hat{\boldsymbol{x}}$ instead of $\hat{\boldsymbol{p}}$, we obtain the second formula $\frac{\mathrm{d}\langle \boldsymbol{x}\rangle}{\mathrm{d}t} = \frac{1}{m}\langle \boldsymbol{p}\rangle$, which says that the temporal change of a system's average spatial coordinates corresponds to an average velocity, again just as in classical mechanics.

This seems to indicate that QM indeed merely applies to entities on the smallest scales and that things only become 'weird' at these very scales—or maybe even just *appear* weird in virtue of the *impact* of the large on the tiny in measurement-interactions—whereas huge accumulations of these weird tiny entities behave just as classical mechanics predicts. Ballentine et al. (1994, p. 2854), however, have argued that only under the condition $-\langle \nabla V(\boldsymbol{x}) \rangle \approx -\nabla V(\langle \boldsymbol{x} \rangle)$ this reasoning from Ehrenfest's theorem is sound, because only then will a classical force act on the average (center of mass) coordinate of all small systems, and this requirement is only fulfilled in special cases (cf. Joos et al. 2003, p. 87; Ballentine 2000, p. 391). And they also (1994, p. 2858) make a (strong) case "that Ehrenfest's theorem is neither necessary nor sufficient to characterize the classical regime in quantum theory", by demonstrating how the predictions from Ehrenfest's theorem and those from classical physics differ significantly in many cases.

Additionally, Joos et al. (2003, p. 2) have argued that

> it remains unexplained why macro-objects come only in narrow wave packets, even though the superposition principle allows far more 'nonclassical' states[...]. Measurement-like processes would necessarily produce nonclassical macroscopic states as a consequence of the unitary Schrödinger dynamics.

In summary, Ehrenfest's theorem on its own does not really provide robust and straightforward reasons to treat the quantum peculiarities as confined to a 'microscopic realm', inaccessible to the unaided senses, and the 'weirdnesses' as arising merely from the problems of accessing that domain.

Another general strategy for connecting the micro- and macroscopic is the so called $\hbar \to 0$ *approximation* or *quasi-classical limit*, which means taking into account that $\hbar$ is almost negligible compared to (action) scales of macroscopic systems. So for instance, $\Delta x \Delta p \approx 0$, if viewed from these scales. This limiting procedure has proven its importance in many practical applications, the earliest one probably being Einstein's and Planck's observation that the classical equipartition law follows from Planck's radiation formula for $\hbar\omega / k_B T \to 0$ (cf. Landsman 2009, p. 626).

Ballentine (2000, p. 388) nevertheless objects that in this limiting strategy the "limit is not well defined mathematically unless one specifies what quantities are to be held constant during the limiting process", which implies a certain *arbitrariness*, and that one has to be guided by experimentally or theoretically informed expectations. But there is a more important *conceptual* problem with the $\hbar \to 0$ approximation.

Compare the $\hbar \to 0$ strategy for connecting QM and classical mechanics to case of *special relativity* and classical mechanics. Considering $c \to \infty$ (i.e. only velocities v such that $v \ll c$), with c the speed of light, one recovers the predictions of classical mechanics from (special) relativity. Ballentine (2000, p. 389) here points out that there is a *conceptual continuity* between classical and relativistic mechanics that is missing between QM and classical mechanics. Classical mechanics and relativity both treat of the spacetime trajectories of material objects in terms of 'point particles'. But QM ultimately does no such thing; the state vector, as we have demonstrated above, cannot be interpreted as describing the trajectory of an individual 'point particle' (whence our notion of a particle). And neither does it describe waves such as those assumed to exist in classical electrodynamics.

Maybe the strongest reasons to be dissatisfied with considering QM to be simply a matter of size, however, stem from the fact that the scales to which a QM treatment is applicable and even indicated have become larger and larger over time. We already mentioned the impressive experiments of Arndt et al. (1999, 2001), demonstrating interference effects with larger molecules. But the most striking examples are the 'mesoscopic superpositions' of current-density states. In certain arrangements involving superconductors, so called 'supercurrents', consisting of numbers of electrons in the order of 10^9, can be brought into states of the form $\frac{1}{\sqrt{2}}\left(\left|\boldsymbol{j}_s^{\circlearrowleft}\right\rangle \pm \left|\boldsymbol{j}_s^{\circlearrowright}\right\rangle\right)$, where $\left|\boldsymbol{j}_s^{\circlearrowright}\right\rangle$ and $\left|\boldsymbol{j}_s^{\circlearrowleft}\right\rangle$ represent states of the supercurrent $(\boldsymbol{j}_s)$ in which it moves in a clockwise or counterclockwise fashion respectively, through a ring of superconducting material. A short (pointed) presentation of how this comes about will be given in Appendix B, in which one profits, among other things, form our discussion of double-well potentials and tunneling; some of the concepts underlying the theory of superconductors however also require formal methods established only in Sect. 2.2, whence it may be indicated to read that section first.

Finally, another thought that might have crossed one's mind in the discussion should be dispelled at this point: That *superselection rules, by themselves*, can help to solve the OP. Assume that we simply stipulate that certain kinds of macroscopic superpositions, of cats being dead and alive, say, are prohibited by a superselection rule. This would also mean that for all physically realizable observables $\hat{O}$, eigenstates $|Z_j\rangle$ of the 'pointer observable' $\hat{Z}$, indicating the cat's state, would disallow transitions of the form $\langle Z_j|\hat{O}|Z_k\rangle$ $(j \neq k)$. But then either $\hat{O}$ would have to be *codiagonal* (and commuting) with $\hat{Z}$ or would otherwise not be a physically realizable observable after all (if it *did* possess matrix elements $\langle Z_j|\hat{O}|Z_k\rangle \neq 0$, for $j \neq k$). This also goes for the Hamiltonian $\hat{H}_{\text{int}}$, describing the interaction of cat and pointer—which implies that $\hat{Z}$ would either be *unchanged* by the interaction ($\hat{Z}\hat{H}_{\text{int}} = \hat{H}_{\text{int}}\hat{Z} \Rightarrow \hat{Z}e^{-\frac{i}{\hbar}\hat{H}_{\text{int}}\Delta t} = e^{-\frac{i}{\hbar}\hat{H}_{\text{int}}\Delta t}\hat{Z}$, i.e. interaction first and then measurement = measurement and then interaction), or that the interaction Hamiltonian is not physically realizable (cf. d'Espagnat 1995, p. 176). Both options seem absurd. What we will see, however, is how what may be thought of as *dynamically created* superselection rules *do* have *some* say in the interpretation of QM and solutions to the OP after all.

2.1.5 A Technical Note on Measurements and Properties

So far we have generally identified a quantum state with a vector $|\psi\rangle$ in Hilbert space $\mathcal{H}$. But this abstract description is still not sufficiently general for all purposes, since it does not admit of a *statistical weighting* of different (possibly non-orthogonal) quantum states $|\psi_1\rangle, \ldots, |\psi_n\rangle$. Such a weighting may sometimes be indicated, since a given state preparation procedure might be known to have imperfections and result in either of n quantum states. For instance, one could have an imperfect preparation of spinful particles that results in either the spin up state or in some state that corresponds to spin up along an axis tilted by a small angle θ to the intended axis of preparation. To take such possibilities into account, it is customary to define a *statistical* or *density operator* by

$$\hat{\rho} = \sum_{j=1}^{n} p_j \, |\psi_j\rangle\langle\psi_j| \,, \tag{2.34}$$

where the p_j correspond to statistical weights representing the (non-quantum) probability of either state resulting.

In case there is only one such ψ in question, the density operator will simply coincide with a projector

$$\hat{\rho}_\psi = \hat{P}_\psi = |\psi\rangle\langle\psi| \,,$$

and the state will be termed a *pure (quantum) state.* It satisfies the condition that $\hat{\rho}_\psi^2 = \hat{\rho}_\psi$. If this is not the case, the state will be referred to as a *mixed state.* It is important to note that the p_j are *not* the quantum probabilities, but are deemed to stem from a source of uncertainty external to QM, as explained above. Thus a mixed state involves quantum and 'classical' probabilities at the same time.

Ubiquitous talk of density *matrices* stems from the fact that for finite spaces, the prescription $\rho_{ij} := \langle i|\hat{\rho}|j\rangle$ defines (the elements of) an actual matrix; but one can also use prescriptions such as $\rho(\boldsymbol{x}, \boldsymbol{x}') := \langle \boldsymbol{x}|\hat{\rho}|\boldsymbol{x}'\rangle$ to define (elements of) an abstract 'density matrix'.

As an illustration, consider a preparation method which uses a *randomizer* to produce spins either definitely up or down along the x axis. The density matrix can then be given by

$$\hat{\rho} = \frac{1}{2}(|\uparrow_x\rangle\langle\uparrow_x| + |\downarrow_x\rangle\langle\downarrow_x|) = \frac{1}{4}\left(\begin{pmatrix} 1 & 1 \\ 1 & 1 \end{pmatrix} + \begin{pmatrix} 1 & -1 \\ -1 & 1 \end{pmatrix}\right) = \frac{1}{2}\mathbb{1}.$$

Incidentally, a state of this particular sort, i.e. of the form $1/n \cdot \mathbb{1}$ for n different states, is called *completely* or *maximally mixed* (cf. Audretsch 2007, p. 79; Spekkens 2007, p. 4). This is to be contrasted with the pure state density matrix of $|\uparrow_z\rangle$ in the x-spin basis, which reads:

$$\hat{\rho}_{\uparrow_z} = \frac{1}{2}\Big[(|\uparrow_x\rangle + |\downarrow_x\rangle)(\langle\uparrow_x| + \langle\downarrow_x|)\Big] = \frac{1}{2}\Big[\,|\uparrow_x\rangle\langle\uparrow_x| + |\downarrow_x\rangle\langle\downarrow_x| + |\uparrow_x\rangle\langle\downarrow_x| + |\downarrow_x\rangle\langle\uparrow_x|\,\Big].$$

Still, a question that will require some considerable attention in subsequent chapters is whether the quantum probabilities *are* in fact different *in principle* from the probabilities p_j. According to Spekkens (2007, p. 1), it is a widespread view that the kinds of probability *do* differ, and that mixed quantum states reflect a *lack of knowledge*—as to the pure quantum state pertaining to the system in question that is—whereas pure states are somehow descriptive of the actual state of the system. We will see, however, that this is not as uncontroversial as it may seem, and that a lot depends on one's stance towards these issues.

Joint states of multiple systems as described by density operators come in three different 'flavors': *factorized* states are of the form $\hat{\rho}_1 \otimes \hat{\rho}_2$ (with $\hat{\rho}_1$ and $\hat{\rho}_2$ operating on spaces $\mathcal{H}_1$ and $\mathcal{H}_2$ respectively), *separable* states are convex combinations $\sum_j \lambda_j \hat{\rho}_{1_j} \otimes \hat{\rho}_{2_j}$ of factorized states (where 'convex' means that $\sum_j \lambda_j = 1$ and $\lambda_j \geq 0, \forall j$), and *entangled* states are such that they are not separable (cf. Heinosaari and Ziman 2012, p. 262).

Additionally, density operators can be equipped with dynamical equations such as the *von Neumann equation*

$$\frac{\partial}{\partial t}\hat{\rho}(t) = -\frac{i}{\hbar}[\hat{H}, \hat{\rho}(t)], \qquad \text{(vNE)}$$

which can be easily derived by taking into account the Schrödinger equation for a bra $\langle\psi(t)|\,(i\hbar\frac{\partial}{\partial t})^\dagger = \langle\psi(t)|H\rangle^\dagger \Leftrightarrow -i\hbar\frac{\partial}{\partial t}\langle\psi(t)| = \langle\psi(t)|H\rangle$ (cf. Schwabl 2007, p. 382).[60] One can use a density operator to compute the average of an observable by

$$\langle O\rangle = \mathrm{Tr}(\hat{\rho}\hat{O}) \qquad (2.35)$$

where $\mathrm{Tr}(\hat{X})$ is the *trace* of operator $\hat{X}$ (as defined in Appendix A). Crucially, for any density operator $\hat{\rho}$ it must hold that $\mathrm{Tr}(\hat{\rho}) = 1$, which effectively says that probabilities must sum to one. Pure states can also be classified by appeal to their density operator in that they must satisfy $\mathrm{Tr}(\hat{\rho}^2) = 1$, whereas for a mixed state $\mathrm{Tr}(\hat{\rho}^2) < 1$ (cf. Nielsen and Chuang 2010, p. 100).

Moreover, we can generalize Born's rule (postulate (VI)) in terms of the density operator by realizing that[61]

$$\mathrm{Tr}(\hat{\rho}\hat{P}_{o_i}) = \sum_j \langle j|\sum_k p_k\,|\psi_k\rangle\langle\psi_k|\;|o_i\rangle\langle o_i||j\rangle = \ldots = \mathrm{Pr}_O^{\hat{\rho}}(o_i), \qquad (2.36)$$

[60] It is also a direct consequence of time evolution in the Heisenberg picture introduced later. A generalization to cases with complex environment-interactions is possible in terms of so called *master equations* (cf. Schlosshauer 2007, p. 154 ff.).

[61] The omitted computation is somewhat similar to that subsequent to equation (2.38).

with $\hat{P}_{o_i} = |o_i\rangle\langle o_i|$ a projector onto the ray of eigenvalue o_i for some $\hat{O}$,[62] and the Hilbert space taken to be separable or finite. As an example, consider a system in the pure state $|\uparrow_x\rangle$ or equally $\hat{\rho}_{\uparrow_x}$. Then the probability for measuring 'spin up' along z is simply given by

$$\Pr_{\sigma_z}^{\hat{\rho}_{\uparrow x}}(\uparrow_z) = \mathrm{Tr}(\hat{\rho}_{\uparrow_x}\hat{P}_{|\uparrow_z\rangle}) = \mathrm{Tr}(|\uparrow_x\rangle\langle\uparrow_x||\uparrow_z\rangle\langle\uparrow_z|) =$$
$$= \mathrm{Tr}\left(\frac{1}{2}\begin{pmatrix}1 & 1\\ 1 & 1\end{pmatrix}\begin{pmatrix}1 & 0\\ 0 & 0\end{pmatrix}\right) = \frac{1}{2},$$

which is identical to the value computed by the version of Born's rule introduced before.

With the density operator-formalism one can also generalize the projection postulate.[63] But as a (historically interesting) matter of fact, what von Neumann called the 'first kind of intervention' he formally described in terms of density operators as

$$\hat{\rho} \longmapsto \sum_j \mathrm{Tr}(\hat{\rho}\hat{P}_{o_j})\hat{P}_{o_j}, \tag{2.37}$$

with $\hat{P}_{o_j}$ projectors onto the eigenvalues o_j of a non-degenerate observable O (cf. von Neumann 1932, p. 418; Jaeger 2007, p. 36). Thus in terms of density operators von Neumann actually modeled this process as the passage from some arbitrary quantum state into a mixture of eigenstates of a given observable—*not* as the passage from a superposition into *one* such eigenstate. These two passages are arguably quite different: in the latter case one obtains a state which indicates the presence of a given value for a given observable on some observed systems, in the latter case one does not. Von Neumann's formula (2.37) also comes with serious restrictions, as it is not suitable to describe measurements of degenerate observables ('non-maximal measurements'; e.g. Khrennikov 2010, p. 4 on this point).

Thus it is customary to appeal, as a generalization of the PP at any rate, to what is called *Lüders' rule* (after Lüders 1951), the general form of which is

---

[62] In case of a degenerate spectrum $\sigma(\hat{O})$, one can sum over the n projectors onto eigenvectors spanning the subspace with eigenvalue o_i, and then use a projection operator $\hat{P}_{\{o_{i_j}\}} = \sum_{j=1}^{n} |o_i\rangle\langle o_i|_j$ hence defined instead (cf. Redhead 1987, p. 15). It is, however, crucial that the projectors $|o_i\rangle\langle o_i|_j$ are *orthogonal* in the sense that $|o_i\rangle\langle o_i|_j\,(|o_i\rangle\langle o_i|_k\,|\psi\rangle) = |o_i\rangle\langle o_i|_k\,(|o_i\rangle\langle o_i|_j\,|\psi\rangle) = 0, \forall\,|\psi\rangle \in \mathcal{H}, j \neq k$, in order for $\hat{P}_{\{o_{i_j}\}}$ to be a projector itself. This also goes for a countably infinite subspace ('$n = \infty$') corresponding to the same eigenvalue (cf. Heinosaari and Ziman 2012, p. 23).

[63] The details of mathematically generalizing the projection postulate to continuous spaces are quite intricate though (cf. Srinivas 1980), and we shall here restrict our attention to separable or finite dimensional spaces.

$$\hat{\rho} \longmapsto \frac{\hat{P}_{o_j}\hat{\rho}\hat{P}_{o_j}}{\mathrm{Tr}(\hat{\rho}\hat{P}_{o_j})}, \tag{2.38}$$

with $\hat{P}_{o_j}$ a projector onto the (not necessarily $1D$) subspace corresponding to the eigenvalue o_j of O. Measurements modeled by a family of projection operators $\left\{\hat{P}_j\right\}_{j\in J}$ (J some suitable indexing set) are called *projective* (e.g. Wiseman and Milburn 2010, p. 10) for obvious reasons. In case one of the projectors is being selected in the process, as in (2.38), the measurement is called *selective* (cf. Jaeger 2007, p. 37). Note that

$$\hat{P}_j\hat{\rho}\hat{P}_j = |j\rangle\langle j|\Big[\sum_k p_k\,|\psi_k\rangle\langle\psi_k|\Big]|j\rangle\langle j| = |j\rangle\,\langle j|\Big[\sum_k p_k\Big(\sum_{\ell,m}\alpha_\ell^{(k)}\alpha_m^{(k)*}\,|\ell\rangle\langle m|\Big)\Big]|j\rangle\,\langle j| =$$
$$= |j\rangle\Big[\sum_k p_k\Big(\sum_{\ell,m}\alpha_\ell^{(k)}\alpha_m^{(k)*}\underbrace{\langle j|\ell\rangle}_{=\delta_{j\ell}}\underbrace{\langle m|j\rangle}_{=\delta_{mj}}\Big)\Big]\langle j| = |j\rangle\Big(\sum_k p_k|\alpha_j^{(k)}|^2\Big)\langle j| \propto |j\rangle\langle j|,$$

where we have used an expansion $|\psi_k\rangle = \sum_\ell \alpha_\ell^{(k)}\,|\ell\rangle$, and the proportionality is obviously given by $\mathrm{Tr}(\hat{\rho}\hat{P}_j)$. This shows that the process described by (2.38) is indeed selective, and Lüders' rule is easily seen to provide a generalization of the projection postulate as introduced in Sect. 2.1.4.

Projective measurements are always *repeatable* in the sense that if a system exhibits some value corresponding to a projector, it will yield the same value again when measured shortly after for the same observable (think of our initial Stern-Gerlach examples). According to Nielsen and Chuang (2010, p. 91) however,

> many important measurements in quantum mechanics are not projective measurements. For instance, if we use a silvered screen to measure the position of a photon we destroy the photon in the process. This certainly makes it impossible to repeat the measurement of the photon's position!

It thus seems desirable to generalize, for these occasions, the quantum representation of *observables* so that one can also model non-repeatable measurements. Such a generalization can be provided in terms of *positive operator valued measures* (POVMs). For all practical purposes[64] a POVM can be thought of as a set of *positive* Hermitian operators, $\{\hat{E}_m\}_{m\in J}$, where the positivity means that $\langle v|\hat{E}_m|v\rangle \geq 0, \forall\,|v\rangle \in \mathcal{H}, \forall m \in J$, and where it holds that $\sum_m \hat{E}_m = \mathbb{1}$. Since the POVM is intended to represent a generalized observable, m can in principle be replaced by a continuous variable ω with the identity-resolution then given by $\int_\Omega \hat{E}(\mathrm{d}\omega) = \mathbb{1}$ (Ω the set of all ω; cf. Hayashi 2006, p. 14; Peres 2002, p. 386).[65]

[64] For a few technical details see Appendix A.

[65] The notation may differ though, as can be seen from the discussion of the spectral decomposition in Appendix A. The example in Peres (2002, p. 386) is instructive w.r.t. the use of the above notation.

One now has $\mathrm{Tr}\left(\hat{E}_m\hat{\rho}\right) = \mathrm{Pr}^{\hat{\rho}}_M(m)$ instead of (2.36), where m is some measurable value and M refers to some measurement procedure (cf. Nielsen and Chuang 2010, p. 90). Thus, the crucial thing about POVMs is that they define a probability measure.

Comparing this to what has been established about Born's rule above, we may conjecture (correctly) that projectors themselves represent a (special case of) POVM. The expectation value of any projector will be ≥ 0, and they also sum to the identity operator. Hence when all the $\hat{E}_m$ in $\{\hat{E}_m\}_{m\in J}$ correspond to (orthogonal) projectors, the measure is *projector valued* (PVM). The elements of a more general POVM crucially need not equal their own squares, in contrast to those of a PVM (cf. Busch et al. 1995, p. 8).

POVMs also allow to generalize Lüders' rule, since one can—always, in *separable* spaces, as a consequence of the so called *square root lemma* (cf. Heinosaari and Ziman 2012, p. 19)—define operators $\hat{M}_m$, usually referred to as *measurement operators* (e.g. Nielsen and Chuang 2010, p. 81), such that for every $\hat{E}_m$ of a POVM it holds that $\hat{E}_m = \hat{M}^\dagger_m\hat{M}_m$. With these, Lüders' rule generalizes to

$$\hat{\rho} \longmapsto \frac{\hat{M}_m\hat{\rho}\hat{M}^\dagger_m}{\mathrm{Tr}(\hat{M}_m\hat{\rho}\hat{M}^\dagger_m)} \tag{2.39}$$

(cf. Jaeger 2007, p. 42).

Moreover, POVMs can usually be understood as coarse grained PVMs, often by appeal to the known states of an auxillary system or 'ancilla' to which the system of interest is coupled (cf. Peres 2002, pp. 282–283). As an example of the graining, take some (non-degenerate) observable $\hat{O} = \sum^n_j o_j\,|o_j\rangle\langle o_j|$ with n eigenstates $\{|o_1\rangle, \ldots, |o_n\rangle\}$, but assume that the measurement outcomes m, observable on the measuring-device (lights flashing up, pointers pointing on a scale), cannot be related one-to-one to the values o_j. In case one can identify at least a probability (which may arise from the experimental setup in a suitable way; cf. the example in Busch et al. 1995, pp. 9–10) of m occurring on the device, given that the system takes on (or is assumed to take on) value o_j, a POVM can be defined by $\hat{F}_m = \sum_j p(m|o_j)\,|o_j\rangle\langle o_j|$, where $p(m|o_j)$ provides the probability of obtaining m given that o_j obtains on the system. The measurement will then give *some* outcome m, given that the system is in state $|o_j\rangle$, but no outcome m uniquely reveals a particular pure state $|o_j\rangle$. POVMs can thus used to represent the aforementioned *unsharp measurements*.

In case of the *simultaneous* measurement of two *incompatible* observables, the *only possible* measurements are (of course) all unsharp. As an example (cf. Wallace 2012, p. 18), take a state $|q, p\rangle$ with position representation as a wavepacket $\langle x|q, p\rangle = \psi_{p,q}(x) = \frac{1}{\sqrt{\sigma\sqrt{\pi}}}e^{-\frac{i}{\hbar}px}e^{-\frac{(x-q)^2}{2\sigma^2}}$, centered around some fixed position value q. Here a suitable POVM, which gives at least an approximate position *and* momentum in some range Σ_i (a phase space cell) around q and p, can be given as $\hat{E}_i = \frac{1}{2\pi\hbar}\int_{\Sigma_i} \mathrm{d}p\,\mathrm{d}q\,|q, p\rangle\langle q, p|$.

But in concert with all our previous concerns, the question arises whether the system thus measured *does not have* either a precise value of position q *or* momentum p, but *is* at best 'located in the phase space cell Σ_i' if the ith of a range of possible measurement outcomes occurs. Essentially, we are here reissuing questions about the 'uncertainties' appearing in QM, in virtue of noncommutative observables and the uncertainty relations, rephrased in terms of sharp and unsharp *measurements*. Due to the lack of clarity about the role of the projection postulate in the orthodox interpretation, we are still left with the question of whether these 'uncertainties' and the associated need to retreat to POVMs (or unsharp measurements) for two incompatible observables reflect something more than merely an expression of our *incomplete knowledge* of the system's 'true state'.

We can get a better grip on these issues by first going a little deeper into the formal aspects of unsharp measurements. With Busch et al. (1996, p. 10), call the set of all operators $\hat{O}$ which are elements of some POVM the set of *effects* $\mathcal{E}(\mathcal{H})$ on a Hilbert space $\mathcal{H}$, the intuition being that they describe the events which may occur as a result of some measurement. Then the (proper) subset of these effects such that their spectrum extends both below and above $\frac{1}{2}$ are called *properties*.[66]

To understand the intuition behind this use of 'property' better, first note that the spectrum of the operators in a POVM will always be a subset (proper or improper) of $[0, 1]$, and for a PVM it is just $\{1, 0\}$. So projectors can be interpreted as representing 'definite answers' to some yes-no question that one could pose about a given value of a given observable on a given system. Intuitively, they hence provide a way to represent the properties of a system as exhibited in measurements: if the system's state is an eigenvector of some projector with eigenvalue 1 the property represented pertains to it, for 0 it does not.

Using the formal distinction between general POVMs and PVMs, a property (in the present use of the word) may thus be called *unsharp* in case the corresponding operator does not satisfy $\hat{E}_m = \hat{E}_m^2$ (i.e. is not a projector), and, accordingly, an associated *measurement* is called unsharp in case it has at least one unsharp property in its range. As we haven seen, Busch et al. (1996, p. 10) only require for operators to count as properties that their spectrum extend both below and above $\frac{1}{2}$, and this can be non-trivially the case, of course, for the elements of a POVM.[67]

But now if a given operator $\hat{E}_m$ has value $< \frac{1}{2}$ on some vector this must mean something to the effect that the property in question is 'rather absent' in the state represented by the vector, whereas if it has value $> \frac{1}{2}$, it is 'rather present'. Should this be taken to mean that a property can *de facto* also have only an *approximate* degree of presence or absence instead of being definitely there (1) or not there (0)? Does neither of the two possibilities have to be always realized?

---

[66] For technical reasons, the two operators $\mathbb{O}$ and $\mathbb{1}$ are also included in the set of properties as 'trivial cases' (cf. Busch et al. 1996, p. 10).

[67] In fact, these weak requirements preserve the *orthocomplementation* property, which can be interpreted as a form of negation in the context of quantum logic (cf. Jaeger 2009, p. 268; Redhead 1987, p. 160).

Busch and Jaeger (2010, p. 1352) in fact use the suggestive term "approximately real" instead of 'approximately present', and Busch et al. (1995, p. 3) similarly write:

> The unsharpness in question should not in general be taken as an imperfect perception of an underlying more sharply determined property. On the contrary, this term is intended to describe possible elements of reality whose preparation and determination are subject to inherent limitations.

What are we to make of this? It is far from clear what it *means*, exactly, that a property should be 'unsharp', beyond the *operationalistic* specifications given in the above quote. And Busch et al. (ibid.) also concede that unsharp measurements "may or may not admit the kind of *ignorance* interpretation familiar from classical physical experimentation" (my emphasis—FB). Hence 'unsharp property' might be considered merely as a *technical term* here, fully defined by the *formal* considerations given above. Whether this assessment is suitably exhaustive depends on whether QM *more generally* allows for the kind of ignorance-interpretation in question. This will be a point we essentially return to in Chap. 4.

2.2 Can Fields Help to Solve the Riddles? A Glimpse at Quantum Field Theory

2.2.1 Relativity and Many Bodies

So far we have not concerned ourselves at all with QM's connections to the relativity theories. It is important here to distinguish rather sharply between connecting QM to the *general* theory of relativity (GR), which is Einstein's theory of gravity, and connecting it to the *special* theory of relativity (SR), which might best be characterized as Einstein's investigation of the consequences of the constancy of light's velocity in vacuum. The quest to unite GR with QM may still be considered pretty much as an open and active field of research, whereas SR is considered by many to live in a (more or less) happy marriage with QM in the form of relativistic QFT. We will mostly constrain ourselves here to a discussion of some of the basic ingredients of QFT, and of how it relates QM to SR. We will presuppose, however, a basic understanding of SR in what follows.[68]

To recall, in SR the (vacuum) speed of light is a constant, c, as was suggested by theory and experiment in optics and electromagnetism even before the advent of SR. c also constitutes an upper limit to possible speeds at which carriers of matter and energy can travel. More precisely, "it is not possible to take a body travelling at less than the speed of light and *accelerate* it to a velocity greater than the speed of

---

[68] Maudlin's (2011) book on QM and relativity may be a good starting point for philosophers not acquainted with SR at all, alongside any good textbook on the subject such as Walker et al. (2012, chapter 37) or, for a more technical treatment, Rindler (2006, chapter 1).

light. In fact it is not possible for a massive body even to reach the speed of light [...]." (Adams 1997, p. 79; my emphasis—FB)[69]

Usually, this restriction to (sub-)luminal speeds is understood in the following sense:

> The value of c is the limit for the speeds of material bodies or of the processes that could be used for the transmission of a signal. Under the term 'signal' we mean the transmission of a certain amount of *energy* that carries *information* about an event at a point $\boldsymbol{r}_1$ at the moment t_1 and can change the state of a certain physical system at a point $\boldsymbol{r}_2$ at the moment t_2. (Fayngold 2002, p. 142; my emphasis—FB)

Of course this reading of the restriction induces worries about the aptness of the involved understanding of 'signal' in the unweary philosopher. But the idea is intuitive enough; if person A wants to send a message to person B, she needs to alter the state of B's sense organs, which is—supposedly—only possible through physical interaction. And the only kinds of physical interaction we *know of*, and know how to *control* in order to get our message through, are those involving (sub)luminal carriers of energy (air molecules propagating sound waves, photons making up flashes of light, electrons traveling through a wire, and so forth). We can hence accept this understanding of the constraints set up by the constancy of the speed of light here as sufficiently plausible and intuitive. But note that the point is ultimately not as uncontroversial as it may seem.[70]

To establish one particular kind of conflict that traditional QM has with the speed limit, imagine some quantum system, a 'single particle', located around $\boldsymbol{x} = 0$ at some initial time $t = 0$, say. We take it that SR demands that the particle should not be able to reach places at such a distance $|\boldsymbol{x}|$ to 0 in such times that it would have to travel at a velocity greater than c in some frame. In fact, from the set of transformations able to deal with the constancy of c, the *Lorentz*-, or more generally: *Poincaré transformations*, it follows that talk about 'places' and 'times', as if these two categories were entirely separate, is somewhat ill-founded.

Within the context of SR both concepts cannot be thought of separately but must be unified into *one* common structure, the *Minkowski space-time* (e.g. Maudlin 2011, p. 40 ff.). Next to the constancy of c, it is a postulate of SR that in every so called *inertial frame of reference*, the physics is the same, where an inertial frame of reference is usually defined as one in which Newton's first law applies, i.e. in which systems that are not subject to external forces remain at rest or have constant velocities (cf. Fließbach 2009, p. 9; Sexl and Urbantke 1992, p. 1). For each system one can define a so called *past and future light cone* by centering a space-time coordinate system on it; the space-time points which can then be reached at

[69]This does not preclude, however, the possibility of particles which *always* travel at superluminal velocities, so called *tacyhons*, which could not be *de*celerated to subluminal velocities instead (e.g. Maudlin 2011, p. 65 ff. for discussion). So far, however, there is no evidence for the existence of these; and there is even a sense in which they are incompatible with QFT *qua* localizable entities (cf. Sexl and Urbantke 1992, p. 27).

[70]Cf. in particular Maudlin (2011, p. 93 ff.) for an extensive discussion of different kinds of possible signals, and p. 2 ff. therein for a statement of the more general potential controversy.

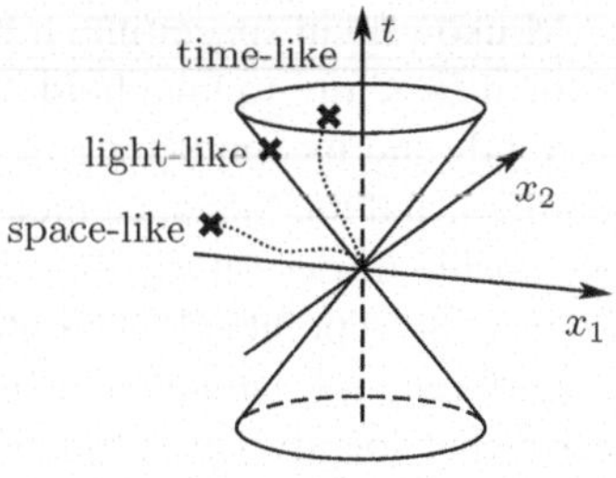

Fig. 2.9 Past and future light cone of an event. The curved lines indicate possible trajectories in space-time which a particle may take to reach the respectively separated space-time points. We have chosen units in which $c = 1$

speeds $v \le c$ from the system's own space-time position (here the point $(\boldsymbol{x}, t) = 0$) constitute its *future light cone*, those from which it could have been reached at those very velocities define its *past light cone* respectively. This light cone can be easily visualized if one restricts the spatial degrees of freedom to 1 or 2 instead of 3 (cf. Fig. 2.9).

Let us make things slightly more precise. In SR already, *spatial distances* become *frame-dependent*, i.e. depend on one's own state of motion. Phrased more technically, the *Euclidean metric* $d_E(p, p') = \sqrt{(x - x')^2 + (y - y')^2 + (z - z')^2}$, the 'natural' distance between two points $p = (x, y, z)$ and $p' = (x', y', z')$ in a $3D$ (Euclidean) space, is not *Lorentz-invariant*, i.e., not invariant under changes $x \mapsto \tilde{x} = \gamma(x - vt)$, say, where $\gamma = 1/\sqrt{1 - (\frac{v}{c})^2}$ (v a constant, unidirectional velocity).[71] The 'distance' that *is* invariant under the Lorentz transformations, is the quantity $\Delta s^2 = -c^2(t - t')^2 + (x - x')^2 + (y - y')^2 + (z - z')^2$, the *Minkowski metric*, where the signs on the right may be reversed, depending on convention (e.g. Carroll 2004, p. 7; Rindler 2006). This leads to the $4D$ *Minkowski spacetime*, which can be described as the pair $(\mathbb{R}^4, \eta)$, where $\mathbb{R}^4$ is endowed with a vector space structure for four-component (column) vectors, $(ct, x, y, z)^T$, 'pointing to' respective *spacetime points*, and where $\eta = (\eta_{\mu v})$ is called a *metric tensor*, which may be represented (depending on convention and basis) as the matrix

$$\eta = \begin{pmatrix} -1 & 0 & 0 & 0 \\ 0 & 1 & 0 & 0 \\ 0 & 0 & 1 & 0 \\ 0 & 0 & 0 & 1 \end{pmatrix}. \tag{2.40}$$

$\eta(u, v) = u^T \cdot \eta \cdot v$ for two four-vectors u, v defines the generalization of a scalar product.[72]

[71] Whence the Lorentz transformations? A nice heuristic introduction can be found in Walker et al. (2012, chapter 37).

[72] Strictly speaking, η should be called a *pseudo-metric tensor*, since $\eta(u, v)$ may yield values < 0. It is otherwise symmetric and linear in both arguments, and if $\eta(u, v) = 0, \forall u \in \mathbb{R}^4$, then v is the null vector (e.g. Nakahara 2003, p. 244).

Minkowskian spacetime, however, is 'flat'. It does not have the required mathematical structure to take hold of the *curvature* introduced by masses that we know from GR and its confirming observations. Thus a more general sort of spacetime is required, and the 'distance measure' provided by the (pseudo-) metric tensor should be capable of describing the 'bending of distances', i.e. should be allowed to vary from point to point. The essential ingredient for such a generalization is the theory of (differentiable) *manifolds*. Intuitively, such a manifold is a possibly oddly shaped space that locally (in sufficiently small regions) 'looks like' $\mathbb{R}^n$ (in the relevant cases: $n = 4$).[73]

The appeal, of course, is that the successful predictions of SR and non-relativsitc physics be preserved in such a space, so that it may locally (and under suitable conditions) be approximated by any of the more traditional spacetimes. GR hence appeals to a more general *Lorentzian* manifold (of which Minkowski spacetime is a particular example) with a metric tensor $g = g(p)$ that may change across points (p) on the manifold and otherwise has the same *signature* (number of -1s in its diagonal matrix form) as the Minkowskian metric tensor η (e.g. Straumann 2004, p. 22). Of course on such a manifold, the algebraic and analytic notions such as 'derivative' and 'vector' have to be generalized as well. Vectors, for instance, are construed as tangent vectors to curves in the manifold, and are mathematically constructed as directional derivatives of smooth functions of equivalence classes of such curves, where curves may formally be thought of as parametrized (e.g. time-coordinate-dependent) sets of points, and the directional derivatives are taken w.r.t. coordinates that points are equipped with in terms of so called *charts* (cf. Footnote 73). We need not really bother with the details any further here though, and the interested reader may be referred e.g. to Nakahara (2003, p. 178 ff.) instead.

Returning, thus, to the single particle that we considered to raise a problem for relativistic 'single particle' QM, we can now say that it should only be able to access spacetime regions which lie *inside* the (future) cone. Any point beyond the boundary of the light cone should not be accessible, and for massive particles the same goes for the points *on* the boundary. This induces a threefold classification of separations from the origin: points within the light cone are called *timelike separated* from it, points which lie somewhere on the boundary are called *lightlike separated*, and those outside are called *spacelike separated*. In this terminology, spacelike separations cannot be traversed by light or (all known) matter. In terms of the metric tensor, the notions of spacelike ($g(u, u) > 0$), lightlike ($g(u, u) = 0$),

[73] More precisely, a differentiable manifold is a *topological space* that is locally *homeomorphic* to $\mathbb{R}^4$. A topological space is a set X, together with a collection of subsets of X which contains X, $\varnothing$ and is closed under infinite unions and finite intersections. A homeomorphism is a continuous invertible map between topological spaces whose inverse is also continuous. A differentiable manifold M now is a topological space, endowed with a collection $\{(U_i, \varphi_i)\}_{i \in I}$ (called an 'atlas') of pairs (called 'charts') of open sets U_i which jointly cover M ($\bigcup_i U_i = M$) and homeomorphisms φ_i that map the U_i into open sets $V_i \subseteq \mathbb{R}^n$, and for which, if $U_i \cap U_j \neq \varnothing$ ($i \neq j$), $\varphi_i \circ \varphi_j^{-1} : \varphi_j(U_i \cap U_j) \to \varphi_i(U_i \cap U_j)$ is smooth. The φ_j equip the manifold *locally* (in the sets U_j) with *coordinates* (cf. Nakahara 2003, pp. 81, 85, and 171–172).

and timelike ($g(u, u) < 0$) as applied to *vectors* can also be made precise (e.g. Nakahara 2003, p. 246; Rindler 2006, p. 101). Understanding these as signifying displacements between points, so that $u = (c(t - t'), (x - x'), (y - y'), (z - z'))^T$ (where $(ct', x', y', z')^T$ could just be $(0, 0, 0, 0)^T$), the prescription $\eta(u, u)$ then computes Δs^2 in Minkowski spacetime.

But it turns out that single particle QM cannot always respect the requirement of predicting only timelike trajectories (those inside the light cone). For the probability amplitude $\langle \boldsymbol{x} | U \rangle (t) | \boldsymbol{x} = 0 \rangle$ of finding the particle initially located at $\boldsymbol{x} = 0$ after some time t at $\boldsymbol{x}$ so that $(\boldsymbol{x}, t)$ (or $(ct, \boldsymbol{x})$, in the standard 4-vector notation) lies *outside* the light cone can be computed to be *non-zero* (cf. Lancaster and Blundell 2014, pp. 75–76; Peskin and Schroeder 1995, p. 14). In other words, there are instances where ordinary QM yields results which *prima facie* make it *incompatible* with SR.

QFT, in contrast, avoids this difficulty and reconciles the two theories, QM and SR, in many further respects, by choosing a completely different starting point (as will become clear below). Most notably, it includes a so called *microcausality condition*, which states that any two self-adjoint operators that represent observables measured at a space-like distance to one another must commute, i.e.

$$[\hat{O}(x), \hat{O}(y)] = 0 \qquad \text{if} \qquad (x - y)^2 > 0 \tag{2.41}$$

(x, y spacetime points). This can be understood as saying that two spacelike separated measurements should not influence one another (cf. Teller 1995, p. 84 ff; Greiner and Reinhardt 1993, p. 103 ff. for more details). Some (e.g. Gottfired and Weisskopf 1986, p. 579) take this condition to constitute a basic postulate of QFT *in general*, comparable in status to the postulates of QM we have discussed in the last section. However, 'ordinary QFT', as *used* by physicists in calculating magnitudes and predicting outcomes of experiments, proceeds in a rather *heuristic* fashion, as we will see below; only in 'axiomatic' (algebraic) approaches is the microcausality condition (or rather: its adaptation to the algebraic program) therefore usually really introduced as an explicit postulate.

From where we are standing now, the simplest route to QFT[74] is that *via second quantization* and the *Fock space* formalism. However, the Fock space formalism cannot give a full picture of the structure of QFT, the main reason being that there will generally be *unitarily inequivalent* Fock spaces in QFT (cf. Friebe et al. 2015, pp. 250 and 259; Ruetsche 2011, p. 69 ff.), and the case can be made that unitarily inequivalent representations should be taken to correspond to *physically* inequivalent theories (cf. Ruetsche 2011, pp. 24–29 and pp. 70–71).[75] An expression

[74]Note that we will sometimes use 'QFT' to mean the theoretical field in general, but also occasionally talk of '*a* QFT', thereby meaning a concrete (heuristically) quantized field theory, such as e.g. QED.

[75]However, things are somewhat subtle here. Ruetsche presupposes the semantic view of theories (cf. Sect. 1.2), as did van Fraassen (1991) before her. On this account it is possible—at least in principle—to make two unitarily inequivalent theories come out physically equivalent, by assign-

of this is the so called *Unruh effect*, according to which observers in different states of accelerated motion should experience differences in the presence or absence of certain particles (cf. Friebe et al. 2015, p. 250 and below), which arguably suggest a clear *physical* difference. But to make things precise, we should first introduce the Fock space formalism and return to the Unruh effect and unitary inequivalence below.

Now consider the (entangled) wave functions for two or more indistinguishable particles that we had met with in Sect. 2.1.3. For one whole collection of N such particles, one can define the space of all wave functions which satisfy the appropriate (anti-)symmetrization requirement, a subspace of the total N-fold tensor-product Hilbert space $\mathcal{H}^{\otimes N}$. Call this appropriately symmetricized subspace of $\mathcal{H}^{\otimes N}$ '$\mathcal{F}_N$'. Then by summing up all the spaces for different values of N, one obtains the so called *Fock space* $\mathcal{F} = \mathcal{F}_0 \oplus \mathcal{F}_1 \oplus \mathcal{F}_2 \oplus \ldots$ Crucially, we have included a space $\mathcal{F}_0$ which amounts to including states with no particle at all.

The general mathematics used in this Fock space formalism can best be understood by appeal to the QM treatment of *harmonic oscillators*. A harmonic oscillator is a system which swings back and forth in a simple periodic motion. Take, for instance, a little mass, a block of lead or something, attached to a spring with no friction at all (e.g. in a remote region of outer space and constructed out of some amazing material with no inner friction). Then extending the spring once will lead to a sinusoidal motion in time, as the mass will bounce back, contract the spring beyond the point of rest, and then bounce back again and again. The force thus exerted on the spring must be proportional to the length of elongation $\Delta x = x - x_0$ and the material it is made of, so that we can write $F = -k\Delta x$, where k is a constant associated with the spring. If we choose the point of rest as $x_0 = 0$, we have $\Delta x = x$, and the formula simplifies to $F = -kx$. Since $F = ma = m\frac{\mathrm{d}^2x}{\mathrm{d}t^2}$ (by Newton's second law), one obtains a differential equation $\frac{\mathrm{d}^2x}{\mathrm{d}t^2} = -\frac{k}{m}x$ whose solutions are trigonometric functions $\cos(\omega t)$ and $\sin(\omega t)$, with $\omega = \sqrt{\frac{k}{m}}$ the frequency of oscillation (e.g. Walker et al. 2012, p. 414 ff. for a more comprehensive introduction). Since the negative gradient of a potential energy is a force, one can define the harmonic oscillator *potential* (generalized now to three dimensions) as $V(\boldsymbol{x}) = \frac{1}{2}m\omega^2\boldsymbol{x}^2$.

This potential can be used for a QM Hamiltonian, with $\boldsymbol{x}$ replaced by $\hat{\boldsymbol{x}}$, so that the corresponding SSE becomes

$$E\psi = -\frac{\hbar^2}{2m}\Delta\psi + \frac{1}{2}m\omega^2\hat{\boldsymbol{x}}^2\psi.$$

This SSE has 'complicated looking' solutions called *Hermite polynomials* (cf. Basdevant and Dalibard 2002, p. 454; Schwabl 2007, p. 51), and solutions and

ing to them appropriately gerrymandered interpretations. Thus, physical equivalence becomes a question about *fully interpreted* physical theories (cf. Ruetsche 2011, p. 29).

energy eigenvalues turn out to be parametrized by a natural number n in each spatial dimension. Allowing, for generality's sake, for different oscillation frequencies (ω_j) in all three spatial directions, the energy E can be worked out to be

$$E = \sum_{j\in\{x,y,z\}} (n_j + \tfrac{1}{2})\hbar\omega_j,$$

so that for $n_j = 0$ ($\forall j$), one still obtains an energy of $\sum_{j\in\{x,y,z\}} \frac{\hbar\omega_j}{2}$. This incidentally means that the oscillator will always possess a *ground state*- or *zero point energy*, even for the lowest permitted values of all the ns.

There exists, however, an equivalent but much more elegant algebraic treatment of the quantum harmonic oscillator (e.g. Basdevant and Dalibard 2002, p. 148), which incidentally serves as a formal paradigm for the Fock space treatment of multi-particle systems. The idea here is to 'cleverly' define (vector-)operators (assuming all ω_js to be equal now, for simplicity)

$$\hat{\boldsymbol{a}} = \sqrt{\frac{m\omega}{2\hbar}}\left(\hat{\boldsymbol{x}} + \frac{i\hat{\boldsymbol{p}}}{m\omega}\right) \tag{2.42}$$

$$\hat{\boldsymbol{a}}^\dagger = \sqrt{\frac{m\omega}{2\hbar}}\left(\hat{\boldsymbol{x}} - \frac{i\hat{\boldsymbol{p}}}{m\omega}\right) \tag{2.43}$$

with which the Hamiltonian can be rewritten as

$$\hat{H} = (\hat{\boldsymbol{a}}^\dagger\hat{\boldsymbol{a}} + \tfrac{1}{2}\mathbb{1})\hbar\omega. \tag{2.44}$$

Working in the eigenbasis of this Hamlitonian, one can use the compact notation $|\boldsymbol{n}\rangle = |n_x, n_y, n_z\rangle$, which is also independent of the choice of representation (momentum or position). By 'playing around' with the operators, one finds some important relations between their components ($j, k \in \{x, y, z\}$):

$$\begin{aligned}
\hat{a}_j |n_j\rangle &= \sqrt{n_j}\,|n_j - 1\rangle \\
\hat{a}_j^\dagger |n_j\rangle &= \sqrt{n_j + 1}\,|n_j + 1\rangle \\
\hat{a}_j^\dagger\hat{a}_j |n_j\rangle &=: \hat{n}_j |n_j\rangle = n_j |n_j\rangle \\
\hat{a}_j |0\rangle &= 0
\end{aligned} \tag{2.45}$$

$$\begin{aligned}
\left[\hat{a}_j, \hat{a}_k\right] &= \left[\hat{a}_j^\dagger, \hat{a}_k^\dagger\right] = 0 \\
\left[\hat{a}_j, \hat{a}_k^\dagger\right] &= \delta_{jk} = -\left[\hat{a}_k^\dagger, \hat{a}_j\right] \\
\left[\hat{n}_j, \hat{a}_k\right] &= -\delta_{jk}\hat{a}_k, \quad \left[\hat{n}_j, \hat{a}_k^\dagger\right] = \delta_{jk}\hat{a}_k^\dagger
\end{aligned} \tag{2.46}$$

These relations, especially (2.46), turn out to be of utmost importance for the Fock space formalism and QFT in general. They also motivate the names *raising* and *lowering operator* for $\hat{a}_j^\dagger$ and $\hat{a}_j$ respectively (since these raise and lower the energy), *number operator* for $\hat{n}_j$ (since it determines the value of the number n_j), and *ground state* for $|0\rangle$. Note that any state $|n_j\rangle$ can now be written as $\frac{(\hat{a}_j^\dagger)^{n_j}}{\sqrt{n_j}}|0\rangle$ (cf. Basdevant and Dalibard 2002, p. 148 ff. and Schwabl 2007, p. 49 ff. for more details).

Now to establish a connection with the Fock space formalism, consider a symmetricized state function of three bosons, where (for instance) two of them occupy the first of an ordered number of possible states, ψ_a, and the remaining one occupies the third of these states, ψ_c. The properly symmetricized state function then is

$$\Psi(\boldsymbol{x}_1, \boldsymbol{x}_2, \boldsymbol{x}_3) = \frac{1}{\sqrt{3}}\Big[\psi_a(\boldsymbol{x}_1)\psi_a(\boldsymbol{x}_2)\psi_c(\boldsymbol{x}_3) + \psi_a(\boldsymbol{x}_3)\psi_a(\boldsymbol{x}_2)\psi_c(\boldsymbol{x}_1) + \\ + \psi_a(\boldsymbol{x}_1)\psi_a(\boldsymbol{x}_3)\psi_c(\boldsymbol{x}_2)\Big].$$

But since nothing is being said about *which* boson occupies *which* of the states, one can equally resort to a much shorter representation:

$$|\Psi\rangle = |2, 0, 1\rangle\,. \tag{2.47}$$

Here the possible states $a, b, c, \ldots$ are simply listed as an ordered sequence of numbers indicating how often any of them is occupied. The 2 hence means that ψ_a is occupied twice, the 0 that ψ_b is unoccupied, and so forth. For fermions, multiple occupations of the same state are obviously prohibited when all degrees of freedom are taken into consideration (in accord with the Pauli principle). This kind of representation is called the *occupation number representation* and (2.47) represents a state in a Fock space for three bosons.

The inclusion of a 0 into the ordered sequence of possibly occupied quantum states is obviously necessary to include the possibility of *un*occupied states, e.g. unoccupied orbitals in an atom. But what if there is *no* particle present at all in some region or configuration under consideration—what if the atom, say, is fully ionized? This leads to the concept of the so called *vacuum state*, $|0, 0, 0, \ldots\rangle$, reminiscent of the ground state of the harmonic oscillator. The best analogy so far would be that of a whole bunch of uncoupled oscillators, all in their ground state, i.e. in the lowest possible state of excitation. But since we are talking about particles occupying or not occupying states, this analogy is somewhat flawed, and this 'vacuum' requires proper interpretation.

It should also be clear that a countable sequence of definite occupation numbers can only be obtained for subspaces of a Hilbert space of *countable* dimension, unlike the spaces for free particles we have appealed to above. This could be brought about e.g. by imposing *periodic boundary conditions* (e.g. Lancaster and Blundell (2014, pp. 28–29) or Basdevant and Dalibard (2002, p. 75 ff.)). To wit, imagine some quantum system confined to a ($1D$) box of length L. 'Imposing periodic boundary conditions' now means that the wave function must behave at $x = L$ as it does at $x = 0$, i.e. $e^{\frac{i}{\hbar}p\cdot 0} = 1 = e^{\frac{i}{\hbar}pL}$. But this in turn means that $pL = 2\pi n,\ n \in \mathbb{Z}$, because of the properties of the complex exponential (think Euler formula and properties of sine and cosine). So one obtains a countable basis in terms of quantized (discretized) momenta. Solutions to the SSE which are normalized inside the box are of the form $\psi(x) = \frac{1}{\sqrt{L}}e^{\frac{i}{\hbar}px}$ and have eigenvalues $\hat{p}\psi(x) = p\psi(x) = \frac{2\pi n}{L}\psi(x)$.

The next key step in the developing the Fock space formalism further is to introduce analogues of the raising and lowering operators from the harmonic oscillator into the occupation number representation. Take one of the above states of definite momentum, in $3D$ abstractly written as $|\boldsymbol{p}\rangle$. Then in analogy to the harmonic oscillator, we can take this to be $\hat{a}^{\dagger}_{\boldsymbol{p}}\,|0\rangle$, with $|0\rangle$ the single particle vacuum state. The operator $\hat{a}^{\dagger}_{\boldsymbol{p}}$ is now rather called a *creation operator*, as it creates a particle in state $|\boldsymbol{p}\rangle$, and its adjoint $\hat{a}_{\boldsymbol{p}}$ is accordingly called an *annihilation operator* ($\hat{a}_{\boldsymbol{p}}\,|\boldsymbol{p}\rangle = |0\rangle$). But now consider that (in our countable basis)

$$|\boldsymbol{x}\rangle = \sum_i |\boldsymbol{p}_i\rangle\langle\boldsymbol{p}_i|\boldsymbol{x}\rangle = \sum_i \hat{a}^{\dagger}_{\boldsymbol{p}_i}\,|0\rangle\,\langle\boldsymbol{p}_i|\boldsymbol{x}\rangle = \sum_i \langle\boldsymbol{p}_i|\boldsymbol{x}\rangle\,\hat{a}^{\dagger}_{\boldsymbol{p}_i}\,|0\rangle\,. \tag{2.48}$$

In our so far non-relativistic treatment we have $\langle\boldsymbol{p}|\boldsymbol{x}\rangle = \frac{1}{\sqrt{V}}e^{-\frac{i}{\hbar}\boldsymbol{p}\boldsymbol{x}}$, with V the 'box volume' (a generalization of the length L from above), or more generally we can write $\langle\boldsymbol{p}|\boldsymbol{x}\rangle = \phi^*_{\boldsymbol{p}}(\boldsymbol{x})$. Thus we identify the mathematical object $\sum_i \phi^*_{\boldsymbol{p}_i}(\boldsymbol{x})\hat{a}^{\dagger}_{\boldsymbol{p}_i} =: \hat{\psi}^{\dagger}(\boldsymbol{x})$, acting on $|0\rangle$, as a *field operator*.[76]

Multiplying $\hat{\psi}^{\dagger}(\boldsymbol{x})$ by the adjoint $\hat{\psi}(\boldsymbol{x})$ from the right gives an expression $\hat{\varrho}(\boldsymbol{x}) := \hat{\psi}^{\dagger}\hat{\psi}(\boldsymbol{x})$ for a *number density operator*; integrating over some volume V', $\int_{V'} \mathrm{d}^3\boldsymbol{x}\,\hat{\varrho}(\boldsymbol{x})$, will result in an operator $\hat{N}_{V'}$ which has integer eigenvalues and determines the number of 'particles' present in V' (cf. Teller 1995, p. 50). But of course we do not generally have to restrict ourselves to the countable case; in analogy to our free wave packet-story above, we can define $\hat{\psi}^{\dagger}_{\text{free}}(\boldsymbol{x}) := \int \mathrm{d}^3\boldsymbol{p}\,\phi^*_{\boldsymbol{p}}(\boldsymbol{x})\hat{a}^{\dagger}_{\boldsymbol{p}}$ as a representation of the 'creation of a particle' at an arbitrary point $\boldsymbol{x}$ in free space (with uncountable values of $\boldsymbol{p}$) out of a vacuum $|0\rangle$, and with all other expressions adapted accordingly (cf. also Teller 1995, pp. 55–56).

From considerations of multiple applications of different or equal annihilation and creation operators to the vacuum state, one can develop *commutation relations* for these. Crucially, one finds that these relations are different for fermionic and

[76]Note that without the hat on $\hat{a}^{\dagger}_{\boldsymbol{p}_i}$ and $\dagger$ replaced by $*$ this would just be a (countable) Fourier expansion of a kind of quantum wave packet ψ^*. In virtue of this, the operator $\hat{\psi}^{\dagger}(\boldsymbol{x})$ can be viewed as a quantization of an object that is already the solution to a quantum mechanical equation (the ψ-function), which provides at least an intuitive (though supposedly historically inaccurate) explanation for the name 'second quantization'.

bosonic operators. Call $[\hat{A}, \hat{B}]_+ := \hat{A}\hat{B} + \hat{B}\hat{A}$ the *anti-commutator*. Then for bosonic creation and annihilation operators it holds that

$$\left[\hat{a}_{\boldsymbol{p}_j}, \hat{a}_{\boldsymbol{p}_k}\right] = \left[\hat{a}^\dagger_{\boldsymbol{p}_j}, \hat{a}^\dagger_{\boldsymbol{p}_k}\right] = 0 \tag{2.49}$$

$$\left[\hat{a}_{\boldsymbol{p}_j}, \hat{a}^\dagger_{\boldsymbol{p}_k}\right] = \delta_{jk}, \tag{2.50}$$

very much in analogy to relations (2.46), and where the δ_{jk} can be replaced by $\delta^3(\boldsymbol{p} - \boldsymbol{p}')$ in the continuous case. For fermionic operators, we instead need

$$\left[\hat{c}_{\boldsymbol{p}_j}, \hat{c}_{\boldsymbol{p}_k}\right]_+ = \left[\hat{c}^\dagger_{\boldsymbol{p}_j}, \hat{c}^\dagger_{\boldsymbol{p}_k}\right]_+ = 0 \tag{2.51}$$

$$\left[\hat{c}_{\boldsymbol{p}_j}, \hat{c}^\dagger_{\boldsymbol{p}_k}\right]_+ = \delta_{jk}, \tag{2.52}$$

where we have used $\hat{c}$ to distinguish the two cases, and where the continuous generalization works analogously. The same behavior also carries over to the field operators in both cases. This anti-commutating behavior of the fermionic operators implements the Pauli principle in this representation, because $\left[\hat{c}^\dagger_{\boldsymbol{p}_k}, \hat{c}^\dagger_{\boldsymbol{p}_k}\right]_+ = \hat{c}^\dagger_{\boldsymbol{p}_k}\hat{c}^\dagger_{\boldsymbol{p}_k} + \hat{c}^\dagger_{\boldsymbol{p}_k}\hat{c}^\dagger_{\boldsymbol{p}_k} = 0 \Rightarrow \hat{c}^\dagger_{\boldsymbol{p}_k}\hat{c}^\dagger_{\boldsymbol{p}_k} = 0$, i.e. no state with two indistinguishable fermionic particles in it can be created (cf. Lancaster and Blundell 2014, pp. 31–38).

So far we have only discussed space-dependent field operators, but we have advertised QFT as a kind of unification of QM and SR, and (as noted above) according to SR space and time cannot be properly thought of in separation. The time dependence of the field operators is established by appeal to the so called *Heisenberg picture*. This picture rests on the fact that viewing the state vector as evolving in time, as suggested by the Schrödinger equation, is not the only perspective one can take in QM. To see this, recall that $|\psi(t)\rangle = \hat{U}(t)\,|\psi(0)\rangle$. But then

$$\langle O\rangle_{\psi(t)} = \langle\psi(t)|\hat{O}|\psi(t)\rangle = \langle\psi|\hat{U}^\dagger(t)\hat{O}\hat{U}(t)|\psi\rangle\,, \tag{2.53}$$

and we could equally think of the object $\hat{O}_H(t) := \hat{U}^\dagger(t)\hat{O}\hat{U}(t)$ as representing the thing which evolves over time, instead of the ket vector $|\psi\rangle$. Put frankly, QM is neutral about what it actually is that evolves in time, the state the theory attributes to a system, or the observables on it, i.e. that which we can determine about it in terms of experiments (in itself is an interesting point).

Taking the time derivative of the operator $\hat{O}_H(t)$, one finds the *Heisenberg equation*

$$\frac{\mathrm{d}\hat{O}_H(t)}{\mathrm{d}t} = \frac{i}{\hbar}[\hat{H}, \hat{O}_H(t)] + \frac{\partial\hat{O}_H(t)}{\partial t}, \tag{2.54}$$

where the last term only occurs in case the Operator $\hat{O}$ is *explicitly* time dependent (i.e. also in the Schrödinger picture; cf. Schwabl 2007, p. 177 for discussion and examples). A time dependent field operator is thus given, in the Heisenberg picture, by $\hat{\Psi}^{\dagger}(\boldsymbol{x},t) = \hat{U}^{\dagger}(t)\hat{\psi}^{\dagger}(\boldsymbol{x})\hat{U}(t)$ (e.g. Lancaster and Blundell 2014, p. 100).

To thoroughly connect QFT to SR now, it seems desirable to look for relativistic dynamics for our new object $\hat{\Psi}^{\dagger}(\boldsymbol{x},t)$. Two equations which provide such dynamics have been developed in the twentieth century. The first one is called the *Klein-Gordon equation* (KGE) and it can be 'derived' very easily in the same heuristic way as the SE. To this end, consider the square of the relativistic energy, $E^2 = \boldsymbol{p}^2c^2 + m_0^2c^4$, and multiply both sides by some function ϕ. Then replace $E \rightsquigarrow i\hbar\frac{\partial}{\partial t}$ and $\boldsymbol{p} \rightsquigarrow -i\hbar\nabla$ to obtain:

$$-\hbar^2\frac{\partial^2}{\partial t^2}\phi = -c^2\hbar^2\Delta\phi + m_0^2c^4\phi$$

$$\Leftrightarrow \left(\frac{1}{c^2}\frac{\partial^2}{\partial t^2} - \Delta + \frac{m_0^2c^2}{\hbar^2}\right)\phi = 0. \qquad \text{(KGE)}$$

$ƛ_C := \frac{\hbar}{m_0c}$ is also called the (reduced) Compton wavelength, since it incidentally occurs in Compton's scattering theory (e.g. Demtröder 2010, pp. 86–88); so we could replace $\left(\frac{m_0c}{\hbar}\right)^2 = 1/ƛ_C^2$.

So far we are treating the KGE as an equation which applies to (wave) functions $\phi = \phi(\boldsymbol{x},t)$, which in virtue of their *space-time dependence* and their scalar values are called *scalar fields*. Moreover, the KGE corresponds exactly to the wave equation of the electric and magnetic fields mentioned in Sect. 2.1.1, if one sets $m_0 = 0$, and otherwise it merely includes an 'inhomogeneity' $1/ƛ_C^2$. All of this seems to be indicative of waves again, but of course we know already from Sect. 2.1.3 that 'it's all about waves!' cannot be the final verdict.

Moreover, there are good reasons to replace ϕ by an operator $\hat{\phi}(\boldsymbol{x},t)$ to make sense of it in a *quantum* context. A first reason to this effect is that for a scalar field, the general solution to the KGE is of the form

$$\phi(\boldsymbol{x},t) = \int\frac{d^3\boldsymbol{p}}{(2\pi)^{3/2}\sqrt{2E_{\boldsymbol{p}}}}\left(a(\boldsymbol{p})e^{\frac{i}{\hbar}(\boldsymbol{px}-E_{\boldsymbol{p}}t)} + b(\boldsymbol{p})e^{\frac{i}{\hbar}(\boldsymbol{px}+E_{\boldsymbol{p}}t)}\right) \qquad (2.55)$$

with $a(\boldsymbol{p})$ and $b(\boldsymbol{p})$ respective amplitudes (e.g. Teller 1995, p. 67; Friebe et al. 2015, pp. 236).[77] Since derivatives occur in second order for place *and* time, one will generally obtain $E^2\phi = (\boldsymbol{p}^2c^2 + m_0^2c^4)\phi$, which is independent of the + and − in the exponent. But then both square roots, $\pm\sqrt{\boldsymbol{p}^2c^2 + m_0^2c^4}$, yield values for E,

[77] The factor $\frac{1}{\sqrt{2E_{\boldsymbol{p}}}}$ is included to ensure Lorentz-invariance (cf. Lancaster and Blundell 2014, p. 101 ff. for discussion).

so that there are also *negative energy solutions*. For classical waves, i.e. with $E/\hbar$ replaced by ω and the connection between the two ignored, this would simply imply waves traveling in opposite directions. In the QM treatment, however, the form of the solution allows for systems having *arbitrarily large* negative energies.

In principle, negative energy is not a problem. For instance, negative energy for an electron in some atom means that such and such an amount of energy is needed to ionize the atom (i.e., 'kick the electron out of the orbit') due to the binding potential of the nucleus; in hydrogen the ground state energy of the electron is (approximately) -13.6 eV. But we are here considering *free* fields, not subject to any binding potential, and if there is no lower bound to this energy, this of course creates much more of a problem and the former interpretation becomes hardly tenable. Historically these problems ultimately lead to the introduction of *antiparticles*, with positive energies and charge and momentum opposite to that of the particles corresponding to the positive energy solutions (e.g. Lancaster and Blundell 2014, p. 61 ff.; Teller 1995, p. 79). But this requires second quantizing the solution of the KGE, i.e. replacing ϕ by $\hat{\phi}$ by using appropriate operators in the place of the of amplitudes in (2.55), the first one being an annihilation operator ($\hat{a}(\boldsymbol{p})$) for particles, the second one a creation operator ($\hat{b}^\dagger(\boldsymbol{p})$) for corresponding antiparticles (and *vice versa* in $\hat{\phi}^\dagger$).

The need for a second quantization of the solutions to the KGE is hence intimately connected to the negative energy solutions. But it is also connected to the fact that the above solution does not allow for a probability interpretation: Consider the so called *probability current density*, which in nonrelativistic QM is defined as

$$\boldsymbol{j}(\boldsymbol{x},t) = \frac{\hbar}{2mi}[\psi^*(\nabla\psi) - (\nabla\psi^*)\psi](\boldsymbol{x},t). \tag{2.56}$$

This probability current density of norelativistic QM satisfies a *continuity equation* $\frac{\partial}{\partial t}\varrho(\boldsymbol{x},t)+\nabla\boldsymbol{j}(\boldsymbol{x},t) = 0$, with $\varrho(\boldsymbol{x},t) = \psi^*\psi(\boldsymbol{x},t)$, expressing the conservation of probability (e.g. Schwabl 2007, p. 31). But the quantity which satisfies an analogous continuity equation in the case of the KGE can become *negative* in virtue of the negative energies (cf. Lancaster and Blundell 2014, p. 61), whence the ordinary QM interpretation of the Klein-Gordon field as a probability amplitude cannot be applied.

Even though one can put these difficulties of the KGE aside by second quantizing its solutions, it is ultimately too restrictive: it cannot handle spin degrees of freedom and is thus only an equation for (massive) spinless particles. But we mentioned that there are *two* notable relativistic equations, and it was Dirac who developed the other one in 1928. This *Dirac equation* (DE) 'automatically' included terms acknowledging the existence of spin, as we shall see below. Historians have it (e.g. Cantor et al. 1990, p. 473) that Dirac was concerned with finding an equation for relativistic quantum mechanics that was not quadratic in time, i.e. an equation of the form $\hat{H}\psi = i\hbar\frac{\partial}{\partial t}\psi$ with relativistic Hamiltonian. To this end, Dirac sought for a way to quantize the non-quadratic relativistic energy $E = \sqrt{\boldsymbol{p}^2c^2 + m_0^2c^4}$.

Unfortunately, the heuristic substitutions $E \rightsquigarrow i\hbar\frac{\partial}{\partial t}$ and $\boldsymbol{p} \rightsquigarrow -i\hbar\nabla$ in this case yield $i\hbar\frac{\partial}{\partial t} = \sqrt{-\hbar^2c^2\Delta + m_0^2c^4}$, and one might be lead to wonder what the square root of the Laplacian is. This is not really a problem though, since the square root of a differential operator can be defined in terms of Fourier transformation, so the essential problem was indeed rather connected to the order of derivatives, namely that "due to the asymmetry of space and time derivatives Dirac found it impossible to include external electromagnetic fields in a relativistically invariant way." (Thaller 1992, p. 2)

Dirac's 'trick' ultimately was to analyze the expression under the square root as the square of something else by introducing operators $\hat{\boldsymbol{\alpha}}$ and $\hat{\beta}$, so that

$$\begin{aligned}\hat{\boldsymbol{p}}^2c^2 + m_0^2c^4 &= (c\hat{p}_x\hat{\alpha}_x + c\hat{p}_y\hat{\alpha}_y + c\hat{p}_z\hat{\alpha}_z + \hat{\beta}m_0c^2)^2 = \\ &= (c\hat{\boldsymbol{\alpha}}\hat{\boldsymbol{p}} + \hat{\beta}m_0c^2)^2,\end{aligned} \tag{2.57}$$

and one finally obtains

$$\left(c\hat{\boldsymbol{\alpha}}\hat{\boldsymbol{p}} + \hat{\beta}m_0c^2\right)\boldsymbol{\Psi} = i\hbar\frac{\partial}{\partial t}\boldsymbol{\Psi}. \tag{DE}$$

After studying the commutation properties of $\hat{\beta}$ and the $\hat{\alpha}_i$, one also finds that

$$\hat{\alpha}_i = \begin{pmatrix} \mathbb{O} & \hat{\sigma}_i \\ \hat{\sigma}_i & \mathbb{O} \end{pmatrix} \quad \text{and} \quad \hat{\beta} = \begin{pmatrix} \mathbb{1} & \mathbb{O} \\ \mathbb{O} & -\mathbb{1} \end{pmatrix}$$

are possible representations, with $\mathbb{O}$ a 2×2 zero-matrix, and $i \in \{x, y, z\}$ (for details e.g. Shankar 1994, p. 565; Schwabl 2008, p. 123).

For the given choice of $\hat{\boldsymbol{\alpha}}$ and $\hat{\beta}$, the occurrence of the Pauli spin matrices immediately explains how the DE with (free) Hamiltonian $c\hat{\boldsymbol{\alpha}}\hat{\boldsymbol{p}} + \hat{\beta}m_0c^2 = \hat{H}$ acknowledges the existence of spin. Since we are now concerned with 4×4-matrices rather than 2×2-ones (as in our previous treatment of spin), the solution $\boldsymbol{\Psi}$ to the equation must be given in terms of a *Dirac spinor*, containing two two-component (Weyl-) subspinors. These two subspinors each correspond to different *chirality* and *helicity* states, which means that they represent an intrinsic *handedness* of particles.[78]

Solutions to the DE are compatible with a first quantized reading, i.e. with $\boldsymbol{\Psi}$ as a four component (spinor-)*wavefunction* that has a probabilistic interpretation as does ψ in non-relativistic QM (e.g. Schwabl 2008, p. 121). Different niceties become

[78]In principle the meaning of the intrinsic handednesses is itself worth philosophizing about, since it may be taken to have implications, say, for substantivalism or relationalism about space (e.g. Earman 1991; Lyre 2005). But we will here rather concern ourselves with other matters more relevant for the discussion to come.

possible though when one phrases things in terms of operator valued fields instead (e.g. Peskin and Schroeder 1995, p. 52 ff.), but we wont bother with details.

Note also that our notation here is quite non-standard, and we have merely chosen it to remain within the delimiters of what has been introduced so far. In textbooks (e.g. Lancaster and Blundell 2014, p. 322 ff.; Peskin and Schroeder 1995, p. 40 ff.), you will rather find $(i\gamma^\mu \partial_\mu - m)\psi(x) = 0$, for natural constants set to 1 and a *summation convention* in place according to which the same upper and lower index appearing in immediate succession is shorthand for a sum over the index ($\sum_\mu \gamma^\mu \partial_\mu$). These upper and lower indices ($\mu \in \{0, \ldots, 4\}$) are called *contra-* and *covariant indices* respectively, additionally taking track of the transformation behavior of an object so indexed under coordinate changes. $x = (t, \boldsymbol{x})$ is a spacetime point (for $c = 1$), and ψ is interpreted as a 4-component spinor (operator) without signifying this by using bold face font or the like.

The γ^μ have representations as 4×4 matrices. They can be defined in terms of $\hat{\beta}$ and the $\hat{\alpha}_i$ by multiplying the DE through (from the left) by $\hat{\beta}/c$ (e.g. Schwabl 2008, p. 123 ff.). Acknowledging also that $\partial_0 := \frac{1}{c}\frac{\partial}{\partial t}$, this fully explains the latter mentioned form of the DE.

2.2.2 *Canonical Quantization and the Concept of a Quantum Field*

We started off with the Heisenberg picture as providing the time dependence of field operators, but now we have switched to overtly relativistic dynamical equations. This does not, however, lead to a conflict with the Heisenberg picture; one 'rederives', for instance, the KGE from the Heisenberg equation upon presupposing some connections between Lagrangian and Hamiltonian theory. So far we have only talked about Hamiltonians in a very loose manner, and with that we have hidden a whole lot of classical mechanics from plain sight. A *few* details, at least, must now be uncovered.[79] First note that it is a quite general fact of classical mechanics that the Hamiltonian *function* H of a problem, representing the sum of kinetic and potential energy, is connected to the *Lagrangian* (function) L, representing the *difference* of potential from kinetic energy, *via* a so called *Legendre transformation*, $H = \sum_j p_j \dot{q}_j - L$. Here $\dot{q}_j$ is called a *generalized velocity* which could, e.g., be $\frac{\partial}{\partial t}\varphi =: \dot{\varphi}$, with φ an angle, and hence need not be a position on a Euclidean straight line. p_j is a corresponding *generalized momentum*, which is derived by taking the (partial) derivative of the Lagrangian w.r.t. a given generalized velocity, such as in $\frac{\partial L}{\partial \dot{\varphi}} = p_\varphi$ (p_φ is then *angular* momentum along φ). Thus the Hamiltonian and Lagrangian formulations of classical mechanics are intimately connected, and they offer a most useful tool to compute mechanical problems in terms of problem-

[79]For a very gentle start cf. Susskind and Hrabovsky (2013).

suited coordinates. In QFT the Lagrangian formulation often is of particular interest because it ensures Lorentz invariance (cf. Peskin and Schroeder 1995, p. 16).

To establish a Lagrangian and Hamiltonian theory of *fields*, one takes things a step further and understands the Hamiltonian and Lagrangian functions as integrals over Hamiltonian and Lagrangian *densities*, $\mathscr{H}$, $\mathscr{L}$. Here the relation $\pi(\boldsymbol{x}) = \frac{\partial \mathscr{L}}{\partial \phi}$ is used, in analogy to the above, to define the *momentum density* $\pi(\boldsymbol{x})$, where $\mathscr{L}$ is the Lagrangian density with $\int \mathrm{d}^3\boldsymbol{x}\, \mathscr{L} = L$, and $\phi(\boldsymbol{x})$ the field under consideration. Using the quantized, operator valued analogues of both, one then stipulates *cannonical commutation relations* $[\hat{\phi}(\boldsymbol{x}), \hat{\pi}(\boldsymbol{x}')] = i\hbar\delta(\boldsymbol{x} - \boldsymbol{x}')\mathbb{1}$, in analogy to the momentum and position operator from ordinary QM (cf. Peskin and Schroeder 1995, p. 18).[80]

We have here reverted temporarily to a time-independent view of the field operators, which we were eager to leave behind; the commutation relations between a field operator and its conjugated momentum should thus instead be taken as 'equal-time' commutation-relations (cf. Peskin and Schroeder 1995, p. 20), and the derivation of a Lagrangian from the Lagrangian density presupposes a globally hyperbolic[81] spacetime $\mathcal{M}$, which can be foliated in such a way as to separate it into a spatial and a temporal part, $\mathcal{M} = \Sigma \times \mathbb{R}$. The integration then is over the spatial part only, but w.r.t. a time dependent measure: $L(t) = \int_\Sigma \mathrm{d}^3\mu(t)\, \mathscr{L}$ (cf. Wallace 2006, p. 36).

What we have just sketched are the general steps of what is called *canonical* (or sometimes: *heuristic*) *field quantization*, the algorithm (nay *heuristic*) used to obtain a QFT from some classical field theory. Any operator $\hat{\psi}(\boldsymbol{x}, t)$ that comes out of such a procedure may be called a *quantum field*. We have thus established how 'quantum fields' are *formally* identified by the canonical approach, namely as specific operators that depend on spacetime-coordinates. But what do these spacetime dependent operators *represent*? Are they representative of actual, *physical* fields? If so, what precisely *are* physical fields?

To get a better grip on these questions, contrast the following two entirely different characterizations of what 'field' means from a modern textbook on

80 Indeed, in even closer analogy to non-relativistic QM, $\hat{\phi}(\boldsymbol{x})$ may just be represented as 'multiplication by a scalar field $\phi(\boldsymbol{x})$', when operated on a quantum state $|\phi\rangle$, and $\hat{\pi}(\boldsymbol{x})$ as a functional derivative operator $-i\hbar\frac{\delta}{\delta\phi(\boldsymbol{x})}$ w.r.t. $\phi(\boldsymbol{x})$. This is the *functional Schrödinger representation* (e.g. Hatfield 1992, p. 199 ff.), where the theory is taken to treat of states $|\Psi\rangle$ that have functionals $\langle\phi|\Psi(t)\rangle = \Psi[\phi, t]$ of field configurations ϕ as expansion coefficients (the 'functional analogue' of wave functions).

81 For the notion of global hyperbolicity e.g. Smeenk and Wüthrich (2010, p. 593 ff.) or Ruetsche (2011, p. 107). Roughly, the idea is that there is a spacelike hypersurface Σ which has a future and past domain of dependence ($D(\Sigma)^+$, $D(\Sigma)^-$) such that the union of these is the whole spacetime $\mathcal{M}$. The domains of dependence $D(\Sigma)^\pm$ are defined as sets of points p such that any past (+) or future (−) directed inextendible timelike curve through p has to intersect Σ, where timelike past/future directedness means that at any point the tangent vector of the curve falls into or on the past/future light cone, and inextendibility means that the curve has no endpoints.

QFT: A field is here (i) thought of "as some kind of machine[82] that takes a position in spacetime and outputs an object representing the amplitude of something at that point in spacetime [...]. The amplitude could be a scalar, a vector, a complex number, a spinor or a tensor." (Lancaster and Blundell 2014, p. 2) And (ii) as "an unseen entity which pervades space and time" (ibid.). Both of these characterizations convey important intuitions, but they are both also misleading in some respects.

(i) is a(n informal) characterization of the *formal* aspects of a field as some mathematical entity that depends on spacetime coordinates. But whether the 'output' (value) of this field in the merely *formal* sense really represents the *amplitude of something* at a spacetime point is not necessarily obvious. (ii) informs us rather about the *ontology* of the notion 'field', of what a field is in the *physical* sense. But neither do fields *have to* be "unseen"—a heat distribution on a hot plate might count as a field and could radiate in the visible energy spectrum—, nor is it clear that all fields can be said to 'pervade space and time': the metric tensor $g(x)$ in GR is at least formally a field, but it seems contentious to say that it 'pervades space and time'; it rather represents 'an aspect *of*' spacetime.

However, the distinction between what we have called 'formal fields' (mathematical objects that depend on spacetime coordinates) and 'physical fields' (the extended entities which formal fields purportedly represent) seems valid and important. Auyang (1995, p. 47), in a similar spirit, notes that "'[f]ield' has at least two senses in the physics literature. A field is a continuous dynamical *system* or a system with infinite degrees of freedom. A field is also a dynamical *variable* characterizing such a system or an aspect of the system." (my emphasis—FB) The field as an operator which creates quantum states of fixed position or momentum, i.e. the "machine" or that which we have referred to as a 'formal field', is a field in Auyang's latter sense; the "unseen entity", i.e. that which we have referred to as a 'physical field' and identified as the possible referent of the formal field, is a field in Auyang's former sense.

As we suggested above, in QFT the relation between formal and physical field becomes especially delicate. The formal fields of classical electromagnetism, say, are (in the first place) scalar and vector fields. The physical content of formal vector fields can be represented by imagining little arrows assigned to space-time points which then indicate, at any such point, the *strength* and *direction* of the physical field hence represented. Strength and direction can be accessed empirically by putting test-objects that couple to the field into the region in which it persists; anyone familiar with magnets should have an idea of this procedure. The ontology of a vector field is hence that of a continuous extended entity with a sense of directionality. Quantum fields, as we have seen, are *operator valued*. They seem to defy such a direct interpretation.

[82]The metaphor "machine" for mathematical objects which take something in and give back an output can be traced back to John Wheeler's use of the word for operators in QM (cf. Susskind and Friedman 2014, pp. 52–53).

"But", the reader well-educated in GR may object, "what about *tensor fields*? They are mathematically more complex than vectors!" This is certainly correct, but it is rather beside the point. The tensor fields that really do occur in GR or even in classical electromagnetism typically have (some subtleties aside) a rather straightforward intended interpretation: Maxwell's stress tensor or the field-strength tensor in classical electrodynamics appear as mere conveniences for rewriting equations in a compact form (e.g. Jackson 1990, pp. 261 and 556), and tensorial quantities in GR typically have straightforward *geometrical* interpretations, such as the Riemann curvature tensor $R^{\rho}{}_{\sigma\mu\nu}$, describing the curvature of a manifold (e.g. Carroll 2004, p. 94), or the quantity $h_{\alpha\beta}(x)$ in the linear approximation $g_{\alpha\beta}(x) = \eta_{\alpha\beta} + h_{\alpha\beta}(x)$ of the metric (tensor), describing local disturbances of the 'flat' Minkowski metric, $\eta_{\alpha\beta}$ (i.e. "metric perturbations" or "gravitiational waves"; cf. Hartle 2003, p. 332).

For operator-valued fields in QFT, the situation is arguably different. Physicist often times help themselves to a pragmatically healthy but all too simplistic attitude. Steven Weinberg (1997, p. 2), for instance, puts things in a way that certainly captures the intuitions of many (if not most) physicists: "the idea of quantum field theory is that quantum fields are the basic ingredients of the universe, and particles are just bundles of energy and momentum of the fields." And similarly Carroll (2004, p. 40): "Upon quantization, excitations of the field are observable as particles."

Rather in contrast to this, Teller (1995) has suggested to make sense of quantum fields in terms of *determinables*, i.e. *collections* of properties, "such that anything that can have one of the properties in the collection must have exactly one of the properties." (Teller 1995, p. 95) This notion can be applied to classical fields; a mass density, say, is a continuous determinable that associates a mass to every spacetime point. But the quantum field operators are not of this kind; they do not assign values to spacetime points, they assign operators. Hence Teller suggests to liken them to *field determinables* instead, which he defines as determinables "the values of which are the full field configurations." (1995, p. 99) Put frankly, Teller (ibid.) thinks that:

> At any given space-time point, the associated operator corresponds not to the value of some physical quantity but to the full spectrum of values of some quantity, which value being applicable being determined by the state that happens to obtain.

Field operators (representing formal quantum fields) assign, according to Teller, whole collections of values to spacetime points, and are thus *at every spacetime point* more like *full determinables*, instead of representing determinate field configurations over spacetime as a whole or the amplitude of something at any spacetime point. Since formal quantum fields are operators, this is *prima facie* quite reasonable, given what we have established about ordinary QM and the operators occurring therein so far.

But Teller's view of quantum fields has been met with quite some criticism, for instance by Fleming (2002) or Wayne (2002), and on various grounds. While Fleming (roughly) criticizes the revisionary character of Teller's writings and the underappreciation of certain other interpretative possibilities, Wayne (2002, p. 130), in contrast, positively suggests to identify "[a]n *actual* state of a physical system containing a quantum field" with a "specific state vector/operator combination." (emphasis in original) "The determinate state of a quantum field", Wayne believes, "is given by the association of a set of quantum field operators with a specific quantum state vector", and "this relationship between operators and state vector [...] fully specifies physically measurable quantities in a quantum system." (ibid.) Hence the very *notion* of a 'quantum field' is here straightforwardly construed in a *physical* manner, and to that end *separated* somewhat from the formal fields or field operators. Wayne's interpretation rests on the fact (cf. Wightman 1956) that a *set*[83] of vacuum expectation values *uniquely* determines a given field operator $\hat{\Psi}(\boldsymbol{x}, t)$, whence he believes that vacuum expectation values "for field operators and products of field operators in models of heuristic QFT correspond to field values in physical systems containing quantum fields." (ibid.)

However, ultimately Wayne comes up with the (complete) set of vaccuum expectation values "corresponding to a Lorentzian immaterial ether, rich in structure, which contributes to the production and explanation of QFT phenomena" (p. 131), and this is certainly quite a stretch from the conception of classical fields as well. It involves, as Wayne (2002, p. 130) himself notes, "widening the concepts of field value and field configuration." We can see that ontologically meaningful interpretations of the term 'quantum field' come with radical differences and are rather 'metaphysically thick'.

Wallace (2012, p. 320) has indeed argued that talk of the operator-valued fields in QFT as representatives of actual fields is *misleading*:

> Quantum field theory uses operator-valued fields for exactly the same purpose that ordinary quantum mechanics uses position, momentum, and spin operators: the Hamiltonian of the theory is defined in terms of them, and the structural properties of the system are given by the expectation values of those operators with respect to the quantum state. And in quantum field theory, just as in quantum mechanics, we can formally speaking transfer the dynamics from states to operators, shifting to the Heisenberg picture.

Moreover, Wallace (2006, p. 39) also—*critically*—remarks that according to many interpreters of QFT, the (field-)"operators represent physical operations which can be performed, by the observer, on the QFT state [...]." Similarly Streater (1988, p. 144) recalls: "When I was a PhD student I asked my supervisor what the quantized field $\phi(\boldsymbol{x}, t)$, is. He said: 'It is the operator in Hilbert space assigned by the physicist to the classical nuclear field at $\boldsymbol{x}$ at time t according to the correspondence principle.'" And finally Cao (1997, p. 176) thinks that

> the localized excitation described by a local field operator $O(x)$ acting on the vacuum means the *injection* of energy, momentum, and other special properties *into* the vacuum at a spacetime point. [...] As a result, the field operators no longer refer to the physical particles and become abstract dynamical variables, with the aid of which one constructs the physical state. (my emphasis—FB)

[83] Why a set? The reason is ultimately to be found in the non-local correlations between field values at different points (cf. Wayne 2002, p. 130; Fleming 2002, p. 141).

On these views, the concept of a quantum field obtains a decidedly *anthropocentric* spin, and the general question arises whether QFT really indicates any serious departure from the basic themes of ordinary QM. We can see that interpretations of quantum fields or field operators are truly diverse and that there seems to be little agreement (aside from use). We will certainly not be able to settle the matter here. What we note is that it is not unambiguously clear that QFT *is* concerned with 'physical fields' (extended entities) in the same sense as is (say) classical electrodynamics. The belief that this is so may rest on a similar confusion as does the belief that wavefunctions in QM describe waves.

Recall, however, that we had also seen the notion of a particle in need of revision already in the light of ordinary QM. In *relativistic* QFT (and relativistic QM, as provided by the DE, for that matter) this need is furthered by such results as Malament's (1996) theorem or that of Halvorson and Clifton (2002a), both of which roughly state that in any relativistic quantum theory, *localizability* becomes particularly iffy.[84]

So QFT introduces a new worry about the proper interpretation of 'fields' in quantum mechanical terms and reissues and deepens the problems associated with a particle-notion. Does it, however, at least offer new perspectives on solving the OP in terms of quantum fields or field configurations (whatever they be)?

Unfortunately, the answer here is 'no', since, as we saw, a direct interpretation of the field operators in terms of physical fields is forestalled, and the quantum states to which the operators are applied play a substantial role in the interpretation of the theory. And, as Teller (1990, p. 606) puts it, "most of the states are superpositions, the components of which correspond not only to particles in different states but to different numbers of particles." Thus, since measurements in particle physics usually reveal rather definite particle numbers (or the impression thereof), there occurs the need for a "transition from a superposition to one of the components of the superposition" (Teller 1990, p. 607), and the *OP remains*. Virtually the same judgment is made by Barrett (2002, pp. 168–169), who writes that

> relativistic quantum field theory provides no account whatsoever for how determinate measurement records might be generated. The problem here is analogous to the problem that arises in nonrelativistic quantum mechanics. If the possible determinate measurement records are supposed to be represented by the elements of some set of orthogonal field configurations, then there typically are no determinate measurement records since (given the relativistic unitary dynamics) the state of the field in a given space-time region will typically be an entangled superposition of different elements of the orthogonal set of field configurations.

[84]Halvorson and Clifton (2002a, p. 207; emphasis in original) conclude the discussion of their results with the remark that relativistic QFT "does permit *talk* about particles—albeit, if we understand this talk as really being about the properties of, and interactions among, quantized fields." But they concede that QFT only gains a capability of "explaining the appearance of macroscopically well-localized objects" in virtue of talk about the interactions of 'quantized fields' "*modulo the standard quantum measurement problem*[...]" (ibid.; my emphasis—FB), whose persistence in QFT is, of course, intimately connected to the mathematical representation of 'quantum fields' as operators. Our phenomenological particle concept is untouched by all this, as will be explained in more detail later.

To sum up: QFT possibly introduces *further* interpretational issues, *complicates* old ones, and the central interpretational issue (the OP) is *retained* from ordinary QM.

2.2.3 A Note on the Vacuum State

One of the further intrepretational issues that QFT raises, and that we had already hinted at above, is the meaning of the *vacuum state*. For later reference, we should spend at least a few thoughts on this concept. First we note that, in analogy to the zero point energy of the harmonic oscillator, one in principle also always has non-vanishing energies present in vacuum sates of QFT, and their occurrence ultimately leads to remarkable consequences. In principle, the vacuum energy here becomes *infinite*. Requiring normalization in a finite volume, the KGE-Hamiltonian can be rewritten as $\hat{H} = \sum_{\boldsymbol{p}}(\hat{a}^{\dagger}_{\boldsymbol{p}}\hat{a}_{\boldsymbol{p}} + \frac{1}{2})\hbar\omega_{\boldsymbol{p}}$ (cf. Peskin and Schroeder 1995, p. 19 ff.; Teller 1995, p. 69 ff.), which looks exactly like the sum over Hamiltonians for a bunch of (uncoupled) harmonic oscillators of different momentum $\boldsymbol{p}$. But the sum $\sum_{\boldsymbol{p}}$ ranges unrestrictedly over all values of $\boldsymbol{p}$, so that the adding up of the $\frac{1}{2} \cdot \hbar\omega_{\boldsymbol{p}}$ diverges. According to Teller (1995, p. 72), "[a]ll standard presentations treat the zero-point energy with the remark that only energy differences are significant, so that a constant can always be discarded." A concise way to effect this 'discarding' is by appeal to a so called *normal ordering* of creation and annihilation operators. Normal ordering simply means that in a given product of operators, one puts all the annihilation operators to the right. The term $\frac{1}{2} \cdot \hbar\omega_{\boldsymbol{p}}$ now vanishes because in the last intermediate step of bringing the KGE Hamiltonian into the form above, the expression for each given $\boldsymbol{p}$ reads $\frac{\hbar\omega_{\boldsymbol{p}}}{2}\left(\hat{a}^{\dagger}_{\boldsymbol{p}}\hat{a}_{\boldsymbol{p}} + \hat{a}_{\boldsymbol{p}}\hat{a}^{\dagger}_{\boldsymbol{p}}\right)$[85] so that a normal ordering gives $\hbar\omega_{\boldsymbol{p}}\hat{a}^{\dagger}_{\boldsymbol{p}}\hat{a}_{\boldsymbol{p}}$, and the diverging term disappears.

This normal ordering has the nice consequence that expectation values of the form $\langle 0|\hat{a}^{\dagger}_{\boldsymbol{p}}\hat{a}_{\boldsymbol{p}}\hat{a}_{\boldsymbol{p}'}\hat{a}^{\dagger}_{\boldsymbol{p}'}|0\rangle$ are reordered into $\langle 0|\hat{a}^{\dagger}_{\boldsymbol{p}}\hat{a}^{\dagger}_{\boldsymbol{p}'}\hat{a}_{\boldsymbol{p}}\hat{a}_{\boldsymbol{p}'}|0\rangle$, so that the average number of particles in the vacuum is (obviously) zero. Normal ordering certainly seems very much like an ad hoc move, but it need not be viewed so: Teller (1995, p. 131) reminds us that "it applies in the process of initially choosing the form of a specific theory or model. Once the choice is made, one calculates with the theory, living with whatever operator ordering arises." With the zero-point energies discarded and vanishing vacuum expectation values, we could be lead to think that the vacuum state is just a vacuum in the intuitive sense of the word, a 'mere nothing', an emptiness, or just the absence of anything at all (in an ontologically slim and deflationary sense of these words). After all, it appears as a state with no particles in the Fock space formalism.

[85] Since $\hat{a}_{\boldsymbol{p}}\hat{a}^{\dagger}_{\boldsymbol{p}} = [\hat{a}_{\boldsymbol{p}}, \hat{a}^{\dagger}_{\boldsymbol{p}}] + \hat{a}^{\dagger}_{\boldsymbol{p}}\hat{a}_{\boldsymbol{p}} = \delta_{\boldsymbol{pp}} + \hat{a}^{\dagger}_{\boldsymbol{p}}\hat{a}_{\boldsymbol{p}}$, this immediately gives the desired expression.

But since normal ordering is used only in the initial stages of theory formation, there *are* operators which will *not* have zero expectation value[86]; the value of the square $\hat{\Psi}^2$ of some field operator $\hat{\Psi}$ does not vanish in case the value of $\boldsymbol{p}$ is definite (cf. Teller 1995, p. 108 ff.). Take, for instance, the quantized version of the electric field in a volume V, which may be written as $\hat{\boldsymbol{E}}(\boldsymbol{x}, t) = i \sum_{\boldsymbol{p},\pi} \sqrt{\frac{\hbar\omega_p}{\epsilon_0 V}} \left(\hat{a}_{\boldsymbol{p},\pi}(t)e^{\frac{i}{\hbar}\boldsymbol{px}} - \hat{a}^{\dagger}_{\boldsymbol{p},\pi}(t)e^{-\frac{i}{\hbar}\boldsymbol{px}}\right)\boldsymbol{n}_{\boldsymbol{p},\pi}$, where π is an index for different polarizations, $\boldsymbol{n}_{\boldsymbol{p},\pi}$ a normalized vector for each component, and the time dependence is absorbed into the operators (cf. Milonni 1994, p. 45; Vedral 2005, p. 138). Then in $\hat{\boldsymbol{E}}^2$, occurring in the energy density of the electromagnetic field, we have terms proportional to $\hat{a}_{\boldsymbol{p},\pi}(t)\hat{a}^{\dagger}_{\boldsymbol{p}',\pi'}(t)$ which do not vanish when $\langle 0|\hat{\boldsymbol{E}}^2|0\rangle$ is evaluated. For many practicing physicists this certainly represents nothing worth worrying about, and it might (again, pragmatically healthily) be thought of in terms of "statistical fluctuations of the electric and magnetic fields." (Milonni 1994, p. 42) But above we have given reason to worry about the meaning of 'electric and magnetic fields', appearing as *operators* in QFT, whence this characterization does not really help.

Kuhlmann (2010, p. 81) has noted that "[s]ince expectation values come about by averaging over all possible measurement outcomes according to their respective probabilities, non-vanishing [...] expectation values seem to indicate that there is something happening in the vacuum without there being anything to which this activity could be predicated." But what *is* the appropriate interpretation of the non-vanishing expectation values or the vacuum *itself*, of which this activity seems to be predicated? Teller and Kuhlmann, e.g., both give very different analyses. Again, this seems like a matter that we cannot possibly settle here, as was the case with the physical meaning of 'field' in QFT. So let us make a few *meta-theoretic* remarks about the vacuum state instead.

What we should first ask is: are these remarkable consequences of the vacuum state even empirically accessible? In many cases the answer is 'yes'. A prominent example is the *Casimir effect*, according to which two closely spaced uncharged but conducting metal plates in a vacuum will exert an attractive force on each other.[87] The Casimir effect exploits exactly the existence of a zero point energy between the two plates, which comes about by the periodic boundary conditions that the plates impose (they count as a 'box'). The gradient of this energy is an attractive force, and it has been tested and confirmed experimentally (more recently e.g. by Sushkov et al. 2011) that two such plates will indeed attract each other to the predicted amount under appropriate conditions. Another example is the *Lamb shift* in atomic spectra that was known experimentally long before the completion

[86]Teller (1995, p. 131), in particular, highlights that these non-vanishing vacuum expectation values (or 'vacuum fluctuations') are a consequence of the theory *with* normal ordering in place.

[87]E.g. Lancaster and Blundell (2014, p. 111) for a toy calculation and Ballentine (2000, p. 533 ff.) for a more detailed treatment.

of quantum electrodynamics (QED), and unexplained until the acknowledgement of non-vanishing expectation values in the quantum vacuum (cf. Milonni 1994, p. 82 ff.).

Even more astonishing predictions arise from explicitly *relativistic* considerations. The aforementioned Unruh effect, according to which an accelerated observer in Minkowskian spacetime will be able to detect a thermal bath of particles instead of the vacuum, whose energy is proportional to the magnitude of acceleration (e.g. Kuhlmann 2010, p. 111), is just such an example. The Unruh effect is relativistic in that it draws on the fact that solutions to the KGE are not form invariant under a switch to coordinates parametrizing the rest frame of a uniformly accelerated observer outside one's own light cone. In a $2D$ Minkoski spacetime, this observer will have a spacetime-trajectory that looks like a hyperbola (placed under the left or right exterior of the light cone) and can be parametrized by setting $x = \xi \cosh(\eta), t = \xi \sinh(\eta)$ (with ξ and η so called *Rindler coordinates*). The Klein-Gordon field that solves the KGE based on these coordinates then includes creation and annihilation operators $\hat{a}^{\dagger}_{\hbar\omega_p}, \hat{a}_{\hbar\omega_p}$ with an expectation value $\langle 0_{\mathcal{M}}|\hat{a}^{\dagger}_{\hbar\omega_p}\hat{a}_{\hbar\omega_p}|0_{\mathcal{M}}\rangle = 1/(e^{\frac{2\pi c\omega_p}{a}} - 1)$, which corresponds to an average number of Bose-Einstein particles at temperature $T = \frac{\hbar a}{2\pi c k_B}$, where $|0_{\mathcal{M}}\rangle$ is the Minkwoski-vacuum (for the unaccelerated observer), a the proper acceleration, and k_B Boltzmann's constant (cf. Crispino et al. 2008, for details). As we mentioned above, this incidentally demonstrates the unitary inequivalence of the Fock spaces for these two observers, reflected in the fact that the accelerated observer will now have a *different vacuum* $|0_{\mathcal{R}}\rangle \neq |0_{\mathcal{M}}\rangle$,[88] i.e., a different state of lowest energy.

The vacuum in QFT (relativistic or not) appears as a radically new concept whose interpretation requires much philosophical caution. More precisely, we should think of it is a *theoretical concept*, meaning that what a 'vaccum state' is "cannot be fully specified by a single definition, but only by the joint effect of the core axioms of a *theory*." (Schurz and Gebharter 2016, p. 1075; emphasis in original)

To make a case, consider the following. In their discussion of theoretical concepts, Schurz and Gebharter (2016) use the concept of *force* in Newtonian physics as a primary example, which is defined, according to them, only by the joint axioms of Newtonian mechanics. Each vacuum state is equally only defined by its role as the state of lowest eigenvalue for a given energy operator of some particular quantized field theory, and hence by the joint assumptions of the theory instead of one single theory-independent definition. This should be compared also to *individual* forces being defined in terms of particular differential equations and initial conditions for given physical problems. This dependence of the vacuum state on a given field theory goes so far as to lead to two physically inequivalent vacua, in the case of the Unruh effect, for two observers who are non-inertially related to one another. This, in turn, may be compared to the 'ficticious forces' in Newtonian

[88] $\mathcal{R}$ for 'Rindler', since the spacetime region in which the accelerated trajectory is located for the stationary observer is also called a *Rindler wegde* (cf. Crispino et al. 2008, p. 792 ff.).

mechanics, which equally result from coordinate transformations between relatively non-inertial frames. Hence the situation of the quantum vacuum is quite comparable to that of Newtonian force, and no general, all-applying definition can be given. This should give some credibility for considering 'vacuum state' as a theoretical term of QFT in the aforementioned sense.

2.2.4 *From Renormalization to the Algebraic Approach (and Back)*

We have left untouched, in the previous section, interactions as they occur in the QFT-formalism. But of course interactions play a crucial role e.g. in particle physics, one of QFT's most important applications. A simple example of an *interacting* QFT is the so called $\hat{\phi}^4$*-theory*, described by a Lagrangian density (which occurs e.g. in the description of the Higgs mechanism; cf. Peskin and Schroeder 1995, p. 77)

$$\hat{\mathscr{L}} = \frac{1}{2}\left[\dot{\hat{\phi}}^2 - (\nabla\hat{\phi})^2 - m^2\hat{\phi}^2\right] - \frac{\lambda}{4!}\hat{\phi}^4, \tag{2.58}$$

where (for convenience) we let $\hbar = c = 1$, λ is a dimensionless *coupling constant*, and the scaling factor $\frac{1}{4!}$ may also be thought of as a 'mere convenience'. Now from a so called *variational principle*, that the action $S = \int \mathrm{d}^4\boldsymbol{x}\, \mathscr{L}$ be stationary, i.e. that $\delta S = 0$ (S is a *functional* of $\mathscr{L}$), one can derive[89] the so called *Euler-Lagrange equation*,

$$\begin{aligned} &\frac{\partial\mathscr{L}}{\partial\phi} - \left(\frac{\partial}{\partial t}\frac{\partial\mathscr{L}}{\partial\dot{\phi}} + \sum_x \frac{\partial}{\partial x}\frac{\partial\mathscr{L}}{\partial(\partial\phi/\partial x)}\right) = 0 \\ \Leftrightarrow &\frac{\partial\mathscr{L}}{\partial\phi} - \partial_\mu\left(\frac{\partial\mathscr{L}}{\partial(\partial_\mu\phi)}\right) = 0, \end{aligned} \tag{2.59}$$

which equally applies to classical fields (whence we have omitted the hats), and where in the last line we have appealed to the notation introduced at the end of Sect. 2.2.1 (sum convention and covariant derivative). It provides the *equation of motion* for the field ϕ, and applied to our above interacting Lagrangian density, it yields

$$-m^2\hat{\phi} - \frac{\lambda}{3!}\hat{\phi}^3 + \Delta\hat{\phi} - \frac{\partial^2}{\partial t^2}\hat{\phi} = 0$$

---

[89]Cf. Peskin and Schroeder (1995, pp. 15–16) for details; for a non-rigorous but intuitive treatment in the context of non-field mechanics cf. also Susskind and Hrabovsky (2013, p. 111 ff.).

$$\Leftrightarrow \left(\frac{\partial^2}{\partial t^2} - \Delta + m^2 + \frac{\lambda}{3!}\hat{\phi}^2\right)\hat{\phi} = 0. \tag{2.60}$$

When one reinserts $\hbar$ and c in appropriate places, this is just the KGE with an additional term $\frac{\lambda}{3!}\hat{\phi}^2$ that may be interpreted as describing the 'self-interaction' of $\hat{\phi}$ (cf. Lancaster and Blundell 2014, p. 67; Peskin and Schroeder 1995, p. 77). Such equations, however, are not strictly solvable, and in principle one could also construct all kinds of interactions with powers higher than four in the field $\hat{\phi}$. But it is typically *stipulated*, as a "simple and reasonable axiom" (Peskin and Schroeder 1995, p. 79), that such interacting theories be at least "*renormalizable*" (ibid.; emphasis in original), which rules out a lot of possible interaction terms.

Now what exactly does that mean? That is kind of a 'long story', but here are some highlights. First of all, recall that we related the Hamiltonian to the Lagrangian in Sect. 2.2.1 *via* the legendre transformation $H = \sum_j p_j \dot{q}_j - L$. This treatment generalizes to the densities as $\mathscr{H} = \pi\dot{\phi} - \mathscr{L}$, where $\pi = \frac{\partial \mathscr{L}}{\partial \dot{\phi}}$ is the momentum density. Our (quantum) Hamiltonian density will thus be

$$\hat{\mathscr{H}} = \left[\dot{\hat{\phi}}^2 + (\nabla\hat{\phi})^2 + m^2\hat{\phi}^2\right] + \frac{\lambda}{4!}\hat{\phi}^4. \tag{2.61}$$

The interaction part of this density is clearly $\frac{\lambda}{4!}\hat{\phi}^4$, so an interaction Hamiltonian will be defined as $\hat{H}_{\text{int}} = \int \mathrm{d}^3\boldsymbol{x}\, \frac{\lambda}{4!}\hat{\phi}^4$. When integrated over space, the rest of our density will similarly define the free Hamiltonian, which we may call $\hat{H}_0$.

One of the most important applications of QFT certainly are *scattering* scenarios, in which one typically models the initial and final states, $|i\rangle$, $|f\rangle$ to be *asymptotically free*, i.e. to be eigenstates of the free Hamiltonian $\hat{H}_0$ at $t \to \pm\infty$ respectively (e.g. Binney and Skinner 2014, p. 328 ff.; Lancaster and Blundell 2014, pp. 166–194). Now to say something about the time evolution in these scattering interactions, first note that both $\hat{H}_{\text{int}}$ and $\hat{H}_0$ will typically be time dependent, which, as we briefly mentioned at the end of Sect. 2.1.1, will imply the need for time ordering and integration when one uses them to define a time evolution operator. Thus let

$$\hat{U}_I(t_1; t_2) = \mathbb{T}\exp\left(-i\int_{t_1}^{t_2} \mathrm{d}t\, \hat{H}_{\text{int}}(t)\right), \tag{2.62}$$

where we still set $\hbar = 1$, and $\mathbb{T}$ is the *time ordering symbol* which orders the terms in the exponential from right to left according to increasing time (this will become clear a little below). If one now uses the limits $t_{1/2} \to \pm\infty$ respectively, one obtians the so called *S-operator*

$$\hat{S} = \mathbb{T}\exp\left(-i\int \mathrm{d}^4x\, \hat{\mathscr{H}}_{\text{int}}(x)\right) = \mathbb{T}\exp\left(-\frac{i\lambda}{4!}\int \mathrm{d}^4x\, \hat{\phi}^4(x)\right), \tag{2.63}$$

where we have appealed to the spacetime-coordinate notation $x = (x^0, x^1, x^2, x^3)^T$, x^0 being a timelike coordinate, and integration is understood over all of spacetime.

By appeal to the asymptotically free states, one can use the matrix element $\langle f|\hat{S}|i\rangle$ to evaluate the effect of the interaction on $|i\rangle$, or rather $|\langle f|\hat{S}|i\rangle|^2$ as a probability that a free state $|f\rangle$ evolves out of an initially free state $|i\rangle$ after the intermediate interaction described by $\hat{S}$. More precisely, if there is no interaction, i.e. $\hat{H}_{\text{int}} = 0$, we have $\hat{S} = \mathbb{1}$. This motivates the introduction of the *transition operator* $i\hat{T} = \hat{S} - \mathbb{1}$ (where the i is basically 'for convenience') that evaluates how much the evolution described by $\hat{S}$ deviates from the identity (i.e. 'how much really happens'). Equivalently, we thus have $\hat{S} = \mathbb{1} + i\hat{T}$, so that the quantity of interest is rather $\langle f|\hat{T}|i\rangle =: \mathcal{M}_{fi}$, the *transition matrix element*.

This matrix element figures, in particular, in empirically meaningful quantities such as (differential) *scattering cross sections*: In scattering events, i.e. the imagined 'bouncing off' of imagined little 'particles' (our particles) of each other, the magnitude $\frac{\mathrm{d}\sigma}{\mathrm{d}\Omega}$ will describe the ratio between the number of events per unit time in a 'solid angle' $\mathrm{d}\Omega$ (the detection surface, construed as a fraction of a sphere's surface) and the number of projectiles per unit time and scattering centers per unit area (e.g. Povh et al. 2006, p. 46). Since this scattering cross section thus basically provides a rate of measured particles as a consequence of the scattering, it is unsurprising that a QM-informed treatment will include the transition probability $|\mathcal{M}_{fi}|^2$, i.e. the probability that an initially (asymptotically) free particle will in fact transition into the (asymptotically free) state that has it travel into the detector (e.g. Povh et al. 2006, p. 48 ff.).

However, recall that an exponential of operators is fully defined by its series expansion. So $\hat{S}$ will be of the form[90]

$$\hat{S} = \mathbb{T}\left[\mathbb{1} - \frac{i\lambda}{4!}\int \mathrm{d}^4x\,\hat{\phi}^4(x) - \frac{\lambda^2}{2!(4!)^2}\int \mathrm{d}^4y \int \mathrm{d}^4z\,\hat{\phi}^4(y)\hat{\phi}^4(z) + \ldots\right], \tag{2.64}$$

which is called a *Dyson expansion* (cf. Lancaster and Blundell 2014, p. 170), and $\mathcal{M}_{fi} = \langle f|\hat{T}|i\rangle$ becomes

$$\langle f|\mathbb{T}\left[-\frac{\lambda}{4!}\int \mathrm{d}^4x\,\hat{\phi}^4(x) + \frac{i\lambda^2}{2!(4!)^2}\int \mathrm{d}^4y \int \mathrm{d}^4z\,\hat{\phi}^4(y)\hat{\phi}^4(z) + \ldots\right]|i\rangle\,. \tag{2.65}$$

The effect of the time ordering symbol on two spacetime-dependent operators $\hat{O}(x)$, $\hat{O}(y)$ can here be written as $\mathbb{T}\hat{O}(x)\hat{O}(y) = \Theta(x^0 - y^0)\hat{O}(x)\hat{O}(y) +$

[90]We here emphasize again that the integrals are definite and x, y, z represent four-vectors, not Cartesian coordinates. The square of an integral means multiplying the integral by itself; but the integration variables can always be renamed individually (in each integral), whence in the square one will meet with an integration over two sets of variables. You can also think of it like this: A simple integral computes the area under a curve, so multiplying two integrals means computing a volume under two curves along different dimensions.

$\Theta(y^0 - x^0)\hat{O}(y)\hat{O}(x)$, with x^0, y^0 the timelike coordinate of the four-vectors x, y respectively, and $\Theta(z)$ the Heaviside-theta distribution that gives zero unless $z \geq 0$ and 1 otherwise (cf. Schwabl 2007, p. 295; Lancaster and Blundell 2014, p. 156).

Using further methods such as propagators and contractions (e.g. Lancaster and Blundell 2014, pp. 154 ff. 171 ff.), one can rephrase the integral terms in (2.65) as 'tidier' momentum integrals. However, these terms are typically *divergent*, tidied up or not, and the formula therefore yields no physically meaningful expression. The first general strategy to handle these infinities is to introduce an arbitrary *momentum cut-off*, i.e.,"some large but finite momentum Λ" which then constitutes the upper limit of the integration (the lower limit will be zero in the momentum integrals), where "[a]t the end of the calculation one takes the limit $\Lambda \to \infty$, and hopes that physical quantities turn out to be independent of Λ." (Peskin and Schroeder 1995, p. 80) Theories that can be handled in this way are the ones called *renormalizable* (cf. ibid.).

For $\hat{\phi}^4$-theories of the sort considered above and other interacting QFTs such as QED (e.g. Peskin and Schroeder 1995, p. 330 ff.), this can be done by introducing additional *counterterms* into the Lagrangian, which then eliminate divergences to some order when $\Lambda \to \infty$, and can be understood in terms of corrections to the coupling constants (cf. Lancaster and Blundell 2014, pp. 289–291). Renormalizable theories may then be identified, more precisely, as "those in which a finite number of counterterms cancel all divergences." (Lancaster and Blundell 2014, p. 293) Renormalization methods by the introduction of cut-offs were 'revolutionized' with the occurrence of the so called *renormalization group theory*,[91] whose basic idea is that "rather than hiding the cut-off, we live with it." (Lancaster and Blundell 2014, p. 303) More precisely, this has the effect that "the parameters of a renormalizable field theory can usefully be thought of as *scale-dependent entities*." (Peskin and Schroeder 1995, p. 393; my emphasis—FB) In turn, this means that one eliminates the $\Lambda \to \infty$ limits for the cutoffs and rather investigates how the behavior of the corrected coupling constants changes as a function of Λ, i.e. when one changes the *scale* of interest.[92] A given renormalizable QFT may now, for a given cut-off Λ, be described be described by a set of N coupling 'constants' $g_i(\Lambda)$ so that $(g_1(\Lambda), \ldots, g_N(\Lambda))$ constitutes a 'point' in a high-dimensional 'parameter space', and the renormalization (semi-)group then describes the transformation of that point under changes $\Lambda \mapsto \Lambda/b,\ b \in \mathbb{R}$ (cf. Lancaster and Blundell 2014, p. 304).

However, it seems quite 'weird' to base the soundness of a theory on '*variations* of *constants*' according to a *scale of interest*, and the corrections to the coupling constants that provide counterterms in the Lagrangian and eliminate the divergences

[91] Strictly speaking, this 'renormalization group' of scaling transformations is only a *semigroup* because there is not an inverse to every transformation (cf. Kadanoff 2013, p. 167). One also encounters talk of the 'Wilsonian revolution' (e.g Kadanoff 2013, p. 162), due to the importance of Kenneth Wilsons's contributions to the field (e.g. Wilson 1971a,b).

[92] Λ is a momentum, so $\hbar/\Lambda$ defines a length scale that becomes smaller as Λ. $\Lambda \to \infty$ means 'increasing the fineness of grain' of the QFT.

may be *infinite*, as is the case for the 'bare' couplings corrected (cf. Lancaster and Blundell 2014, p. 291). This has tipped off e.g. Haag (1996, p. 55) to judge that "the fields in an interacting theory are more singular objects than in the free theory", and he also notes that "we do not have the canonical commutation relations [...]." Why "more singular"? Because, mathematically speaking, even our free (canonical) quantum fields $\hat{\phi}$ are "not really" operators on a Fock space $\mathcal{F}$, but only provide *finite matrix elements* $\langle\psi_1|\hat{\phi}|\psi_2\rangle$ for vectors $|\psi_1\rangle$, $|\psi_2\rangle$ from a subspace $\mathcal{D} \subset \mathcal{F}$ that is dense in $\mathcal{F}$ and "characterized by the property that the probability amplitudes for particle configurations decrease fast with increasing momenta and increasing particle number." (Haag 1996, p. 45) And there are further complaints about renormalization in canonical QFT, such as its conflicts with the Euclidean symmetries of interest in solid state physics (e.g. Wallace 2011, p. 116).[93] In spite of the overall *practical* success of renormalization and renormalization group techniques, all of this leaves kind of a foul taste to the aforementioned methods and considerations.

All in all these complaints hence reasonably motivate an *alternative research program*, i.e. to go "back to basics, and look[...] for an axiomatised, fully rigourous quantum theory" (Wallace 2011, p. 116) wherein these difficulties do not occur. Such an alternative research project is *axiomatic-algebraic* QFT (AQFT). So let us briefly review some of its fundamentals, implications, and limitations.

AQFT, as we mentioned, proceeds from 'axoims' or postulates like those of ordinary QM (cf. Sect. 2.1.4), which are here formulated, however, with an eye on the possibility of a mathematically rigorous theory that preserves the insights from renormalization theory and the other advances of heuristic QFT. One set of such postulates are the so called *Wightman axioms*, discussed e.g. by Haag (1996, p. 56). Among these is the requirement that field operators be replaced by *operator valued distributions*

$$\hat{\phi}(f) = \sum_j \int \mathrm{d}^4x\, \hat{\phi}^j{}_\sigma(x) f^{j\sigma}(x), \tag{2.66}$$

where x is a spacetime point, j, σ are indices for different particle types and spin components repsectively, and the $f^{j\sigma}$ are suitably 'well behaved' ($C_0^\infty(\mathbb{R}^4)$; cf. Appendix A) functions. In other words: the operators are being '*smeared out*'. It

[93]More precisely, it is a consequence of *Haag's theorem* that if one holds that there be unique vacuum states $|\Omega_0\rangle$, $|\Omega_\lambda\rangle$ for free and interacting theories respectively, and if one requires that these be invariant under unitaries such as $\hat{U}(\boldsymbol{a})$ representing Eucildean symmetries such as a translation by $\boldsymbol{a}$, then given that one also allows initial and final scattering states to be asymptotically free, an interacting theory such as the $\hat{\phi}^4$-theory discussed above can be demonstrated to be *unitarily equivalent* to the free theory, which one can take to mean that it is no interacting theory after all, i.e. that (given the previous assumptions) there *are no interacting theories* (cf. Haag 1996, p. 55; cf. also Ruetsche 2011, pp. 251–253 and Teller 1995, pp. 115–116 and 122–123 for further discussion).

is then assumed that one can approximate *any operator* on a suitable Hilbert space $\mathcal{H}$ in terms of linear combinations and products of such operators $\hat{\phi}(f)$ (cf. Haag 1996, pp. 57–58).

More precisely, an operator $\hat{O}(\mathcal{O})$, *local* to some (open, bounded; we will generally omit the qualifier below) region $\mathcal{O} \subset \mathcal{M}$ of the underlying spacetime manifold $\mathcal{M}$ can then be defined (neglecting the j, σ-indices) by

$$\hat{O}(\mathcal{O}) := \sum_n \int \prod_k \mathrm{d}^4 x_k \, \hat{\phi}(x_1) \ldots \hat{\phi}(x_n) f^n(x_1, \ldots, x_n) \tag{2.67}$$

in case all f^n vanish for all points x_j outside $\mathcal{O}$. This generates a *polynomial algebra* $\mathcal{A}(\mathcal{O})$ of such operators localized to $\mathcal{O}$ (cf. Haag 1996, p. 84; and cf. Appendix A, Definition A.9 for the notion of an algebra). Since there will be multiple such algebras, one can speak of a *net of algebras* (cf. Haag 1996, p. 105).

Now these polynomial algebras $\mathcal{A}(\mathcal{O})$ come with certain restrictions and their elements are *unbounded* operators, which induces further mathematical complications. This is why Haag (1996, p. 106) suggests to switch "without loss of generality", to local algebras $\mathfrak{A}(\mathcal{O})$ of *bounded operators*, from which e.g. Ruetsche (2011, p. 104) starts immediately in her exposition of AQFT.

In fact, bounded operators $\mathfrak{B}(\mathcal{H})$ on a (separable) Hilbert space $\mathcal{H}$ that we know already from *ordinary* QM are an instructive example. As noted in Appendix A, they constitute a special kind of algebra, a *unital *-algebra* (cf. Definition A.10), where $\mathbb{1}$ is the unit element. The *involution* $*$ on $\mathfrak{B}(\mathcal{H})$ is, as may have been guessed, the *adjoint* operation † (cf. Ruetsche 2011, p. 75). More precisely, $\mathfrak{B}(\mathcal{H})$ with † is a *C^*-algebra*, meaning that it is closed w.r.t. a norm (cf. Definition A.11). In the case of $\mathfrak{B}(\mathcal{H})$ the norm in question is $\|\hat{A}\|_{\mathfrak{B}} := \sup_{\psi \in \mathcal{H}} \frac{\|\hat{A}\psi\|_{\mathcal{H}}}{\|\psi\|_{\mathcal{H}}}$, where $\| \cdot \|_{\mathcal{H}} = \sqrt{\langle \cdot | \cdot \rangle}$ is the Hilbert space norm (cf. Haag 1996, p. 112–118; Ruetsche 2011, p. 76).

The key idea of AQFT now is to start from a general local C^*-algebra $\mathfrak{A}(\mathcal{O})$ and then ('axiomatically') build a suitable quantum theory from it. These algebras (for respective $\mathcal{O}$) are typically referred to as algebras of *local observables* (e.g. Haag 1996, p. 105), and their elements are, in the generally anthropocentric sort of reading of QFT, sometimes thought of "as representing physical operations performable within $\mathcal{O}$ [...]." (Haag 1996, ibid.) However, Ruetsche (2011, pp. 104–105) argues that things need not be viewed so, and she outlines a way to associate a local algebra with the (local) solution space of a corresponding classical theory, leaving "open the questions of how or whether to further interpret the association [...]." (ibid.)

Considering abstract local algebras of observables is certainly somewhat of a liberation from the Hilbert space formalism in ordinary QM. But Hilbert spaces of state vectors can still be fit into this picture. The key 'bridging principle' between an abstract C^*-algebra and a Hilbert space $\mathcal{H}$ and its elements $|\psi\rangle$ is the *Gelfand-Naimark-Segal-construction* (GNS-construction; cf. Haag 1996, p. 122 ff.). For completeness' sake, we will give a pointed review of this construction in Appendix C, based on the expositions in Haag (1996) and Ruetsche (2011, pp. 73 ff. and 104 ff.).

So from purely algebraic considerations one can recover the (essential) formal methods introduced before (Hilbert spaces and state vectors). The 'deeper' connection to canonical QFT is, however, provided exactly by the *axioms* or *postulates* of AQFT, which are supposed to capture the underlying intuitions while avoiding the aforementioned difficulties. Ruetsche (2011, p. 105 ff.) discusses one set of such postulates, and to make the connection, we will briefly sketch at least the central implications of these.

Among the postulates are multiple *causality postulates* such as a generalization of the microcausality condition (2.41), namely that for two spacelike separated regions $\mathcal{O}, \mathcal{O}'$, all elements of the respective algebras $\mathfrak{A}(\mathcal{O}), \mathfrak{A}(\mathcal{O}')$ commute; or the *primitive causality* requirement that if some region $\mathcal{O}$ is a subset the *domain of dependence* $D(\mathcal{O}') = D^-(\mathcal{O}') \cup D^+(\mathcal{O}')$ (cf. Footnote 81) of another region $\mathcal{O}'$, i.e. $\mathcal{O} \subset D(\mathcal{O}')$, then the algebras will also satisfy $\mathfrak{A}(\mathcal{O}) \subset \mathfrak{A}(\mathcal{O}')$. The upshot of the latter is that in virtue of the causal constraints set up by the 'speed limit' c (the speed of light), the measurable quantities in a region will be causally connected *only* to measurable quantities in the areas that can be reached with speeds $v \leq c$. Put in an operationalistic parlance, the upshot of the former is (as it was above, in the simple, heuristic version) that spacelike separated operations such as measurements can have no direct bearing on one another (we will later see in more detail why this is important).

There are also constraints on the 'behavior' of the local algebras under transformations of the underlying spacetime $\mathcal{M}$ (or rather $(\mathcal{M}, g)$, g the metric tensor), namely that isometries on $\mathcal{M}$ (transformations that do not change the metric) will be reflected by automorphisms on the respective local algebras $\mathfrak{A}(\mathcal{O})$ (bijective maps from the algebra onto itself) on whose defining regions the isometries act. These automorphisms will then provably be represented as suitable unitaires in a representation (cf. Ruetsche 2011, pp. 105–106). Moreover, it is assumed that there exists an irreducible faithful representation of the algebra $\mathfrak{A}(\mathcal{M})$ over the entire spacetime, and that there is a (Lorentz-invariant) *vaccum state* ω_0 whose representation is equally faithful, irreducible, and, basically and informally, will not predict energy-momentum transfers backwards in time or outside the light-cone (cf. Ruetsche 2011, pp. 107–108).

This is quite 'nice and tidy', and AQFT allows, in particular, to prove a range of theorems (one of which we will encounter in Chap. 4) any QFT axiomatized in this or a relevantly similar fashion must satisfy. Moreover, Wallace (2011, p. 119) informs us that "non-interacting quantum field theories [...] can readily be incorporated into the formal framework", and that there is also an AQFT-based "two-spacetime-dimensional scalar quantum field theory with an interaction term ($\lambda\phi^4$) which was exactly definable without cutoffs."

But a sobering realization is that "[d]espite 40 years of work [...] the only known physically realistic algebraic quantum field theories in four dimensions are *free-field theories*." (Wallace 2011, ibid.; my emphasis—FB) In other words: *some* progress can be made with nice and tidy AQFTs. But generally speaking, the successes based on the 'unpleasant' introductions of cutoffs and startling scale-consideration of canonical QFT *cannot* be reproduced by an AQFT to date.

AQFT, although sometimes favored by philosophers for its mathematical clarity (e.g. Ruetsche 2011; Friederich 2015; Halvorson 2007), should be viewed rather as a field of active research that so far cannot help to remove or clarify the confusing *additional* worries introduced by QFT, let alone the OP. So we *still* have the OP *and* the complaints about notions such as 'quantum field', 'vacuum state', and suspect procedures such as introducing infinities that cancel out, or cut-offs to live with. To repeat the point: QFT *in general* at best seems to make the pressing interpretational issues of QM *worse*—AQFT does not really help this point.

2.3 Outlook: Demands for an Interpretation

We have outlined how the OP persists in QFT, the most up-to-date version of QM, and how the reference to fields introduces further problems at QM's foundations. Moreover, we had explained at the end of Sect. 2.1.4 how the limits of the applicability of QM have been pushed back into the 'macroscopic realm' ever further, such as to give rise to at least 'mesoscopic' superpositions. Mesoscopic superpositions bear a certain importance in that they make the OP especially pressing. If it is possible to gather evidence for superpositions occuring in such comparatively large systems, why do they not also appear in everyday life objects (such as the cat in Schrödinger's thought experiment or the friend in Wigner's)? Why are we not well-acquainted with them? What, if anything, triggers the collapse of the quantum state? How does it take place? And what do superposition states actually *represent*? Should we literally think of them as 'both options occurring at the same time', in some sense to be specified? And how *do* we consistently *make sense* of that?

We here need to finally face the task of sketching possible options, possible directions in which to proceed, in order to find a suitable interpretation of QM, one that (re)solves as many of the conceptual problems raised by the theory as possible. We have highlighted the OP as the central problem of QM. Arguably, it is ultimately the problem of the unfamiliarity with 'quantum superposition' in the domain of everyday life experience and QM's implication of there potentially being such a thing, if we do not artificially restrict it to the 'microcosm' or to 'exotic domains'. Thus for the task of interpreting QM, we may formulate the following minimal criterion:

Minimal Adequacy Criterion (MAC) Any interpretation of QM must either solve or avoid the OP without contradicting experimental evidence.

Above, we had discussed kinematical and dynamical aspects of the OP. On the one hand, the difficulty is to understand what quantum superposition actually *is*, and how it relates to experience. Thus we need to figure out what exactly the (ontological) status of quantum states or wave functions is. But a word of caution from a philosopher is due at this point, as to what is meant when we talk about 'the status of the wave function':

> The topic we are trying to address is, more often than not, referred to as 'the ontological status of the wave function.' But one obvious thing about the wave function is that it is a *function*, that is, a mathematical object. [...] In this sense, the question of the ontological status of the wave function is trivial: it is a piece of mathematics. [...] Of course, that is not at all the question we have in mind. Rather, the question is what, if anything, does this particular piece of mathematics *represent*, what physical entity (if any) corresponds to it, and how is that correspondence to be understood? (Maudlin 2013, p. 129; emphasis in original)

There is indeed a certain problem of confusion of semantic levels in the interpretational debate on QM, as indicated in Maudlin's quote, which is not as trivial as may seem. Often suggestions to the solution of interpretational problems are made that only refer to the *mathematics* and hardly have an immediate bearing on the conceptual issues (we shall see this e.g. in the discussion of collapse models later, and the need to supply them with an ontology). This is due, at least in part, to a lack of clarity about *what is being asked.*

The ontological status of mathematical entities is itself certainly an exciting and controversial topic, but it is not the kind of question we are concerned with here. We are assuming that the wave function is used to *represent something*, or at least has a unique, identifiable *role* in scientific conduct that gives a specific *meaning* to it. Otherwise QM would be an empty play of symbols and experimental applications would be impossible. But *what* does it represent? How 'seriously' do we have to take the wave function? Does it represent the actual physical situation of something? And if so, how? Or does it merely represent something about the *experimenter*, maybe his *knowledge* of the actual situation? Or about the experimental setup...? And so forth. In short: we are searching for the *referents* and *application conditions* of QM wave functions or state vectors.

On the other hand, depending on our interpretation of quantum states, we have to tell a story about the dynamics. We saw above that the unitary dynamics at any rate amplifies the OP, if the measurement process is described in QM terms. Thus one should also tell a story as to how we come to experience definite outcomes—as does the orthodox interpretation, in a somewhat incomprehensive fashion, by introducing the projection postulate. The task of fulfilling the MAC thus is twofold:

(i) Explain what the state vector refers to, thereby elucidating the meaning of the kinematical superposition principle and all associated issues (*kinematical task*).
(ii) Provide a suitable modification of the unitary dynamics if indicated by (i), and if not, explain why not (*dynamical task*).

To the 'associated issues' of superposition we here count quantum interference, entanglement, and the peculiar probabilistic structure, because all of these are ultimately related to quantum superposition: Probability evaluations may presuppose specific superpositions of quantum states that give rise to interference terms. And (pure) entangled states are of the form of superimposed product states.

How can an interpretation accomplish fulfilling the two tasks of the MAC? To accomplish the kinematical task, there seem to be two major options: one can either provide an ontology of the wave function that is compatible with scientific evidence *and* everyday life experience, or one can explain why the wave function should *not* constitute part of the ontology of the physical world at all, and then give an alternative account of what it represents or how it relates to scientific practice instead.[94] The former class of interpretations we will refer to as *ontological* interpretations of QM, the latter class as *non-ontological* ones (suitable specifications will follow when indicated).

As for the dynamical task, we will call interpretations that modify the dynamics so as to lead from the unitarily evolving quantum state to definite outcomes *collapse* interpretations, in analogy to the wave-collapse suggestions appealed to in the discussion of Conjecture 0, and thereby following standard terminology. The remainder will be called *non-collapse* interpretations. Note that collapse and non-collapse interpretations may exist in both domains, ontological and non-ontological interpretations, as will become evident subsequently.

To devise a third coarse classificatory dimension, we note that some interpretations suggest stronger modifications, whereas some rather avoid these. We will call the former ones *revisionary*, the latter ones *conservative*. But the (non-)conservativeness may appeal either to the conceptual side of the theory or to its formal side, so that we will distinguish between *formally* conservative/revisionary and *conceptually* conservative/revisionary interpretations. Our standard of evaluation for *conceptual* conservativeness will be the orthodox interpretation. But note that we will have to allow for *some* variation, on conceptual and formal grounds, since otherwise nothing but the orthodox interpretation could fit into the 'conservative' category.

To take hold of these aspects, we will use a three dimensional classificatory scheme (cf. Fig. 2.10) to sort out where each of the interpretations discussed in the following fits. To take hold of the formal/conceptual distinction, each of the 'revisionary'-cells is subdivided. Interpretations or interpretational schemes that propose both, significant conceptual and formal revisions, will be placed on the boundary. With this scheme in hand, we can proceed to discuss constructive proposals to solve or avoid the OP (with its long rat tail of issues associated with quantum superposition), and hence to interpret QM.

---

[94]Talk of a 'physical world' may suggest that there is a 'non-physical world' as well, maybe a 'mental' one. But we do not intend to take sides on this issue here at all. Purely on the level of scientific *description*, it is not the case that physical and mental phenomena do coincide. That is, to date there is no fully physical theory of the mind, and if 'the hard problem' (Chalmers 1996, p. xii) of why there even *is* an experienced inner life accompanying neurophysical processes is indeed as hard as it seems, it is not clear that there ever will be. But again, we are here only suggesting a *prima facie* non-identity of physical and mental (or social, or economical, or...) phenomena, not serious mind-body dualism.

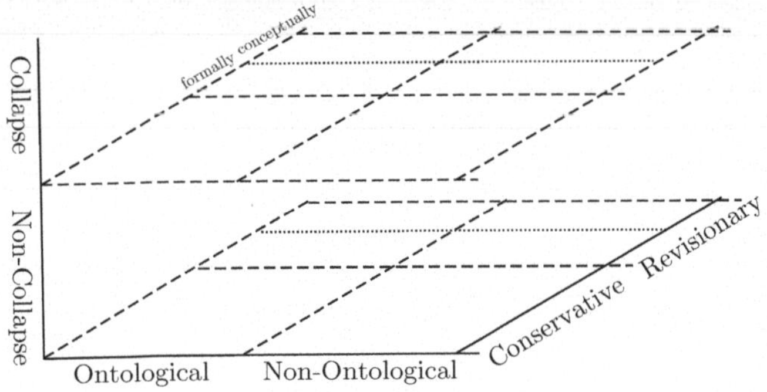

Fig. 2.10 $3D$ scheme for classifying interpretations of QM

However, classifying and fulfilling minimal criteria is only half the battle. Given that we are trying to escape the minimality of the Dirac-von Neumann orthodoxy and are striving for at least a non-minimal interpretation, we formulate the following desideratum for an interpretation:

Desideratum of Ontological Clarity (DOC) Any non-minimal interpretation of QM should be ontologically as clear as possible. This means that it should

(i) explain the *appearance* of 'classical' *objects* that seem to exhibit simultaneously definite but quantum mechanically incompatible properties at all times,
(ii) explain the precise *relation* between the 'quantum' and 'classical' realms,
(iii) and specify the ontological *significance* of all formal ingredients.

By 'classical' we here simply mean something along the (more innocent) lines of Bohr (1938, p. 269), i.e. objects describable by "everyday concepts, eventually refined by the terminology of classical physics[...]", thereby emphatically not encouraging "the mistaken thought that any use of such a description carries with it the *full* content of classical physics [...]." (Healey 2012d, p. 740; my emphasis—FB) Point (iii) of the DOC may urge us, depending on where we stand on the kinematical task of the MAC, to provide an ontology of the high-dimensional space on which the wavefunction is defined, and points (i) and (ii) may then also urge us to specify how $4D$ spacetime, wherein the classical objects 'live', and the high-dimensional ('configuration'-)space wherein the wavefunction 'lives' relate to one another. We will see that fulfilling the DOC on these grounds is a major hurdle.

Chapter 3
Philosophical Interlude I: 'Probability' and 'Realism'

So far we have talked about probabilities for finding a certain value for a certain observable, or for a system to collapse into some definite state, without specifying at all what we *mean* by 'probability'. Famously, there is a whole host of differing views of probability, and possibly the broadest dichotomy[1] one can draw between probability-concepts is that between *epistemic* and *objective* probability, as advocated e.g. by Gillies (2000, p. 2). With an eye on the discussion to follow later, it seems desirable to further subdivide both categories, epistemic and objective probability, into suitable subcategories.[2]

Objective conceptions of probability include *relative frequencies* (of occurrences of certain types of events in finite domains), *limits* of relative frequencies (in random sequences defined over infinite domains), long-run *propensities* (as dispositions of *types of experiment* to bring about a certain outcome with a certain rate), and *single-case* propensities (as dispositions of *individual* experiments to bring about a certain outcome or of single events to occur). Epistemic conceptions of probability, on the other hand, include degrees of *belief* in some hypothesis (of a real *or* ideal agent), and degrees of *confirmation* of a hypothesis (by evidence).

But of course none of these views is entirely free of problems, and the notions are interrelated in various ways. For statistical probabilities in the sense of the limit of a relative frequency, for instance, it is necessary to provide an account of what counts as a *random* sequence of events, because otherwise all kinds of probabilities could be *fabricated* by a suitable ordering of the elements of the sequence. Such

[1] As is well known, Carnap, in particular, also defended a *two*-concept view of probability, e.g. in his 1945 and 1955 papers. But Carnap's distinctions between degree of confirmation and relative frequency in the long run (1945), or statistical and inductive probability (1955), are ultimately too narrow for our purposes. So is the distinction between credences and chances, which Lewis (1987, p. 83 ff.) put in their place.

[2] We appeal, in what follows, to the expositions given in Gillies (2000), Mellor (2005, p. 8 ff.), and Schurz (2014, p. 129 ff.).

F. J. Boge, *Quantum Mechanics Between Ontology and Epistemology*,
European Studies in Philosophy of Science 10,
https://doi.org/10.1007/978-3-319-95765-4_3

an account was, indeed, worked out by von Mises (1957), who first assumed some ground sequence s of all realizations of a certain kind of experiment continuing on indefinitely into the future. In this ground sequence, there should be definite limits for all frequencies of all occurring events, which should be invariant under *arbitrary* selections of partial sequences s' from s (where the s' should be extendable indefinitely as s). And to provide a notion of *arbitrariness*, von Mises introduced *admissible* place selections for occurences of events in the sequence by requiring an outcome-independence, i.e. that "the question whether or not a certain member of the original sequence belongs to the selected partial sequence should be settled *independently of the result* of the corresponding observation, i.e., before anything is known about this result." (von Mises 1957, p. 25, emphasis in original)

Von Mises' account has, however, also been met with criticism, most notably that "there is not just only one physically possible infinite random sequence; there are many, indeed uncountably many. It seems arbitrary to pick out one of them and declare it to be the 'ground sequence.'" (Schurz 2014, p. 156; cf. also Hájek 2009b, p. 217 for the original argument).

Moreover, it is worthy of question how to thoroughly *connect* an ideal mathematical concept such as the limit value of some sequence to actual *experience*. Hence, as Howson and Urbach (2006, p. 47, emphasis in original) put it,

> it has [...] elicited from positivistically-minded philosophers and scientists the objection that we can never in principle, not just in practice, observe the infinite n-limits. Indeed, we know that in fact (given certain plausible assumptions about the physical universe) *these limits do not exist*.

Schurz (2014, p. 156) refers to this as the *problem of empirical content*.

A particular solution that has been suggested to this problem is to take the limit of some relative frequency of occurrence of a certain outcome under certain (experimental) conditions "as an exact measure of the *tendency* of those conditions to deliver the outcome in question [...]." (Howson and Urbach 2006, p. 46; my emphasis—FB) And arguably that is also what von Mises had in mind in assuming experimental setups to produce genuinely random sequences of outcomes[3]:

> The probability of a 6 is a physical property of a given die and is a property analogous to its mass, specific heat, or electrical resistance. Similarly, for a given pair of dice (including of course the total setup) the probability of a 'double 6' is a characteristic property, a physical constant belonging to the experiment as a whole and comparable with all its other physical properties. (von Mises 1957, p. 12)

This obviously links the frequentist approach to the long-run propensity approach and gives the unobservable (and in some cases maybe even physically nonexistent) limits a meaning. But propensities can be understood as a special kind of *disposition* (cf. Mumford 1998, p. 11; Popper 1957, p. 67) which is simply not "sure fire" (Mumford 1998, ibid.). Hence they inherit all the metaphysical and analytical difficulties associated with the notion of a disposition.

---

[3]Cf. also Howson and Urbach (2006, p. 46) and Schurz (2014, p. 154) on this point.

The epistemic degrees-of-belief and degrees-of-confirmation accounts are usually also jointly referred to as *Bayesianism* (after Reverend Thomas Bayes 1763). The latter of these, broadly speaking, has to do with an agent's *evidence* (something *objective, external*), whereas the former need not. This broadly coincides with Williamson's (2010, p. iii) characterization of the distinction between *subjective* and *objective* Bayesianism: "Subjective Bayesians hold that it is largely (though not entirely) up to the agent as to which degrees of belief to adopt. Objective Bayesians, on the other hand, maintain that appropriate degrees of belief are largely (though not entirely) determined by the agent's evidence."

Moreover, Williamson characterizes objective Bayesianism as a *normative* theory. I.e., it is not that any actual observer's beliefs may be represented by probabilities without difficulty, but rather:

> The strengths of an agent's beliefs *should* behave like probabilities: they should be representable by real numbers in the unit interval and one should believe a disjunction of mutually exclusive propositions to the extent of the sum of the degrees of belief of the disjuncts. Moreover, these degrees of belief should be shaped by empirical evidence: for example, they should be calibrated with known frequencies. (Williamson 2010, p. 1; my emphasis—FB)

But Bayesian views of probability, sometimes also thought of as *epistemological* hypotheses about how agents (should) form, update, and manage beliefs (e.g. Williamson 2010, p. 10), can also be subdivided into 'finer' categories, depending on the *degree* of objectivity they embrace: All forms of Bayesianism, Williamson (ibid.) tells us, hold a *probability norm*, meaning that "one's degrees of belief at a particular time must be probabilities if they are to be considered rational." *Empirically based subjective* Bayesians add a *calibration norm*, meaning that "one's degrees of belief [...] should also be calibrated with known frequencies." (ibid.) *Objective* Bayesians differ from this in that they assume an *equivocation norm* as well, meaning that "one's degrees of belief at a particular time are rational if and only if they are probabilities, calibrated with physical probability and otherwise *equivocate* between the basic possibilities." (Williamson 2010, p. 16; my emphasis—FB) In the evidence/belief divide, empirically based subjective Bayesians should hence rather be grouped together with objective ones than with subjective ones.

Williamson's is a useful distinction between three 'flavors' of Bayesianism which we may appeal to, on occasion, in what follows. But it should be stressed that it is ultimately *pointless* to demand too sharp and robust distinctions between different Bayesianisms: By a simple combinatorial argument alone, I.J. Good (1983, p. 20) makes out 46,656 varieties of Bayesianism, depending on how one stands on a few details, and he (consciously) ignores, in his calculation, that one might even construct a *continuum* of intermediate positions (i.e. an uncountably infinite number of Bayesianisms). The most one should hence hope for is a suitable distinction between 'more objective' views and 'more subjective' ones.

Of course epistemic approaches to probability also come with *problems* that in principle deserve equal attention as those raised for objective ones; but we will defer the discussion of some of these problems (for radically subjectivist positions at least) to Chap. 7, where the context demands it.

Now which category is appropriate to the quantum probabilities, as they emerge from the formalism? This is all but a simple and uncontroversial question, and a significant part of the task of interpreting QM concerns what one believes on these issues. Assume, for instance, that one aims for a *thoroughly operationalist* understanding of QM. Then the probabilities must of course be regarded as relative frequencies of occurrences of a certain type of outcome in *actual* (finite) sequences of experiments, because there is otherwise no possibility of 'translating them' or their test-conditions into laboratory operations. But this is obviously highly problematic, since it is unclear *how long* a given such sequence must be to deliver the *correct* probability (frequency), and hence when QM—or any probabilistic theory for that matter—should ever count as *confirmed* (not to mention the difficulties with translating non-rational probabilities into laboratory operations). Adding a few more runs which all happen to have one particular outcome will change the frequency again, so there is a major arbitrariness or selectiveness about when to consider the sequence 'long enough'. Even operationalists, it seems, must tacitly appeal to something like von Mises account of limiting frequencies, and hence, in fact, possibly embrace a metaphysics of dispositions.

Popper actually refrained from his originally frequentist understanding of probability, and developed his account of propensities—first in the long run sense, and later in the single case sense (cf. Popper 1990, p. 12 ff.)—in part due to his worries about the quantum probabilities, as he explicitly states e.g. in Popper (1959, p. 27). *His* main reason was that the frequentist account is hardly capable of making sense of *single case* probabilities, which QM seems to assert all the time (cf. Popper 1957, p. 66).

But assume now that one has a large grid, i.e. essentially a 'multi-slit'-arrangement, a generalization of the double-slit, and assume that a photon is incident on this grid. The wavefunction should now assign a *propensity* for the photon being measured in any of these slits, should one use detectors to determine which path it takes. As we already know, only in such an arrangement, *with* detectors behind slits, will the photon behave as if it took exactly one path. This means that detection at any one slit should alter *immediately* and spontaneously what happens in all the other slits, if the wavefunction (or the associated probability amplitude) is somehow indicative of a 'single case propensity'. This seems hard to reconcile with relativity in which there is no preferred frame and hence no over-arching simultaneity, i.e. no 'definite moment' in which the propensity 'manifests' in one slit and 'ceases' in the others. And we will eventually see, subsequently to Chap. 4, how QM raises even worse conflicts with relativity when construed along these lines.

A partially connected issue that has surfaced in the previous chapter on multiple occasions is that of '*realism*' in QM. For instance: QM does not, it seems, warrant the assumption of pre-existing well defined properties on investigated systems, including such important ones as positions and velocities. The assumption of such is of course highly intuitive on the basis of everyday-life experience and its 'relative stability'; closing one's eyes for a second and making one's visual perceptions of a fastly seated table and laptop vanish, one may tend to firmly believe that these perceptions will re-appear when one opens one's eyes again. However, as Russell put it in *Our Knowledge of the External World*:

> We naturally believe, for example, that tables and chairs, trees and mountains, are still there when we turn our backs upon them. I do not wish for a moment to maintain that this is certainly not the case, but I do maintain that the question whether it is the case is not to be settled off-hand on any supposed ground of obviousness. (Russell 1914, p. 77)

This is an attitude that we should equally entertain, as a matter of philosophical caution if you will, in the present context, and especially w.r.t. to the 'states' that we would like to attribute to microscopic (or 'quantum') systems in certain (experimental) situations on the basis of our successful practice of doing so with unobserved everyday-life objects.

Quite general talk of 'realism' is, however, rather uninformative, as 'realism', not unlike 'probability', is not an unequivocal notion: there is a wide variety of differing meanings in different contexts, and it has been criticized, notably by Travis Norsen (2007), that in certain applications in the context of QM (namely, the violations of Bell-type inequalities to be discussed later) the discussion is *blurred* by the use of the word 'realism'; because "it is almost never clear what exactly a given user means by the term [...] and [...] none of [the] possibly-meant senses of 'realism' turn out to have the kind of relevance that the users seem to think they have." (Norsen 2007, pp. 311–312) Whether Norsen's assessment as to the relevance of 'realism' is correct should be evaluated on the basis of later discussion. But we concede that there is a crucial terminological problem here that one should try to fix.

To confront a concrete example, consider that Clauser and Shimony (1978, p. 1883) characterize 'realism' in a particular (actually: the same) QM context as "a philosophical view, according to which external reality is assumed to exist and have definite properties, whether or not they are observed by someone." Denying the existence of the external may be construed as a thorough idealism along the lines of Berkeley (2009 [1710-13]). But denial of the second conjunct is just the most straightforward reading of the position of Busch et al. (1995), outlined in Sect. 2.1.5. On the one hand we can immediately see that the issue is nontrivial for the interpretation of QM, not just for philosophy in general. On the other hand, we can also see that there may be radically differing intuitions as to what is *meant* by 'realism', since it is far from clear that 'realism' implies 'definiteness of properties'.

A related point of criticism towards uses of the word 'realism' in the context of QM is expressed in a more recent paper by Maroney and Timpson (2014). The paper specifically concerns the implications of violations of so called *Leggett-Garg inequalities* (Leggett and Garg 1985), but some of the arguments carry over to the present context. The aforementioned class of inequalities was introduced by Leggett and Garg with the aim of testing a metaphysical position which they refer to as "macroscopic realism" (Leggett and Garg 1985, p. 857), and flesh out as the supposition that "[a] macroscopic system with two or more macroscopically distinct states available to it will at all times be in one or the other of these states." (ibid.) Together with the assumption of "[n]oninvasive measurability at the macroscopic level", i.e. that "[i]t is possible, in principle, to determine the state of the system with arbitrarily small perturbation on its subsequent dynamics" (ibid.), Legget and Garg believe to be able to show (by appeal to a SQUID-example as discussed in Appendix B) that "[a] direct extrapolation of quantum mechanics to the macroscopic level denies this [macroscopic realism]." (ibid.) Our point of contention here is not the assessment of the experimental details, but the *notion* of 'realism' involved and its connection to the notion of realism employed by Clauser and Shimony.

Setting aside questions of macroscopicity (as do Maroney and Timpson 2014, p. 9), the question remains whether the assumption of being in one out of a range of possible states at all times is a necessary condition for 'being real'. To this Maroney and Timpson (2014, p. 10) utter the concern that the notion of 'state' employed in the definition requires clarification, since such a notion usually "comes as part of a theory, or as part of a general framework of theories." (ibid.) The background theory they identify in the Leggett-Garg case is (ordinary Hilbert space) QM, and the 'states' in question are meso- or macroscopic current states, mentioned briefly at the end of Sect. 2.1.4 and discussed at some length in Appendix B. But against this, Maroney and Timpson (2014, p. 10) counter with the following slogan:

> *There is nothing realist about denying the existence of superpositions macroscopic or otherwise.*

This slogan they motivate (their p. 11) by the observation that

> one can seek to incorporate superposition, including macroscopic superposition, into one's realist, descriptive, account of how the mind-independent world is—incorporate it, moreover, in such a way as to recover the determinate nature of our experience, and of the macroscopic world.

This is obviously correct, since, as we shall see later (and as is elaborated by Maroney and Timpson), there are well-known ontological interpretations of QM which do just that. How does this argument carry over as an objection to Clauser and Shimony's notion of realism, accroding to which "reality is assumed to [...] have definite properties, whether or not they are observed by someone"? The answer obviously is that analogous reasoning can be applied to the notion of a *property*, instead of 'state'. The notion of property that Clauser and Shimony have in mind is that of a 'definite' one, which we may equally consture as 'sharp', in contrast to the 'unsharp' ones we introduced in Sect. 2.1.5. That is, the definite properties in question, those whose observation-independent existence is construed as constitutive of realism, are those that could be represented as *projectors* in QM. Because the existence of *these* as always present observer-independently is what is put in question by QM.

But in close analogy to Maroney and Timpson, we can dispute the intuition that this has anything to do with realism, which we summarize in the following slogan:

> *There is nothing anti-realist about accepting the existence of unsharp properties.*

Maybe the world just happens to be such that entities can only have an approximate degree of localization, an approximate velocity, an approximate spin... and so forth. Given everyday life experience, this certainly is highly counter-intuitive. But *prima facie* that has nothing to do with the *existence* of a mind independent reality altogether.

Above (Sect. 2.1.5) we indicated that the assumption of 'unsharp properties' is in need of some explanation though. So to make the argument plausible, we should here provide a rough sketch of what a 'realism' including unsharp properties could look like.

An obvious possibility would be to entertain the conviction that the *linguistic expressions* we form on the basis of everyday-life experience do not *refer* properly to mind-independent reality, but only in an imprecise or ambiguous way, and that this is reflected in the appeal to QM concepts such as POVMs and non-commuting self-adjoint operators. More precisely, the 'true constitution' of reality could be hidden from us altogether, thus not sanctioning the *attribution* of 'properties' in any 'fundamental', non-pragmatic sense. And through sense perception and conceptualization based (partly) on our experience, we might only get access to this reality in a semi-definite manner, i.e. by means of sometimes unsharply applying concepts such as 'position' or 'momentum'.

We see that certain applications or understandings of 'realism', often employed in the QM literature, involve criteria worthy of clarification. So let us review some philosophically more well-grounded terminology. Norsen (2007, p. 316), first of all, appeals to a notion of *naïve* realism, which to him means that "whenever an experimental physicist performs a 'measurement' of some property of some physical system [...] the outcome of that measurement is simply a passive revealing of some pre-existing intrinsic property of the object." (emphasis omitted) This, Norsen thinks, is the physics-appropriate generalization of "the view that all features of a perceptual experience have their origin in some identical corresponding feature of the perceived object." (his p. 315)

Classical physics is viewed by many to entertain exactly the former sort of epistemological position, as it seems to endorse that at least in principle measurements can be as subtle and non-invasive as desired, whence to a good degree of approximation, a pre-existing intrinsic property of the system *is* being revealed (almost) passively (think again of Heisenberg's microscope and "destruction of knowledge", which he apparently considered as a kind of philosophical revelation).

The most fundamental notion of 'realism' in the philosophical tradition is certainly that of *metaphysical realism*, typically identified also with *external realism*, the basic statement of which can be phrased as "[t]he world [being] (largely) made up of objects that are mind-, language-, and theory-independent." (Button 2013, p. 8) Since languages and theories (arguably) depend on minds, we can identify metaphysical realism simply as the view that there exists a mind-independent 'outside world'. Norsen (2007, p. 330), who equally acknowledges a notion of 'metaphysical realism', puts it in slightly different terms; metaphysical realism, according to him, "accepts the existence of a single, objective, external world 'out there' whose existence and identity is independent of anyone's awareness [...] of it." (ibid.)

More illuminatingly, there is the much stronger notion of *scientific* realism in the philosophy of science, which, following Putnam (1975b, p. 179) and Psillos

(1999, p. xvii) and in *contrast* to Norsen's (2007, p. 320) less detailed treatment, we can summarize as endorsing, *on top of* a general metaphysical realism, the central tenets that (i) mature and well confirmed *theories* are *capable* of being true, and that (ii) the concepts of these very theories typically *do* refer to entities in the external world, in *all* domains (including unobservable microstates, say). Condition (i) may be thought of as *semantic*; it says something about the *meaning* of scientific notions (including formal-mathematical ones). Condition (ii), on the other hand, may be viewed as *epistemic*; it says that we can *access*, or *know about*, the external world by means of scientific theorizing.

Both of these tenets are nontrivial parts of the scientific realist stance,[4] as different kinds of anti-realism have challenged them separately; Dummett's (1982) semantic anti-realism, e.g., representing a challenge to the first statement, van Fraassen's (1980) constructive empiricism a challenge to the second one.[5] It is also worth pointing out that one could deny *scientific* realism without denying external realism altogether. I.e., one could express reservations about science's capability to represent the (outside) world while reserving that there is an outside world that we can truthfully represent by means other than science. At any rate, mystics or some scientific layman do just that.

Notably, on a more careful view of things, the above identification of *external* and *metaphysical* realism seems misguided. Even on basic metaphysical grounds, more subtle versions of realism are available that do posit that there is "a [...] world 'out there' whose existence and identity is independent of anyone's awareness [...] of it", as Norsen has it, but which are more careful about the existence of "*objects* that are mind-, language-, and theory-independent" (Button's formulation; my empahsis—FB). A contrasting position to the latter is what Putnam (1977) calls *internal realism*. Internal realism "has been summarized by Putnam in several different places and in a number of different ways" (Conant 1990, p. xix), with implications not necessarily coincident (cf. also Putnam 1992, p. 353 ff.; Button 2013, p. 74). Moreover, internal realism has been linked closely to Kant's (1781) *transcendental idealism* (cf. in particular Brown 1988), and Conant (1990, p. xix) even finds that "Putnam discerns a version [of internal realism—FB] in Kant's work[...]." We here agree with (Conant's reading of) Putnam that Kant's transcendental idealism can be understood as a *version* of internal realism.

Despite the fact that there may be multiple versions, can we say what internal realism *generally* implies and how it is distinguished from external realism? To this end, it is instructive to review (relevant) commonalities between Kant and

[4]This careful notion of a 'stance' is embraced, in particular, by van Fraassen (2002, pp. 47–48) who identifies it with an "attitude, commitment, approach, a cluster of such—possibly including some propositional attitudes such as beliefs as well", but which "cannot be simply equated with having beliefs or making assertions about what there is".

[5]We should note though that van Fraassen (1980, p. 8) goes further in even disputing that the *aim* of science is "a literally true story of what the world is like", and that "acceptance of a scientific theory involves the belief that it is true". More on this later.

Putnam first and then discern the central distinguishing elements in their respective positions. Here is Brown's (1988, p. 146) analysis:

> Both consider, somewhat uncomfortably, the view that there is *an unknowable noumenal world behind the phenomena*. Both are motivated in part by the threat of scepticism: Kant by scepticism about our ability to *know the external world*, Putnam by scepticism about our ability to *refer to it*. Both Kant and Putnam hold that the world we know and talk about is *empirically real*, but both hold also that it is *mind-dependent*. [...] Putnam, like Kant, stresses the pervasive importance of causation, and argues that *causation is partly our own imposition on the world*. (my emphasis—FB)

This characterizes internal realism reasonably well: In its broadest understanding, it is the position that there *is* a mind independent world, but that the way the world *appears* to us, the way 'empirical reality' *is*, is largely dependent on our *minds*; or as we shall here prefer to phrase it: on our *cognitive interior*. Notably, this extends even to such fundamental concepts as *causation* (about which we will have to say more later).

However, from the quote we can also discern differences between Kant and Putnam: Putnam (1990, p. 41; my emphasis—FB) describes his internal realism as driven by a rejection of the correspondence theory of *truth*, i.e. as stating, in the first place, "that truth comes to no more than *idealized rational acceptability*" and that "what is supposed to be 'true' be warrantable on the basis of experience and intelligence for creatures with 'a rational and a sensible nature.'" Kant, on the other hand, was famously concerned with

> objects of sense as mere appearances,[...] based upon a thing in itself, though we know not this thing in its internal constitution, but only know its appearances, viz., the way in which our senses are affected by this unknown something. (Kant 1783, §32)

It seems that while the former denies the *semantic condition* of scientific realism, but also w.r.t. to *pre-scientific* domains of inquiry and *everyday life* conduct, the latter does the same w.r.t. the epistemic condition.

Neither denies the very *existence* of a mind-independent—albeit "noumenal" (cf. Putnam 1977, p. 492; Kant 1781, p. 248 ff.)—world though, so we identify (in contrast to Putnam's use of these words) internal realism as a brand of metaphysical realism. But of course the 'realism' portion in internal realism is considerably weaker than in external realism, whence Sankey (2008, p. 115; my emphasis—FB) has it that "internal realism is [...] an *inherently idealist doctrine*." Idealist, maybe. But only in the "more complex" Kantian sense that there is "an important distinction between the mental and the physical, but that the structure of the empirical world depend[s] on the activities of minds." (Brown 1988, p. 145) We hence identify internal realism as a *weak* metaphysical realism and external realism as a *strong* one, and we will use the items in these pairs interchangeably respectively in the following.[6] Scientific realism in our present understanding presupposes *strong* metaphysical realism.

---

[6]We completely forego the role of the a priori in Kant's thought and the details of Putnam's theory of reference at this point, since we are here only interested in the role of 'reality' in both conceptions. But both these aspects will make their way into Chap. 7.

This clarifies how Clauser and Shimony (1978) can think that having “definite properties, whether or not they are observed by someone” could be constitutive of realism: They do not seem to acknowledge the possibility of an internal realism, which, in some version or other, is perfectly compatible with our option for embracing ‘unsharp’ properties outlined above.

The terminology introduced in this interlude will prove useful in what follows, as we can use it to see to what extent and in what sense QM ‘threatens’ or even concerns realism—depending, of course, on its interpretation. We may note here that all interpretations that we classify as ‘ontological’ (to be discussed in Chap. 6) obviously endorse scientific realism; QM, certainly being an instance of a *mature* theory, is assumed to refer (successfully) to a mind-independent reality, although its way of referring has to be pointed out by precisification of notions, rewriting of equations, and/or the introduction of further formal elements. The same (embracing of scientific realism) can be said about the interpretational program investigated in the following chapter, although *not* w.r.t. QM. Since QM is, however, a well-confirmed, mature theory, it clearly needs some sorting out *how* exactly scientific realism is endorsed therein. In this connection, the different views of probability introduced above will play a substantial role. The same goes for the interpretations discussed in Chap. 7.

Chapter 4
Just a Matter of Knowledge?

The field in a many-dimensional coordinate space does not smell like something real.

—A. Einstein, in a letter to Ehrenfest, 1926 (cf. Howard 1990, p. 83)

4.1 Prelude: Ensemble Interpretations and Hidden Variables—The Historical Background

There is, it seems, a rather natural response to the conceptual problems raised by QM. This response, put frankly, is to say that 'it's all just epistemic!' More precisely this would mean to deprive the quantum state of its ontological significance and to construe the theory *not* as a description of the actual, real situation of physical systems, but rather as a representation of the *knowledge* an actual or ideal observer or agent has about these. So for instance, when a quantum system passes a double slit, we cannot know exactly where it is going to turn up. But we can try to make precise predictions about its future behavior, based on our previous experience with similar situations and systems, and hence quantify, in a sense, our knowledge about its future behavior and about the occurrence of spots in various regions on the screen behind the slit. Again given past experience, a quantum mechanical state function may be the tool of choice to accomplish this task. But in the instant we see a dot appearing on the screen we can update our knowledge about the system's actual state, since we can now be rather certain (assuming that we have not visited too many epistemology classes) that the system has occupied exactly that region on the screen at the moment of the appearance of the dot.

But there should be no doubt, the intuition might go, that every system always has a unique, definite state, a unique way of how it 'actually is', despite our ignorance of this 'actual how'. This should also—a slightly stronger assumption—at least *in*

F. J. Boge, *Quantum Mechanics Between Ontology and Epistemology*,
European Studies in Philosophy of Science 10,
https://doi.org/10.1007/978-3-319-95765-4_4

principle enable us to give a description of this 'how'. The 'weirdnesses' elaborated on above, the uncertainty relations, the alleged unsharpness of properties, quantum superposition, and with it the OP, are all just expressions of our inability to properly *access* the true states of certain (typically microscopic) systems, and hence they vanish when properly construed in terms of incomplete knowledge. This natural response, as we have coined it, we can formulate as a first serious conjecture as to how to interpret QM while satisfying the MAC:

> **Conjecture 1 (The natural response)** Quantum mechanical systems are (typically tiny) objects which always possess a true, definite state with precise values for all observable physical magnitudes. The need for a quantum mechanical description of these systems is just an indication of our lack of knowledge about their true states.

Thus, the need for a 'collapse of the wave function' in any *physical* sense, i.e. for the system to be considered as a weirdly dislocalized entity inhabiting a high-dimensional configuration space when in isolation, just to collapse into a 'point-particle' in physical space(time) upon certain kinds of 'measurement-like' interactions, is removed. The 'collapse' in the sense of the PP is just a sort of informational update for the experimenter upon registration of a given result, which may be compared to Bayesian updating (cf. also Wiseman and Milburn 2010, pp. 10–11). In this sense, the interpretations we are about to investigate should be considered as *non-ontological collapse interpretations*, using the terminology we have introduced above.

A central tenet behind this general approach is that QM is in fact an *incomplete* theory, which will (hopefully) be replaced by a more complete and comprehensive one in the future. This tenet is, indeed, a rather old one which quickly surfaced when the peculiar features of QM became apparent in its early development. But it has also gained attention again in recent years and our primary concern will be with the more recent debate. The most prominent proponent of an early epistemic view of quantum sates was Einstein. This is reflected vividly, for instance, in his 1939 correspondence with Schrödinger, where he writes:

> I am as convinced as ever that the wave representation of matter is an incomplete representation of the state of affairs, no matter how practically useful it has proved itself to be. The prettiest way to show this is by your example with the cat (radioactive decay with an explosion coupled to it.)[1] At a fixed time parts of the ψ-function correspond to the cat being alive and other parts to the cat being pulverized.

[1]At that time, Einstein knew the cat thought-experiment from a letter Schrödinger had sent him, in which the latter described it in terms of an explosion rather than poisoning (cf. Mehra and Rechenberg 1987, p. 743).

> If one attempts to interpret the ψ-function as a complete description of a state, independent of whether or not it is observed, then this means that at the time in question the cat is neither alive nor pulverized. But one or the other situation would be realized by making an observation.
> If one rejects this interpretation then one must assume that the ψ-function does not express the real situation but rather that it expresses the contents of our knowledge of the situation. (Einstein 1939, p. 43)

The intuition here is clear enough, and the motivation should be clear as well. QM generally only predicts probabilities and averages, statistical magnitudes which are familiar from contexts in which the actual conditions are simply unknown or at least supposed to be so. QM, interpreted according to orthodoxy, arguably has 'unbearable' consequences such as superpositions of dead and alive cats, and hence cannot possibly refer to the world as it really is. Thus (early) Heisenberg's 'destruction of knowledge' due to the measurement interaction is exactly apt, and it is *all there is* to the quantum weirdnesses. This is certainly due to some feature of physical particles or fields or whatever the theory 'really' treats of that has yet to be understood (or so the intuition might go).

But how does one actually spell out such an *epistemic* interpretation of QM in detail? Einstein's own views are often also referred to as an *ensemble interpretation*[2] of quantum states because of his continuing appeal to *statistical ensembles*, as witnessed e.g. in his reply to criticisms in Schilpp's volume on his life and work (cf. Einstein 1949b, p. 668), or in his 1936 article on physics and reality: "The ψ-function does not in any way describe a condition which could be that of a single system; it relates rather to many systems, to 'an ensemble of systems' in the sense of statistical mechanics." (Einstein 1936, p. 375)

A view of this kind was later also explicitly defended and extended by Ballentine (1970), who describes it (his p. 361) in the following terms:

> For example, the system may be a single electron. Then the ensemble will be the conceptual (infinite) set of all single electrons which have been subjected to some state preparation technique (to be specified for each state), generally by interaction with a suitable apparatus. Thus a momentum eigenstate (plane wave in configuration space) represents the ensemble whose members are single electrons each having the same momentum, but distributed uniformly over all positions.

Note that the "state preparation technique" must in fact be viewed rather as an *equivalence class* of such techniques (cf. Busch et al. 1995, p. 5). The possibility of using either a calcite crystal or a yes-no polarizer to prepare a photon in a certain polarization state makes this immediately obvious, and completely different sets of preparation procedures can be used to prepare the same *mixed* state, due to the latter's multiple decompositions (more on this in Chap. 7).

---

[2]Fine (1984), however, gives a detailed critical analysis of the historical data concerning Einstein's view of the quantum state and contends that the standard representation of Einsteins views is incorrect. In essence, the point is that Einstein seems to have used talk of ensembles merely as a means for grounding the belief in the incompleteness of QM, not as a serious suggestion for an alternative (cf. also Whitaker 1996, p. 239, on this point).

Now it may not be immediately obvious why this view of the quantum state should count as 'epistemic', but it is obvious from the above that at least Einstein was aiming for an epistemic interpretation. So he may have had in mind an *ideal* or *conceptual* ensemble, as suggested in Ballentine's quote, construed merely as a cognitive tool for determining probabilities of experimental outcomes in experiments on equally prepared systems. This is also the viewpoint of Harrigan and Spekkens (2010, p. 150) who write that

> [...] the ensembles Einstein mentions are simply a manner of grounding talk about the probabilities that characterize an observer's knowledge. [...] Ultimately, then, the only difference we can discern between the ensemble view and the epistemic view concerns how one speaks about probabilities [...].

Similarly, Bartlett et al. (2012, p. 4) have it that "the thesis that quantum states describe the statistical properties of a virtual ensemble of systems [...] is equivalent to saying that it describes one's limited information about a single system drawn from the ensemble." And distinguishing, as Ballentine (1970, p. 361) suggests, the *probability* of each outcome as determined by the aforementioned conceptual ensemble from the actually observed *statistical frequency* is equally compatible with this understanding of an ensemble approach as epistemic: The conceptual, infinite ensemble may be invoked here as a means to determine a probability, a mathematical expression of what frequencies to *expect* in an experimental situation, an epistemic tool for making headway in a situation of uncertainty.

But *what* exactly are these ensembles *composed* of? From what we have established so far, it is obvious that they cannot be composed of (conceptual) electrons, say, *in the sense of QM*. Because all QM assigns is the state vector, and this state vector, as we saw, does not—and *cannot*—attribute definite values for all observables at all times. So the conceptual ensembles must be about something else, something not exhaustively described by QM, about some additional set of *hidden variables*.

This notion of 'hidden variables' is a widespread term used to classify a wide range of interpretations of QM. But the name *hidden* variable may create confusion, as has been objected e.g. by Belinfante (1973, p. 8). Why are these variables 'hidden'? Hidden by what? From whom? A nice clarification is offered by Pearle (1968, pp. 464–465):

> The 'hidden' in the phrase 'hidden-variable theories' refers to the fact that at present the variables in these theories have not been experimentally detected, so that the variables must be averaged over in some way, in order to produce predictions which agree with experiment and with quantum theory.

The general contention of such theories or interpretations of QM hence is that there are some additional features in nature, typically supposed to be more in line with the concepts of classical physics, which then allow for a more 'complete' theory, capable of explaining away the perplexing consequences of QM (cf. Belinfante 1973, p. xvii ff., for a similar assessment). The hidden variables are supplemented so that the more complete theory may refer to more than does

QM. Applied to the ensemble in Ballentine's quote above, described by a particular eigenstate ϕ_p of momentum $\boldsymbol{p}$, this would mean the following:

> *all* members of the ensemble will have the corresponding value of momentum, but in addition each has a precise value of position, though these values will all be different. The values of position must be called hidden variables, because they are not related to the wave-function. (Whitaker 1996, p. 284; emphasis in original)

As should be clear by now, and as is laid out by Home and Whitaker (1992, p. 263 ff.) and d'Espagnat (1995, p. 297 ff.) as well, the assumption of such hidden variables has to figure in all *meaningful* ensemble interpretations in the sense of Conjecture 1. But there is also a rather trivial sense in which the term 'ensemble' plays a role in QM itself, without the hidden variables:

> When a quantum theorist wants to make some predictions about a physical system, he will admit at once that in most cases he can make at most some probability predictions. After he has ascertained in what way the system was prepared, he will choose an initial state vector in Hilbert space, which in simple cases often is called the 'wavefunction' of the initial state. This ψ will describe all systems that underwent a similar preparation. These systems taken together form an ensemble (E_ψ), and ψ will describe this entire ensemble. (Belinfante 1973, p. 6)

Obviously, since QM is concerned with probabilistic predictions, the state function must always *also* be allowed to refer to an entire ensemble of equally prepared systems *to test these predictions*. But in orthodox QM, ψ is taken as equally representing ('completely') the state of an *individual* system, which is decidedly *not* the case with probability distributions (or densities) in classical theories. This is actually one decisive aim in advancing an ensemble interpretation of QM, in the sense of Conjecture 1: to dispense with the idea that the quantum state refers to one individual system, and hence with the idea that system should be in a superposition state w.r.t. some basis.

On a classificatory note, the traditional ensemble interpretations discussed above may be viewed as conceptually revisionary but formally conservative. This is so since in order to makes sense, they are committed to the existence of variables not contained in QM, i.e. further *concepts* which differ from those employed in the orthodox interpretation, but neither Ballentine nor Einstein, say, proposed a concrete modification of the formalism. QM *as it is* should be seen as a formal tool for devising statistical predictions, the underlying intuition says; the possibility of a more complete physical theory is merely left open or hoped for.

There are, however, many good reasons why ensemble interpretations of this kind were historically considered as failures quite early on. Schrödinger (1935b, p. 156),[3] for instance, found counter-examples to the conceivability of such simple, formally conservative ensemble interpretations right away. One of his examples was that of a harmonic oscillator with a given fixed value of total energy (say $(n+\frac{1}{2})\hbar\omega$ for some fixed n), where in the corresponding quantum state of the system, an eigenstate of energy, there would be a large uncertainty as to the oscillator's position

[3]Cf. also Whitaker (1996, p. 214) for an analysis of Schrödinger's examples.

x. But according to the ensemble view, kinetic *and* potential energy, depending on velocity and position respectively, should be well defined at all times for each individual member of the ensemble (or rather, the actual objects formally described by the conceptual ensemble), whence there should be a clear cut-off value for the position. To see this, consider that the potential energy increases with position x in an oscillatory system, and kinetic energy (increasing with velocity) cannot be negative since this would mean that the oscillator does 'less than not move'—an obvious absurdity. So the limit $(n + \frac{1}{2})\hbar\omega$ in total energy implies a limit for possible positions. But QM predicts non-vanishing probabilities for positions which should not be reached according to this analysis, and a straightforward ensemble interpretation hence seems untenable in this case.

A further example can be invoked by appeal to quantum tunneling, as encountered in the ammonia molecule-example in Chap. 2 or the SQUID example in Appendix B. In radioactive α-decay, an α-particle has to tunnel through the Coulomb barrier of the nucleus in order to be emitted. This escape is impossible in a classical physical scenario, since the particle would have to have greater potential than total energy at some point, which again implies negative kinetic energies. Thus construing the wave function as a representation of an ensemble of particles with definite kinetic and potential energies makes tunneling phenomena appear impossible to explain (cf. Whitaker 1996, pp. 214–215).

Other arguments can be invoked by appeal to quantum interference. We have already seen that a simple statistical particle interpretation should not predict the observed patterns of distribution in a double slit experiment (interference patterns). Only the incorporation of an active part of the double slit and possible detectors behind both slits could raise hopes for a suitable statistical analysis in terms of particles, since the interference pattern vanishes in case detectors are placed behind the slits. But no explanation of this kind is possible with the formally conservative ensemble interpretations.

In summary, what these and many further examples show is "how far away from the basic [...] ensemble one has to go—[...] as Bohr would have stressed, one must include the measuring device as an active participator in the measurement, not just a recorder of a fixed value." (Whitaker 1996, p. 217) And the failure to do so seems to be the major crux of historical epistemic ensemble interpretations.

Modern epistemic approaches *do* in fact suggest a revised formal inventory that implies the possibility to formally model an active part of the measuring device in producing the outcome statistics, or even 'automatic' transformations between preparation and measurement. And with this, they can claim some *prima facie* successes in reproducing some of the predictions peculiar to QM from merely epistemic restrictions, including examples of the infamous quantum interference phenomena. We should hence give some deeper thought to these approaches that would, if successful, provide a nice and intuitive solution to the OP that leaves most of our common sense convictions untouched.

4.2 Formal Revisions and ψ-Epistemic Models

4.2.1 *General Outline and Classification of 'Ontological Models'*

As mentioned before, today there is renewed interest in an epistemic interpretation of quantum states. In part, this is due to the success of *quantum information theory* (QIT), which is "the study of the information processing tasks that can be accomplished using quantum mechanical systems" (Nielsen and Chuang 2010, p. 1), and incidentally the most direct modern application of the traditional QM formalism, as compared to e.g. QFT. Its technical-implementational successes have lead some physicists to speak of a *second quantum revolution* (e.g. Dowling and Milburn 2003), next to the technological and scientific revolution brought about by the early development of QM, as mentioned in the introduction. And this success has also inspired some physicists to suspect that all of QM is merely about information (cf. Spekkens 2007, p. 2).

Timpson (2013, p. 2), however, objects to this reasoning as follows:

> The conviction that quantum information theory will have something to tell us about the interpretation of quantum mechanics seems natural when we consider that the measurement problem is in many ways the central interpretive problem in quantum mechanics and that measurement is a transfer of information, an attempt to gain knowledge. But this seeming naturalness only rests on a confusion between the two meanings of 'information'.

The 'two meanings' concern "'information' as a technical term which can have a legitimate place in a purely physical language, and the everyday concept of information associated with knowledge, language, and meaning [...]." (ibid.) Timpson then goes on to characterize the technical, communication-theoretic notion of information (which he coins *information$_t$*) in classical as well as quantum contexts, and to distinguish it from the everyday concept. And subsequently, he dispenses with a range of arguments (e.g. from Dretskean semantic naturalism) which suggest a rather intimate relation between both concepts (cf. Timpson 2013, p. 38 ff.).

But if QIT is concerned with information$_t$ and information$_t$ is not (intimately) connected to the notion of knowledge (as Timpson concisely argues), does this not preclude the more recent epistemic approaches from being serious endeavors of resolving QM's foundational issues in terms of knowledge in the first place? The answer here must be 'no', since for one, it may be viewed as a mere historical contingency that QIT has (partly) motivated the renewed interest in an epistemic approach to QM, and the successes of such an approach should be evaluated independently. And secondly, it is not clear that Timpson's criticism applies to *all* uses of the word 'information' in QIT—as he readily acknowledges in stating that

> descriptions of the quantum state in terms of a person's knowledge or information will typically involve [...] both the everyday semantic/epistemic concept of information and at the same time, the distinct technical concept of information$_t$ introduced in information theory. (Timpson 2013, p. 147)

At least *some* consequences of QIT, or rather some of the suggestive uses of the word 'information' in modern applications of QM, may hence be indicative of some observer's knowledge being involved after all.

A particularly influential epistemic view of quantum states in the more recent debate, developed especially in papers by Spekkens (2005) and Harrigan and Spekkens (2010), and very much alike in style to (if not an instance of) the formalism used by Bell (1964), is that of the so called *ontological models* (OMs). What is meant by 'model' in this context? As Hartmann (1996, p. 80) has observed,

> [q]uite often, the term 'model' is used [...] synonymously with 'theory'. By and large, scientists prefer 'model, because [...] it is safer to label one's thought products 'models' instead of 'theories' for they are most likely provisionary anyway, and the term 'model' seems to acknowledge this right from the beginning[...].

The models concerned here should be viewed along these lines: as provisionary, and as defining new paradigms for *alternative theories* that preserve QM's successful predictions but transgress its descriptive boundaries. Spekkens (2005, p. 2), more precisely, describes OMs as "an attempt to offer an *explanation* of the success of an *operational* theory by assuming that there exist physical systems that are the subject of the experiment." (my emphasis—FB) Thus, an *operationalistic* understanding of QM is presupposed in this general approach, which in the words of Harrigan and Spekkens (2010, p. 128) means that "the primitives of description are simply preparation and measurement procedures—lists of instructions of what to do in the lab." The goal of QM, on such an operational reading, is then just to determine outcome probabilities for measurement procedures. The primitives of description in an OM for QM thus construed are the properties of micro-systems, and the goal of the OM is to account for the measurement statistics in terms of these (cf. ibid.).

To match the operational understanding thus envisioned with the quantum formalism, Spekkens (2005, p. 3) associates a *preparation procedure* $P_{\hat{\rho}}$ with a density operator $\hat{\rho}$ and a *measurement M* with a POVM $\left\{\hat{E}_j\right\}_{j\in J}$. Since $\hat{\rho}$ may be a pure state density operator, and hence correspond to an eigenstate of some observable, but is not directly associated with the physical condition of any system, the EE-link, (V), is severed in this kind of interpretation.

But how *precisely* is the quantum state 'associated' with the (equivalence class of) preparation procedure(s), and what, if anything, does it *represent* about the system? Busch et al. (1995, p. 5), for instance, write in their extensive exposition of operational QM: "Any type of physical system is characterised by means of a collection of preparation procedures, the application of which *prepare the system in* a state T. The set of states is taken to be convex, thus accounting for the fact that different preparation procedures can be combined to produce mixtures of states." (my emphasis—FB) Here the state is named *in addition* to the collection of preparation procedures, as that which results from them. In accord with this, we will make sense of the purported 'association' as follows: The quantum state $\hat{\rho}$ of a given system is the state of the system *according to its preparation*, i.e., with the word 'state' read in a decidedly non-ontological fashion. $\hat{\rho}$ does not represent 'how the system actually *is*', but rather what can *pragmatically* be *said* about it for the context of experimentation in virtue of what was *done* to it.

The preparation procedure could, for instance, consist of using an arrangement of DuBois magnets and screens (cf. Fig. 2.7a) to select only quantum systems with their spin up along a chosen axis (say z), and the state according to that preparation would then be $\hat{\rho}_{\uparrow_z} = \hat{P}_{\uparrow_z} = |\uparrow_z\rangle\langle\uparrow_z|$. But this should not be confused with the assertion that the system 'really has its spin up'; there might not even be spins in the OM supplemented to explain the statistics of the experiment. And this emancipation from the EE-link is already a step beyond the ensemble interpretations discussed before.

Outcomes in *selective* projective measurements, however, can be identified with (pure) quantum states as well, whence quantum states should also be allowed to represent 'states according to measurement'. Notably, this reading fits well with the general operationalism about QM, since the measurement is also an operation performed on the system and generally not so much different from the preparation procedure (think of a Stern-Gerlach measurement, where both preparation and measurement involve magnets and screens). Indeed, we thus preserve *half* of the EE-link: that when an observable A is measured to have value a on S, the state of S is given by $|a\rangle$—albeit only in the operational reading of the word 'state'. In accord with this analysis, we will, in what follows, occasionally call $\hat{\rho}$ (or ψ) the *P/M-state* of the system. Given these prerequisites, the *Born rule* $\Pr_M^{\hat{\rho}}(k) = \mathrm{Tr}(\hat{E}_k\hat{\rho})$ should be read, accordingly, as providing a probability of obtaining value k in a *measurement* procedure of type M given some *preparation* procedure resulting in P-state $\hat{\rho}$.

OMs are now defined by appeal to a bunch of formal ingredients not contained in QM. The first ingredient is a *state space* Λ, with elements λ termed *ontic states*.[4] These ontic states are supposed to represent a "complete specification of the properties of a system[...]." (Harrigan and Spekkens 2010, p. 128) Talk of 'ontic states', however, seems rather clumsy, since *every* state is at least a *state*, and hence it *is*, rather than not, regardless of *what* it is (a sate of an observer, or a state of the experimental setup, or...). We will hence prefer to speak of *true states*—i.e. states which are true of respective systems under consideration in a correspondence theoretic understanding of truth, because this is what the OM approach obviously aims for. To avoid confusion, however, one should also keep the standard terminology in mind which will occasionally resurface in the discussion below.

In addition to the space of true states λ, two probability densities are defined. The first one is termed *epistemic state*, and is intended to reflect the knowledge a possible observer might have about the $\lambda \in \Lambda$. It corresponds to a conditional probability density $p(\lambda|\hat{\rho})$ or $p_{\hat{\rho}}(\lambda)$ of obtaining a certain true state λ, conditional on having

[4]Note that things might be phrased more accurately in terms of *random variables* $\underline{\lambda}$—or even random vectors $\underline{\boldsymbol{\lambda}}$—taking on values $\lambda \in \Lambda$ (or $\boldsymbol{\lambda}$ in some suitable Cartesian product-space $\times_{i=1}^{n}\Lambda_i$) on systems S. Accordingly, when we occasionally speak about λ as a variable in what follows, this should be read as short for 'value λ of variable $\underline{\lambda}$'.

prepared P-state $\hat{\rho}$. The second one, denoted by $\xi(k|\lambda, M)$ or $\xi_M^k(\lambda)$, is called an *indicator-* or *response function*, and it is supposed to reflect uncertainties in a given measurement M leading to outcome k, conditional on the fact that state λ obtains on a system (cf. Spekkens 2005, p. 3; Harrigan and Spekkens 2010, p. 128). This conveys some basic intuitions, but a deeper interpretation of these two probability densities is deliberately left unspecified (cf. Harrigan and Spekkens 2010, p. 150). Leifer (2014, p. 70), however, notes that "calling a probability density 'epistemic' [...] presupposes a broadly Bayesian interpretation of probability theory in which probabilities represent an agent's knowledge, information, or beliefs." 'Broadly Bayesian' seems still a little broad though, and below we shall utter a few comments on the potential meaning of the probabilities involved here, when the 'stage is set'.

In addition to epistemic state and response function, *transition matrices* $\Gamma(\lambda', \lambda)$ (elements thereof) may be introduced (cf. Spekkens 2005, p. 3) which describe (possibly automatic) *state transformations*, or rather probability densities for transitions from true state λ to true state λ' between preparation and measurement.

These ingredients allow to tackle the MAC as follows: the *kinematical* task is undertaken by depriving quantum states of their ontological significance and letting them refer only to P/M-states. And it is *such* states that are 'ambiguous', i.e. include superpositions or imply 'unsharp properties', with property read in a *merely* operational sense. This is not much of a problem now, because they are not (or need not be) indicative of the *true* states, λ, of investigated systems which states, in turn, need not include any of the aforementioned properties. The OP *disappears*, because the PP is only Bayesian updating in disguise. The dynamical task must be tackled by each model individually, the dynamics then being provided in terms of transformation matrices Γ, occurring in between preparation and measurement.

It would probably not be fair (or even make sense) to demand that the OM approach in general should even satisfy the DOC; it is merely a *framework* for defining possible models for QM when the latter is construed purely operationally. But an eventual model of the right kind (cf. below) should certainly be capable of satisfying the DOC. And we will see throughout the discussion that this is not really the case in any of the models that are already out there.

Now for any OM to be a model of QM it is required (cf. Spekkens 2005, p. 5; Harrigan and Spekkens 2010, p. 128) that the model's epistemic states and response functions satisfy

$$\int \mathrm{d}\lambda' \, \mathrm{d}\lambda \, \xi_M^k(\lambda')\Gamma(\lambda', \lambda)p_{\hat{\rho}}(\lambda) = \mathrm{Tr}(\hat{E}_k\hat{\rho}), \tag{4.1}$$

for any given measurement $M = \{E_k\}_{k\in K}$, and P-state $\hat{\rho}$.[5] That is: Summing up (integrating) all the probabilities of obtaining a certain outcome k, given a certain measurement M and true state λ', weighted by the probability that the state λ' results

[5]For completeness' sake we note that Leifer (2014, p. 82), in his analysis of the approach, provides a treatment in terms of measure-theoretic notions, using a σ-algebra Σ (cf. Appendix A) over Λ,

via transformations prior to the measurement from state λ and the probability that λ occurs due to some preparation $P_{\hat{\rho}}$ in the first place, must reproduce the quantum probabilities.

This *fully* defines what an OM *is*, whence 'ontological model' should be rather viewed here as a *technical term*; not much ontology is actually conveyed. The OM approach sketches a road to possible modifications of QM's formalism which would then *allow* for the specification of an ontology in which the quantum state need not figure, or at least not fundamentally. It does not *provide* any such ontology, besides the introduction of additional λs.

Ipso facto we are here dealing with hidden variables again, the true states λ, and the 'hiddenness' is expressed, according to formula (4.1), exactly by some averaging out of the λs, just as suggested by Pearle (1968, pp. 464–465). Generally λ *need not* be interpreted as a hidden variable though, since it can be interpreted as the quantum state ψ itself—the OM-approach is formally neutral on this point. In fact, Harrigan and Spekkens (2010, p. 129 ff.) draw a multifold distinction between different classes of OMs (cf. Fig. 4.1).

The major division here is between *ψ-onitc* and *ψ-epistemic* OMs, which intuitively concerns whether the quantum state be construed as something really pertaining to some system after all, or just something we ascribe to that system in virtue of a lack of knowledge about its actual physical state. Within the first category they distinguish *ψ-supplemented* from *ψ-complete* models, where the former category simply consists of OMs in which the quantum state is something that pertains to reality, but is still not *all* there is. The notion of *ψ-complete* models should be self-explaining. For obvious reasons, ψ-epistemic and ψ-supplemented models are jointly termed *ψ-incomplete.*

More precisely, Harrigan and Spekkens (2010, p. 131) define ψ-completeness, as follows.[6]

Definition (ψ-completeness) An ontological model is ψ-complete if the space of true states Λ is isomorphic to the projective Hilbert space $\mathcal{P}(\mathcal{H})$ (the space of rays of Hilbert space) and if every preparation procedure P_ψ associated in quantum theory

---

equipped with a σ-additive measure μ that maps from Σ into the interval [0, 1]. The epistemic state is then viewed as the (Radon-Nikodym) derivative of μ w.r.t. λ, which is only well-defined under certain conditions (cf. below) that not all conceivable models satisfy (cf. Leifer 2014, p. 90; Leifer and Maroney 2013, p. 4). Leaving out transition matrices, Leifer requires that $\int \mathrm{d}\mu_{\hat{\rho}}\, \xi_M^k(\lambda) = \mathrm{Tr}(\hat{E}_k \hat{\rho})$, where $\mu_{\hat{\rho}}$ is the probability measure over Σ induced by preparation $P_{\hat{\rho}}$. This is only equivalent to the above condition (modulo transformation matrix) in case there is a measure λ which dominates all measures $\mu_{\hat{\rho}}$ over the space Λ (cf. Appendix A), whence one can appeal to $p_{\hat{\rho}}(\lambda) = \frac{\mathrm{d}\mu_{\hat{\rho}}}{\mathrm{d}\lambda}$ in the integral (cf. Leifer and Maroney 2013, p. 4). Fortunately, we can here restrict our attention to models in which the assumption is valid and we need not bother with these details any further.

[6]We slightly alter notation and wording, since Harrigan and Spekkens also call λ_ψ and ψ "isomorphic", which is meaningless for *elements* of spaces. We have also used our notion of *true* rather than 'ontic' states.

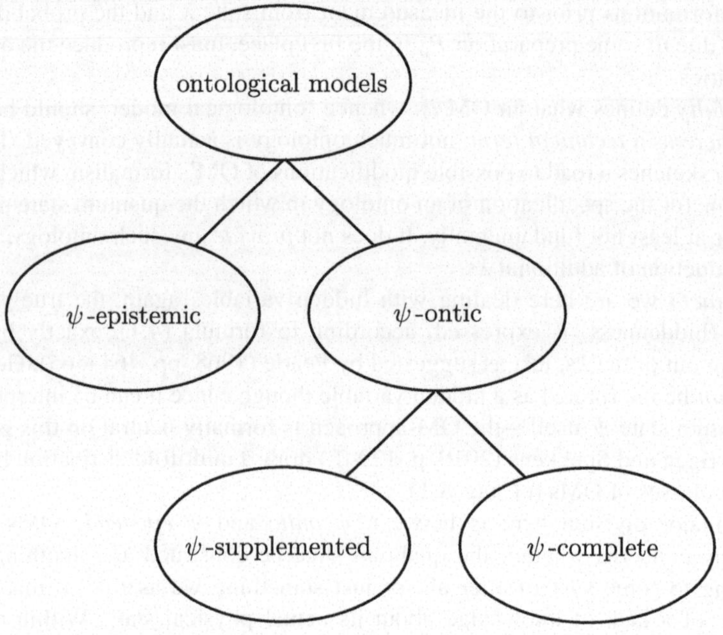

Fig. 4.1 Classification of OMs according to the status of the quantum state (cf. Harrigan and Spekkens 2010, p. 134, for a tabular representation)

with a given ray ψ is associated in the OM with a Dirac delta function centered at the true state λ_ψ that is the value of ψ in the isomorphism, $p_\psi(\lambda) = \delta(\lambda - \lambda_\psi)$.

The appeal to projective Hilbert space is due to the (observational) invariance of quantum states under multiplication by a global phase (a $z \in \mathbb{C}$ s.t. $|z| = 1$); so the isomorphism between Λ and quantum states picks out equivalence classes $[\psi]$ of vectors $|\psi\rangle$ rather than individual vectors, with the equivalence relation defined by multiplication by a(n overall global) phase.[7] Put frankly, the definition tells us that an OM is ψ-complete in case it reproduces QM *tout court*. The true states in Λ are bijectively mapped onto rays in Hilbert space, and the probability of a true state obtaining, given a preparation procedure associated to a ray in $\mathcal{H}$, is such that it is 1 for the true state that is the value of the ray in the isomorphism and 0 for all other true states. The quantum statistics is reproduced in a trivial fashion.

[7]These equivalence classes are then usually called 'rays' by physicists (cf. Heinosaari and Ziman 2012, p. 82), even though a ray is more precisely the set of *all* complex multiples of some vector. So the 'ray' in the sense of $[\psi]$ is basically a ray of *normalized* vectors (cf. Gustafson and Sigal 2011, p. 193). The name 'projective' here obviously stems from the fact that any projector $|\psi\rangle\langle\psi|$ projects equally onto $|e^{i\varphi}\psi\rangle := e^{i\varphi}|\psi\rangle$, since $|e^{i\varphi}\psi\rangle\langle e^{i\varphi}\psi| = e^{i\varphi}|\psi\rangle\langle\psi|e^{-i\varphi} = |\psi\rangle\langle\psi|$ (cf. also Heinosaari and Ziman 2012, p. 82).

As we saw, the notion of ψ-onticity is supposed to allow for supplementation of ψ by elements of the model which do not simply mirror elements of QM, whence a general ψ-ontic model is defined by the following property (cf. Harrigan and Spekkens 2010, p. 131; notation adapted).

Definition (ψ-onticity) An ontological model is ψ-ontic if for any pair of preparation procedures, P_ψ and P_ϕ, associated with distinct quantum states ψ and ϕ, we have $p_\psi(\lambda)p_\phi(\lambda) = 0$ for all λ.

In other words: the supports of two epistemic states, i.e. the sets of points for which they are non-zero, should not overlap, i.e. have intersections of non-zero measure.

Now ψ-ontic models which do not satisfy the first definition are called ψ-supplemented, non-ψ-ontic models are called ψ-epistemic. Since ψ-ontic and ψ-epistemic models also exhaust the classification, the decisive criterion for a model to be ψ-epistemic is *that there be an overlap* in the supports of the epistemic states associated with distinct quantum states.[8]

The intuition here is that, if it may happen that λ is *really* the case in two instances and we have prepared for ψ in the one, for ϕ in the other instance, then ψ and ϕ themselves do not reflect the true state of the system. Given that the probability densities $p_{\psi/\phi}(\lambda)$ were supposed to reflect "what can be known and inferred by observers", this could be translated more crisply into a statement of the form 'I cannot know/infer for sure that λ is not sometimes the case when I prepare for ψ, and sometimes when I prepare for ϕ'. Since we are concerned with ways to fill Conjecture 1 with content, we will here only be interested in the OM approach to the extent that it can facilitate the underlying intuitions, i.e. only with ψ-epistemic models.

4.2.2 A Note on the Philosophical Issues at Stake

The account presented above actually connects quite nicely to the historical issues introduced in Sect. 4.1. First of all, the idea of demonstrating the incompleteness of QM by showing that two different ψ-functions may simultaneously pertain to the very same physical system was also explicitly advocated by Einstein: "[...]

[8]In some receptions (e.g. Lewis et al. 2012, p. 3 or Maroney 2012, p. 2) this requirement is refined such that an overlap between epistemic states is required only for *non-orthogonal* quantum states, as two orthogonal states $|\phi\rangle$, $|\psi\rangle$ are usually construed as indicative of mutually exclusive preparation procedures, which one might (but need not) assume to result in mutually exclusive sets of true states. The negation of ψ-onticity merely implies the existence of two *distinct* quantum states that have densities associated to them with overlapping support. This is comparatively weak and makes a broad range of models possible. Refinements in terms of distance measures between the epistemic states have also been proposed (e.g. Pusey et al. 2012, p. 477; Aaronson et al. 2013, p. 2).

coördination of several ψ functions with the same physical condition of [some—FB] system [...] shows [...] that the function cannot be interpreted as a (complete) description of a physical condition of a unit system." (Einstein 1936, p. 376) These views of Einstein have served as an overt inspiration to Harrigan and Spekkens (2010, p. 147), and since he was so philosophically involved with QM, we will meet Einstein's views again at multiple junctions in this book. In particular, we shall have to say more about the infamous EPR incompleteness argument later.

But in connection to the notion of 'completeness', a general worry may be expressed at this point. Bell, whose work basically laid the foundations for the OM approach (as mentioned earlier) was cautious to talk merely of a "*more* complete specification", for which it is "a matter of indifference [...] whether λ denotes a single variable or a set, or even a set of functions, and whether the variables are discrete or continuous." (Bell 1964, p. 15; my emphasis—FB) 'More complete' may be read in the sense of 'with more descriptive content', and Bell himself certainly had classical physical theories as role models in mind, with their simultaneous 'sharp values' for positions and momenta or energies at precise times, a fact evidenced by his lifelong endorsement of Bohmian mechanics (cf. Chap. 6), in which such simultaneous assertions become bluntly possible.

For Harrigan and Spekkens (2010, p. 128), in contrast, λ is straightforwardly supposed to provide a "complete specification of the properties of a system[...]." But in any philosophically informed reading of 'property', this appears quite *impossible*, for the trivial reason that there is, in principle, an infinity of arbitrary or arbitrarily complex properties (think e.g. of Goodman's (1955) bleen and grue). For the present case suppose, for instance, that we have prepared for ψ in the one case and ϕ in the other, and that in both cases λ is supposed to occur. Then we can say that there is a (complex) *relational* property of being-in-a-P_ψ-situation in the first case, and a relational property of being-in-a-P_ϕ-situation in the second, and hence λ cannot concern strictly *all* properties.

Possibly the completeness in question is meant with regard to a specific set of *kinematical quantities* similar to the phase space coordinates in classical physics (generalized positions and momenta), quantities which are *productive* in the sense that the values of all other quantities of the theory can be derived from their values. Ruetsche (2011, p. 31) makes a point in favor of our interpretation: she identifies generalized coordinates and momenta in classical physics as paradigmatic examples of magnitudes she calls "fundamental in the physicist's sense", for which it is precisely assumed that the value of every other magnitude pertaining to a system can be determined by assigning values to these on the system.

QM itself, however, *has* the complete set of compatible observables, which is fundamental or productive in just that sense. One must hence ask what exactly is sought for in aiming for 'completeness', and, more precisely, what could possibly single out such a set of properties *independently of any background theory*. If 'completeness' in Harrigan and Spekkens' sense is meant as completeness as provided by a list of all quantities that are fundamental-in-the-physicist's-sense, then a background theory other than QM is required to specify what these quantities are, and hence what even *should* be derivable. The suspicion arises that Harrigan and Spekkens do have a possible background theory in mind, namely (not entirely unlike Bell), a suitable prospective adaptation of *pre-quantum physics*. And in fact, when we establish the connection to Bell's theorem, the EPR argument, and the Kochen-Specker theorem later, this will become all the more plausible.

The whole issue is treated, however, in a rather loose and intuitive manner by Harrigan and Spekkens, and without any considerations of background theories. The underlying *intuition* seems to be that there could be such a thing as an exhaustive list of all *natural* properties pertaining to a system, i.e. "an élite minority of special properties" a description in terms of which would "carve reality at the joints[...]." (Lewis 1983, p. 346) Such an assumption, of course, drags us deeply into the metaphysics of natural kinds, properties, universals... and so forth, with all the associated subtleties (e.g. Bird and Tobin 2015, for an overview of the natural kind-debate). The completeness issue, in short, is quite non-trivial and much less innocent than the casual writings of Harrigan and Spekkens make it appear.

These issues are issues of ontology, and we are driven to ask how ψ-epistemic OMs fare w.r.t. the DOC. Now while the details depend on specific models, we can still make some general remarks, given the above considerations. At least partly and at least for the kinematics (virtually nothing is being said about the dynamics in the debate), requirement (iii) is met in a rather trivial way: the quantum state is simply a P/M-state—something to do with lab operations—, as explained above. It does not figure in the underlying ontology. And the formal entities that contain all the ontology are λ or λ and $\xi_M^k(\lambda)$, depending on the interpretation of the latter (cf. below). Since these models assume that there are microsystems with definite states (λ), these systems should be able to account for classical appearances by mere *composition*, and requirement (i) is hence equally met in rather trivial way. Point (ii), the relation between quantum and classical, is obviously tackled quite generally in terms of epistemic restrictions: what appears to be 'quantum' is really only 'inaccessibly classical'. So up to the completeness issues, ψ-epistemic OMs *set a frame* for an ontologically clear interpretation of QM in terms of missing knowledge. But we will see that many difficulties arise in actually spelling out such an ontology, in the context of concrete models.

There is another subtlety involved here, namely, that the assumption of 'definite states' makes for a connection to the interpretive questions about probability that we raised in interlude I. With Leifer (2014, p. 70) we already noted that "calling a probability density 'epistemic' [...] presupposes a broadly Bayesian interpretation of probability[...]." More precisely, the epistemic state, $p_{\hat{\rho}}(\lambda)$, is supposed to represent "what can be known and *inferred* by observers." (Harrigan and Spekkens 2010, p. 129; my emphasis—FB) This means that the 'broad Bayesianism' in question cannot be *radially subjective*, because, as we saw in interlude I, that would mean that agents would be largely at liberty to assign any probability they want instead of being bound to 'infer' from evidence (observed frequencies). To use

Williamson's coarse grained classification, the Bayesian view in question is at least *empirically based*.[9]

All of this fits quite well with the general 'objectivist flavor' of the approach: It presupposes that there is a mind-independent reality that determines, in a rationally compelling way, how we must adjust our credences. And given the above considerations on completeness, the hope underlying the construction of ψ-epistemic models seems to be that there might be a future theory that allows us to carve nature at its joints in a way that we can (more or less) *understand*, just as classical theories were once taken to do. That this is ultimately hoped for is also evidenced by the fact that the probabilities proposed here are all 'classical' in the sense that they allow for a *joint* probability space for *all* the components of λ (or rather: components of the values of $\underline{\lambda}$). As is well known, such a joint probability space does *not* exist for non-commuting observables in QM (e.g. Bub 1974, p. 35).

The epistemic state, to recall, is not the only probability density assumed in such models though. Regarding the *response function* Harrigan and Spekkens (2010, p. 128) have it that "the model *may be such* that the ontic state λ *determines only* the probability $\xi_M^k(\lambda)$ of different outcomes k for the measurement M." (notation adapted; my emphasis—FB) This use of "determines" rather smells like *propensities* being involved, because a model in which the very same apparatus may respond differently every time to the same kind of state but in such a way that at least frequencies are foreseeable would exactly implement such propensities. The "may be" in conjunction with the fact that Harrigan and Spekkens (2010, p. 129) also hold that *both* "$p_{\hat{\rho}}(\lambda)$ *and* $\xi_M^k(\lambda)$ specify what can be known and inferred by observers" (notation adapted; my emphasis—FB), however, tells a different story.

If $\xi_M^k(\lambda)$ is interpreted in terms of propensities, the latter statement is of course trivially true as well. But by conversational implicature we infer from the quotes that it is at least hoped for that $\xi_M^k(\lambda)$ turns out 'just as epistemic' as $p_{\hat{\rho}}(\lambda)$.[10] In any such model, a Laplacian demon (cf. de Laplace 1814, p. 4; Gillies 2000, p. 14) could avail himself of a completely deterministic description of reality taking precise note of all micro-components, and $\xi_M^k(\lambda)$ would become obsolete.

These considerations obviously shift the debate towards questions of *(micro-)determinism*, a point in favor of our comparison to intuitions underlying classical theories. But the general accusation of assuming such determinism is guarded against by the fact that the response function is allowed to play a non-trivial role after all, and the central question of ψ-epistemic OMs remains one of

[9] In fact, in a concrete ψ-epistemic toy model investigated later, there will even be an equivocation norm operative, whence it may even be seen as presupposing *objective* Bayesiansim in Williamson's sense.

[10] This reading is supported by further textual evidence: Harrigan and Rudolph (2007, p. 4), for instance, concede that the response functions "could occur because of our failure to take into account the precise ontological configurations of either [preparation or measurement]"; and similarly Spekkens constantly refers to an "un*known* disturbance" (my emphasis—FB) of the system caused by the measurement in his 2007 paper, with obvious similarities to Heisenberg's original formulation of the microscope thought experiment.

micro*definiteness*—of the very *existence* of true states λ that do not exhibit such strange features as superposition or 'unsharp properties'.

This, of course, connects to broader questions of realism as considered in interlude I, because the assumption of such true states—any one of which, to recall, is supposed to provide a "complete specification of the properties of a system"—is obviously an expression of strong metaphysical realism. ψ-epistemic models, however, are *not* naïvely realist in Norsen's sense as long as the response function is non-trivial, because then pre-existing intrinsic properties of systems are not just passively revealed in measurements. By judging thusly we appear to be disagreeing with Norsen (2007, p. 316), who thinks that naïve realism underlies all non-contextual hidden variable models. But the apparent disagreement is simply rooted in the understanding of 'non-contextual', as shall become clear later.

Moreover, since *scientific methods* are clearly endorsed here—a probability calculus, higher order mathematics, and rough specifications of application to experiment—these models certainly also express some sort of *scientific* realism. But a remarkable thing to note is that this scientific realism is of a peculiar kind: QM, the *most well-confirmed* mature theory is regarded as merely operational, and its success is regarded to be in need of explanation 'from the outside', i.e. in terms of the formalism of an entirely *different* theory.

It seems that scientific realism is at play here only *selectively*. But this immediately raises the question of what the 'selection criteria' are. Consider e.g. Peters' (2014, p. 377) understanding of selective scientific realism, who characterizes it as the view "that not all the propositions of an empirically successful theory should be regarded as (approximately) true but only those elements that are *essential for its success*." (my emphasis—FB) There is an obvious drawback to this reading of 'selective', which Peters (ibid.) readily acknowledges: "It is [...] not obvious how a term like 'essential' is to be understood." Plausibly, for the proponents of ψ-epistemic OMs, an extra-empirical standard is at play in figuring out what is essential for the success of a scientific theory: that it provide a somewhat *intuitable* view of the world, 'complete' by standards close to common sense intuitions as employed (in extended and modified form) in classical theories. Why else would one read QM as merely operational and demand an explanation for its success?

The realism issues thus raised are of course *philosophically* non-trivial; but we need to stress that they are also non-trivial for the particular discussion at hand, because a fully-fledged (strong) metaphysical *or* scientific realism is *not* shared across *all* interpretations of QM, not even all 'epistemic' ones. Leifer (2014, p. 72) e.g. maintains that

> it is important to distinguish two kinds of ψ-epistemic interpretation. The most popular type are those variously described as anti-realist, instrumentalist, or positivist.[...] The second type of ψ-epistemic interpretation are those that are realist, in the sense that they do posit some underlying ontology. They just deny that the wavefunction is part of that ontology. Instead, the wavefunction is to be understood as representing our knowledge of the underlying reality, in the same way that a probability distribution on phase space represents our knowledge of the true phase space point occupied by a classical particle.

As should be clear by now, we are so far only concerned with epistemic interpretations of the second type in Leifer's classification. ψ-epistemic models in the sense of the OM approach will (mostly) serve as our paradigm example for such an epistemic interpretation, since they constitute the most 'up-to-date' attempt. And they also make possible a quite far-reaching discussion of limitations in terms of formal results.

What we must first ask, though, is: *Are* there even any models which fit the definitions above? Indeed, Harrigan and Spekkens provide examples of models for each of their categories. To bring some substance to the discussion, we will hence now consider two examples of ψ-epistemic models, one of which is formally precisely suited to fit the definitions, the other one being conceptually more elaborate and having more intuitive appeal. Both models have a limited domain of application. But the first one is a perfectly fine example of how *general* ψ-epistemic models can be constructed *formally*, and the second one for how more *meaningful* ψ-epistemic models can be constructed when *intuitions* are allowed to play a role. Both will ultimately help to understand the limitations of the project.

4.2.3 A Formal Example: The Kochen-Specker Model

The one model that Harrigan and Spekkens consider as an example of a ψ-epistemic model, which is hence of greatest interest here, is the so-called *Kochen-Specker model*.[11] As indicated above, the model is utterly *formal*, i.e. of *low conceptual value*; it hardly serves to explain *how* QM is just a reflection of incomplete knowledge of hidden true states. The model can be straightforwardly formulated in the language of the OM approach, and it resembles in style other models such as the 'Bell model' (e.g. Lewis et al. 2012) or Lewis et al.'s own one, which is just a modified and generalized ψ-epistemic version of the former. The Kochen-Specker model was originally devised by Kochen and Specker (1967) to make a general formal point (namely the existence of a non-contextual hidden variable model in *two* dimensions; cf. later). So recapitulating it under the same premise seems quite appropriate.

As indicated above, the model is thus limited in scope; more precisely, it is only concerned with systems describable by a two-dimensional Hilbert space ($\mathcal{H} = \mathbb{C}^2$) of which we had introduced three examples above: spins, polarizations, and atoms or molecules constrained to two possible (energy- or coarse-grained position-) states. In the context of QIT these systems are usually referred to as *qubits* (quantum-bits), in analogy to the *bit* as the fundamental unit of classical information theory (cf. Nielsen and Chuang 2010, p. 13), which "corresponds to a single binary digit, or to the answer to a yes/no question." (Maudlin 2011, p. 153)

[11] We will treat their exposition of it as the only relevant reference for our purposes.

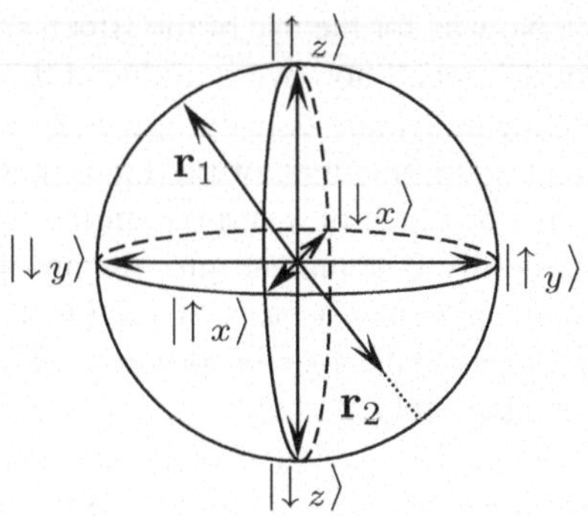

Fig. 4.2 Exemplary Bloch sphere for spins. $\boldsymbol{r}_1$ corresponds to a pure state, $\boldsymbol{r}_2$ to a mixed one

To describe such qubits, it is customary to make use of the so-called *Bloch sphere* (cf. Fig. 4.2). To understand this conceptual tool, think of a three dimensional (solid) sphere of radius one. A spin-up state, which we have represented so far by $|\uparrow_z\rangle$, or more concretely by a 2-entry column vector $\binom{1}{0}$ in $\mathbb{C}^2$, is then associated with a unit vector pointing in positive z-direction, the sphere's north-pole. The spin-down state $|\downarrow_z\rangle$ is accordingly represented as a unit vector pointing in $(-z)$-direction.

This is a quite natural, intuitive way to picture spins, given the treatment of intrinsic angular momenta in classical physics. One might hence wonder, at this point, why spins or qubits are not exclusively represented in this fashion. But subtleties arise as follows: Spins, as represented by elements of $\mathbb{C}^2$, are transformed or 'rotated' by unitary operators that can be defined in terms of some unit vector $\boldsymbol{n}$ and an angle θ via $\hat{U}(\boldsymbol{n}, \theta) = e^{i\frac{\theta}{2}\hat{\boldsymbol{\sigma}}\cdot\boldsymbol{n}}$, with $\hat{\boldsymbol{\sigma}}$ the vector of Pauli matrices. This is phrased in mathematicians terms as the Pauli matrices being the infinitesimal generators of the (transformation) group $SU(2)$ of unitary 2×2 matrices with determinant 1. The group of transformations which rotate arrows inside the sphere is the group $SO(3)$ of 3×3 matrices A with $\det(A) = 1$ and $A^T = A^{-1}$, and the group homomorphism is 2 to 1 (e.g. Chen et al. 2007, p. 524). The crucial point being that transformations of spinors by elements of $SU(2)$ induce a phase of -1 for the choice $\theta = 2\pi$, whereas a rotation of an arrow by 2π gives back the original configuration.

For a *single qubit* this can be regarded as an overall phase with no observable consequences, so no harm comes from the identification. More technically speaking, this is expressed by the fact that one finds an isomorphism between the quotient group $SU(2)/\{\mathbb{1}, -\mathbb{1}\}$ and $SO(3)$ (e.g. Chen et al. 2007, p. 524), i.e. if one identifies states $\pm |\psi\rangle$. Still, for composite systems the analogy breaks down due to empirical consequences of the relative phase when 'rotated' and 'non-rotated' systems are joined together and their states are superposed (cf. Kiefer 2003, p. 44; Werner et al. 1975). The temptation to think of spins in terms of little pointers which indicate the direction of rotation of some tiny charged sphere (hence giving rise to a magnetic field) should vanish in virtue of this disanalogy, and the difference in transformation behavior is sometimes also advanced as an invocation of the 'decided nonclassicality' of the concept of spin (cf. Sect. 2.1.2).

Now as for the use of the Bloch sphere in the description of the Kochen-Specker model, recall how we introduced the Pauli matrices to describe observables on $\mathbb{C}^2$. Incidentally, any density operator on the space $\mathbb{C}^2$ can also be defined in terms of linear combinations of the Pauli matrices and the unit matrix $\mathbb{1}$, which should not come as much of a surprise in the light of how operators representing observables spectrally decompose into eigenprojectors (e.g. Appendix A). In fact, these four matrices define a basis of the (abstract vector-)space of complex 2×2 matrices, equipped with a scalar product $\frac{1}{2}\mathrm{Tr}(\hat{A}\hat{B})$ (cf. Heinosaari and Ziman 2012, p. 62; Shankar 1994, p. 383).

In (2.19), we also demonstrated how to construct a spin observable for an arbitrary direction *via* $\hat{\boldsymbol{\sigma}} \cdot \boldsymbol{n}_{\theta\varphi}$. But since any direction of $3D$ space can be defined in terms of the polar and azimutal angles, θ, φ, this equally amounts to just giving an arbitrary linear combination of the three Pauli matrices (for fixed θ, φ). Thus using pointers $\boldsymbol{r}$ that lie inside the three-dimensional (solid) sphere, a general description of a density operator in $\mathbb{C}^2$ can be given as

$$\hat{\rho} = \frac{1}{2}(\mathbb{1} + \hat{\boldsymbol{\sigma}} \cdot \boldsymbol{r}). \tag{4.2}$$

The factor $\frac{1}{2}$ here stems from the fact that this operator can then be shown to have eigenvalues $\mu_\pm = \frac{1}{2}(1 \pm |\boldsymbol{r}|)$, which are $\mu_+ = 1$ and $\mu_- = 0$ for unit vectors ($|\boldsymbol{r}| = 1$) (cf. Heinosaari and Ziman 2012, p. 63). Hence $\frac{1}{2}$ figures as a normalization, and unit vectors can be used to represent *pure* states, whereas vectors with a norm smaller than 1 represent *mixed* states (cf. Fig. 4.2).

In this representation, we can obtain the Born-probability for pure states and projective measurements by

$$\begin{aligned}
\mathrm{Tr}(\hat{\rho}_\psi \hat{P}_\phi) &= \mathrm{Tr}\left(\frac{1}{2}(\mathbb{1} + \hat{\boldsymbol{\sigma}} \cdot \boldsymbol{n}_\psi)\frac{1}{2}(\mathbb{1} + \hat{\boldsymbol{\sigma}} \cdot \boldsymbol{n}_\phi)\right) = \\
&= \mathrm{Tr}\left(\frac{1}{4}(\mathbb{1}+\hat{\sigma}_1 n_{\psi_1}+\hat{\sigma}_2 n_{\psi_2}+\hat{\sigma}_3 n_{\psi_3})(\mathbb{1}+\hat{\sigma}_1 n_{\phi_1}+\hat{\sigma}_2 n_{\phi_2}+\hat{\sigma}_3 n_{\phi_3})\right) = \\
&= \frac{1}{4}\left(\mathrm{Tr}(\mathbb{1}) + \mathrm{Tr}(\mathbb{1} n_{\psi_1} n_{\phi_1}) + \mathrm{Tr}(\mathbb{1} n_{\psi_2} n_{\phi_2}) + \mathrm{Tr}(\mathbb{1} n_{\psi_3} n_{\phi_3})\right) = \\
&= \frac{1}{2}(1 + \boldsymbol{n}_\psi \cdot \boldsymbol{n}_\phi)
\end{aligned} \tag{4.3}$$

where $\boldsymbol{n}_\psi$ corresponds to a unit vector in ψ-direction, and n_{ψ_j} is its j-th component.[12]

[12]We have here appealed to a few properties such as $\mathrm{Tr}(\hat{A} + \hat{B}) = \mathrm{Tr}(\hat{A}) + \mathrm{Tr}(\hat{B})$, $\mathrm{Tr}(\mathbb{1}\sigma_j) = 0$, and $\sigma_k \sigma_j = \mathbb{1} \cdot \delta_{kj}$, $j, k \in \{1, 2, 3\}$ (cf. Heinosaari and Ziman 2012, p. 62 ff.). These properties are not difficult to prove; so the interested reader is encouraged to prove them herself.

In the Kochen-Specker model, the points on the surface $\mathbb{S}^2$ of the unit (Bloch) sphere are used as the state space Λ, whence unit vectors $\boldsymbol{n}_\lambda$ can be used to represent true states λ. The epistemic state is given by $p_\psi(\lambda) = \frac{1}{\pi}\Theta(\boldsymbol{n}_\psi \cdot \boldsymbol{n}_\lambda)\, \boldsymbol{n}_\psi \cdot \boldsymbol{n}_\lambda$ with Θ the Heaviside step function (cf. Appendix A), and the response function by $\xi_M^\phi(\lambda) = \Theta(\boldsymbol{n}_\phi \cdot \boldsymbol{n}_\lambda)$. We can immediately see that the model is ψ-incomplete; $p_\psi(\lambda)$ is not a Dirac-δ.

Now one can demonstrate that $\Pr_M^\psi(\phi) = \int \mathrm{d}\lambda\, p_\psi(\lambda)\xi_M^\phi(\lambda) = \frac{1}{2}(1 + \boldsymbol{n}_\phi \cdot \boldsymbol{n}_\psi)$ which is exactly the Born probability $\mathrm{Tr}(\hat{\rho}_\psi \hat{P}_\phi)$, as given by (4.3). The calculation is a little tedious though, and we refer the reader to De Zela (2008, p. 6) or Leifer (2014, p. 14 ff.) for proofs. So the model is indeed an OM for QM in $\mathbb{C}^2$, as it reproduces the quantum probabilities in the required fashion. Moreover, for two non-orthogonal states ψ and ϕ it holds that $p_\psi(\lambda)p_\phi(\lambda) > 0$, as can easily be seen from the definition of the epistemic states. Hence this is in fact a ψ-epistemic OM.

But beyond this, the model is of little conceptual value: It does not help much in clarifying how the probabilities arise and come to distribute in the way they do. This is the case with many models discussed in the debate. Another qubit model which additionally includes *transformations* and at least *prima facie* comes much further in offering clarification on a conceptual level is Spekkens' (2007) toy model, which we will now turn to.

4.2.4 Gathering Evidence: Spekkens' Toy Model

What we here call a 'toy model' was originally developed by Robert Spekkens (2007) under the name "toy theory", but it can be made to fit into the OM approach, as shown by Leifer (2014, p. 84) and below. The toy model is also, like the Kochen-Specker model, only concerned with qubits, but can be expanded to include systems of multiple, coupled qubits. The analogues of qubits in the toy model are called *elementary systems* (cf. Spekkens 2007, p. 3). For these elementary systems, Spekkens postulates four possible true states, simply denoted by $\{1, 2, 3, 4\}$. There is a 'foundational principle' at the heart of this model, called the *knowledge balance principle*:

Knowledge Balance Principle (KB) If one has maximal knowledge, then for every system, at every time, the amount of knowledge one possesses about the [true—FB] state of the system at that time must equal the amount of knowledge one lacks. (Spekkens 2007, p. 3)

This, of course, immediately raises the question of how to *measure* knowledge. To provide a measure, Spekkens first defines what he calls *canonical sets* (cf. ibid.):

Definition (Canonical set) A *canonical set* is a set of yes-no questions that is sufficient to fully specify the true state, and that has a minimal number of elements.

This means that if one only knows that the state of the system under investigation is in the set $\{1, 2, 3, 4\}$ and one wants to find out in which of the states it actually is, one could ask "Is it in state 1?", "Is it in state 2?", and so forth. Or one could be smart instead, and just ask, say, "Is the system's state in the set $\{1, 2\}$?", and "Is the system's state in the set $\{2, 3\}$?" Two nos will give assurance that it is in state 4, two yeses that it is in 2, and one yes and one no that it is either in 1 or 3, depending on the order. Now the amount of knowledge one has is defined within the toy model as "the maximum number of questions for which the answer is known, in a variation over all canonical sets of questions." (ibid.).

(KB) then dictates that one can always only know half the answers in such a set, and this is somewhat reminiscent of an epistemic reading of the uncertainty relations (Eq. 2.22). Applied to physical systems, e.g. once more spinful particles traveling, say, in z-direction, we can understand it such that "if we know the x-coordinate with certainty then we cannot know anything about the y-coordinate." (Leifer 2014, p. 73)

This immediately implies that the epistemic states of simple systems in this model must be distributions which assign probability $1/2$ to two states, and probability 0 to two others. That is, we can always know that the state is in a subset like $\{1, 2\}$ but nothing more. To elaborate, consider the following six quantum states, which are the ones that may be prepared and measured in the model:

$$
\begin{aligned}
&|0\rangle\,, && |1\rangle\,,\\
&|+\rangle = \frac{1}{\sqrt{2}}(|0\rangle + |1\rangle), && |-\rangle = \frac{1}{\sqrt{2}}(|0\rangle - |1\rangle),\\
&|+i\rangle = \frac{1}{\sqrt{2}}(|0\rangle + i\,|1\rangle), && |-i\rangle = \frac{1}{\sqrt{2}}(|0\rangle - i\,|1\rangle).
\end{aligned}
$$

These could, of course, be the spin states for z, x, and y, or equally the vacuum state and one excited mode,[13] and linear combinations of them prepared by suitable equipment. Accordingly, our epistemic states will be of the form $p_P(\lambda)$, $\lambda \in \{1, \ldots, 4\}$, $P \in \{0, 1, +, -, +i, -i\}$.

Since we are only concerned with a discrete set of possible true states, the probability distributions can be represented by n-tuples (or, if desired, column-vectors). This also means that condition (4.1) which connects the QM probabilities with the probabilities in the OM must be changed to a sum:

$$\mathrm{Tr}(\hat{E}_k \hat{\rho}) = \sum_{\lambda \in \Lambda} \xi_M^k(\lambda')\Gamma(\lambda', \lambda) p_P(\lambda). \tag{4.4}$$

[13]One should not generally confuse $|0\rangle$ with the vacuum state in this context though. 0 is merely a label here, reminiscent of the binary bit-language of 1s and 0s.

Note that we can here equally use the vector formula for the Born probabilities, as we will be concerned only with pure states and projective measurements. Spekkens (2007, p. 4) uses a different, convenient notation for the epistemic states, which we will also make use of in what follows. We hence make the following identifications[14]:

$$\boldsymbol{p}_0 = (\tfrac{1}{2}, \tfrac{1}{2}, 0, 0) \leftrightsquigarrow 1 \vee 2, \qquad \boldsymbol{p}_1 = (0, 0, \tfrac{1}{2}, \tfrac{1}{2}) \leftrightsquigarrow 3 \vee 4,$$

$$\boldsymbol{p}_+ = (\tfrac{1}{2}, 0, \tfrac{1}{2}, 0) \leftrightsquigarrow 1 \vee 3, \qquad \boldsymbol{p}_- = (0, \tfrac{1}{2}, 0, \tfrac{1}{2}) \leftrightsquigarrow 2 \vee 4,$$

$$\boldsymbol{p}_{+i} = (0, \tfrac{1}{2}, \tfrac{1}{2}, 0) \leftrightsquigarrow 2 \vee 3, \qquad \boldsymbol{p}_{-i} = (\tfrac{1}{2}, 0, 0, \tfrac{1}{2}) \leftrightsquigarrow 1 \vee 4,$$

where we have used curvy arrows to denote the correspondence between different notations, and boldface to indicate that the 4-tuples may be treated as 'probability vectors'.

As for the response functions, these turn out to be *deterministic*. For instance $\mathrm{Pr}^{|0\rangle}_{+/-}(+) = |\langle +|0\rangle|^2 = 1/2 \stackrel{!}{=} \sum_{\lambda\in\Lambda} p_0(\lambda)\xi^+_{+/-}(\lambda)$, where '+/−' refers to the measurement associated with outcomes + and −. But this must mean that $\xi^+_{+/-}(\lambda)$ has to give 1 for the first of the λs, and cannot also give 1 for the second one. Equally, $\sum_{\lambda\in\Lambda} p_1(\lambda)\xi^+_{+/-}(\lambda) \stackrel{!}{=} |\langle +|1\rangle|^2 = 1/2$, $\sum_{\lambda\in\Lambda} p_+(\lambda)\xi^+_{+/-}(\lambda) \stackrel{!}{=} |\langle +|+\rangle|^2 = 1$, $\sum_{\lambda\in\Lambda} p_-(\lambda)\xi^+_{+/-}(\lambda) \stackrel{!}{=} |\langle +|-\rangle|^2 = 0$, and so forth. All in all, we get $\boldsymbol{\xi}^+_{+/-} = (1, 0, 1, 0)$, so that the $\boldsymbol{\xi}$ for outcome + mirrors the $\boldsymbol{p}$ which is conditional on +, but with 1s instead of $\frac{1}{2}$s. All the $\boldsymbol{\xi}$s can be worked out to look this way.[15] So $\boldsymbol{\xi}$ actually does not do any real work here at all, and could be omitted altogether.

With this simple setup, Spekkens is *prima facie* able to reproduce a bunch of quantum phenomena. To this end, measurements are assumed to be "*reproducible* in the sense that if repeated upon the same system, they yield the same outcome." (Spekkens 2007, p. 9; emphasis in original.) In other words: they are like the *projective* measurements of QM. And since all relevant quantum outcomes are associated with an epistemic state that is assigned in consequence of the measurement, the QM measurements mirrored or modeled are also *selective*. But as noted before, due to (KB) measurements cannot reveal the true state λ, but can only change what one knows about the system.

Before any measurement one is considered, in the model, to be in a state of total ignorance about the $\lambda \in \Lambda$, i.e. a state where one only knows that $\lambda \in \{1, 2, 3, 4\}$.

[14] Actually, Spekkens lets quantum states directly correspond to the probability distributions; but we are here using the OM framework, whence they should be representative of preparations P instead.

[15] Note that it is not in contradiction to the Kolmogorov axioms that the entries in $\boldsymbol{\xi}$ sum up to 2 instead of 1, as $\boldsymbol{\xi}$ expressed in this way is variable in λ, i.e., in the true state on which it is conditional, not in the outcome. Only the sum over all *outcome probabilities*, given *fixed* parameters (λ, M) must sum to one.

Accordingly, one's epistemic state is $p(\lambda) = 1/4, \forall\lambda \in \Lambda$, i.e. $\boldsymbol{p} = (\frac{1}{4}, \frac{1}{4}, \frac{1}{4}, \frac{1}{4})$ or $1 \vee 2 \vee 3 \vee 4$ (Spekkens 2007, p. 4).[16] Upon measurement, however, $\boldsymbol{p}$ will be changed so that one knows one of the (symbolic) disjunctions $1 \vee 2$, $3 \vee 4$, $1 \vee 3, \ldots$ This is represented in the model as the measurement 'inducing a partition' (cf. Spekkens 2007, p. 9). But frankly speaking, it amounts to a *probability update* reminiscent of *Bayesian conditionalization*. In fact it can even be *reconstructed* in terms of Bayesian conditionalization perfectly well.

To see this, let us say that some experimenter has no prior knowledge about the true state of a system. Equivocating between the alternatives, her epistemic state should hence be $\boldsymbol{p} = (\frac{1}{4}, \frac{1}{4}, \frac{1}{4}, \frac{1}{4})$. Upon measuring the value $+$ (say), she will instantaneously think, however, that the system must be in one of the states 1 or 3, but she can still give no preference to any of the two. Thus her knowledge about the system would have to be represented as $\boldsymbol{p}_+ = (\frac{1}{2}, 0, \frac{1}{2}, 0)$.

Now using the parameters of the model, we can reconstruct this situation in terms of Bayesian conditionalization as follows. Given her knowledge of the measurement, the experimenter will believe beforehand that the result will be either $+$ or $-$. So equivocating between these possibilities, she will assign priors $p(+) = p(-) = 1/2$. Given also the nature of the response function, there will only be four possible joint events from $\Lambda \times \{+, -\}$, i.e. of true state and measurement result that can jointly occur in consequence of the measurement, namely $(1, +)$, $(3, +)$, $(2, -)$, and $(4, -)$. Equivocating between these as well, she will assign $1/4$ to all of them and 0 to all other events in $\Lambda \times \{+, -\}$. Then according to Bayesian conditionalization, her beliefs after measuring $+$ will be $p_+(1) = p(1|+) = \frac{p(1,+)}{p(+)} = \frac{1/4}{1/2} = \frac{1}{2}$, $p_+(2) = p(2|+) = \frac{p(2,+)}{p(+)} = \frac{0}{1/2} = 0$ and so forth. Hence all in all the change is $\boldsymbol{p} = (\frac{1}{4}, \frac{1}{4}, \frac{1}{4}, \frac{1}{4}) \mapsto \boldsymbol{p}_+ = (\frac{1}{2}, 0, \frac{1}{2}, 0)$. This is the picture of what happens in a measurement provided by the model's formal setup.

It is important to note that (KB) is restricted to the knowledge about a system *at a given time*. This is so because given that one knows $1 \vee 2$, a measurement which partitions Λ into $\{\{1, 3\}, \{2, 4\}\}$ will lead to definite knowledge of the state of the system *prior* to the measurement; in case one measures $1 \vee 3$, the state must have been 1, in case of $2 \vee 4$, it must have been 2. The fact that one still lacks complete knowledge about the system's state *after* the measurement is accounted for by an "unknown disturbance" of the state, caused by the measurement (Spekkens 2007, p. 10). This is certainly strongly reminiscent of the 'disturbance interpretation' of the uncertainty relations in Heisenberg's electron microscope.

Given these prerequisites, the first remarkable achievement of the model is that these measurements can be demonstrated to exhibit *non-commutativity*, just as quantum measurements do. Consider two measurements A and B inducing partitions $\{\{1, 2\}, \{3, 4\}\}$ and $\{\{1, 3\}, \{2, 4\}\}$ respectively, and performed on a system in state $1 \vee 2$. Performing the A-measurement first will keep the system in $1 \vee 2$ and

[16] This is the equivocation norm that we had claimed was operative in the model.

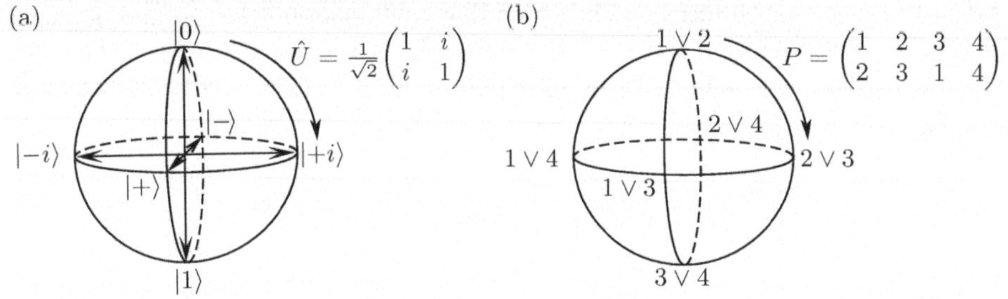

Fig. 4.3 (**a**) Is a regular Bloch sphere for the qubit states, (**b**) is an analogous diagram for the epistemic states. Two of these can be combined by operations $+_1, \ldots, +_4$ to yield one of the respective other states, just as two quantum states can be superimposed to yield a third one. In (**a**), transformations are represented by unitary operators (which can be mapped to rotations), in (**b**) permutations are used instead (the elements of the upper row are replaced by those immediately below them) (Cf. Spekkens 2007, p. 6 for a similar illustration)

the B-measurement will then yield $1 \vee 3$ and $2 \vee 4$ with equal frequencies. Performing them the other way around, the B-measurement will first update the epistemic state to either $1 \vee 3$ or $2 \vee 4$; but now the A-measurement will yield $1 \vee 2$ and $3 \vee 4$ with equal frequency. This could be viewed as a model of the non-commutativity of spin measurements along orthogonal axes in Stern-Gerlach experiments, as discussed in Sect. 2.1.2.

The next interesting achievement of the toy model is the (partial) reproduction of quantum superposition. This is accomplished by defining different rules for combining the epistemic states.[17] For instance, we could combine two states such as $1 \vee 2$ and $3 \vee 4$ by taking the true states of lowest index from each and combining them into a new state, i.e. $1 \vee 3$. This could be symbolized by writing $(1 \vee 2) +_1 (3 \vee 4) = 1 \vee 3$. Equally, we could take the true states of highest index to obtain $2 \vee 4$, which could be written as $(1 \vee 2) +_2 (3 \vee 4) = 2 \vee 4$. Taking one of higher and one of lower index from both epistemic states respectively will yield two further possibilities; $+_3$ could be chosen to be high-low, and $+_4$ to be low-high (cf. Spekkens 2007, p. 6). With these four combination rules, the interrelations of all six quantum sates which we have considered in this context can be mirrored, which is best illustrated in terms of Bloch spheres (or Bloch sphere-like diagrams), as in Fig. 4.3.

There are, however, a few subtleties about this analogy which lead into a first kind of trouble. Combining, say, $(2 \vee 3) +_4 (1 \vee 4) = 2 \vee 4$ in the toy model should, according to the Bloch sphere-image, be analogous to superposing $|+i\rangle$ and $|-i\rangle$ to get $|-\rangle$ in QM, i.e. developing $|-\rangle = \langle +i|-\rangle\,|+i\rangle + \langle -i|-\rangle\,|-i\rangle =$

[17] For completeness' sake, note that Spekkens (2007, p. 5) also introduces a notion of convex combination for the model so that $1 \vee 2 \vee 3 \vee 4$ comes out as the toy-analogue of a completely mixed state which can be decomposed into $1 \vee 2$ and $3 \vee 4$ or $1 \vee 3$ and $2 \vee 4$ or...

$\frac{1+i}{2}\,|{+i}\rangle + \frac{1-i}{2}\,|{-i}\rangle$. It turns out, however, that combination rules $+_3$ and $+_4$ have an *ordering sensitivity*, so that $(1 \vee 4) +_4 (2 \vee 3) = 1 \vee 3 \neq (2 \vee 3) +_4 (1 \vee 4)$. One can model this situation instead by superimposing $\frac{1}{\sqrt{2}}(|{+i}\rangle - i\,|{-i}\rangle)$, which is equal to $e^{-i\frac{\pi}{4}}\,|{-}\rangle$, because

$$e^{-i\frac{\pi}{4}} = \cos\left(-\frac{\pi}{4}\right) + i\sin\left(-\frac{\pi}{4}\right) = \cos\left(\frac{\pi}{4}\right) - i\sin\left(\frac{\pi}{4}\right) = \frac{1}{\sqrt{2}}(1-i),$$

so that

$$\begin{aligned} e^{-i\frac{\pi}{4}}\,|{-}\rangle &= \frac{1}{\sqrt{2}}(1-i)\,|{-}\rangle = \frac{1}{\sqrt{2}}\left(\frac{1}{\sqrt{2}}(|0\rangle - |1\rangle) - i\frac{1}{\sqrt{2}}(|0\rangle - |1\rangle)\right) = \\ &= \frac{1}{\sqrt{2}}\left(\frac{1}{\sqrt{2}}(|0\rangle + i\,|1\rangle - i\,|0\rangle - |1\rangle)\right) \\ &= \frac{1}{\sqrt{2}}\left(\frac{1}{\sqrt{2}}(|0\rangle + i\,|1\rangle) - i\frac{1}{\sqrt{2}}(|0\rangle - i\,|1\rangle)\right) = \\ &= \frac{1}{\sqrt{2}}(|{+i}\rangle - i\,|{-i}\rangle) \end{aligned}$$

This is just $|{-}\rangle$ up to a(n empirically meaningless) global, overall *phase*. But the superposition rule thus also induces a *relative* phase of $\frac{3\pi}{2}$ ($e^{i\frac{3\pi}{2}} = -i$) between the two states superimposed, which accounts for the ordering sensitivity and, as will become clear a little below, *does* matter empirically. In fact, the four combination rules above can all be understood in terms of quantum superpositions with a relative phase, and Spekkens (2007, p. 7) makes the following identifications:

$$+_1 \leftrightsquigarrow +e^{i\cdot 0}, \qquad +_2 \leftrightsquigarrow +e^{i\pi},$$

$$+_3 \leftrightsquigarrow +e^{i\frac{\pi}{2}}, \qquad +_4 \leftrightsquigarrow +e^{i\frac{3\pi}{2}}.$$

These identifications, however, reveal the subtleties mentioned above, and show that the analogy between combinations of epistemic states and quantum superpositions is not—and *cannot* be made—perfect. In the given choice, one obtains $(1 \vee 3) +_3 (2 \vee 4) = 2 \vee 3$ and $(1 \vee 3) +_4 (2 \vee 4) = 1 \vee 4$, but $\frac{1}{\sqrt{2}}(|{+}\rangle + e^{i\frac{\pi}{2}}\,|{-}\rangle) = e^{i\frac{\pi}{4}}\,|{-i}\rangle$ and $\frac{1}{\sqrt{2}}(|{+}\rangle + e^{i\frac{3\pi}{2}}\,|{-}\rangle) = e^{-i\frac{\pi}{4}}\,|{+i}\rangle$, which, given the identifications between combination rules and epistemic- and quantum states, should be exactly the other way around. Exchanging identifications in the latter case will always only shift the problem (cf. Spekkens 2007, p. 7). According to Spekkens (ibid.), "[t]his curious failure of the analogy shows that an elementary system in the toy theory is not simply a constrained version of a qubit."

So here the model already fails to correctly reproduce the QM toolkit from epistemic restriction, and is *bound* to do so. This need not be seen as a strong objection yet, because it should not be required that any successful alternative to QM must mirror the quantum *formalism* isomorphically; a successful replacement of, or alternative to QM should only be required to preserve QM's successful *predictions*. If we construe the model, however, as a means to *reduce* the (exact) rules of QM, in a limited domain, to a theory about incomplete knowledge, then it must still appear as a drawback that the model fails to do so.

Be that as it may, there are further interesting phenomena which the toy model can reproduce to some extent, and to introduce a particularly interesting one, we should consider how *state transformations* are represented in the model. We have seen that transformations in QM are represented (mostly) by unitary operators[18]; the time evolution operator, the spatial propagation operator, and the (pre-)measurement operator discussed in the context of the measurement problem being decisive examples. In the toy model, transformations are represented by *permutations* of the true states in an epistemic state. These correspond to *resamplings* of the probabilities in the epistemic state which, in the present case, in turn amount to changes in *knowledge*. This means that the true state of a system does not have to change *at all* when the epistemic state of an observer does, i.e. that one may always find out some new piece of information even though the system this information pertains to remains entirely unchanged.

The analogy between permutations and unitary transformations can be visualized in a Bloch representation where, as we have seen, unitaries correspond to rotations up to an overall phase, and permutations in the toy-analog of the Bloch sphere will equally appear as rotations by integer multiples of $\pi/2$ (cf. Fig. 4.3b). But more interestingly, permutations can be put to use in the reproduction of *interference* examples within the toy model. Given what we have said about the *meaning* of permutations as toy-replacements of unitaries, this, if successful, should obviously count as a major achievement of the model: quantum interference, Feynman's "only mystery", is explained away in terms of mere ignorance about the underlying reality and the dynamics of our knowledge about it.

But *can* these toy examples of interference count as successful? This is a subtle question which will require considerable discussion below. First consider the following example, based on a setup called *Mach-Zehnder interferometer* (Fig. 4.4), a widely used tool of quantum optics which is also frequently discussed in introductory level books to explain many quantum peculiarities (e.g. Albert 1992, p. 2 ff.; Baaquie 2013, p. 154; Jaeger 2007, p. 20 ff.; Thaller 2005, p. 184 ff.).

Imagine a beam of photons which is of very low intensity, so that only one photon at a time enters the setup. These photons are moving one by one from a source (S) towards a device called a *beam splitter* (BS_1), which can be implemented by a half silvered mirror, and has the effect that each individual photon either passes right

[18] In fact, unitary operators together with *antiunitary* ones exhaust the state-automorphisms or symmetries on the set of all density matrices on a separable Hilbert space (e.g. Heinosaari and Ziman 2012, pp. 29 and 92 ff.). Antiunitary operators, however, "can describe only abstract symmetries (e.g. time inversion), not physically realizable symmetries such as rotations or translations." (Heinosaari and Ziman 2012, p. 91)

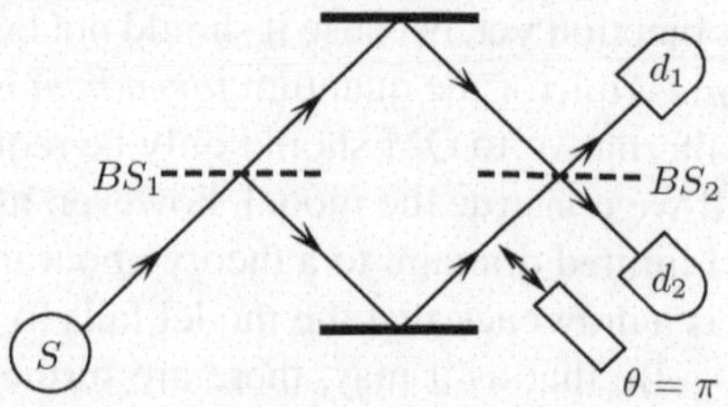

Fig. 4.4 Mach-Zehnder interferometer with an optional phase

through or gets reflected. In case the beam of photons is incident on BS_1 at an angle of 45° and the beam splitter is fabricated in an appropriate fashion, this will result in a 50/50 chance for each photon of either moving right through the mirror or being reflected at a right angle to its original trajectory. One can also in principle make sure that there is always only one photon in the setup by inserting detectors in both arms, right after BS_1, *either* of which will then (ideally) always give off some signal indicating the presence of a photon, but (ideally) never *both* at the same time.

Neglecting polarization etc., we can model this situation as a simple spatial qubit, with one state the 'moving up state' $|\nearrow\rangle \doteq \binom{1}{0}$, and the other one the 'moving down state' $|\searrow\rangle \doteq \binom{0}{1}$. Now take a photon prepared in the up state $|\nearrow\rangle$. The beam splitter BS_1 will change the state into a quantum superposition of moving up and moving down, which can be represented (in our chosen basis) by a unitary matrix

$$\hat{U}_H \doteq \frac{1}{\sqrt{2}} \begin{pmatrix} 1 & 1 \\ 1 & -1 \end{pmatrix}, \tag{4.5}$$

also called the *Hadamard gate* (cf. Thaller 2005, pp. 176 and 186). It is easy to verify that

$$\hat{U}_H |\nearrow\rangle = \frac{1}{\sqrt{2}}(|\nearrow\rangle + |\searrow\rangle) =: |\psi\rangle\,. \tag{4.6}$$

Imagine now that behind each beam of photons emanating from BS_1 there are mirrors (indicated by the thick black lines in Fig. 4.4), so that both paths any single photon could take run towards one another again. We can represent the transformation effected by the mirrors by the $\hat{\sigma}_x$ Pauli-matrix, since Pauli matrices are not only Hermitian but also unitary. $\hat{\sigma}_x$ will only exchange the flying up- and down-components of $|\psi\rangle$ and hence essentially leave it untouched:

$$\begin{pmatrix} 0 & 1 \\ 1 & 0 \end{pmatrix} \frac{1}{\sqrt{2}} \begin{pmatrix} 1 \\ 1 \end{pmatrix} = \frac{1}{\sqrt{2}} \begin{pmatrix} 1 \\ 1 \end{pmatrix}.$$

Now we could also insert an 'obstacle' into one of the paths, with the effect of shifting the phase of a photon on it (or 'delaying the wave', in this beloved metaphor).[19] Let us say that the phase shifter is inserted in the lower path but after the mirrors, so that it will only affect the flying-up part of the spatial superposition state. If we choose $\theta = \pi$ as our phase, we will obtain a transformation which can be represented by the matrix

$$\hat{\Phi}(\theta) \doteq \begin{pmatrix} e^{i\theta} & 0 \\ 0 & 1 \end{pmatrix} \overset{\theta=\pi}{=} \begin{pmatrix} -1 & 0 \\ 0 & 1 \end{pmatrix},$$

(which is just $(-1) \cdot \hat{\sigma}_z$) so that we get

$$\hat{\Phi}(\pi) |\psi\rangle \doteq \begin{pmatrix} -1 & 0 \\ 0 & 1 \end{pmatrix} \begin{pmatrix} \frac{1}{\sqrt{2}} \\ \frac{1}{\sqrt{2}} \end{pmatrix} = \begin{pmatrix} -\frac{1}{\sqrt{2}} \\ \frac{1}{\sqrt{2}} \end{pmatrix} \doteq \frac{1}{\sqrt{2}}(|\searrow\rangle - |\nearrow\rangle) =: |\psi'\rangle. \tag{4.7}$$

But in case we insert a second beam splitter at the point where the two paths cross (BS_2 in the figure, and again represented by $\hat{U}_H$ as in (4.5)), the photon described by $|\psi'\rangle$ will experience a change as

$$\hat{U}_H |\psi'\rangle \doteq \frac{1}{\sqrt{2}} \begin{pmatrix} 1 & 1 \\ 1 & -1 \end{pmatrix} \begin{pmatrix} -\frac{1}{\sqrt{2}} \\ \frac{1}{\sqrt{2}} \end{pmatrix} = \begin{pmatrix} 0 \\ -1 \end{pmatrix} \doteq -|\searrow\rangle. \tag{4.8}$$

That is: our simple qubit model predicts that we will always find a down moving photon in this setup, which has picked up an unobservable phase of π ($e^{i\pi} = -1$).

Computing the probability for detecting an up- or downward traveling photon at the end of this setup is, of course, $|-\langle\nearrow|\searrow\rangle|^2 = 0$ and $|-\langle\searrow|\searrow\rangle|^2 = 1$. We now also see why *relative phases* between two kets in a superposition state do matter: Here the phase is entirely responsible for the resulting behavior at BS_2,[20] since without it we would instead have

$$\hat{U}_H |\psi\rangle \doteq \frac{1}{\sqrt{2}} \begin{pmatrix} 1 & 1 \\ 1 & -1 \end{pmatrix} \begin{pmatrix} \frac{1}{\sqrt{2}} \\ \frac{1}{\sqrt{2}} \end{pmatrix} = \begin{pmatrix} 1 \\ 0 \end{pmatrix} \doteq |\nearrow\rangle, \tag{4.9}$$

i.e. only photons moving *up* at the end of the setup.

Whether the predictions of this simple model are essentially correct could be checked by installing detectors (d_1 and d_2 in Fig. 4.4) which would amplify the energy deposited by incoming photons so as to give off a humanly perceivable signal (e.g. a click or the increase of a number on some electronic counter's display). With

[19] Such a phase shifter can, for instance, be implemented by a piece of matter with a refraction index different from that of air, in which light would travel at an altered velocity (e.g. Walker et al. 2012, p. 1050).

[20] Note that we have assumed both arms of the interferometer to be of equal length, so that none of the two states can pick up a phase due to a spatial delay. In fact, the spatially induced phase difference is what accounts for the interfence pattern in the double slit experiment (Sect. 2.1.1).

the phase shifter in place this would (ideally) mean only detections in d_2 and without it (ideally) only in d_1. And indeed, the successful execution of such experiments is reported in the literature, for instance by Grangier et al. (1986) where the predictions are confirmed quite clearly.[21]

Since we assumed that only one photon at a time enters the setup, it seems surprising that it should matter to photons traveling along the upper path whether there is a phase shifter in the lower one. The effect observed here hence constitutes another example of single particle quantum interference, as in the advanced double-slit setups discussed in Sect. 2.1.1.

But, as was suggested above, the example can be reconstructed entirely within the toy model by appeal to permutations instead of unitary matrices. The simplest type of permutation is a *swap* of two elements in some ordered n-tuple (the epistemic state in our case, being a 4-tuple of probabilities), and we will describe all permutations occurring in the example in terms of such swaps here. Thus, let (jk) represent the swap of elements j and k in an n-tuple. Then in the toy model we start out with $1 \vee 2 \leftrightsquigarrow p_{\nearrow}(\lambda)$ as the epistemic state corresponding to the preparation of $|\nearrow\rangle$ $(= |0\rangle$, accroding to the formerly used nomenclature). The first beam splitter is represented by a permutation (23), which results in $1 \vee 3$ (i.e. 3 will now be assigned the probability previously assigned to 2, which is $\frac{1}{2}$). The mirrors can be represented by (13), yielding $3 \vee 1 = 1 \vee 3$, so that not much happens here, just as in the QM treatment. In case the phase shifter is in, this can be modeled as a permutation corresponding to two successive swaps (12) (34) which then yield $2 \vee 4$. And the second beam splitter will again correspond to (23), so that the final state is $3 \vee 4$. But this distribution is the one corresponding to the quantum sate $|1\rangle = |\searrow\rangle$ so that the quantum predictions are indeed preserved. Equally, if the phase shifter is not inserted, this means that the permutation (12) (34) is left out, whence $1 \vee 3$ will just be transformed into $1 \vee 2$ at the second beam splitter, and we obtain the state that we started off with, again just as in QM.

All of these swaps can, of course, be represented also in the form of transformation matrices, as indicated in formula (4.4). In particular, (23) on $1 \vee 2$, say, may be written as

$$\Gamma_{(23)} \boldsymbol{p}_{\searrow}^{T} = \begin{pmatrix} 1\,0\,0\,0 \\ 0\,0\,1\,0 \\ 0\,1\,0\,0 \\ 0\,0\,0\,1 \end{pmatrix} \begin{pmatrix} \frac{1}{2} \\ \frac{1}{2} \\ 0 \\ 0 \end{pmatrix} = \begin{pmatrix} \frac{1}{2} \\ 0 \\ \frac{1}{2} \\ 0 \end{pmatrix} \leftrightsquigarrow 1 \vee 3, \tag{4.10}$$

[21] In fact, varying the phase somewhat more than just $\theta \in \{0, \pi\}$, one can appeal to probabilities of detection in either d_1 or d_2, where (say) $\Pr_x^{\psi_\theta}(d_1) = |\langle\nearrow|\psi_\theta\rangle|^2 = \cos^2(\frac{\theta}{2})$ for $|\psi_\theta\rangle := \frac{1}{2}\Big((1 - e^{i\theta})\,|\searrow\rangle + (1 + e^{i\theta})\,|\nearrow\rangle\Big)$, as results from the setup with a general phase shift. One can equally use a difference in path length, as mentioned in Footnote 20, and this is what Grangier et al. (1986) actually did to confirm that the number of counts would conform to the predicted $\cos^2$-regularity (cf. their p. 178).

and the probability (say) of obtaining M-state $|\nearrow\rangle$, given P-state $|\searrow\rangle$ and transformation $\Gamma_{(23)}$ as above (effected by BS_1) can be computed as

$$\xi_M^{\nearrow}\Gamma_{(23)}p_{\searrow}^T = (0,0,1,1)\begin{pmatrix}1&0&0&0\\0&0&1&0\\0&1&0&0\\0&0&0&1\end{pmatrix}\begin{pmatrix}\frac{1}{2}\\\frac{1}{2}\\0\\0\end{pmatrix} = 1/2, \tag{4.11}$$

(where $M = \{\searrow, \nearrow\}$). The entire example can be rewritten in this fashion (sandwiching matrices between rows and columns) which looks strikingly similar to our matrix representation of the original quantum example.

So the toy model can *prima facie* reproduce interference examples with the aid of resamplings of probability distributions by permutations of state labels, and such permutations can result in the toy-analog of certain superpositions just as (unitary) QM transformations can result in the corresponding quantum superpositions. We have here considered only a limited example with a certain fixed phase, but a mathematical generalization of Spekkens' work exists (Garner et al. 2013) which can also handle arbitrary phase arguments in terms of probability vectors and transformation matrices. This achievement has lead several authors to conclude that "a whole host of Mach-Zehnder interferometry experiments can be qualitatively reproduced by the theory[...]." (Leifer 2014, p. 79; cf. also Hardy 2013, p. 3 or Fuchs 2014, p. 388 for similar judgments)

"But hold on!", you may interject, "How can a lack of knowledge account for the fact that what I do in the lower arm of the interferometer will influence *all* photons in the setup, even if they take the upper route?" And as well you should. We have here rather 'blindly' applied the *formal* tools of the toy model, which then appeared to nicely mirror some features of QM. But that permutations can be made to look like unitary operations on qubits is a long shot from accepting that resamplings of a probability distribution, a formal representation of knowledge on a Bayesian view, can possibly account for what goes on in a Mach-Zehnder interferometer. *How* is it that our knowledge *should* be affected by the putting in of the phase shifter? *Why* should it be affected in *this* way?

More precisely, in any reasonable model where we can talk about something taking this or that route, the true states should be representative of local degrees of freedom of systems located somewhere in the setup. Hence many of the true states—those representing something moving through the upper route—should, for all we know, *not* be affected at all, whence possibly neither should our knowledge of them. While we had claimed above that permutations can *generally* amount to changes in knowledge about the true state of a system *without* real changes in this very state, in the present case an appeal to this fact seems far fetched: putting in a phase shifter physically alters the setup, and any changes in the statistics would of course *intuitively* result as an effect of a change in the true states of the systems affected by this. Building the example bottom-up, we would certainly not have guessed that putting in a phase shifter must result in interference, in case only one photon enters the setup. It is only background knowledge of QM and the confirming experiments that allows one to concoct a toy model this way, and this leaves us without any explanation as to what is going on in these experiments.

While Spekkens (2007, p. 2) claims to develop the toy model in order to "identify phenomena that are characteristic of states of incomplete knowledge regardless of what this knowledge is about", this entire toy-interference example is hence in conflict with the overall aim of the OM approach (as would be others like it): to provide an *explanation* of the operational/empirical success of QM. What do $1, 2, 3, 4$ represent? How are they affected by the setup in such a way that the kind of probability update exemplified above is indicated? How are *local* degrees of freedom in any one path represented in the model and why should we accept the kind of 'nonlocal' probability update of the epistemic state $1 \vee 3$, *jointly* representing everything that can be said (known/inferred) about the goings on *in both arms*

In fact, this need for explanation seems to be widely acknowledged, whence there *is* a kind of (*ex post*) explanation out there, discussed e.g. in Spekkens (2008), Hardy (2013), and Leifer (2014). But we will only be able to assess this explanation properly, and then also demonstrate difficulties with it, when we concern ourselves more deeply with questions of *locality* later. So we shall defer the discussion to the end of this chapter.

We should now also take a look at the *combined* states of two (or more) simple systems as provided by the model. Spekkens represents the simultaneous occurrence of two true states, on two distinct systems a, b respectively, simply by a (symbolic) conjunction, which we here choose to symbolize by $1(a) \wedge 2(b)$, say, for true state 1 pertaining to system a and true state 2 pertaining to system b.[22] Of course having such an epistemic state is prohibited by (KB) since it would correspond to complete knowledge of the true states of both systems. But combinations of epistemic states, i.e. states of the form $[j(a) \vee k(a)] \wedge [\ell(b) \vee m(b)]$, with $j, k, \ell, m \in \Lambda$, and $j \neq k, \ell \neq m$, are of course possible. These mimic simple *product states* of QM for two separate systems, such as $|\psi^{(a)}\rangle\ |\phi^{(b)}\rangle$ (the bracketed upper index referring to the respective system here).

A second possibility are states of the form $[j(a) \wedge k(b)] \vee [\ell(a) \wedge m(b)] \vee [n(a) \wedge o(b)] \vee [p(a) \wedge q(b)]$ with $j \neq \ell \neq n \neq p,\ \ k \neq m \neq o \neq q$, and all these letters still representing numbers from the set $\{1, 2, 3, 4\}$. I.e.: it could be known that both systems are in the same state, but not in *which* state. Or it could be known that both are in different states, related by a certain specified permutation, but not which is in which. States of this form are supposed to mimic *entangled states* and, *prima facie*, they do capture the essence of such states quite well. This will become evident from the following example.

Take two systems which have been prepared in an entangled state, say $|\pi\rangle = \frac{1}{\sqrt{2}}(|0, 0\rangle + |1, 1\rangle)$ (we will give concrete examples of similar states later). Then this state implies that there is a probability of $1/2$ for each (sub)system to exhibit

[22]Spekkens uses '$\cdot$' instead and refrains from labeling the systems, i.e. lets the conjunction be ordering sensitive (cf. 2007, p. 11).

either of the two measurable values (0,1), but both systems are bound to exhibit the same value if the same observable is measured on them. Now consider a situation in which the two systems are spatially (spacelike) separated and two agents, A and B or 'Alice' and 'Bob', as they are usually called, perform measurements on them. Then at the very moment Alice measures '1', she will know that Bob will measure '1' as well, as long as he measures the same observable.

If we were to endorse the orthodox interpretation, with its sudden change in the system's actual state due to the measurement, then Alice would be capable of 'steering'[23] Bob's system into some definite state by choosing a certain kind of measurement on *her* system—and *instantaneously* so at *arbitrarily* large distances. But from the point of view of the toy model, this surprising consequence dissolves. Alice's state prior to measurement should be represented as $[1(a) \wedge 1(b)] \vee [2(a) \wedge 2(b)] \vee [3(a) \wedge 3(b)] \vee [4(a) \wedge 4(b)]$, since she knows that both systems are in the same state, even though she cannot know in which one. Accordingly, the measurement must result in something like $[1(a) \vee 2(a)] \wedge [1(b) \vee 2(b)]$, say. Treating the connectives in these symbolic formulae as actual conjunctions and disjunctions from propositional logic for the moment, the latter statement straightforwardly follows from $[1(a) \wedge 1(b)] \vee [2(a) \wedge 2(b)]$ by case distinction and adding disjuncts. But the other direction is *not* valid, since $[1(a) \vee 2(a)] \wedge [1(b) \vee 2(b)]$ is also true if $[1(a) \wedge 2(b)]$ holds, and taking into account that states pertaining to the same system mutually exclude each other,[24] $[1(a) \wedge 1(b)] \vee [2(a) \wedge 2(b)]$ would actually be *false*. Only with the epistemic state as given above (the fourfold disjunction) and mutual state exclusion on the same system can Alice draw the appropriate conclusion.

However, which *epistemic* state will result for *both* of the two systems together depends on which *measurement* is performed. Let us say that Alice chooses to measure $\{1 \vee 3, 2 \vee 4\}$ on her system and finds $1 \vee 3$. Then she will come to know that *both* systems must be in either of *those* two true states (1 or 3). If she decides to measure $\{1 \vee 2, 3 \vee 4\}$ instead and finds $1 \vee 2$, then she comes to know that both systems must be in one of these states. So in fact, performing both measurements in a row and obtaining these respective results, Alice can come to the conclusion that both her *and* Bob's system must have been in state 1 all along. She thus instantaneously obtains information about the distant system. But since the act of measurement effects an unknown disturbance, the states of both systems may now (after both measurements) be different: the state of her system (a) could have changed to 2, in virtue of the disturbance effected by the second measurement. And assuming Bob performs the same protocol, he need not even obtain outcome $1 \vee 2$ in the second measurement, since his system's state could have been changed to 3 in the first measurement and then $3 \vee 4$ would result in the second case. All that Alice can hence come to know in the second measurement is that during the *first* measurement both systems *were* in state 1; and this setup can obviously not be used as a means of *communication*.

[23] This is the much-used term that Schrödinger (1935a, p. 556) introduced to describe the situation.

[24] It would hence be more appropriate to use exclusive disjunction $\dot{\vee}$ instead of $\vee$.

This is reminiscent of a family of results from QM called *no-signaling theorems*, one version of which goes as follows (cf. Dickson 2007, pp. 393–394). Consider two systems in the entangled state $|\pi\rangle$ from above, with density matrix $\hat{\rho}_\pi = \frac{1}{2}[(|0,0\rangle + |1,1\rangle)(\langle 0,0| + \langle 1,1|)]$, and consider a measurement of an observable with operator $\hat{O} = \sum_i o_i \hat{P}_{o_i}$, performed e.g. on the second system (with two system-operator $\mathbb{1} \otimes \hat{O}$). Then in a *selective* measurement in which the k-th outcome is measured on the second system, the state would change, according to Lüders' rule, as

$$\hat{\rho}_\pi \mapsto (\mathbb{1} \otimes \hat{P}_{o_k})\hat{\rho}_\pi(\mathbb{1} \otimes \hat{P}_{o_k}) / \mathrm{Tr}\Big(\hat{\rho}_\pi(\mathbb{1} \otimes \hat{P}_{o_k})\Big), \tag{4.12}$$

whereas in a non-selective one, it will change as

$$\hat{\rho}_\pi \mapsto \sum_j (\mathbb{1} \otimes \hat{P}_{o_j})\hat{\rho}_\pi(\mathbb{1} \otimes \hat{P}_{o_j}), \tag{4.13}$$

resulting in what can be read as a sum over (possible PVM) outcomes, weighted by their respective probabilities $\mathrm{Tr}\Big(\hat{\rho}_\pi(\mathbb{1} \otimes \hat{P}_{o_j})\Big)$. Since we assume (as is implied by the state $|\pi\rangle$) that one cannot control which of the quantum systems will end up in which of the two states, any setup which could be used as a kind of signaling between the two remote systems in state $\hat{\rho}_\pi$ would have to make use of multiple different measurements on a bunch of equally prepared pairs of systems and an observable change in the *statistics* of the behavior of (say) Alice's system in virtue of Bob's measurements on his system. Alice and Bob could, for example, agree that if Bob does perform a measurement (at certain evenly spaced points in time), this counts as a 1 and if he does not, this counts as a 0. This sequence of 1s and 0s could then be used to encode a message.

Since Alice cannot know *what* Bob has measured if he did indeed measure, for her Bob's occasional measurements will result in a mixed state of form (4.13). The fact of the matter is that Alice, measuring system (*a*), will *not* detect a change of statistics depending on whether Bob has or has not measured his system (*b*), because if she measures an observable $\hat{Q} = \sum_j q_j \hat{P}_{q_j}$ on her system where the joint state is $\hat{\rho}_\pi$ (i.e. no measurement has occurred on Bob's side), she will obtain result q_ℓ with probability $\mathrm{Tr}\Big(\hat{\rho}_\pi(\hat{P}_{q\ell} \otimes \mathbb{1})\Big) = \mathrm{Tr}\Big((\hat{P}_{q\ell} \otimes \mathbb{1})\hat{\rho}_\pi\Big)$; but in case she measures it in state $\sum_j (\mathbb{1} \otimes \hat{P}_{o_j})\hat{\rho}_\pi(\mathbb{1} \otimes \hat{P}_{o_j})$ (i.e. a measurement *has* occurred on Bob's side), she will obtain result q_ℓ with the *same probability*:

$$\mathrm{Tr}\left(\left[\sum_j (\mathbb{1} \otimes \hat{P}_{o_j})\hat{\rho}_\pi(\mathbb{1} \otimes \hat{P}_{o_j})\right](\hat{P}_{q\ell} \otimes \mathbb{1})\right) = \mathrm{Tr}\left((\hat{P}_{q\ell} \otimes \mathbb{1})\sum_j (\mathbb{1} \otimes \hat{P}_{o_j})\hat{\rho}_\pi(\mathbb{1} \otimes \hat{P}_{o_j})\right) =$$

$$= \sum_j \mathrm{Tr}\Big((\hat{P}_{q\ell} \otimes \mathbb{1})(\mathbb{1} \otimes \hat{P}_{o_j})\hat{\rho}_\pi(\mathbb{1} \otimes \hat{P}_{o_j})\Big) = \sum_j \mathrm{Tr}\Big((\mathbb{1} \otimes \hat{P}_{o_j})(\hat{P}_{q\ell} \otimes \mathbb{1})\hat{\rho}_\pi(\mathbb{1} \otimes \hat{P}_{o_j})\Big) =$$

$$= \sum_j \mathrm{Tr}\Big((\mathbb{1} \otimes \hat{P}_{o_j})(\mathbb{1} \otimes \hat{P}_{o_j})(\hat{P}_{q\ell} \otimes \mathbb{1})\hat{\rho}_\pi\Big) = \sum_j \mathrm{Tr}\Big((\mathbb{1} \otimes \hat{P}_{o_j})(\hat{P}_{q\ell} \otimes \mathbb{1})\hat{\rho}_\pi\Big) =$$

$$= \mathrm{Tr}\left(\sum_j (\mathbb{1} \otimes \hat{P}_{o_j})(\hat{P}_{q\ell} \otimes \mathbb{1})\hat{\rho}_\pi\right) = \mathrm{Tr}\Big((\hat{P}_{q\ell} \otimes \mathbb{1})\hat{\rho}_\pi\Big),$$

where we have used a few properties of the trace, the fact that $\mathbb{1} \otimes \hat{P}_{o_j}$ and $\hat{P}_{q\ell} \otimes \mathbb{1}$ commute, that $\hat{P}^2_{o_j} = \hat{P}_{o_j}$, and that $\sum_j (\mathbb{1} \otimes \hat{P}_{o_j}) = \mathbb{1} \otimes \mathbb{1}$. This no-signaling or no-communication feature is retained in the toy model in virtue of the 'unknown disturbance', as we have seen above.

The truly crucial thing to realize, however, is that, in the toy model, even if there is no real change in the *true* state of Alice's system due to Bob's measurement (or *vice versa*)—because all that happens is that Bob updates his knowledge about Alice's system in virtue of the information obtained on his system—it may still *appear* this way, in case one *confuses* the epistemic state with the true state of the system. And if this confusion were what happens in orthodox QM, this could account for a lot of the apparent weirdness.

This example is indeed suggestive, and *prima facie* the ψ-epistemicist has a major advantage here. But the example is also quite *selective*, and we will argue in detail below that quantum entanglement should in fact rather count as the strongest case *against* an epistemic view of quantum states in the sense of Conjecture 1, not for it. For 'dialectical' purposes however, we will postpone 'dropping the big bomb' of Bell's famous theorem and start off with an analysis of more recent theorems to a similar effect—all of which ultimately have to do with entanglement, as the discussion will show.

4.2.5 A Brief Look at Spinoffs

For completeness' sake, it should not go unmentioned that Bartlett et al. (2012) have worked out a model similar in spirit to Spekkens' original toy model, which reproduces a bunch of phenomena in continuous-range systems (not qubits).[25] A thorough discussion of this model exceeds the scope of this work, whence we only give a brief review. In short, the authors show that putting an epistemic restriction (similar to (KB)) on *Liouville mechanics* (the statistical version of classical Hamiltonian mechanics), one obtains a theory which is "operationally equivalent" (p. 2) with a subtheory of QM, which they spell out to mean that

> there is a one-to-one mapping between the preparations, measurements, and transformations that are allowed in the first theory and those that are allowed in the second and [that] the

[25] It is however not clear that the model fits into the OM approach or whether it can be made to do so. This does not really pose a problem for us, though, since we are in principle more generally concerned with epistemic approaches to QM here.

statistics predicted for every possible experiment in the first theory are precisely the same as those predicted for the corresponding experiment in the second theory. (p. 15)

Because the model is a restricted version of classical statistical mechanics, the true states of systems in question are points $z = (q_1, \ldots, q_{3n}, p_1, \ldots, p_{3n})$ in phase space (for n mass points with 3 position and 3 momentum coordinates q_i, p_j).

The epistemic restriction is twofold. First of all, Bartlett et al. (2012, p. 5) define the set $L_+(\Gamma) := \left\{\mu | \mu : \Gamma \to \mathbb{R}, \mu \geq 0, \int_\Gamma \mu(z)\, \mathrm{d}^{6n}z = 1\right\}$ of (Liouville) probability distributions (or rather densities) on phase space Γ. Then for any such μ to be considered a *valid* density for their model, it is required that (i) the covariance matrix $\gamma(\mu)$ satisfies the 'classical uncertainty principle' $\gamma(\mu) + i\lambda\Sigma \geq 0$, where λ is a free parameter and $\Sigma := \begin{pmatrix} 0 & -1 & 0 & 0 & \cdots \\ 1 & 0 & 0 & 0 & \\ 0 & 0 & 0 & -1 & \\ 0 & 0 & 1 & 0 & \\ \vdots & & & & \ddots \end{pmatrix}$, and that (ii) μ has maximum entropy $S(\mu) = -\int_\Gamma \mu(z) \ln(\mu(z))\, \mathrm{d}^{6n}z$ over Γ among all phase space distributions with the same covariance matrix. The covariance matrix of a distribution that depends on multiple coordinates z_i, z_j (in phase space, in this case) describes, in components γ_{ij}, (twice) the covariance $\left\langle \left(z_i - \langle z_i \rangle\right)\left(z_j - \langle z_j \rangle\right)\right\rangle_\mu$, i.e. the correlation of departures from the mean values $\langle z_i \rangle_\mu, \langle z_j \rangle_\mu$ according to μ (cf. Jaynes 2003, p. 361). The bite of (i) is that it parallels an actual formulation of the uncertainty relation, (2.22), and thus ensures that in restricted Liouville mechanics, relations such as $\Delta p_x \Delta x \geq \lambda/2$ hold (for adjustable λ). (ii), on the other hand, "ensures that an agent should have the maximum uncertainty about the physical state of the system consistent with knowing the means and the covariance matrix." (Bartlett et al. 2012, p. 5) The valid distributions satisfying (i) and (ii) are all of Gaussian form.

The theory which results is thus operationally equivalent (in their sense) to what they call "Gaussian quantum mechanics" (p. 2), the part of QM "including only those preparations, measurements, and transformations that have Gaussian Wigner representations [...]." (ibid.) A Wigner function is a function

$$
\begin{aligned}
w(\boldsymbol{q}, \boldsymbol{p}) &= \frac{1}{(2\pi\hbar)^{3n}} \int \left\langle \boldsymbol{q} + \frac{\boldsymbol{s}}{2} \middle| \hat{\rho} \middle| \boldsymbol{q} - \frac{\boldsymbol{s}}{2} \right\rangle e^{i\boldsymbol{s}\boldsymbol{p}/\hbar}\, \mathrm{d}^{3n}\boldsymbol{s} \\
&= \frac{1}{h^{3n}} \int \sum_j p_j \left\langle \boldsymbol{q} + \frac{\boldsymbol{s}}{2} \middle| \psi_j \right\rangle \left\langle \psi_j \middle| \boldsymbol{q} - \frac{\boldsymbol{s}}{2} \right\rangle e^{i\boldsymbol{s}\boldsymbol{p}/\hbar}\, \mathrm{d}^{3n}\boldsymbol{s} \\
&= \frac{1}{h^{3n}} \sum_j p_j \int \psi_j^*(\boldsymbol{q} - \boldsymbol{s}/2)\psi_j(\boldsymbol{q} + \boldsymbol{s}/2) e^{i\boldsymbol{s}\boldsymbol{p}/\hbar}\, \mathrm{d}^{3n}\boldsymbol{s}\,,
\end{aligned}
\tag{4.14}
$$

of the positions and momenta of n particles ($\hat{\rho}$ some density operator). 'Smearing out' an operator $\hat{A}$ over $\boldsymbol{q}, \boldsymbol{p}$-coordinates in the same fashion (i.e. with $\hat{A}$ in place of $\hat{\rho}$), one obtains its so called *Weyl transform* $\tilde{A}(\boldsymbol{q}, \boldsymbol{p})$, and together with a given Wigner function $w(\boldsymbol{q}, \boldsymbol{p})$, one obtains

$$\langle A \rangle_{\hat{\rho}} = \int \tilde{A}(\boldsymbol{q}, \boldsymbol{p}) w(\boldsymbol{q}, \boldsymbol{p}) \, \mathrm{d}^{3n}\boldsymbol{q} \, \mathrm{d}^{3n}\boldsymbol{p} \,, \tag{4.15}$$

and equally

$$\mathrm{Pr}_{\boldsymbol{q}}^{\hat{\rho}}(\boldsymbol{q} \in \Delta) = \int_{\Delta} \int w(\boldsymbol{q}, \boldsymbol{p}) \, \mathrm{d}^{3n}\boldsymbol{p} \, \mathrm{d}^{3n}\boldsymbol{q} \,, \tag{4.16}$$

and similarly for $\boldsymbol{p}$ (cf. Basdevant and Dalibard 2002, p. 442; Case 2007). So $w(\boldsymbol{q}, \boldsymbol{p})$ shares many properties of, or 'almost looks like' a phase space distribution. But Wigner functions cannot per se be considered as probability distributions on phase space as they may become negative. Gaussian ones, however, do not. And using Wigner representations of POVMs as representations of measurements, and Wigner representations for completely positive (positive for composite systems; cf. Heinosaari and Ziman 2012, p. 176) nonincreasing linear maps as the most general sorts of transformations, Bartlett et al. (2012, pp. 19–20) establish their Gaussian QM. With these tools in hand, they then demonstrate the claimed sort of equivalence between the epistemically restricted version of Liouville mechanics and Gaussian QM. And they also show, in analogy to what is being done in Spekkens' toy model, how some quantum phenomena such as quantum teleportation or no-cloning constraints can be captured, as well as a case similar to the Alice-Bob scenario (which we are going to discuss in more detail later).

However, we shall argue below that even in this more elaborate model some quantum phenomena—and arguably the most important ones—*cannot* be reproduced. There *is* (yet another) model (van Enk 2007) which *can* reproduce them, but only on the pains of accepting *negative probabilities*. Given everything that was said in interlude I, it is hard to see how one could ever make sense of such a notion on epistemic grounds, and we hence choose to dismiss this model as 'too implausible' without further consideration.

What we have seen, in summary, is that there exist some models which can more or less reproduce a bunch of quantum phenomena[26] purely from epistemic considerations. We have here selected but a fraction, and according to Fuchs's (2014, p. 388) count, there is over a dozen of such phenomena that are reproduced by Spekkens' model alone. This is certainly suggestive of limited knowledge being involved *in some sense*, in the appropriate interpretation of quantum states.

But we have also seen that many questions arise when one tries to introduce a 'more complete' (or more intuitive, or more classical...) description which could help us to find an *explanation* (and hence: *interpretative elimination*) of 'weird' quantum phenomena. What we must ask, then, is: are these weirdnesses of QM *just* a matter of knowledge, as promoted by Conjecture 1? Well... things do not seem to be so simple. In the next section, we will introduce more thorough arguments against

[26]Recall that, in concert with the Bogen-Woodward understanding of 'phenomena' which we endorsed from Chap. 2 on, this may simply mean 'implications of QM or QIT'.

an epistemic view as introduced in this chapter, and hence in against Conjecture 1. We will begin with more recent arguments, (more) peculiar to the OM approach. We will then shed some light on two 'classics', and demonstrate how they put the natural response very much in question as well.

4.3 Restrictions for the ψ-Epistemic Approach

4.3.1 The PBR Theorem

In 2012, Matthew Pusey, Jonathan Barrett, and Terry Rudolph (PBR) published a theorem in *Nature Physics*, aimed at showing that the very assumption of ψ-epistemic models leads to predictions which contradict those of QM, and testably so (cf. Pusey et al. 2012). In a preprint-version of their paper, the authors proposed an error-tolerant version of the experimental conditions described in the proof, which allows for an actual test of the diverging predictions (cf. Pusey et al. 2011). Such tests have been implemented and are reported to confirm the QM predictions (cf. Nigg et al. 2012). The theorem constitutes what is usually called a *no-go theorem*, supposed to demonstrate, in this case, that a ψ-epistemic interpretation of QM is not possible or at least faces serious restrictions. The theorem is formulated by appeal to the OM-framework, and thus somewhat specific to epistemic interpretations of QM in the sense of this approach. Of course the proof is not free of presuppositions and we should hence discuss these carefully. The theorem is first demonstrated for quantum states with overlap $\langle\phi|\psi\rangle = \frac{1}{\sqrt{2}}$ and then generalized to states with arbitrary overlaps. We shall restrict ourselves to a discussion of the former case and only briefly sketch how the generalization is established.

To show the incompatibility of QM with ψ-epistemic models, PBR consider two qubit systems which are supposed to be prepared independently. The first crucial assumption is thus that systems *can* be prepared entirely independently of one another, in the range of situations of interest. The states which are assumed to result from the preparation procedure are $|0\rangle$ and $|+\rangle$, where $\mathcal{B} = \{|0\rangle, |1\rangle\}$ is a basis of $\mathbb{C}^2$, and $|+\rangle = \frac{1}{\sqrt{2}}(|0\rangle + |1\rangle)$. Note that these states are *non-orthogonal*, whence, in line with the discussion above, they are plausible candidates for P-states with overlapping associated probability densities, signifying ψ-epistemicity.

An important thing to realize is that the independence assumption translates into two different formal requirements in the two different formalisms (QM and OMs), whose intertranslation hence requires a bridging assumption. In QM, independence can be represented by the use of product states; thus, if $|\Psi\rangle$ denotes the total quantum state of the two systems, we can translate the assumption of preparation independence into

$$|\Psi\rangle \in \{\underbrace{|0\rangle\,|0\rangle}_{=:|\Psi_1\rangle}, \underbrace{|0\rangle\,|+\rangle}_{=:|\Psi_2\rangle}, \underbrace{|+\rangle\,|0\rangle}_{=:|\Psi_3\rangle}, \underbrace{|+\rangle\,|+\rangle}_{=:|\Psi_4\rangle}\} =: \mathcal{P} \qquad \text{(Prod. 1)}$$

(with $i, j \in \{0, +\}$, and where we use $\mathcal{P}$ for 'preparation'). Now in the OM-approach *two* true states λ_1 and λ_2 need to be specified, since two systems are concerned. The next non-trivial assumption (which basically functions as a bridging principle) then is that the state space is *separable* or *factorizable* in an appropriate manner, which could be represented by using a Cartesian product. I.e.:

$$\Lambda_\Psi = \Lambda_1 \times \Lambda_2, \tag{Sep.}$$

with Λ_1 and Λ_2 the state spaces of the two systems respectively (cf. also Spekkens 2012). This separability assumption amounts to assuming that, "when modeling independent local preparations, there are no additional properties of the joint system that are not derived from the properties of the individual systems." (Leifer 2014, p. 100) It is hence basically the ontological assumption which justifies the next step (cf. also Emerson et al. 2013, p. 2). Namely, given (Sep.), the independence-assumption can be translated into a classical probabilistic language, suitable for the OM-approach, as

$$p_j(\lambda_1, \lambda_2) = p_k(\lambda_1)p_\ell(\lambda_2), \quad j \in \{1, \ldots, 4\}, \quad k, \ell \in \{0, +\} \tag{Prod. 2}$$

(cf. Pusey et al. 2012, p. 477; Drezet 2012, p. 14), with $p_j(\lambda_1, \lambda_2) := p(\lambda_1, \lambda_2|\Psi_j)$ and $p_k(\lambda) := p(\lambda|k)$ $(j \in \{1, \ldots, 4\}, k \in \{0, +\})$.[27] The two conditions (Prod. 1) and (Prod. 2) are neither logically equivalent, nor does (Prod. 1) straightforwardly imply (Prod. 2). But we can make the case that (Prod. 1) *conceptually* implies (Sep.), and that (Sep.) conceptually implies (Prod. 2): If we can *prepare* two systems in (sufficient) isolation from one another, we use a tensor product in QM to represent the (P-)state of a composite system. But if we use such a product state, we assume both component systems to *be* (sufficiently) independent of one another. And given that we hence assume their respective (*true*) states to be independent of one another, i.e. given (Sep.), we would also model this very situation in a classical probabilistic framework by letting the joint probability distribution for both systems be a mere product-distribution over the true states of each individual system.[28] Hence it fully suffices to claim that (Prod. 1)→(Sep.), and that (Sep.)→(Prod. 2) to get the central premise:

$$\text{(Prod. 1)} \rightarrow \text{(Prod. 2)} \tag{P.-Indep.}$$

[27] Indeed, this definition is not maximally general again, since we have appealed directly to probability densities. Leifer (2014, p. 99) instead uses the condition that the probability measure on the space $\Lambda_\Psi = \Lambda_1 \times \Lambda_2$ is the product measure $\mu_1 \times \mu_2(\Lambda_\Psi) = \int_{\Lambda_2} \mu_1(\Omega_{\lambda_2})\, d\mu_2(\lambda_2)$, where $\Omega_{\lambda_2} = \{\lambda_1 \in \Lambda_1 | (\lambda_1, \lambda_2) \in \Lambda_\Psi\}$. For our discussion, no harm comes from using the simpler definition above.

[28] This of course means that $p_j(\lambda_1|\lambda_2, k^{(2)}) = p_j(\lambda_1)$, $j, k \in \{0, +\}$, with $k^{(2)}$ the preparation for the *second* system, and analogously for $p_k(\lambda_2)$.

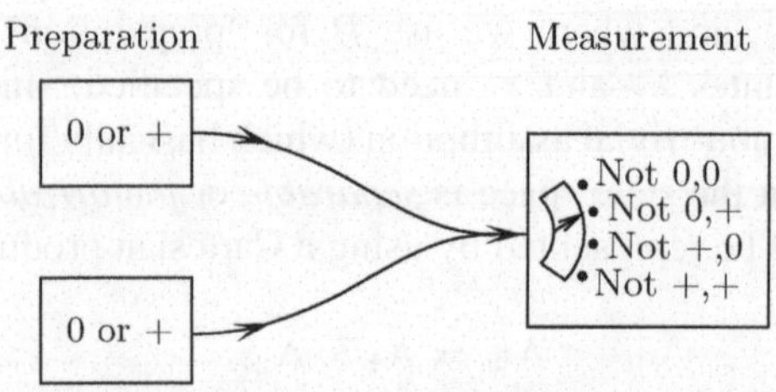

Fig. 4.5 Each system is prepared in one of two quantum states; the entangled measurement performed on both systems simultaneously then finds out which of the four possible product states was not prepared (Cf. Pusey et al. 2012, p. 477 for a similar illustration)

Suppose now that there is a Δ such that $\lambda_1, \lambda_2 \in \Delta$ is not excluded, i.e. that both systems can assume true states in some common range. Also, fix some lower limit $q > 0$ such that $p_k(\lambda_1) \geq q, p_l(\lambda_2) \geq q$ for $k, l \in \{0, +\}$ and $\lambda_1, \lambda_2 \in \Delta$. Then by (P.-Indep.), we get that

$$p_\Psi(\lambda_1, \lambda_2) \geq q^2, \quad \forall \Psi \in \mathcal{P}\, \forall \lambda_1, \lambda_2 \in \Delta \tag{Δ}$$

We call this intermediate result '(Δ)' because the existence of some Δ (i.e. the positivity of the product density p_Ψ on some set of non-zero measure), regardless of the specific preparation on each system, is crucial. It is also crucial to realize that the preparation procedures on both systems do the same thing, i.e., prepare either $|+\rangle$ or $|0\rangle$, whence the (total) range of true states λ possibly resulting from the preparations should be identical for the two systems. This (in concert with (P.-Indep.)) justifies why it even makes sense to consider this setup for *two systems* as a possibility to check for the possibility of a ψ-epistemic model, where the assumption of an overlap is formulated w.r.t. to the states of *one and the same* system.

Now the measurement executed on the two systems is performed by bringing them together in one measurement-device and measuring them jointly (cf. Fig. 4.5). A measurement of this kind is called *global*, since all the systems in some total state $|\Phi\rangle$ are measured together. This should be contrasted with a *local* measurement, where each of a bunch of systems described jointly by $|\Phi\rangle$ is measured individually, whence information about each of them is acquired independently. But among the global measurements, one can further distinguish measurements which have only product states as possible outcomes from such which have at least one entangled state among their outcomes. Meaning that in the latter case, the operators used to describe the measurement have entangled eigenvectors. These are then (unsurprisingly) called *entangled measurements* (cf. Wootters 2006, pp. 219–220; Giovannetti et al. 2004, p. 1333).

The measurement considered by PBR is exactly such an entangled (global) measurement. Furthermore, it is projective, resulting in one of four states from the set

$$\mathcal{R} := \left\{ |\phi_1\rangle = \frac{1}{\sqrt{2}}(|0\rangle\,|1\rangle + |1\rangle\,|0\rangle), \qquad |\phi_2\rangle = \frac{1}{\sqrt{2}}(|0\rangle\,|-\rangle + |1\rangle\,|+\rangle), \right.$$
$$\left. |\phi_3\rangle = \frac{1}{\sqrt{2}}(|+\rangle\,|1\rangle + |-\rangle\,|0\rangle), \qquad |\phi_4\rangle = \frac{1}{\sqrt{2}}(|+\rangle\,|-\rangle + |-\rangle\,|+\rangle) \right\}$$

(cf. Pusey et al. 2012, p. 476), which we call '$\mathcal{R}$' for 'result'.[29] What we now see is that for each of the $|\phi_j\rangle \in \mathcal{R}$ there is a $|\Psi_k\rangle \in \mathcal{P}$ which is *orthogonal* to it (whence the global property to be measured is which of the states was *not* prepared; cf. Fig. 4.5). For instance,

$$\langle\phi_1|\Psi_1\rangle = \frac{1}{\sqrt{2}}((\langle 0|) \otimes \langle 1| + \langle 1|) \otimes \langle 0|)\,|0\rangle \otimes |0\rangle = \frac{1}{\sqrt{2}}(\langle 0|0\rangle\,\langle 1|0\rangle + \langle 1|0\rangle\,\langle 0|0\rangle) =$$
$$= \frac{1}{\sqrt{2}}(1 \cdot 0 + 0 \cdot 1) = 0, \tag{4.17}$$

and (because of the way we have indexed the states) in general $\langle\phi_j|\Psi_j\rangle = 0$.

But recall that the connection between the Born probabilities and the probability densities in the OM was established by an integral over the product of the epistemic state with a response function (formula (4.1)). This integral must now take the form

$$\Pr_{\mathcal{R}}^{|\Psi_k\rangle}(\phi_j) = \int d\lambda_1 \int d\lambda_2\, p_k(\lambda_1, \lambda_2)\xi_{\mathcal{R}}^{\phi_j}(\lambda_1, \lambda_2)$$

(cf. Pusey et al. 2012, p. 477; Drezet 2012, p. 14), with $\xi_{\mathcal{R}}^{\phi_j}(\lambda_1, \lambda_2)$ the response function for outcome ϕ_j.

Moreover, it is plausible to require that

$$\sum_{j=1}^{4} \xi_{\mathcal{R}}^{\phi_j}(\lambda_1, \lambda_2) = 1, \quad \forall (\lambda_1, \lambda_2) \in \Lambda_\Psi, \tag{Outc.}$$

i.e. that there will always be *some* outcome for *all* the states that may result from the preparation (cf. Aaronson et al. 2013, p. 1; Schlosshauer and Fine 2014, p. 1). Of course this is quite an idealization, and we may assume that (Outc.) is only required to hold up to expected experimental noise and error.

But since $p_1(\lambda_1, \lambda_2)$ is at least q^2 on a set Δ of non-zero measure, i.e. in virtue of (Δ), at least after error correction it must hold that

[29] For notational simplicity we will later also use this letter to refer to the measurement (POVM) associated with the outcome states in $\mathcal{R}$.

$$\exists k \forall j : \Pr_{\mathcal{R}}^{|\Psi_j\rangle}(\phi_k) = \int d\lambda_1 \int d\lambda_2\, p_j(\lambda_1, \lambda_2) \xi_{\mathcal{R}}^{\phi_k}(\lambda_1, \lambda_2) > 0$$

$$\stackrel{!}{=} |\langle \phi_k | \Psi_j \rangle|^2 = 0 \quad \text{for } j = k \quad \lightning \tag{PBR}$$

(with $j, k \in \{1, \ldots, 4\}$). This is the PBR contradiction. (Prod. 1), (P.-Indep.), and (Outc.) taken together with the definition of ψ-epistemicity and the probabilistic assumptions of the OM framework (short: {OM}), lead to a contradiction; hence PBR conclude:

$$\{\text{OM}\}, (\text{Prod. 1}), (\text{P.-Indep.}), (\text{Outc.}) \vdash \neg(\psi\text{-epistemicity}) \tag{4.18}$$

Stated differently, this means that any ψ-epistemic OM can not maintain (Prod. 1), (P.-Indep.), and (Outc.) together, all of which are *prima facie* reasonable assumptions.

We have restricted our attention to the two-system case, but the result of PBR is generalized (2012, p. 476 ff.) using tensor-product states $|\Psi\rangle = |\psi_1\rangle \otimes \ldots \otimes |\psi_n\rangle$ of arbitrary finite cardinality n, where each system is prepared in either $|0\rangle$ or $|+\rangle$ ($\psi_j \in \{0, +\}$, $\forall 1 \leq j \leq n$). This allows for states with an overlap different from that between $|0\rangle$ and $|+\rangle$ to be used in the preparation.

But how deep is the impact of PBR's result really? Should it be taken to rule out ψ-epistemic OMs *tout court*? Obviously, the fact that all the aforementioned additional assumptions have to be made in order for the proof to go through limits the scope and depth of the PBR-theorem as a no-go result. Each of the assumptions of the proof could well be the culprit, whence it is worth looking at each of them separately.

Detailed criticism toward the other premises of the PBR theorem can be found especially in an article by Schlosshauer and Fine (2012). Notably, they first of all refrain from even using the terminology of 'ψ-epistemic' and 'ψ-onitc' models, and refer to these classes of models as 'mixed' and 'segregated' instead (which they find "less charged" (2012, p. 4)). Thus, the general aptness of the very *definition* of a ψ-epistemic model used in the OM approach may of course be put into question (and hence the premise {OM}), and a whole other set of criteria for understanding the wave function as a representation of knowledge may of course be available (a thought that we should keep in mind). Schlossauer and Fine then also show a way of transforming mixed models into segregated ones and vice versa, thus lessening the appeal of the definitions from Sect. 4.2 as indeed reflecting a distinction between something that represents knowledge and something that represents something real.[30]

Beyond that, Schlosshauer and Fine suggest to augment the spectrum of outcome values associated with the measurement with so called 'no-shows', i.e. to allow

[30]These charges of transformability between the two types of models are, however, challenged by Leifer (2014, p. 113–114).

for measurements with no discernible outcome at all, and hence to modify the connection between the Born probabilities and the probability densities in the OM-framework accordingly. We have seen above that one crucial step of PBR's theorem is to require (Outc.) and that (Outc.) is somewhat idealized. Modifying this requirement in such a way that, given that the true state is in the overlap region, there will be a probability of obtaining no outcome at all, determined *by the true state itself*, obviously blocks the inference to $\neg$(ψ-epistemicity). Schlosshauer and Fine (2012, p. 2) refer to this as a "built-in inefficiency", since the assumption is that there is something about the measured system itself which lets the probability of a (discernible) outcome dip in the appropriate region—i.e. not just regular sources of experimental error.

A bit more precisely, the general recipe goes like this: Determine some probability $\xi_{\mathcal{R}}^{\emptyset}(\lambda_1, \lambda_2)$ of getting a *null-outcome* $\emptyset$ (i.e. something that cannot be recognized properly as an outcome on the measuring device), sufficiently high for the $\lambda \in \Delta$, so that the QM statistics are reproduced, but from probabilities *conditional* on the fact that a discernible outcome was measured at all (i.e. by 'postselecting' the statistics for runs in which there was a determinate outcome). Then for the set of outcomes $\{\phi_1, \ldots \phi_4, \emptyset\}$, the resulting version of (Outc.) is *not* violated and no contradiction arises. Under these assumptions, all that the PBR-result shows is "how inefficiencies arise as a fundamental property of certain hidden-variables models [...]." (Schlosshauer and Fine 2012, p. 2)

This is a kind of 'prism model', which the reader may be familiar with from the context of Bell inequalities (see also later). However, there is a certain unpleasant ad hocness to assuming that the true states from the overlap mysteriously sabotage the measurement procedure just to recover the quantum statistics. Thus we may be inclined, at this point, to put more doubt on the justifyability of Schlosshauer and Fine's no-show assessment than on PBR's own one.

The various assumptions underlying (P.-Indep.) are also under scrutiny in Schlosshauer and Fine's article. They think that "[c]orrelations [...] cannot be ruled out, even if the preparations appear to be independent, because procedures for preparing the individual subsystems may occur together closely in spacetime or share common sources of energy, as well as a common past." (Schlosshauer and Fine 2012, p. 3) In our reconstruction, we may take this criticism to aim at the validity of the implication (Prod. 1)$\rightarrow$(Sep.), and so indirectly at the validity of (P.-Indep.). But (Sep.) can be weakened to the condition (call it '(Sep.*)') that, if there is a λ in the support of each of the epistemic states associated with the multiple systems and respective quantum states, then there is also some λ_c in the support of the common density p_Ψ associated with the product state $|\Psi\rangle = |\psi_1\rangle \otimes \ldots \otimes |\psi_n\rangle$ (cf. ibid.). The exact nature (and structure) of λ_c can then be left completely unspecified. From this one neither gets the condition (P.-Indep.), because (Sep.*) does not imply (Prod. 2), but rather that $p_\Psi(\lambda_c) > 0$ (call this '(Pos.)'). Nor does one get the (exact) intermediate q^2-result (Δ), which follows from (Prod. 2), not (Pos.). But since the weaker (Pos.) is obviously sufficient to derive a contradiction (i.e. (Outc.) would still be violated) it appears that PBR's conclusion $\neg$(ψ-epistemicity) is not really warranted, and that the theorem need not be considered as applying to ψ-epistemic models after all. Notably this move of Schlosshauer and Fine is only possible on the pains of replacing (Sep.) by (Sep.*) and hence by denying (P.-Indep.), or in other words: by assuming that the systems in question *cannot* be prepared (sufficiently) independently of one another.

Another assumption also scrutinized by Schlosshauer and Fine, which PBR *tacitly* make and which we have not yet discussed so far, is that the response functions $\xi_{\mathcal{R}}^{\phi_j}(\lambda_1, \lambda_2)$ do not depend on Ψ. Here Schlosshauer and Fine (2012, p. 2 ff.) propose that models which avoid the problem raised by PBR can be constructed in case the ξs are allowed to depend on Ψ. They call the class of models presupposed by PBR *state-independent*. Leifer (2014, p. 111), in contrast, thinks that "this criticism is simply a misunderstanding of what is meant by the term 'ontic state' in the ontological models framework", and goes on to demonstrate an example of how models can *trivially* reproduce the Born probabilities in case state dependence is allowed (that is, in case ξ is also conditional on the prepared quantum state Ψ). In a similar vein, Ballentine (2014, p. 6) refers to such models as "functionally ψ-ontic", because

> [t]he most important structure of the model is the separation of *preparation* from *measurement*, with information passing only via the ontic state variables. If the state ψ has a direct effect on the measurement outcome, then ψ should be classified as an *ontic* variable. (emphasis in original)

Hence the assumption of state independence may be considered as justified (or -fiable); the introduction of state dependence *conceptually undermines* the very idea behind ψ-epistemicity in the OM-approach.

Nevertheless, Schlosshauer and Fine's (2012, p. 4) conclusions on the impact of the PBR theorem are overall quite de-emphasizing:

> PBR show that state-independent models of composites formed using systems with mixed [ψ-epistemic – FB] models face restrictions. It is vital to see that those restrictions do not imply any difficulty for models of the *components* themselves. The PBR theorem is not a no-go theorem for the component systems[...]. (my emphasis – FB)

And indeed, the theorem is *not* concerned with several quantum states of a *single* system, but only has an impact on overlapping epistemic states *via* the detour of using product states of compound systems. One may jump in at any point and criticize the assumptions that bridge the gap, as we have just seen. Moreover, Lewis et al. (2012) actually have provided two variants of a ψ-epistemic model which become possible in case (P.-Indep.) is dropped. But these models are utterly formal—or, to use a term beloved by philosophers, appear completely *gerrymandered*—and Lewis et al. (2012, p. 4) themselves also concede:

> None of these models is intuitive or motivated by physical principles or considerations. The primary motivation for exploring the possibility of ψ-epistemic models is to understand the formal limitations of reproducing quantum theory from a deeper theory.

Their conclusion w.r.t. the latter aim is that "any similar no-go theorem will also require nontrivial assumptions beyond those required for a well-formed ontological model." (Lewis et al. 2012, p. 1) We can take from this that, while restricting the possibility of ψ-epistemic models, the PBR theorem and similar results should not count as a *full* no-go theorems for these models, in the sense of demonstrating their *impossibility*. They all rely on additional assumptions and can hence maximally limit the attractiveness of ψ-epistemic hidden variable models, or more precisely, show their incompatibility with these very assumptions.

Regarding the existence of other such theorems, the PBR paper has indeed caused a whole landslide of publications which put forward theorems purportedly showing the impossibility of ψ-epistemic models (so in fact, their incompatibility with other plausible assumptions and QM).

To name a few: Patra et al. (2013) derive a no-go theorem based on a "continuity assumption", which they think ψ-epistemic models should satisfy. The motivation is that "we assign an ontic status to ψ if a variation of ψ necessarily implies a variation of the underlying reality λ, and we assign it an epistemic status if a variation of ψ does not necessarily imply a variation of λ." (p. 2) The continuity assumption then rests on interpreting ψ as associated with an ensemble of possible λs and says that "there are real states λ in the initial ensemble that will remain part of the perturbed ensemble, no matter how we perturb the initial state, provided this perturbation is small enough." (ibid.) In their words, this "captures the intuition that in a model where the quantum state is epistemic, a small variation of ψ does not necessarily imply a variation of the underlying real state λ." (p. 4) Subsequently, models satisfying (the formalization of) this condition are demonstrated to be incompatible with QM.

Colbeck and Renner (2012) provide another theorem which they purport to show that "the quantum wave function can be taken to be an element of reality of a system based on two assumptions: the correctness of quantum theory and the freedom of choice for measurement settings." (p. 3) And the list continues.[31]

One such theorem that we should also take a closer look at is that of Hardy (2013). This will give us a first chance to directly confront one of the purported achievements of existing ψ-epistemic models (and specifically of Spekkens' toy model). In the course of demonstrating the impact of Hardy's theorem, we will thus highlight some defects in reasoning that lead to the conclusion that there even *is* a true achievement of the model(s), and hence a *plausibility argument* for the ψ-epistemic approach.

4.3.2 Hardy's Theorem

The gist of Hardy's theorem can best be captured by appeal to an interferometry example like the ones we had met with in Sect. 4.2.4.

In Hardy's own words, the argument based on the following example amounts to a "version of the popular argument for something going both ways[...]." (2013, p. 6)

[31]Leifer (2014) gives a detailed overview of at least some of the recent development.

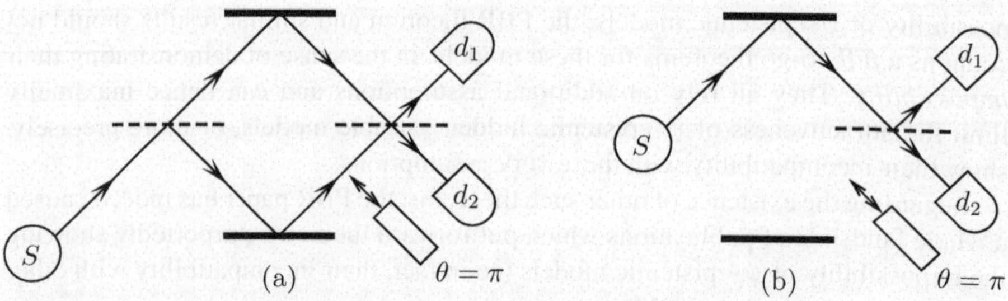

Fig. 4.6 (**a**) is the Mach-Zehnder setup as discussed in Sect. 4.2.4. In (**b**) the photon is emitted somewhere along the upper trajectory, whence the phase shifter in the lower trajectory should have no effect

Consider, in contrast to the Mach-Zehnder example we had discussed in Sect. 4.2.4, an altered setup where the source of photons is placed somewhere along the upper route (cf. Fig. 4.6b). In this altered setup, it should not matter whether the beam splitter is inserted or not; whereas in the original Mach-Zehnder example we would obtain either $-|\searrow\rangle$ or $|\nearrow\rangle$ at the end of the interferometer, depending on whether the phase shifter was in or not, we will here simply have

$$\hat{U}_H \hat{\sigma}_x |\nearrow\rangle = \hat{U}_H |\searrow\rangle = \frac{1}{\sqrt{2}}(|\nearrow\rangle - |\searrow\rangle), \tag{4.19}$$

whence detection at d_1 and d_2 will be equiprobable.

Now consider the state $|\psi\rangle = \frac{1}{\sqrt{2}}(|\nearrow\rangle + |\searrow\rangle)$ as prepared by the first beam splitter in the setup of Fig. 4.6a, and the state $|\phi\rangle = |\nearrow\rangle$ as prepared by the source in the setup of Fig. 4.6b. These two states are non-orthogonal and hence could well be taken to have overlapping supports in a ψ-epistemic model. In this context, we can understand this claim such that it is not impossible for the first beam splitter to prepare a photon which is *actually* traveling up, and that $|\psi\rangle$ is again just indicative of our lack of knowledge about the true state, i.e. the true path that the photon takes.

But then it should make no difference for the photons actually traveling up whether the phase shifter is inserted in the lower path or not. Thus, denote the full set of true states associated with $|\phi\rangle$ by $\Lambda_{|\phi\rangle}$, and the subset of those resulting in a click from detector d_1 or detector d_2 by $\Lambda^{d_1}_{|\phi\rangle}$ and $\Lambda^{d_2}_{|\phi\rangle}$ respectively. One can also associate a given setting of the phase shifter (in or out) to these sets, which we indicate by the notation $\Lambda^{d_j}_{|\phi\rangle}[\theta]$ ($j \in \{1, 2\}, \theta \in \{0, \pi\}$). But since the choice of θ as 0 or π should not alter the behavior of the photon going along the upper path, we obtain a kind of *invariance*:

$$\Lambda^{d_j}_{|\phi\rangle} = \Lambda^{d_j}_{|\phi\rangle}[\theta = 0] = \Lambda^{d_j}_{|\phi\rangle}[\theta = \pi], \quad j \in \{1, 2\}. \tag{INVAR}$$

Assume that the photon is bound to end up in one of the detectors, thereby neglecting experimental errors, i.e., photons getting absorbed somewhere along the way or detectors not responding upon incidence. Then it should hold that

$$\Lambda_{|\phi\rangle} = \Lambda^{d_1}_{|\phi\rangle} \cup \Lambda^{d_2}_{|\phi\rangle}, \tag{TOT}$$

irrespective of the choice of θ. Now consider the set of true states $\Lambda_{|\psi\rangle}$ associated with $|\psi\rangle$ (the state prepared by the first beam splitter). We had established above that in case the phase shifter is in ($\theta = \pi$), the state $|\psi\rangle$ will not result in any clicks from detector d_1. Thus it should hold that

$$\begin{aligned} &\Lambda_{|\psi\rangle} \cap \Lambda^{d_1}_{|\phi\rangle}[\theta = \pi] = \varnothing \\ \Leftrightarrow\ &\Lambda_{|\psi\rangle} \cap \Lambda^{d_1}_{|\phi\rangle} = \varnothing, \end{aligned} \tag{4.20}$$

where the equivalence follows from (INVAR). Analogously, in case the phase shifter is out ($\theta = 0$), there will be no clicks in detector d_2 if $|\psi\rangle$ is prepared, so that

$$\begin{aligned} &\Lambda_{|\psi\rangle} \cap \Lambda^{d_2}_{|\phi\rangle}[\theta = 0] = \varnothing \\ \Leftrightarrow\ &\Lambda_{|\psi\rangle} \cap \Lambda^{d_2}_{|\phi\rangle} = \varnothing. \end{aligned} \tag{4.21}$$

But from (TOT), (4.20), and (4.21) it now follows that $\Lambda_{|\psi\rangle} \cap \Lambda_{|\phi\rangle} = \varnothing$, whence there is no intersection in the sets of true states associated with the two non-orthogonal states $|\psi\rangle$ and $|\phi\rangle$. This in turn means that the epistemic states for the two preparation methods associated with $|\psi\rangle$ and $|\phi\rangle$ cannot have overlapping supports. Thus, it seems, this situation *cannot* be understood ψ-epistemically.

Of course this is not yet a no-go theorem for ψ-epistemic OMs but merely an example. In the remainder of his paper, Hardy provides a generalization, first for finite Hilbert spaces, for which it is shown that non-orthogonal states with a certain lower bound quantum probability $|\langle\phi|\psi\rangle|^2$ (which depends on the dimension of the Hilbert space) will result in distributions with non-overlapping supports (cf. his pp. 9–13). For an infinite dimensional Hilbert space, the result is then shown to hold regardless of the quantum probability (cf. his p. 12). For a rigorous, general proof one of course needs to abstract from beam splitters, mirrors, and phase shifters. The phase shifter, for instance, is replaced by a general unitary transformation with some general parameter m (instead of the phase shift θ) to be varied (cf. his p. 10 ff.).

But of course, a few crucial assumptions also have to be made to run this proof, just as in the PBR case. For the proof of Hardy's theorem, the following two principles have to be assumed (cf. Hardy 2013, pp. 4–5):

Possibilistic Completeness (PC) The ontic state, λ, is sufficient to determine whether any outcome of any measurement has probability equal to zero of occurring or not.

Restricted Ontic Indifference (ROD) Any quantum transformation on a system which leaves a particular given pure quantum state, $|0\rangle$, unchanged can be implemented in such a way that it does not affect the underlying ontic states, $\lambda \in \Lambda_{|0\rangle}$, in the ontic support of $|0\rangle$.

Note that Hardy first assumes a stronger principle of ontic indifference, which is supposed to hold for any arbitrary quantum state $|\psi\rangle$ instead of a particular one ($|0\rangle$). He then demonstrates that the weaker principle (ROD) is sufficient to run the proof (cf. Hardy 2013, p. 12). The 'ontic support' is of course the support of the epistemic state, i.e., the set of true states λ which may result from the preparation procedure associated with $|\psi\rangle$. (PC) is also a rather weak principle, since it has the true state only determine whether an outcome has probability zero or not, instead of determining the exact probability.

We have seen both of these principles at work in the example considered above. (PC) is used to define the sets of states which may give rise to a click from d_1 or d_2 respectively. (ROD) is invoked in assuming that (INVAR) holds, i.e. that it does not make a difference to the photon traveling in the upper path whether the phase shifter is inserted or not. The assumption is, as Hardy (2013, p. 3) also notes, akin to a kind of *locality* or *local causality* constraint, which informally means that whether something is done over here should not *immediately* influence what happens somewhere else. As we will see in the next section, it is open to debate whether or in what sense QM respects such a principle, and hence whether any hidden variable model which purports to reproduce QM's predictions should.

The critical reader will object that we have seen Spekkens' toy model reproduce interferometer examples like the one considered in this section. Is the toy model 'non-local'? *Prima facie* the answer here is 'no', but only on the price of accommodating a non-trivial 'vacuum state', akin to that of QFT, into the ontology presupposed by the model. Thus Hardy (2013, pp. 14–18) writes[32]:

> [T]here are ontic variables associated with the occupation number of the path (take this to be 0 or 1) and a phase associated with the path (take this to be 0 or π). Even if the occupation number is 0 there is still the phase variable which will be affected by a phase shifter. Thus a path with no particle in it still has nontrivial degrees of freedom associated with it. This allows the model to violate ontic indifference in a local way.

And similarly Leifer (2014, p. 121) thinks that it is possible to save the interference example from the consequences of Hardy's theorem in this fashion:

> From quantum field theory, we know that the vacuum is not a featureless void, but has some sort of structure. Therefore, it makes sense that, at the ontological level, there might be more than one ontic state associated with the vacuum, and a transformation that does not affect things localized [in one arm of an interferometer – FB] might still act nontrivially on these *vacuum ontic states*. [...] A transformation acting locally on [one arm – FB] can then switch the ontic states, in violation of ontic indifference, whilst leaving the distribution invariant. (my emphasis – FB)

[32]Here he is referring especially to elaborations from a talk given by Spekkens (2008).

So not: 'something goes both ways', but rather: 'for each way there is something which goes it'. As we already noted, Spekkens (2007, p. 2) describes one central aim of finding an epistemic model for QM as identifying "phenomena that are characteristic of states of incomplete knowledge regardless of what this knowledge is about." The interferometer example does obviously *not* constitute an example of incomplete knowledge *regardless* of what this knowledge is about. For this example to make sense, one has to admit either a direct (causal) influence between the two arms of the interferometer (which would violate the otherwise 'local' spirit of the model), or construct a specific kind of true state, capable of carrying phase information while being otherwise hidden from detection. We should now put some more weight on these worries by concerning ourselves with two (in)famous, 'classic' theorems in the context of QM.

4.3.3 EPR, Einstein, and Bell's Theorem

The theorem we now unfold is arguably the most profound of the four discussed here, and Strapp (1975, p. 271) has even gone so far as to call it "the most profound discovery of science." In virtue of the fact that Bell's theorem makes some of the most fundamental intuitions of many practicing physicists accessible to experiment, Abner Shimony (1984, p. 35) has coined the term "experimental metaphysics" for it.[33] And it is certainly also one of the most thoroughly discussed results in the philosophy of QM, whence we can only cover a tiny fraction of the literature here.

To elaborate the details, we should once more concern ourselves with an experimental example, again possibly implemented with two spin-$\frac{1}{2}$ atoms and Du Bois magnets. Consider the following setup as described by Bohm (1951, p. 614 ff.), and essentially a (crucial) refinement of a thought experiment originally suggested by Einstein et al. (1935), hence usually called an *EPRB* experiment. Two systems with spin-$\frac{1}{2}$ are prepared together at a common source, which leaves them in an entangled state, similar to those encountered in Sect. 2.1.3 or the PBR theorem. In Bohm's example these are two atoms, produced as a result of molecular decay, but similar situations are constructible with protons from scattering process, de-excitations of nuclei or atoms with cascades of temporally coincident photon emissions, or particle-antiparticle pair annihilation, which makes different experimental niceties possible (more on this later).

Now in the present two atom-setup, these atoms will travel in opposite directions after the emission and with their spins anti-aligned, so that the total spin of the pair is zero and the spin of the molecule is conserved. But there is seemingly nothing that predetermines which system will have its spin up and which one will have it

[33] It is of course open to debate whether one prefers to call something that is testable 'physics', and reserves the term 'metaphysics' for a priori investigations. But that is rather a matter of linguistic taste and intuition.

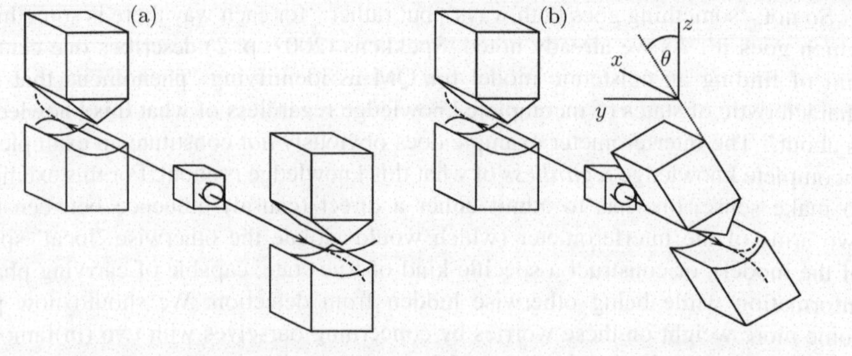

Fig. 4.7 **(a)** A (close to) perfect anti-correlation between two entangled spin-$\frac{1}{2}$ systems is observed when both Du Bois magnets are aligned along a common axis. **(b)** If the magnets are tilted relative to one another, the correlation will depend on the angle θ of relative tilt

down along a given axis. So we should have no particular expectations as to where we will find 'spin up' and where 'spin down' if measure both spins after some time of flight along a common direction. We saw, in our discussion of the Stern-Gerlach experiments, that for any given axis, it is apparently totally random whether a system will fly down or up in the magnet, or that we can at least not predict an individual system's behavior beyond the fact that it will do either of the two things. The randomness carries over to the present case, but the two systems will always end up having their spins anti-alinged, as long as we measure them along one common axis (cf. Fig. 4.7a).

Now something is already fishy here, since the systems could have their spins anti-aligned along an axis *different* from the one along which we are measuring. So why would they always anti-align along the axis that we *chose* to measure for as well? Imagine, more precisely, that we perform an alternative experiment by rotating both magnets simultaneously and by the same angle, possibly even after the emission. We will then find that in each run, no matter which axis we have chosen, somehow the two atoms always anti-align their spins and show the appropriate behavior for a total spin of 0.

The situation here is obviously similar to the Alice-Bob scenario that we discussed at the end of Sect. 4.2.4; we could have Alice sit at the left magnet and monitor the left atom's behavior, while Bob does the same thing on the right. But the invariance of the results under a rotation of the total setup should strike us as odd. How can the systems 'know' in advance which axis we are going to chose? Assume that they come out in a state where their spins are perfectly anti-aligned along an axis exactly perpendicular to the axis of the magnets. Should it then not be possible, and even happen on occasion, that the spins both flip into a spin-up configuration in the magnets, since for each individual atom the measurement has a random effect?

The quantum state appropriate to describe the spin degrees of freedom of the two atoms is the singlet state, $|\chi\rangle = \frac{1}{\sqrt{2}}(|\uparrow\rangle\,|\downarrow\rangle - |\downarrow\rangle\,|\uparrow\rangle)$, which we are familiar with from Sect. 2.1.3, and which we found to have value 0 for the square $\hat{\boldsymbol{S}}^2 = (\hat{\boldsymbol{s}} \otimes \mathbb{1} + \mathbb{1} \otimes \hat{\boldsymbol{s}})^2$ of the total spin operator $\hat{\boldsymbol{S}}$. Of course, we have to assign an axis to be able to say relative to what these spins will end up being down or up respectively. But as we saw, it does not matter *which* axis we choose; the state is rotation invariant. For now let us stick to the convention of taking the z-axis of some chosen coordinate system as our axis of measurement.

We can also do a second kind of experiment and rotate the magnets *relative* to one another (cf. Fig. 4.7b). What we will then observe is that the *rate* of systems with anti-aligned spins will go down (more generally: vary) as a function of the angle θ of relative tilt. We can quantify this by appealing (once more) to the spin observable for some arbitrary axis at angles θ and φ to the z-axis, $\hat{\boldsymbol{\sigma}} \cdot \boldsymbol{n}_{\theta\varphi}$ (cf. Eq. 2.19). But in our given example we are only tilting along the θ-angle, so we obtain (for $\varphi = 0$):

$$\hat{\sigma}_\theta = \begin{pmatrix} \cos\theta & \sin\theta \\ \sin\theta & -\cos\theta \end{pmatrix}.$$

It is not a difficult exercise (e.g. McIntyre 2012, p. 38 ff., for guidance) to verify that this matrix has eigenvectors $|{\uparrow_\theta}\rangle = \cos\frac{\theta}{2}\,|{\uparrow_z}\rangle + \sin\frac{\theta}{2}\,|{\downarrow_z}\rangle$ and $|{\downarrow_\theta}\rangle = \sin\frac{\theta}{2}\,|{\uparrow_z}\rangle - \cos\frac{\theta}{2}\,|{\downarrow_z}\rangle$ with eigenvalues ± 1 respectively, representing spin up/down along the tilted axis respectively. We can use this fact to calculate the probabilities for both spins being up in the tilted setup, as

$$\begin{aligned} \Pr{}^{\chi}_{z,\theta}(+_z, -_\theta) &= |\langle{\uparrow_z}, {\downarrow_\theta}|\chi\rangle|^2 = \frac{1}{2}\cos^2(\theta/2) = |\langle{\downarrow_z}, {\uparrow_\theta}|\chi\rangle|^2 = \Pr{}^{\chi}_{z,\theta}(-_z, +_\theta), \\ \Pr{}^{\chi}_{z,\theta}(+_z, +_\theta) &= |\langle{\uparrow_z}, {\uparrow_\theta}|\chi\rangle|^2 = \frac{1}{2}\sin^2(\theta/2) = |\langle{\downarrow_z}, {\downarrow_\theta}|\chi\rangle|^2 = \Pr{}^{\chi}_{z,\theta}(-_z, -_\theta), \end{aligned} \tag{4.22}$$

where we have abbreviated $\langle x|\langle y| = \langle x, y|$, $\sigma_j = j$, and $\pm_j 1 = \pm_j$ ($j \in \{z, \theta\}$). For now, we let the first argument in the probability function refer to the *left* system and consider it as 'system 1', which will keep the notation simple (we will make suitable adjustments later).

So far we have only varied the orientation of one of the two magnets, but we could equally allow for rotations of both magnets to new axes, defined by unit vectors $\boldsymbol{n}_a$ and $\boldsymbol{n}_b$. Letting one of the two magnets define the 'new z-axis' on each run, it becomes immediately clear that the probabilities depend only on the angle of *relative* tilt, θ_{ab}. This is equivalent to working in the eigenbasis of the operator representing the spin along the axis of one of the magnets.

We could now also arrange things such that the two magnets are sufficiently far apart to make the measurements spacelike separated, and that we rotate them only when both atoms have already been emitted from the source. In this way, we should be able to secure that the atoms have no way of 'communicating what to do'. I.e., drawing on the constraints set by special relativity, we would have to conclude that the two systems have no way of anti-aligning appropriately *after* having left the

Table 4.1 Possible settings for both quantum systems taken together for all three axes

Number	System 1	System 2
N_1	$(+_a, +_b, +_c)$	$(-_a, -_b, -_c)$
N_2	$(+_a, +_b, -_c)$	$(-_a, -_b, +_c)$
N_3	$(+_a, -_b, +_c)$	$(-_a, +_b, -_c)$
N_4	$(+_a, -_b, -_c)$	$(-_a, +_b, +_c)$
N_5	$(-_a, +_b, +_c)$	$(+_a, -_b, -_c)$
N_6	$(-_a, +_b, -_c)$	$(+_a, -_b, +_c)$
N_7	$(-_a, -_b, +_c)$	$(+_a, +_b, -_c)$
N_8	$(-_a, -_b, -_c)$	$(+_a, +_b, +_c)$

source, given that we are free to choose a setting of misalignment after the emission has taken place. Assuming that each atom always has some definite spin value, i.e., that there is a true state λ on each atom that fixes the spin properties, and that the spin state $|\chi\rangle$ is hence an incomplete description of the relevant degrees of freedom, we can now deduce a contradiction with the QM probabilities.[34]

Following Wigner (1970), d'Espagnat (1979), Sakurai (1994), and others, we can give a simple quick and dirty-argument to this effect as follows. Take three possible axes of space, defined by (all real multiples of the) unit vectors $\boldsymbol{n}_a, \boldsymbol{n}_b$, and $\boldsymbol{n}_c$, with different angles of misalignment θ_{ab}, θ_{ac}, and θ_{bc}. Then list all possible configurations which predetermine a given measurement result w.r.t. any of the given axes on each system, and count all such possible configurations (cf. Table 4.1).

This listing corresponds to assuming more definite states than QM ascribes; we are here assuming a hidden configuration λ (a set of hidden variables, the true state of the two systems) which respects the anti-alignment and assigns to each system some definite property that predetermines what the system will do in any measurement. The indices on the $+/-$ signs again refer to the axes. The numbers N_j $(j \in \{1, \ldots, 8\})$ are the cardinalities of the (imagined) sets of possible configurations that predetermine the desired behavior (assumed to be finite). From this list, we can infer the numbers of configurations which will give rise to the appropriate behavior in any given setting of the experiment. For instance, the number of configurations such that the first system has its spin up along the a-axis and the other one has its up along the b-axis is given by $N_3 + N_4$. Similarly, $N_2 + N_4$ is the number of configurations such that the first one has its spin up along a, and the second one up along c, and correspondingly for c and b with numbers $N_3 + N_7$. But since N_3 and N_4 both also occur in the last two expressions and N_2 and N_7 cannot become negative, it must hold that

[34] Of course we could also use a *quantum mechanically* more complete description here, by including spatial degrees of freedom etc. But this has no influence on the relevant predictions; it would only make the description more complicated, since the spatial quantum state for two indistinguishable systems has to be appropriately (anti-)symmetrized as well: two indistinguishable fermions, say, would here have to be described by a state such as $|\Psi\rangle = \frac{1}{\sqrt{2}}(|L\rangle\,|R\rangle + |R\rangle\,|L\rangle) \otimes |\chi\rangle$, where $|L\rangle$ and $|R\rangle$ are two states in position space with non-overlapping supports in $\mathbb{R}^3$, and $|\chi\rangle$ is the singlet state (e.g. Ghirardi et al. 2002, p. 81 ff; Ghirardi and Marinatto 2003, p. 384).

$$N_3 + N_4 \leq (N_2 + N_4) + (N_3 + N_7).$$

We can now straightforwardly evaluate the (a priori) probabilities p of the occurrences of the aforementioned settings by dividing the whole equation by the total number $\sum_{j=1}^{8} N_j$ of possible settings. These probabilities hence satisfy the *Bell-Wigner inequality*

$$p(+_a, +_b) \leq p(+_b, +_c) + p(+_c, +_b) \tag{4.23}$$

(e.g. Hughes 1989, p. 172; Sakurai 1994, pp. 229). But QM predicts differently. Take, for instance, $\theta_{ac} = \theta_{bc} = \pi/4, \theta_{ab} = \pi/2$. Then according to the QM probabilities we have found above (for two spins being up along two different axes), we would have

$$\begin{aligned} & p(+_a, +_b) \leq p(+_a, +_c) + p(+_c, +_b) \\ \overset{\text{QM}}{\Leftrightarrow} & \frac{1}{2}\sin^2(\theta_{ab}/2) \leq \frac{1}{2}(\sin^2(\theta_{ac}/2) + \sin^2(\theta_{bc}/2)) \\ \Leftrightarrow & \; 0,5 \leq 2\sin^2(\pi/8) \approx 0,29 \; \lightning \end{aligned}$$

The inequality which we have derived and shown to be violated according to QM straightforwardly demonstrates the incompatibility between QM and a set of assumptions associated with a certain class of hidden-variable theories. Such an equality is called a *Bell-type inequality*, after Bell (1964), and a whole family of similar such results is generally referred to as *Bell's theorem* (cf. Shimony 2009, p. 1). However, a more thorough derivation with much less contentious assumptions is possible, so we should not rush to conclusions about determinism, finite numbers of possible configurations...and so forth. But before we start analyzing the assumptions really required in more detail, we should first take a look at the theoretical and historical background of the theorem.

First of all, it is worth noting that the incompatibility is again derived in virtue of the existence of *entangled* states. These states have also played a crucial role in the derivation of the PBR theorem, and we can see why Schrödinger (1935a, p. 555) thought that entanglement is "*the* characteristic trait of quantum mechanics, the one that enforces its entire departure from classical lines of thought." (emphasis in original)

But not all entanglement is the same; one can distinguish different *strengths* of entanglement by assigning an *entanglement measure* (e.g. Jaeger 2007, p. 102 ff.). An important subclass of entangled states that can be readily defined without appeal to such a measure are the *maximally* entangled ones. Following e.g. Thaller (2005, p. 247), we can define this notion w.r.t. a compound system, formed of two systems '1' and '2' with respective Hilbert spaces of dimension n, by requiring that the reduced density operator of each system must satisfy

$$\mathrm{Tr}_{i\neq j}(\hat{\rho}_{12}) =: \hat{\rho}_j = \frac{1}{n}\mathbb{1}, \quad i, j \in \{1, 2\}, \tag{4.24}$$

where $\hat{\rho}_{12}$ is the density operator of the compound system, and $\mathrm{Tr}_{i\neq j}(\hat{\rho}_{12})$ means performing the trace operation w.r.t. the respective other system ($i \neq j$) (e.g. Basdevant and Dalibard 2002, p. 444, for more details). In other words: if the reduced density operator for each subsystem is maximally mixed, then the system is maximally entangled. Accordingly, we can also write $|\Psi\rangle = \frac{1}{\sqrt{n}}\sum_j |\phi_j\rangle \otimes |\psi_j\rangle$ for a *pure* maximally entangled state. A system which is made up of two subsystems is usually called *bipartite*; and maximally entangled states of two qubits are often called *Bell states* (e.g. Thaller 2005, pp. 216 and 247).

Based on this existence of different strengths of entanglement, one might hence conjecture that not all entangled states are 'entangled enough' to violate Bell-type inequalities. And indeed, to date "it is unknown whether there exist Bell inequality violations for many nonseparable mixed states." (Jaeger 2007, p. 93) A result which suggests that *not* all entangled states do is the counterexample provided by Werner (1989), in which a (mixed) entangled state is constructed that is 'local' in the sense of admitting a local hidden variable model (cf. also Vértesi and Brunner 2014, p. 2). But regarding this and similar examples, Vértesi and Brunner (2014, ibid.) note that "it turns out that [...] [i]f pre-processing by local operations and classical communication (LOCC) is performed before the local measurements, the 'hidden nonlocality' of some local entangled states can be revealed."

Moreover, for a whole class of mixed entangled states previously conjectured to be *incapable* of violating a Bell-type inequality (cf. Peres 1999, p. 609), so called 'bound entangled states', it has been demonstrated that some of them *do* violate Bell-type inequalities after all (cf. Vértesi and Brunner 2014, pp. 2–3).

Bound sates are such that they are not *distillable*, where

> [a] bipartite entangled state is said to be distillable if, from an arbitrary number of copies, it is possible to extract pure entanglement by LOCC. [...] The main open question now is whether all bound entangled states can give rise to Bell inequality violation, which would imply that entanglement and nonlocality are basically equivalent. (Vértesi and Brunner 2014, p. 2)

By local operations one means operations of the form $\mathbb{1} \otimes \mathbb{1} \otimes \ldots \otimes \hat{O} \otimes \mathbb{1} \otimes \ldots \otimes \mathbb{1}$, i.e. such operations which only affect one system in the total (possibly entangled) state of multiple systems. 'Classical communication' refers to all means of communication which do not straightforwardly require quantum mechanical considerations (such as e.g. telephone calls; cf. Audretsch 2007, p. 144).

Why is this of interest? The significance of Vértesi and Brunner's discovery for us here is that the astonishing implications of entanglement (violations of Bell-type inequalities) are quite widespread, much more so than previously thought. It is hence often—or possibly always—only a matter of experimental ingenuity, not of principled theoretical restriction, whether one can find evidence for the 'weird' features induced by entanglement. This underlines Schrödinger's aforementioned assessment of entanglement's significance.

Initially we noted that the general kind of experiment described in this section (albeit entirely without misalignment considerations) was first conceived of in a paper by Einstein and collaborators, usually referred to as the *EPR-paper*. Since this is instructive for the later discussion, we should spend a few lines on its content. The authors used an entangled wave function $\Psi(x_1, x_2) = \int \mathrm{d}p\, e^{i(x_1 - x_2 + x_0)p/\hbar}$, describing two particles with x_1 and x_2 the coordinates of the respective particles (at a given time), and x_0 a fixed distance. Using the fact that this function can either be viewed as a continuous superposition of products of momentum eigenfunctions $u_p(x_1) = e^{ix_1 p/\hbar}$ and $\psi_p(x_2) = e^{-i(x_2 - x_0)p/\hbar}$, or equally as a superposition of *position* eigenfunctions $v_x(x_1) = \delta(x - x_1)$ and $\varphi_x(x_2) = \int \mathrm{d}p\, e^{i(x - x_2 + x_0)p/\hbar} = h\delta(x - x_2 + x_0)$, they gave an argument that QM must be incomplete. To this end, they assumed the following two premises to hold:

(i) *Necessary condition for completeness* A physical theory can only be considered complete if "every element of the physical reality [has] a counterpart in the physical theory." (Einstein et al. 1935, p. 777)
(ii) *Reasonable criterion for reality* "If, without in any way disturbing a system, we can predict with certainty (i.e., with probability equal to unity) the value of a physical quantity, then there exists an element of physical reality corresponding to this physical quantity." (ibid.)

Since it is possible to predict, in virtue of the total function being an eigenfunction of the two *commuting* observables $\hat{x}_1 - \hat{x}_2$ and $\hat{p}_1 + \hat{p}_2$,[35] with certainty and without in any way disturbing it, the position *or* momentum of particle 2 from the measured result of either the position *or* the momentum of particle 1 respectively, both of these must be elements of reality by (ii). Hence, by (i), QM must be incomplete, since it does not imply nor even allow for the existence of precise values for both quantities, in virtue of the non-commutativity of the single particle operators (cf. also Jammer 1974, p. 182–185).

The EPR paper is sometimes thought of a presenting a 'paradox'. That seems incorrect, since it is just an argument for the incompleteness of QM, and using standard notation from first order logic and a garden-variety logical calculus, we can represent its logical structure quite directly. Thus, let P(λ) denote 'λ is predicted with certainty' (λ ranges over physical quantities), D(S) 'S is being disturbed' (S ranges over physical systems), and C(T) 'T is complete' (T ranges over theories). Let also p_2 be the precise value of momentum of particle 2, say, predicted from $\Psi(x_1, x_2)$ and the measurement result when particle 1 is measured for momentum,

[35] The operators $\hat{p}_{1/2}$ correspond to $-i\hbar\frac{\partial}{\partial x_{1/2}}$ respectively, yielding eigenvalue 0 for the total operator $\hat{p}_1 + \hat{p}_2$; i.e. $p_2 = -p_1$. The respective position operators yield x and $x + x_0$ respectively, so that $\hat{x}_1 - \hat{x}_2$ gives the distance x_0 between the two, and a position measurement on system 1 with result x will imply position $x + x_0$ for system 2 (see also Aharonov and Rohrlich (2005, p. 27) and Schrödinger (1935a, p. 559) on this point).

but which value 2 is supposed to *already have*. And let S_λ be the system to which a given λ pertains. Then the EPR argument (with p_2) has the following structure[36]:

(1)	$\forall T \forall \lambda [\mathrm{C}(T) \wedge \exists x(x = \lambda) \rightarrow \lambda \in T]$	Pr.
(2)	$\forall \lambda \forall S_\lambda [\diamond(\mathrm{P}(\lambda) \wedge \neg \mathrm{D}(S_\lambda)) \rightarrow \exists x(x = \lambda)]$	Pr.
(3)	$p_2 \notin QM$	from QM
(4)	$\mathrm{P}(p_2) \wedge \neg \mathrm{D}(S_{p_2})$	from QM, causal assumptions
(5)	$\diamond(\mathrm{P}(p_2) \wedge \neg \mathrm{D}(S_{p_2}))$	MP $\mathbf{T}_1$, (4)
(6)	$\exists x(x = p_2)$	UI (2), MP (5)
(7)	$\neg \mathrm{C}(QM) \vee \neg \exists x(x = p_2)$	UI (1), MT (3), De M.
(8)	$\neg \mathrm{C}(QM)$	DN, DS (7), (6)

As we can see, two 'lemmas' are needed to substantiate (3) and (4), that we will discuss shortly. $\mathbf{T}_1$ here is the statement that $q \rightarrow \diamond q$ ($\diamond$ being the weak modal operator with standard interpretation 'it is possible that...'), for some proposition q. It is a theorem of the modal system **T**, which is the second weakest discussed in Hughes and Cresswell (1996, cf. p. 42). So the modal commitments are not too worrisome here. That $p_2 \notin QM$ can be inferred from QM itself *in the orthodox interpretation* is due to the specification of p_2 as 'already existent', when the statefunction is $\Psi(x_1, x_2)$ and before the measurement. The EE-link prohibits this existence. This is the first 'lemma' to the proof: $\hat{p}_2\Psi(x_1, x_2) \neq \mu\Psi(x_1, x_2)$ for any $\mu \in \mathbb{C}$, therefore, by the EE-link, $\hat{p}_2$'s corresponding observable does not have a value in $\Psi(x_1, x_2)$.

EPR were also quite aware that "one would not arrive at [their—FB] conclusion if one insisted that two or more physical quantities can be regarded as simultaneous elements of reality only when they can be simultaneously measured or predicted." This would mean denying premise (2) in our reconstruction, the implication from predictability (without disturbance) to *existence*; the mention of the non-simultaneous measurability serves merely as a motivation for denial. The quote reflects EPR's awareness that their sufficient condition for attributing a 'physical reality' (which we should construe rather as a mind- or theory-independent reality) to the quantity that is measured is not without competition.[37] Arguably, many physicists at the time would have rather endorsed the view expressed in the last

[36]The notation '$\lambda \in T$' appealed to below is a bit sloppy, but it should be clear what is meant.

[37]The opinion that this is so however goes contrary to that of Maudlin (2014b, p. 6), who thinks that "the criterion is, in the parlance of philosophers, *analytic*." (emphasis in original) 'Predicting the value of a physical quantity with certainty' could mean to predict the outcome of some experiment which could still not be indicative of what the investigated system 'really did beforehand', which may be the targeted 'element of reality'. Moreover, predicting with certainty *on theoretical grounds* could have no actual experimental counterpart (incompatible experimental setups) and thus not *refer* to anything. These are reasons to doubt that the statement *is* analytic, and in fact we will see how to put these intuitions to work in Chap. 7.

quote, but to EPR "[n]o reasonable definition of reality could be expected to permit this." (Einstein et al. 1935, p. 780)

As we noted, an additional *causal assumption* needs to be made, in order to infer the second conjunct in (4). This causal assumption is the denial of a *direct causal influence* between the two systems: "since at the time of measurement the two systems no longer interact, no real change can take place in the second system in consequence of anything that may be done to the first system." (Einstein et al. 1935, p. 779) These additional premises to make the EPR argument work require, as we will see, some attention. This, however, is the second 'lemma': from $\hat{p}_1 + \hat{p}_2$, $\Psi(x_1, x_2)$, and a measurement on 1, we predict with certainty the value p_2, i.e. we have $\mathrm{P}(p_2)$, and by the causal restriction (no direct causal influence) $\neg\mathrm{D}(S_{p_2})$ follows.

It is on safe grounds now that Einstein was rather dissatisfied with the paper, mostly because the structure of the argument was obscured. In a correspondence with Schrödinger, he expressed his misgivings as follows: "For reasons of language, this was written by Podolsky after many discussions. But still it has not come out as well as I really wanted; on the contrary, the main point was, so to speak, buried by the erudition." (Einstein 1935, as cited in Howard 1985, p. 175) One point which was buried by the erudition is that two *different* ψ functions may correspond to the *same* piece of reality, which, as we saw earlier, is the very basis for considering ψ as epistemic in the sense of Conjecture 1. This point is present in the EPR paper (cf. their p. 779), but was most clearly pointed out in Einstein's 1948 article *Quanten-Mechanik und Wirklichkeit.*

In the above example from the EPR-paper, the two different ψ functions which could result on the *second* system, from choosing at will the measurement to perform on the *first* system, are the position- and momentum eigenfunctions $\varphi_x(x_2)$ and $\psi_p(x_2)$ for particle 2; and in our initial description of the EPRB experiment in this section they are the respective kets $|\uparrow_a\rangle$ or $|\uparrow_b\rangle$, say, given that 'spin down' was observed on the other system, and depending on whether they are jointly measured along either axis a or b. According to the projection postulate (Lüders' rule) some such state should result as a consequence of either of the measurements; again an expression of the 'remote steering' that Schrödinger was concerned about.

In *Quanten-Mechanik und Wirklichkeit* Einstein also made additional assumptions about the *separability*, *locality*, and '*reality*' (if you will) of the two systems quite explicit, which have to be made in order for the argument to go through. In particular, he believed that the 'defenders' of QM would be willing to give up the "requirement [...] of the *independent existence* of the *physically real*, in the *distinct parts of space* [...]." (Einstein 1948, p. 323; my emphasis; my translation—FB).[38] And this conveys a good intuition of what is meant by, and

[38]German original: "Es scheint mir keinem Zweifel zu unterliegen, dass die Physiker, welche die Beschreibungsweise der Quanten-Mechanik für prinzipiell definitiv halten, auf diese Ueberlegung wie folgt reagieren werden: Sie werden die Forderung [...] von der unabhängigen Existenz des in verschiedenen Raum-Teilen vorhandenen Physikalisch-Realen fallen lassen; sie können sich mit

at stake with, the aforementioned additional assumptions. The detailed analysis of these concepts, however, leads us directly back into the discussion of Bell's theorem, since ultimately the very same assumptions are at stake here, and they can be made formally precise.

Some have taken Bell to assume, in his 1964 paper, *determinism* as well as *locality*. But Norsen (2009, p. 274) and Maudlin (2010b, p. 123) emphasize that this is a misunderstanding, since Bell *derives* determinism from locality and the existence of (perfectly) correlated measurement results; and in *Bertlmann's socks and the nature of reality*, Bell (1981a, p. 143) himself laments that "[i]t is remarkably difficult to get this point across, that determinism is not a *presupposition* of the analysis." (emphasis in original)

Determinism can be spelled out as the outcomes of the measurements being *functions* of the settings and the values of the assumed hidden variables (Bell 1964, p. 15; cf. also Wiseman 2014, p. 5), and we have appealed to determinism in our quick and dirty derivation of the Bell-Wigner inequality above, by appealing to sets of configurations predetermining the outcomes on both systems. The locality constraint was formulated by Bell as the assumption "that the result B for particle 2 does not depend on the setting $\boldsymbol{n}_a$, of the magnet for particle 1, nor A on $\boldsymbol{n}_b$." (Bell 1964, p. 15; notation adapted) Einstein must have had in mind basically the same thing in writing that "an external influence on [system 1] has no *immediate* influence on [system 2] [...]." (Einstein 1948, pp. 321–322; emphasis in original; my translation—FB)

However, Bell's statement is more explicit on the nature of the influence on system 1 which should not influence system 2; namely, the *setting* of the other device (magnet) should not influence the remote system's state. Following Jarrett (1984, p. 572 ff., 1989, p. 69) and Wiseman (2014, p. 6), and expanding our notation to include reference to the side on which the outcome occurs, such as in e.g. L_a^+ for 'spin-up is measured on the left side for alignment along axis a',[39] we can reconstruct Bell's locality, which is also sometimes called *parameter independence*[40] after Shimony (1990, p. 35), as

$$p(A_j^x|i, j, \lambda, \chi) = p(A_j^x|j, \lambda, \chi), \ \forall i, j, x, \lambda, \chi, \tag{PI}$$

where $A \in \{L, R\}$, and $i, j \in \{a, b, c\}$ are the settings of the left and right device respectively, λ is an assumed (set of) hidden variable(s), $x, y \in \{+, -\}$

Recht darauf berufen, dass die Quanten-Theorie von dieser Forderung nirgends explicite Gebrauch mache."

[39]This kind of notation is also used by Graßhoff et al. (2005).

[40]The name 'parameter independence' is possibly misleading, since many things should certainly count as causal or probabilistic parameters. The intended 'parameter' here is the distant setting, whence Pawłowski et al. (2010, p. 2), for instance, use the name "setting independence" instead. We will however stick to the more widespread terminology.

are the outcomes, and χ the P-state.[41] But we have thus captured only part of the notion of locality implicit in Einstein's writing. Assuming that the obtainment of an outcome on one side of the experiment is in part determined by the action of the measurement-apparatus on the system, even if allowed to be in principle arbitrarily subtle, an influence between the obtainment of the *remote outcomes* should also be excluded; whence it is in accord with the above Einstein-quote to require an *outcome independence* as well:

$$p(A_j^x|B_i^y, i, j, \lambda, \chi) = p(A_j^x|i, j, \lambda, \chi), \quad \forall i, j, x, y, \lambda, \chi, \tag{OI}$$

where $A, B \in \{L, R\}$ and $A \neq B$ (cf. also Jarrett 1989, p. 69; Shimony 1990, p. 35).

Later (1971, 1976) Bell avoided the appeal to determinism entirely (even as a derived premise) and exclusively relied on probabilistic notions instead, most importantly a notion of *local causality* (cf. also Clauser and Horne 1974, p. 526). As in the discussion of the EPR paper, we here slide into considerations of *causality*, whereas originally we were only concerned with the existence of true states of systems beyond those ascribed by QM, i.e. the structure of mind-independent reality (mostly) at the 'micro-level'.

The tension between these two aspects is also noticed by Cartwright (1989, p. 237), who writes:

> one needs to have a clear idea what purposes a hidden-variable theory is supposed to serve. Why want hidden variables? There are two distinct answers: to restore realism in quantum mechanics; and to restore causality.

But we hold here that the two notions are more intimately connected than Cartwright apparently thinks. Suppes (1998, p. 247), for instance, describes the kind of hidden variables in question as "causes that cannot be observed but that satisfy more classical assumptions in generating quantum mechanical phenomena." Hence questions about the very *existence* of certain unobservable entities that *also* figure as causes of observed phenomena are at stake, i.e. questions regarding what we believe reality 'beyond appearances' to be. This conveys intuition enough so as to see the role of hidden variables as restoring causation *and* a kind of realism: they act as *causes* for *observed phenomena* and observed *correlations*, and they thereby ensure the *interpolability* of correlated phenomena by appeal to unobserved entities or events—Reichenbach's (1944, p. 21) *interphenomena*—that one supposes to be assessable in a suitably 'classical' manner.

Among other things, the question immediately arises whether scientific theories are able to *refer in all domains* (observable and unobservable); and denying this

[41] In contrast to e.g. Wiseman (2014) and Bell (1990b), and in the spirit of our above discussion of OMs, we have omitted direct reference to a preparation procedure P and instead only appealed to the quantum state χ, interpreted as a P-state. P would denote "the values of any number of other variables describing the experimental set-up, as admitted by ordinary quantum mechanics [...]", (Bell 1990b, p. 108) and would hence add no relevant information beyond χ in this context.

would mean denying the semantic condition for scientific realism (cf. interlude I). A subtlety is involved here though: λ *need not* be a 'more classical' cause, but could be *the quantum state* χ *itself*. We will see in interlude II, however, why this is difficult to maintain as well, and the interpretation ultimately settled for in this book will accept neither χ *nor* some λ as a cause of the correlations.

The *probabilistic content* of local causality, however, was originally spelled out by Bell (1976, p. 54) as

$$p(A_j^x|B_i^y, i, j, \lambda, \chi) = p(A_j^x|j, \lambda, \chi), \ \ \forall i, j, x, y, \lambda, \chi, \tag{LC}$$

again using our notation and again $A \neq B$ (cf. also Norsen 2011, p. 1270). Hence, (LC) can be viewed as the conjunction of (PI) and (OI). Since in general $p(A|B) = p(A, B)/p(B) \Leftrightarrow p(A, B) = p(A|B)p(B)$ (assuming $p(B) > 0$), and $p(A_j^x|i, j, \lambda, \chi) = p(A_j^x|j, \lambda, \chi)$ in virtue of (PI), one obtains what has become known as a (local) *factorization condition*:

$$p(R_j^x, L_i^y|i, j, \lambda, \chi) = p(R_j^x|j, \lambda, \chi) \cdot p(L_i^x|i, \lambda, \chi), \ \ \forall i, j, x, y, \lambda, \chi. \tag{FACT}$$

This is the central consequence of (LC) which is often appealed to directly in the discussion of Bell's theorem (e.g. Bell 1990b, p. 109; Clauser and Horne 1974, p. 528; Graßhoff et al. 2005, p. 666; Norsen 2011, p. 1270; Wiseman 2014, p. 13).

Next to the fact that (FACT) plays a key role in a thorough derivation of a Bell-type inequality, it also makes the link to established principles of (probabilistic) causality theory most obvious, to some of which we will return below. To derive a Bell-type inequality, we need to add to (FACT) the assumption that λ itself is independent of the measurement settings:

$$p(\lambda|i, j, \chi) = p(\lambda|\chi), \ \ \forall i, j, \chi, \lambda. \tag{AUT}$$

'AUT' is short for *autonomy* (cf. Friebe et al. 2015, p. 141; van Fraassen 1982b, p. 31),[42] and this merely corresponds to the assumption that there is no 'cosmic conspiracy', as it were, since the settings of the measurement devices on each side could be chosen during flight, and randomized (say) by a computer, so that it would indeed amount to a weird kind of conspiracy or "superdeterminism" (Bell 1990b, p. 110), if λ was dependent on the settings, i and j (cf. also Friebe et al. 2015, p. 168; Shimony 2009, p. 15 ff.). Note also that if $p(\lambda|i, j, \chi) \neq p(\lambda|\chi)$, then $p(i, j|\lambda, \chi) \neq p(i, j|\chi)$, assuming, as must be the case here, that $p(i, j|\chi) \neq 0$. I.e., we would also have dependence of the settings on λ which even more so amounts to a conspiracy or superdeterminism.

[42] In contrast to Friebe et al. (2015, p. 141), we have allowed for λ to depend on χ, since the OM approach requires this to be possible: χ is construed as the P-state therein, a representation of what was done to the system in a preparation procedure.

From (AUT) and (FACT), one of the most important Bell-type inequalities can be derived in a straightforward manner, namely the *Clauser-Horne-Shimony-Holt* (CHSH) *inequality*, named after Clauser et al. (1969).[43] First note that for $p, q, r, s \in [-1, 1]$, it holds that $-2 \leq pr + ps + qr - qs \leq 2$ (you can convince yourself of this fact or look up the proof e.g. in Shimony 2009, p. 5). But all the expectation values of the spin (or photon) EPRB-experiments satisfy the requirement that they lie in the interval $[-1, 1]$ due to the nature of the outcomes ± 1 (in the appropriate units). Hence, taking averages w.r.t. different settings for the respective sides of the experiment, we obtain (by distributivity of sums and the definition of an average)

$$\begin{aligned}-2 \leq &\sum_{x,y} xyp(R_j^x|j, \chi, \lambda)p(L_i^y|i, \chi, \lambda) + \sum_{x,y} xyp(R_j^x|j, \chi, \lambda)p(L_{i'}^y|i', \chi, \lambda)+ \\ &+ \sum_{x,y} xyp(R_{j'}^x|j', \chi, \lambda)p(L_i^y|i, \chi, \lambda) - \sum_{x,y} xyp(R_{j'}^x|j', \chi, \lambda)p(L_{i'}^y|i', \chi, \lambda) \leq 2,\end{aligned} \tag{4.25}$$

with $j', i' \in \{a, b, c\}$ measurement settings, and everything else as before. We can rewrite this, in virtue of (FACT), as

$$\begin{aligned}-2 \leq &\sum_{x,y} xyp(R_j^x, L_i^y|j, i, \chi, \lambda) + \sum_{x,y} xyp(R_j^x, L_{i'}^y|j, i', \chi, \lambda)+ \\ &+ \sum_{x,y} xyp(R_{j'}^x, L_i^y|j', i, \chi, \lambda) - \sum_{x,y} xyp(R_{j'}^x, L_{i'}^y|j', i', \chi, \lambda) \leq 2.\end{aligned} \tag{4.26}$$

We require also that $p(\lambda|\chi)$ is normalized, i.e., that the integral of $p(\lambda|\chi)$ over all λ is 1, which is a trivial requirement since $p(\lambda|\chi)$ is supposed to be a probability density and χ represents a preparation procedure with resulting λs. Thus, multiplying the whole inequality by $p(\lambda|\chi)$, we obtain

$$\begin{aligned}-2p(\lambda|\chi) \leq &\sum_{x,y} xyp(R_j^x, L_i^y|j, i, \chi, \lambda)p(\lambda|\chi) \\ &+ \sum_{x,y} xyp(R_j^x, L_{i'}^y|j, i', \chi, \lambda)p(\lambda|\chi)+ \\ &+ \sum_{x,y} xyp(R_{j'}^x, L_i^y|j', i, \chi, \lambda)p(\lambda|\chi) \\ &- \sum_{x,y} xyp(R_{j'}^x, L_{i'}^y|j', i', \chi, \lambda)p(\lambda|\chi) \leq 2p(\lambda|\chi),\end{aligned} \tag{4.27}$$

[43] For the following proof see also Shimony (1990, p. 34 ff., 2009, p. 5 ff.) or Friebe et al. (2015, p. 142).

which we can rewrite, in virtue of (AUT), as

$$-2p(\lambda|\chi) \leq \sum_{x,y} xyp(R_j^x, L_i^y|i, j, \chi, \lambda)p(\lambda|i, j, \chi)$$
$$+\sum_{x,y} xyp(R_j^x, L_{i'}^y|i', j, \chi, \lambda)p(\lambda|i', j, \chi)+$$
$$+\sum_{x,y} xyp(R_{j'}^x, L_i^y|i, j', \chi, \lambda)p(\lambda|i, j', \chi)$$
$$-\sum_{x,y} xyp(R_{j'}^x, L_{i'}^y|i', j', \chi, \lambda)p(\lambda|i', j', \chi) \leq 2p(\lambda|\chi). \quad (4.28)$$

By appeal to the definition of conditional probability we obtain

$$-2p(\lambda|\chi) \leq \sum_{x,y} xyp(R_j^x, L_i^y, \lambda|i, j, \chi) + \sum_{x,y} xyp(R_j^x, L_{i'}^y, \lambda|i', j, \chi)+$$
$$+\sum_{x,y} xyp(R_{j'}^x, L_i^y, \lambda|i, j', \chi) - \sum_{x,y} xyp(R_{j'}^x, L_{i'}^y, \lambda|i', j', \chi) \leq 2p(\lambda|\chi), \quad (4.29)$$

and from the normalization condition on $p(\lambda|\chi)$, we obtain, by integration of the inequality over all λ (marginalization for λ),

$$-2 \leq \sum_{x,y} xy \int \mathrm{d}\lambda\, p(R_j^x, L_i^y, \lambda|i, j, \chi) + \sum_{x,y} xy \int \mathrm{d}\lambda\, p(R_j^x, L_{i'}^y, \lambda|i', j, \chi)+$$
$$+\sum_{x,y} xy \int \mathrm{d}\lambda\, p(R_{j'}^x, L_i^y, \lambda|i, j', \chi) - \sum_{x,y} xy \int \mathrm{d}\lambda\, p(R_{j'}^x, L_{i'}^y, \lambda|i', j', \chi) \leq 2, \quad (4.30)$$

so that we finally have

$$-2 \leq \sum_{x,y} xyp(R_j^x, L_i^y|i, j, \chi) + \sum_{x,y} xyp(R_j^x, L_{i'}^y|i', j, \chi)+ \quad \text{(CHSH)}$$
$$+\sum_{x,y} xyp(R_{j'}^x, L_i^y|i, j', \chi) - \sum_{x,y} xyp(R_{j'}^x, L_{i'}^y|i', j', \chi) \leq 2.$$

The expressions in this inequality correspond to averages of products of observables *independent* of λ, and may hence be compared with the averages that QM predicts. In virtue of the angle dependence of QM probabilities (and *a fortiori*: averages), this inequality is violated according to QM for the appropriate choices of angles between the three settings (cf. Shimony 2009, p. 8).

We can see how this derivation of the CHSH inequality has much stronger implications than our quick and dirty one of the Bell-Wigner inequality, since it rests on much weaker assumptions. We did not appeal to determinism here, unlike above; we did not appeal to questionable a priori probabilities in the sense of classical probability theory, which was also a contentious point in our above derivation (cf. also Home and Selleri 1991, p. 22 on this point). All we assumed was the absence of cosmic conspiracies and a form of independence of both results from remote (space-like separated) influences; each assumption spelled out in form of a probabilistic requirement, (AUT) and (FACT).

But this, again, raises the question of the *background assumptions* for (AUT) and (FACT), and the validity of connecting the informal assumptions to desired formal premises. We saw that (LC) can be viewed as (is equivalent to) the conjunction of (OI) and (PI). An underlying assumption that in turn motivates these two (pairs of) *probabilistic* formulae is a *causal* notion, sometimes referred to as *causal Einstein locality* (e.g. Friebe et al. 2015, p. 128; Hofer-Szabó and Vecsernyés 2014, p. 1):

Causal Einstein Locality (CEL) Causal influences do not propagate faster than the speed of light.

An attitude of this type is also often found in textbooks on relativity, such as e.g. Adams (1997, p. 138 ff.) or Taylor and Wheeler (1963, p. 39). It can equally be understood as a formulation of *Bell's* own *intuitive* notion of local causality (cf. Bell 1976, p. 54). And, most importantly, (CEL) is not unmotivated because, due to the lack of an overarching notion of *simultaneity* in SR, denying (CEL) would open up the possibility of effects preceding their causes, and ultimately that of contradiction provoking *causal loops* (cf. Maudlin 2011, pp. 142–143 and the discussion in interlude II).

Since we have assumed that the two magnets (measuring devices) may be at a large distance, and the settings may be changed at any time during the flight, one can see how (CEL) motivates (PI) and (OI). It thus seems that one is at liberty to give up either (PI) or (OI) in the light of violations of Bell-type correlations, i.e. the dependence of one outcome on its distant partner or the dependence of outcomes on the distant settings. But this impression has been found wanting for several good reasons.[44]

Dickson (2007, p. 391) e.g. points out that "the claim is often made that a failure of Outcome Independence is somehow consistent with relativity, while a failure of Parameter Independence is not." This claim is then typically substantiated by the belief that "experimenters are in control of parameters—they are in fact normally assumed to be the result of a free choice of the experimenter." (ibid.; emphasis omitted.) This kind of reasoning, however, has been found to be flawed by several authors (cf. Dickson 2007, p. 391; Friebe et al. 2015, p. 151; Maudlin 2011, p. 88 ff.), the main reason being that

[44]Maudlin (2011, p. 87) also demonstrates that (FACT) can equally be derived from two different formulae, which could claim equal right to be called 'parameter-' and 'outcome independence'. So the exact *formalization* of the two intuitive requirements is already questionable.

> the probabilities in Parameter Independence and Outcome Independence are those generated by the hidden state, λ. If the experimenter is not in control of these hidden states, then a failure of Parameter Independence will *also* not imply the possibility of signaling. Moreover, control of the hidden states would mean that in fact a violation of Outcome Independence *also* implies the possibility of signaling, so long as the probabilities for the outcomes generated by different hidden states are different. (Dickson 2007, p. 391; emphasis in original)

In short, the distinction between (OI) and (PI) as the two guiding principles for the derivation of a Bell-type inequality may be misleading, and the case for rejecting (PI) while keeping (OI) is rather weak.

More importantly, Näger (2013b) derives a theorem according to which models in which "at least one of the factors [in some factorization of the joint probability for both systems—FB] involves space-like separated variables, but none of the factors involves both parameters [...] imply Bell inequalities [...]." (p. 10) Thus *some* (probabilistic) dependence on the distant parameter must be assumed once outcome dependence is allowed, on account of Näger's investigation. A similar result is also derived by Pawłowski et al. (2010), who "show that information about a distant setting and outcome is not only sufficient to simulate violation of Bell's inequality but also necessary"; and Maudlin (2011, p. 164 ff.) gives some additional informal arguments to a similar effect.[45]

What these objections conjointly show is that one cannot straightforwardly avoid the consequences of Bell's theorem by letting the outcomes influence each other (causally). This neither invalidates (OI) and (PI) *qua* formulae nor their in-principle capability to reflect independence of outcomes from the remote outcomes or settings. What the complaints show in summary is that "the usual verdict" (Butterfield 2007, p. 828), that denying (OI) and allowing a direct causal influence among outcomes is the appropriate way to view the situation, is at best doubtful.

In spite of this we are still left with the task of thoroughly connecting causal and probabilistic notions. This is so because causation and correlation are, of course, two separate issues. For instance, "the correlation between 'heads up' and 'tails down' will not stand in need of an explanation in terms of a common cause[...]." (Wroński 2014, p. 9) That the correlations in EPRB experiments are not of this sort needs to be postulated, and is a consequence of the Einsteinian separability assumption that we have already met with in the quote above. Following Howard (1989, pp. 226–227), we can state it more precisely as follows:

Einstein Separability (SEP) The contents of any two regions of space-time separated by a non-vanishing spatiotemporal interval constitute separable physical systems in the sense that (1) each possesses its own, distinct physical state, and (2) the joint state of the two systems is wholly determined by these separate states.

---

[45]Näger's result is more straightforwardly concerned with causal influences and the latter results concern information, so to count these arguments as in favor of the same thing, a case has to be made that causation and information are related in an appropriate manner. Cf. Näger (2013a, p. 42 ff.) for discussion on these issues.

Howard (ibid.) construes this as "a fundamental ontological principle governing the individuation of physical systems and their associated states, a principle implicit in many classical physical theories", which seems like an apt characterization. Einstein held this principle dearly and thought that "[w]ithout the assumption of such a mutual independence of existence (of a 'being thus') of the spatially remote things, which stems in the first place from everyday life thinking, physical thinking in the sense familiar to us would not be possible." (Einstein 1948, p. 321; my translation—FB)[46]

(SEP) is obviously also quite similar in content to (Sep.) from the PBR theorem, whence both carry the name of a separability assumption. Yet (SEP) is explicitly concerned with *spatiotemporal* notions, whereas (Sep.) is merely used to proclaim an independence in preparation (thereby however *presupposing* (SEP) for any reasonable preparation; cf. Fig. 4.5).

Using probability theory, a *correlation* between events A and B can be spelled out by requiring that $p(A, B) \neq p(A)p(B)$, because then $p(A|B) \neq \frac{p(A)p(B)}{p(B)} = p(A)$, so that A actually does depend probabilistically on B (and analogously B on A). The events are called *positively correlated* if $p(A, B) > p(A)p(B)$ and negatively correlated if $p(A, B) < p(A)p(B)$ (cf. Wroński 2014, p. 3). We clearly get a correlation from the quantum probabilities in the EPRB case, since $\Pr^{\chi}_{s_j}(R^x_j) = \frac{1}{2} = \Pr^{\chi}_{s_i}(L^y_i)$, which follows from the respective reduced density matrix for $|\chi\rangle = \frac{1}{\sqrt{2}}(|\uparrow\rangle|\downarrow\rangle - |\downarrow\rangle|\uparrow\rangle)$, but the *joint* probability of the two events is given by trigonometric functions, generally unequal to $\frac{1}{4}$, as we have seen above.

The move from correlation to causation however requires the appeal to a bridging principle, and just such a bridging principle is Reichenbach's *Principle of the Common Cause* (PCC), supposedly first suggested by van Fraassen (1982b) to relate to EPRB correlations.[47] Reichenbach himself (1965, p. 157 ff.) gave an informal statement of the PCC as follows: "If an improbable coincidence has occurred, there must exist a common cause." But soon enough in the following, Reichenbach explicates that he is not interested in a treatment of *single* coincidences, but of such cases where "the simultaneous happening of A and B is more *frequent* than can be expected for chance coincidences" (my emphasis—FB), and where, if "the coincidence of A and B [...] has a probability exceeding that of a chance coincidence, we assume that there exists a common cause C." (Reichenbach 1965, p. 157 ff.) Today, differing exact formulations of the principle exist in abundance,

---

[46]German original: "Ohne die Annahme einer solchen Unabhängigkeit der Existenz (des 'So-Seins') der räumlich distanten Dinge voneinander, die zunächst dem Alltags-Denken entstammt, wäre physikalisches Denken in dem uns geläufigen Sinne nicht möglich."

[47]Terminological warning: One sometimes encounters a differentiation between EPR-correlations and Bell-correlations, the former denoting the *perfect* correlation implied for the EPR-state or measurements along the same axis on the singlet, the latter referring to the precise correlations appealed to in Bell-inequalities, i.e. with different misalignment angles and *violated* by QM (e.g. Maudlin 2010b, p. 124). When we use the term 'EPRB'-correlations, the 'B' stands for Bohm; but we allow to include the correlations *predicted by QM* for different *misalignment angles*, i.e. the 'violating' correlations rather than the violated ones.

and with a variety of different implications (cf. Wroński 2014 for discussion). We shall here only sketch the relevant 'nuances'. The simplest basic statement of the PCC is the following (e.g. Butterfield 2007, p. 818):

Principle of the Common Cause (PCC) Given two events A and B which are correlated, i.e. $p(A, B) \neq p(A)p(B)$, and where neither causes the other, there is an event C conditional on which A and B are independent: $p(A, B|C) = p(A|C)p(B|C)$. C is then said to 'screen off' A from B (and vice versa).[48]

In the present context, the common cause should be identified with the two true states of the two systems after the preparation as χ, i.e. after the decay process. Of course for the PCC to apply we have to assume here that the correlation is not explained by some other kind of dependence; but one possible candidate for dependence was ruled out already by the assumption of (SEP). We have assumed that the two events are neither one and the same, nor are they like the two sides of a coin whose up-down correlation need not be explained by a common cause since it is basically analytic.

Now why exactly should the (anti-)correlation between the two remote spin values be of a *causal* nature? A quite natural (and compelling) response here would be to counter: 'Well, what else could it be?' But let us motivate this attitude a little more strongly. Compare the correlation between the distant spins to that between the birds' shadows in *Reichenbach's cube*: Reichenbach (1961, p. 115 ff.) imagined the whole of mankind to live inside a cube with translucent walls. Outside the cube's walls there would be birds whose shadows would be reflected, by a system of mirrors, onto those walls in such a way that the shadow of one and the same bird would always appear simultaneously on two adjacent walls. Reichenbach imagined a brilliant mind, a 'Copernicus', to figure out, by watching the correlated behavior of these pairs of shadows, that they were "nothing but effects caused by one individual thing situated outside the cube within free space." (p. 118) Just as the surprising correlation between the two shadows on adjacent walls is *explained* in a satisfactory way as soon as it is realized that they are *caused* by the absorption of light reflected off a mirror by one and the same bird's body, so, it seems, would the (anti-)correlated spin measurements be explained in a satisfactory way if understood as caused by *something* unobserved which pertains to the two decay products right after the decay, a hidden configuration λ which assigns a definite, local configuration to each system individually and makes them behave in that particular manner.

Suggestive as the image may be, the PCC leads to the local factorization condition (FACT). Because given (CEL) and (SEP), and given that the setup is such that neither settings nor outcomes can influence one another *via* luminal or subluminal signals, (PCC) requires us to assume that there must be some λ (most likely just the trues states of the two decay products) which causes the

---

[48]Reichenbach also required that (i) $\neg C$ would equally screen off A and B, and that (ii) $p(A|C) > p(A)$, $p(B|C) > p(B)$. But as Butterfield (2007, p. 818) remarks, (ii) is simply appealed to by Reichenbach to account for positive correlation, and we are here equally interested in negative correlations. And the screening off by $\neg C$ will be replaced shortly by a more general constraint.

correlation, and given which each measurement only depends (statistically) on the local parameter and the past preparation.

Moreover, a significantly weaker version of (PCC) suffices to prove a Bell-type inequality. In the version we have given above, all the correlated pairs have one and the same cause C. But one could, first of all, think of C as a (random) variable $\underline{\lambda}$ with value space Λ and *multiple* values λ^{ℓ}, or equally in terms of an algebra over the induced partition Λ of an underlying probability space Ω with *multiple* cells λ^{ℓ} (e.g. Butterfield 2007, p. 834). This allows for a generalization of the PCC, which was originally formulated by Reichenbach with regard only to C and $\neg C$, not multiple values (cf. footnote 48). But additionally, it allows to devise the following *weaker* formulation of a common cause principle (cf. Butterfield 2007, p. 834; Portmann and Wüthrich 2007, p. 848):

Weak Principle of the Common Cause (PCC*) Given two sets of events $\{A_m\}_{m\in M}$ and $\{B_m\}_{m\in M}$ which are correlated, i.e. $\exists m \in M : p(A_m, B_m) \neq p(A_m)p(B_m)$, and where neither causes the other, then for every $m \in M$ there is a variable $\underline{\lambda}_m$ with values $\{\lambda^{\ell}_m\}_{\ell\in L}$ conditional on which all A_m and B_m are independent: $p(A_m, B_m|\lambda^{\ell}_m) = p(A_m|\lambda^{\ell}_m)p(B_m|\lambda^{\ell}_m)\ , \forall m \in M, \ell \in L$.

Here M and L are (countable) indexing sets, and the number of ℓs will depend on M. Such a partition or random variable is sometimes also referred to as a *screener system* (cf. Wroński 2014, p. 35). This version is weaker in that we have switched from 'there is a... such that for all' to the reversed order of the quantifiers. Crucially, Portmann and Wüthrich (2007) still derive an error tolerant version of the CHSH inequality even from (PCC*).

One central point in our discussion of the EPRB correlations that has a strong bearing on the assumption of hidden variables *in general* (a fortiori: on the assumption of true states) is that these be *local*, an assumption invoked in particular in deriving (FACT) from (PCC*). Hence if one is in for an explanation of the correlations in EPRB experiments in terms of hidden variables, this explanation must involve *nonlocal* variables (cf. Fig. 4.8). *Prima facie* this provokes a conflict with SR and its locality constraints, and if one believes that this is not so, one has to give an elaborate account of why not. A significant portion of Maudlin's (2011) book may be seen as an attempt to do just that, and we refer the interested reader there for further reference.

We emphasize at this point that a lot more needs to be said about causation, nonlocality, and realism, and in particular how (if at all) these relate to one another; and we shall say at least a bit more in interlude II. An option that we have hitherto not discussed is to give up (AUT), i.e. accept Bell's superdeterminism or a kind of cosmic conspiracy; but we deliberately choose to deem this option 'implausible' (as do many commentators) and not explore it any further. The reader may be referred to e.g. Vervoort (2013) and references therein for views more affirmative of this option.

So far we have discussed the matters at hand on merely theoretical grounds, but of course all of this would be meaningless if there was no experimental evidence for the violation of Bell-type inequalities. The most prominent confirmation of such

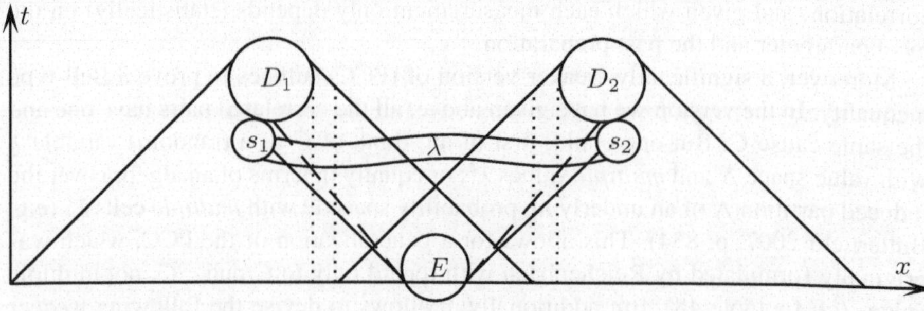

Fig. 4.8 Spacetime diagram of an EPRB experiment with light cones. E is the emission event, $D_{1/2}$ are detection events, and $s_{1/2}$ are setting events. The dashed lines indicate the spacetime trajectories of the two emitted particles. The dotted vertical lines indicate the spatial separation which an at most luminal signal would still have to travel from one side to the other in order to influence the remote detection event. The spread out region labeled λ indicates the nonlocal hidden variable (or cause) which could account for the correlation by instantaneously communicating settings and outcomes (Cf. also Norsen 2009, p. 282 for a similar illustration)

violations is certainly the experiment by Aspect et al. (1982). The authors used the entangled polarizations of coincidentally emitted photons, resulting from the de-excitation of calcium into two subsequent energy levels, and traveling collinearly in opposite directions. Because these photons would be in an entangled state with equal polarizations, this would allow them to count coincidences for given misalignments of analyzers, in order to check for the violation of a Bell-type inequality.

The trick they used to exclude an at most luminal interaction (or signal) was to use spacelike separated *time varying analyzers* (polarization filters), more specifically, a setup in which the photons would be either deflected or transmitted by an optical switch, then ending up in differently oriented analyzers, depending on the setting of the switches. Each of the switches was connected to a different ultrasonic standing wave (with different frequency) used as a quasi-randomizer. The switching occurred so fast (approximately every 10 ns), and the lifetime of the intermediate level of the cascade was so short (approx. 5 ns) that the switching events happened during the flight of the photons (approx. 40 ns). In this way, Aspect et al. found violations of a Bell-type inequality by five standard deviations.[49]

This experiment certainly constitutes an impressive and well thought-out scheme for confirming the violations of Bell-type inequalities. But of course, there are still restrictions. One obvious restriction is the randomization through the two standing sound waves, which were "not truly random, but rather quasiperiodic." (Aspect et al. 1982, p. 1807) However, since these were independent and at different frequencies, it would amount to a kind of conspiratorial assumption to suspect them of fabricating the results.

[49]We remark here that today experiments have been realized in which a violation of a Bell-type inequality was reported using photons in a fiber that allowed them to be separated by a distance >300 km (Inagaki et al. 2013).

The more serious threat is the so-called *detection loophole*. Consider a non-idealized setup, where the detectors will not work perfectly. Then we could have a null outcome Ø, just as in the PBR case, and the probabilities used to judge the violation of a Bell-type inequality would have to be 'postselected' probabilities of the form

$$p_{ps}(R_j^x, L_i^y|i, j, \lambda, \chi) = p(R_j^x, L_i^y|R_j^x \neq \emptyset, L_i^y \neq \emptyset, i, j, \lambda, \chi) = \\ = \frac{p(R_j^x, L_i^y|i, j, \lambda, \chi)}{p(R_j^x \neq \emptyset, L_i^y \neq \emptyset|i, j, \lambda, \chi)}, \quad (4.31)$$

i.e. conditionalized on obtaining an outcome at all (e.g. Branciard 2011, p. 2).

Arthur Fine (1982) has suggested two models which exploit this detection loophole. Since we cannot be sure that all photons which are produced are really measured, one could assume that some photons simply do not provoke a response in the detector (or rather provoke the response Ø). If this happens in a systematic way (i.e. as a built-in inefficiency due to the system, just as in the PBR case), it is possible to reconstruct the quantum probabilities without a violation of locality *or* separability. Using response functions with selective responsivity depending on the angular configuration of a given analyzer (or Du Bois magnet), Fine first defines a '*minimal model*', in which "each particle is targeted to be responsive to exactly one analyzer position." (p. 286) He then improves on this restricted model with a '*maximal*' one, in which from a total of four different measurement settings for the two arms R, L of the experiment only three out of four response functions for the given measurements are simultaneously defined on certain measurable sets of λs (cf. Fine 1982, p. 287).[50]

Strictly speaking, excluding the possibility of such models would require a very high efficiency, i.e., making sure that almost all produced pairs of entangled systems would also be measured. There was a period (cf. Grangier 2001, p. 775) when the detection loophole could be closed only on the pains of opening up a locality loophole, i.e. where (sub-)luminal signals would be possible in the experiments. One might wonder if that really matters, though, as it seems highly questionable that photons in one kind of EPRB experiment should refuse to provoke responses in the measurement device just in order 'fake' the quantum predictions that atoms in another kind of experiment create by exchanging (sub-)luminal signals.

Both loopholes have, in fact, been closed for a while now by appeal to so called *steering inequalities*. To understand the gist, consider again Alice and Bob with Alice sending Bob one out a pair of systems which *could* be entangled. If they are, Alice can claim to 'steer' Bob's system into a quantum state by performing a measurement on hers. Assuming that the two systems are *independent* instead, Bob

[50]Both models are also discussed at length and improved in Maudlin (2011, p. 160 ff.), and the elaborations and drawings therein are instructive. See also Maudlin's criticism of this kind of model (his pp. 165–166).

can work out how "well-correlated Alice's prediction can be with his outcome[...]." (Wittmann et al. 2012, p. 3) In doing so, he can then determine a lower bound for quantifying possible correlations, which is known as a *steering bound* (cf. ibid.), and which gives rise to a corresponding inequality. If that inequality is violated, the pairs must be entangled and the 'steering' happens. In the experiment by Wittmann et al. (2012, p. 4), such a steering inequality which already included error-terms (detection malfunction on either side of the experiment) was violated in a setup with spacelike separation.

Steering, however, is weaker (fulfilled by more states) than the violation of Bell-type inequalities (cf. Wiseman et al. 2007). It is hence a welcome addition that recently a bunch of reports on experiments with entangled photons (Giustina et al. 2015; Shalm et al. 2015) or electron spins (Hensen et al. 2015) have been published, in the latter case with measurements separated by a (Euclidean) distance of 1,3 km in the laboratory frame, in which the experimenters claim loophole free violations of Bell-inequalities.

4.3.4 The Kochen-Specker Theorem and Contextuality

After our lengthy discussion of Bell's theorem and associated issues we can now, in as much brevity as possible, turn to another 'classic' theorem of QM. A version of this theorem was *first*, in fact, proven by *Bell* (1966) as well, but independently by Simon Kochen and E. P. Specker in 1967 (short: KS). For this reason, the theorem is sometimes called 'Kochen-Specker-theorem', sometimes 'Bell-Kochen-Specker-theorem' (e.g. Mermin 1993, p. 806). While acknowledging the existence of Bell's earlier proof, we shall simply use the acronym 'KS'.

"The thrust of this theorem", as Peres (1991, p. L175) puts it,

> is that any purported cryptodeterministic theory which would attribute a definite result to each quantum measurement, and still reproduce the statistical properties of quantum theory, must necessarily be *contextual*. Namely, if three operators $\hat{A}$, $\hat{B}$ and $\hat{C}$ satisfy $[\hat{A}, \hat{B}] = [\hat{A}, \hat{C}] = 0$ and $[\hat{B}, \hat{C}] \neq 0$, the result of a measurement of A cannot be independent of whether A is measured alone, or together with B, or together with C [...]. (Emphasis in original; notation adapted – FB)

We can see that multiple issues are at stake here, and we must sort them out carefully as well. At once we can say that (contrary to Peres' opinion) the theorem is *not* concerned specifically with determinism either; rather the issue is once more a kind of *definiteness*. More precisely, the theorem derives a contradiction between QM in Hilbert spaces $\mathcal{H}$ of dimension ≥ 3 and the following two assumptions (cf. Held 2013, p. 4 ff.; Redhead 1987, p. 121 ff.):

Value Definiteness (VD) For each member of a family of observables $\{A_j\}_{j \in J}$ with operators $\{\hat{A}_j\}_{j \in J}$ on $\mathcal{H}$, there always exists a unique definite value $v(A_j)$ on the system S associated with $\mathcal{H}$.

Functional Composition Rule (FUNC) If two (self-adjoint) operators, $\hat{A}, \hat{B}$, representing observables, A, B, satisfy a functional relation of the form $f(\hat{A}) = \hat{B}$, then the same relation is satisfied by the unique definite values of the observables, $f(v(A)) = v(B)$.

For simplicity we here only concern ourselves with discrete spectra (cf. Isham and Butterfield 1998, p. 2681 ff., for some discussion of continuous spectra). Thus, if desired, think of positions or momenta as suitably coarse grained (a particle being in this or that box, or in this or that arm of an interferometer, say). The values v are of course relative to a system S on which they obtain, and should depend on the purported true state λ, since this is the latter's sole purpose. Hence one should actually rather write $v_S^\lambda(A)$. In the random variable-talk, $v_S^\lambda(A)$ is the value a in $A(S) = a$, say, which S takes on for A when it is in state λ. In other words, v is a *value function* that maps $(A, S, \lambda) \mapsto a$. We will generally suppress λ and S, however, in accord with standard notation in the literature.

(VD) can in fact be understood as the assumption of there being hidden true states λ which subsume all the 'hidden' values of observables, i.e., λ would be the kind of state that simultaneously supplies a definite momentum and position, say. The respective values are the (real) numbers representing what *would* be measured (approximately) if λ *could* be accessed perfectly. (FUNC), moreover, has two important instances[51]:

$$\text{If } [\hat{A}_1, \hat{A}_2] = 0 \text{ and } \hat{A}_1 + \hat{A}_2 = \hat{A}_3, \text{ then } v(A_1) + v(A_2) = v(A_3), \qquad \text{(sum rule)}$$

$$\text{if } [\hat{A}_1, \hat{A}_2] = 0 \text{ and } \hat{A}_1 \cdot \hat{A}_2 = \hat{A}_3, \text{ then } v(A_1) \cdot v(A_2) = v(A_3). \qquad \text{(product rule)}$$

The emphasis on *commutativity* of the $\hat{A}_i$ (*compatibility* of the A_i) is not without reason. Von Neumann (1932, p. 305 ff.) had given a 'proof' that no hidden variables whatsoever could be supplemented to QM without contradicting it (a short version can be found in Ballentine 1970, p. 374–375). As criticized by Bell (1966, p. 2 ff.) and long before him Grete Hermann (cf. Jammer 1974, p. 273), von Neumann's proof crucially relied on the assumption that for *all* observables, not just compatible ones, a sum rule would hold, which he motivated by the additivity of *expectation values* in QM. Much like Hermann, Bell stressed that this assumption is unreasonable. As an example, he considered the three operators $\hat{\sigma}_x, \hat{\sigma}_y, (\hat{\sigma}_x + \hat{\sigma}_y)/\sqrt{2}$, whose expectation values are additive, but whose *eigen*values are not, namely: $\frac{\hat{\sigma}_x + \hat{\sigma}_y}{\sqrt{2}} |\psi\rangle = \pm 1 |\psi\rangle$ on an eigenket $|\psi\rangle$, and the same for $\hat{\sigma}_{x/y}$

[51] Cf. Redhead (1987, pp. 121 and 123) for an actual derivation of these from (FUNC).

respectively, but $(\pm 1 \pm 1)/\sqrt{2} \neq \pm 1$. In 1971 (p. 32), Bell made the questionability of the sum rule for incompatible observables very clear[52]:

> It seems therefore that von Neumann considered the additivity [. . .] more as an obvious axiom than as a possible postulate. But consider what it means in terms of the actual physical situation. Measurements of the three quantities σ_x, σ_y, $(\sigma_x + \sigma_y)/\sqrt{2}$ require three different orientations of the Stern-Gerlach magnet, and cannot be performed simultaneously. It is just this which makes intelligible the non-additivity of the eigenvalues [. . .]. That the statistical averages should then turn out to be additive is really a quite remarkable feature of quantum-mechanical states, which could not be guessed *a priori*.

To see, however, how QM *does* provoke a conflict even with our more innocent assumptions (VD) and (FUNC), we first need to review a few formal details. For one, consider that any function of an operator can be written as $f(\hat{A}) = \sum_j f(a_j)\hat{P}_{a_j}$ (cf. Appendix A on spectral decomposition). Now in virtue of the so called *characteristic function*

$$\chi_{a_k}(x) := \begin{cases} 1 \text{ for } x = a_k, \\ 0 \text{ else}, \end{cases} \tag{4.32}$$

we can turn things upside down and write $\hat{P}_{a_k} = \chi_{a_k}(\hat{A}) = \sum_j \chi_{a_k}(a_j)\hat{P}_{a_j}$. In Sect. 2.1.5, we discussed how projectors are standardly viewed as representing properties. Hence the idea here is that a measurement of some observable A really measures whether a certain property is present or absent; e.g., if the value measured for A is a_k then this is taken to indicate that the investigated system has the property represented by $\hat{P}_{a_k}$. This incidentally means that the whole problem can be phrased by appeal to *projectors*.

Since the projectors resolving $\hat{A}$ project onto vectors which span the whole underlying space, their sum must resolve the identity, i.e. $\sum_j \hat{P}_{a_j} = \sum_j \hat{P}_j = \mathbb{1}$. From the product rule we can infer that since $\hat{P}_j^2 = \hat{P}_j\hat{P}_j = \hat{P}_j$ for any projector,[53] we have $v(\hat{P}_j^2) = v(\hat{P}_j) = v(\hat{P}_j)v(\hat{P}_j) = v^2(\hat{P}_j)$. But this implies that $v(\hat{P}_j) \in \{0, 1\}$, since these are the only $x \in \mathbb{R}$ for which $x^2 = x$.

From the sum rule and $\sum_j \hat{P}_j = \mathbb{1}$ we get that $\sum_j v(\hat{P}_j) = v(\text{id})$, and since $\hat{A} = \mathbb{1}\hat{A}$, we get from the product rule that $v(A) = v(\text{id}\, A) = v(\text{id})v(A) \Rightarrow v(\text{id}) = 1$ (id being construed as a 'trivial' identity observable, represented by $\mathbb{1}$). But then only *one* of the projectors $\hat{P}_j$ can have the value 1, and all others must be zero. Since these projectors each project uniquely onto one of the orthogonal rays of the space $\mathcal{H}$, which in turn are defined by the (orthogonal) basis vectors of $\mathcal{H}$, the problem can

[52] In an interview in a popular magazine (Mann and Crease 1988) Bell even went so far as to call von Neumann's proof "silly"—as did (independently) David Mermin in a talk (cf. Mermin 1993, p. 805).

[53] The product rule applies since $\hat{P}_j$ of course commutes with itself. And more generally, any two projectors that project onto *orthogonal* rays commute as well: $[|i\rangle\langle i|, |j\rangle\langle j|] = |i\rangle\langle i|j\rangle\langle j| - |j\rangle\langle j|i\rangle\langle i| = 0$ (for $\langle i|j\rangle = \delta_{ij}$).

be translated into assigning value 1 to only one out of any set of orthogonal vectors (or rays) spanning the space in question. It can equally be translated into assigning one particular *color* (say red) to one of these very vectors and a second particular color (say blue) to the rest of them. The KS theorem then is that this coloration is impossible for any $\mathcal{H}$ such that $\dim(\mathcal{H}) \geq 3$, and this formulation as a coloration problem is the standard way to present its proof (e.g. Held 2013, for an overview).

Note that it is sufficient to establish the impossibility for a space of dimension 3. For consider that if the envisioned coloring was impossible in some dimension $N \geq 3$, but not for $N+1$, then we could remove one of the blue vectors from each such set, i.e., consider an appropriate N-dimensional subspace $\mathfrak{h}$ of the $N+1$ dimensional space $\mathcal{H}$, and the theorem would suddenly not hold anymore in dimension N. Contradiction. Hence, by induction, we at least obtain the result for all separable spaces from a proof in dimension 3. And it is also sufficient to consider a real instead of a complex vector space, since if the coloring were possible in a complex space, then one could construct a substructure isomorphic to the real counterpart from it (cf. Redhead 1987, p. 124), whence by contraposition impossibility in a real space implies impossibility in a complex one.

The following proof of the impossibility is due to Peres (1991) (cf. also Peres 2002, p. 17 ff.). Peres uses a set of 33 vectors in $\mathbb{R}^3$ which belong to 16 different bases in total. We note at this point, that some of the vectors will have to belong to multiple bases, since 33/3 = 11, so that not all 16 bases can consist of entirely distinct vectors. This point is actually more important than may be apparent, since from this property derives the 'thrust' that was initially identified in Peres' quote.

Moreover, the length of the vector does not play any role, and since projectors project onto rays anyway, we can equally think of comparing mutually orthogonal rays in the space. Peres uses rays parametrized by vectors with components $x, y, z \in \left\{0, \pm 1, \pm\sqrt{2}\right\}$. Following his notation, we will write, as an abbreviation, triplets of numbers like $\bar{1}02$, meaning the ray defined by the vector $\begin{pmatrix} -1 \\ 0 \\ \sqrt{2} \end{pmatrix}$. Thus $10\bar{2}$ will represent the same ray (and a fortiori, projector). From these, orthogonal triads can be constructed which each span the whole space. But assigning blue to two of them and red to the remaining one already leads to the desired contradiction. This can be shown in a simple table (cf. Table 4.2).

Following Peres (1991, p. L176), we also write in boldface letters the one ray that is chosen to be red, and in italic letters rays which have occurred before. Moreover, in each step we list other rays that are orthogonal to the chosen red one and will be needed later (in another row). In the third column the reason that a given ray in the row must be red is explained.

It can easily be checked that all the vectors defining the rays in each triad are orthogonal. But so are the vectors $\begin{pmatrix} 1 \\ 0 \\ 0 \end{pmatrix}$, $\begin{pmatrix} 0 \\ \sqrt{2} \\ 1 \end{pmatrix}$ and $\begin{pmatrix} 0 \\ -1 \\ \sqrt{2} \end{pmatrix}$, which, according to lines 1,4 and 10 are all blue. Hence we have found a triad which by necessity does not fulfill our criterion. This in turn proves the theorem. Note that the choice in the first 4 lines comes with no loss of generality, since the theorem is concerned with *all*

Table 4.2 Contradiction derived from the assumption of being able to color one ray red and the others blue in each orthogonal triad formable in $\mathbb{R}^3$ (cf. also Peres 1991, p. L176)

	Triad			Also orthogonal to first		Reason for redness
1	**001**	100	010	110	$1\bar{1}0$	choice
2	**101**	$\bar{1}01$	010			—‖—
3	**011**	$0\bar{1}1$	100			—‖—
4	$\mathbf{1\bar{1}2}$	$\bar{1}12$	110	$\bar{2}01$	021	—‖—
5	**102**	$\bar{2}01$	010	$\bar{2}11$		orthogonal to 2nd and 3rd ray in row
6	**211**	$0\bar{1}1$	$\bar{2}11$	$\bar{1}02$		—‖—
7	**201**	010	$\bar{1}02$	$\bar{1}\bar{1}2$		—‖—
8	**112**	$1\bar{1}0$	$\bar{1}\bar{1}2$	$0\bar{2}1$		—‖—
9	**012**	100	$0\bar{2}1$	$1\bar{2}1$		—‖—
10	**121**	$\bar{1}01$	$1\bar{2}1$	$0\bar{1}2$		—‖—

possible orthogonal triads; the choice corresponds, in the geometric image, basically to a choice of coordinate system and orientation (cf. Peres 1991, p. L176).

Given the preliminary lemmas about the number of dimensions and real and complex spaces, a simple table shows that there is no way of assigning the number 1 to only one out of a set of mutually commuting projectors which jointly resolve the identity in all choices of basis of a given space $\mathcal{H}$ with $\dim(\mathcal{H}) \geq 3$. But this quite directly has the consequence that it cannot *generally* be the case that, if $[\hat{A}, \hat{B}] = [\hat{A}, \hat{C}] = 0 \neq [\hat{B}, \hat{C}]$, then the value on A is independent of whether it is measured together with B or with C (or alone, for that matter); because $\hat{A}$, $\hat{B}$ and $\hat{C}$ may be projectors, which, as we had established above, also represent (yes-no) observables.

Moreover, the operators for three more complex observables A, B and C can each be decomposed into projectors in their eigenbasis, which then forestalls that the value of A be independent of B and C in this case as well: Three sets of mutually orthogonal projectors may resolve operators for observables which satisfy just the required commutation properties. To see this more clearly, note that in deriving a contradiction we have considered the same ray as a member of multiple different orthogonal bases. Now the projectors corresponding to the rays of *one* orthogonal basis all commute (cf. Footnote 53); but the projectors corresponding to the rays of *different* orthogonal bases generally do *not*, since any two (non-collinear) vectors taken from each of the respective bases may be non-orthogonal, and then $[|\psi\rangle\langle\psi|, |\phi\rangle\langle\phi|] = |\psi\rangle \underbrace{\langle\psi|\phi\rangle}_{=:c\in\mathbb{C}} \langle\phi| - |\phi\rangle \underbrace{\langle\phi|\psi\rangle}_{=c^*} \langle\psi| = c\,|\psi\rangle\langle\phi| - c^*\,|\phi\rangle\langle\psi| \neq \mathbb{O}$.[54]

This is the 'thrust' of the KS-theorem, as Peres has it, that observables are *contextual* in the sense that their value will depend on the total context of the measurement, specifically, on what else is being measured. Just as with the other

[54]You can easily convince yourself of this fact by assuming otherwise and multiplying by either $|\psi\rangle$ or $|\phi\rangle$ from the right. You will find a contradiction with the assumption of non-collinearity.

theorems, however, this does not *exclude* the possible existence of hidden variables or the possibility of finding ψ-epistemic models; it merely puts yet another restriction on any such model. To see this more clearly, consider how Bell (1966, p. 9) viewed the situation:

> [A]s well as P_{ϕ_3} say, one might measure either P_{ϕ_2} or P_{ψ_2}, where ϕ_2 and ψ_2 are orthogonal to ϕ_3 but not to one another. These different possibilities require different experimental arrangements; there is no *a priori* reason to believe that the results for P_{ϕ_3} should be the same. The result of an observation may reasonably depend not only on the state of the system (including hidden variables) but also on the complete disposition of the apparatus [...]. (notation adapted – FB)

So *prima facie* the KS-theorem *by itself* does not have all too daunting implications for a ψ-epistemic interpretation of QM. It simply urges one to search for appropriate response functions in purported ψ-epistemic models which then encode the behavior of the apparatus as sensitive to which *joint* measurements are being made.[55] This seems quite harmless. It just means that the measurement cannot suitably *reveal* the true state of the system, but *contributes* something to what is being observed—where the contribution, in turn, depends on which *total* measurement is performed.

On the other hand, assuming that there are such things as definite values for all observables independent of their being measured, the KS theorem would urge us to read quite carefully the postulates of orthodox QM, in particular postulate (III), the connection between observables and operators. (III), then, must *not* be (mis)understood a *uniqueness* claim, but merely as an *existence* claim: for every observable there *is an* operator that represents it. The KS theorem could hence be taken to 'merely' demonstrate that at least *some* operators will represent *multiple* observables, depending on 'context'.

But things can be assessed more systematically, and a good candidate for delineating which of all (self-adjoint) operators may fill out the 'some' are operators with *degenerate spectra*. That this possibility for delineation is available becomes clear from the following observations (cf. Redhead 1987, pp. 20 and 134).

First of all, if an observable A can be expressed by two different functions, $f(B)$ and $g(C)$, of two different, non-degenerate observables B and C respectively, then if A is itself non-degenerate, it holds that $[\hat{B}, \hat{C}] = 0$. Conversely, if $[\hat{B}, \hat{C}] \neq 0$ and $A = f(B)$ and $A = g(C)$, then A must be degenerate. To see this, note first that for any h and $\hat{Q}$, $h(\hat{Q})$ and $\hat{Q}$ always commute (regardless of degeneracy), since $h(\hat{Q}) = \sum_j h(q_j)\hat{P}_{q_j}$ and $\hat{Q} = \sum_j q_j \hat{P}_{q_j}$ in the appropriate basis, and the (orthogonal) projectors satisfy $\hat{P}_{q_j}\hat{P}_{q_k} = \delta_{kj}\hat{P}_{q_k}$. But now suppose that $\hat{A} = f(\hat{B})$ and $\hat{A} = g(\hat{C})$. Then $\hat{B} = f^{-1}(g(\hat{C}))$, where f^{-1} exists since A and B were assumed non-degenerate (so in the spectral resolution each $f(b_i)$ can only be assigned one a_j). Hence now $\hat{B} = k(\hat{C})$, with $k := f^{-1} \circ g$, and therefore $[\hat{B}, \hat{C}] = 0$, by the above considerations on arbitrary $h(\hat{Q})$ and $\hat{Q}$.

[55] E.g. also Spekkens (2005) and Maroney and Timpson (2014, p. 20 ff.) on this point.

This makes it "plausible if we look at how the contradiction was arrived at" (Redhead 1987, p. 134), that any observable falling prey to the contextuality implied by the KS-theorem must be *degenerate*, and suggests that degenerate observables constitute a special case. In fact, even a formal proof exists that no KS contradiction arises if only non-degenerate observables are considered (cf. Mączyński 1971; cf. also Redhead 1987, p. 134). The projectors in the proof, notably, *are* degenerate observables: any projector in dimensions > 2 gives *zero* for at least two vectors (the ones orthogonal to the one projected onto).

As an example, take some $\hat{P}_1$ in a three dimensional space. Then two non-degenerate observables on which its value is contextual can be constructed by considering it first as a part of the commuting set $\left\{\hat{P}_1, \hat{P}_2, \hat{P}_3\right\}$ and then of $\left\{\hat{P}_1, \hat{P}_2', \hat{P}_3'\right\}$, which comes about by rotating around $|1\rangle$ to form a new orthonormal basis. The two non-degenerate, non-commuting operators are then constructed as $\hat{Q} = q_1\hat{P}_1 + q_2\hat{P}_2 + q_3\hat{P}_3$ and $\hat{R} = r_1\hat{P}_1 + r_2\hat{P}_2' + r_3\hat{P}_3'$ (cf. also Redhead 1987, pp. 21–22).

But consider also what it *means* for an observable to be degenerate: it means that several of its eigenvectors correspond to the same eigenvalue, so that a measurement of such an observable may not even be indicative of some unique (pure) *quantum* state. Why would it be indicative of some unique *true* state, assuming such a thing exists?

Let us take this to be saying that if a degenerate observable A can be constructed as a function of some non-degenerate observable B and equally as a function of some incompatible non-degenerate observable C, then we construe these functions as two *different, contextual* observables $f(B) = A_B$ and $g(C) = A_C$, although represented by the same operator $\hat{A}$. We *need not* accept $v(f(B)) = v(g(C))$, as suggested by (FUNC).

Redhead (1987, p. 137) discusses a possible *adaptation*, namely to replace (FUNC) by the following principle:

Contextualized Functional Composition Rule (FUNC*) If $\hat{B}$ is a *non-degenerate* (self-adjoint) operator, representing observable B, and (self-adjoint) operators $\hat{A}$ and $\hat{C}$, representing observables A and C, satisfy functional relations of the form $\hat{A} = f(\hat{B}), \hat{C} = g(\hat{B}), \hat{A} = h(\hat{C})$, then the unique definite values of the observables satisfy $v(A_B) = h(v(C_B))$, where $A_B = f(B), C_B = g(B)$.

We shall occasionally refer to this replacement of (FUNC) by (FUNC*) as a *minimal revision* below, meaning a minimal revision of the basic assumptions underlying a purported interpretation of QM in terms of hidden variables. This notably presupposes that we read quantum postulate (III) as an existence claim only, not one of uniqueness.

For C also non-degenerate, we have $C_B = C$, and then (FUNC*) has the nice property that $v(A_B) = v(A_C)$ (cf. Redhead 1987, p. 137). But this only applies in virtue of the functional relation between B and C, and (FUNC*) hence does not *generally* imply that $v(A_B) = v(A_C)$. So the KS-theorem cannot be derived from it. Moreover, if we set $v(A_B) = f(v(B)), v(C_B) = g(v(B))$, then we have

$v(A_B) = f(v(B)) = h \circ g(v(B)) = h(v(C_B))$. So the consequent (then-part) of (FUNC*) is an appropriate necessary condition for a context sensitive version of (FUNC): functional relations are preserved between the *context dependent* observables. (FUNC*) seems like a reasonable 'contextualization' of (FUNC)

The minimal revision implies that *projectors* cannot always correspond to unique observables. But, again, think about what this means: it means that a quantum state, revealed in a certain (projective) selective measurement, need not be *directly* indicative of a certain value or property 'truly' applying to the investigated system. I.e.: *the quantum state may not be the true state of the system*. This, in fact, makes perfect sense from a ψ-epistemic point of view.

Redhead (1987, p. 133), moreover, thinks that (FUNC) is implied by three assumptions, namely

(i) if a system is not in some eigenstate of some observable, that observable has a definite but unknown value,
(ii) there is a 1:1 correspondence between self-adjoint operators and observables, and
(iii) if there is an operationally well-defined number associated with a self-adjoint operator, then there is something in reality that corresponds to it.

(ii) is targeted by the minimal revision, as we saw. As regards (i), a ψ-epistemic model, in fact, implies something even stronger than (i), namely that for *any* quantum state (even an eigenstate of some observable), the values of *all* actual physical properties measured in terms of observables may be unknown even at the time of measurement: The M-state ψ need not reveal the true state λ. (iii) is certainly also a reasonable, if not *the* underlying assumption for ψ-epistemic models. Something (λ) is already there that accounts for ('explains'; cf. Spekkens 2005, p. 2) why we can use QM so unreasonably well. But it should be stressed that it need *not* be assumed that the operationally well-defined number *accurately reveals* what that something is: *All* quantum states (even eigenstates of measured observables) are construed *operationally*, as *P/M-states*. And Spekkens (2005, p. 3) actually introduces a notion of *measurement-contextuality*, drawing on features of the *response function*, which allows the in-principle possibility of generating ψ-epistemic models that could overcome the difficulties raised by the KS-theorem so far.

Now the '*real* thrust', if you will, is that the KS-theorem applied to *composite* systems also has implications about *locality* and *separability*. The details are quite intricate (cf. Redhead 1987, p. 139 ff.) and we will not elaborate on them here. Note, however, that the assumption of a so called *value rule* is involved (Redhead 1987, p. 120), namely that $\Pr_Q^{\psi}(q) = 0$ implies $v^{\psi}(Q) \neq q$. This rule need not be generally accepted by the ψ-epistemicist: For one, it is doubt worthy that probability zero must be read as the value in question not obtaining at all (compare the definition *via* limit frequency); but more importantly, one could, in this case too, come up with models (similar to Fine's prism models) in which the response of the measuring device to the system's state accounts for the observed statistics, i.e.

where the measurement is 'intrinsically defective'. Moreover, ψ is only the P-state, and $Q(S) = q$ could still be true for S in state λ, since automatic transformations with density $\Gamma(\lambda', \lambda)$ could account for the loss of value q until the measurement.

If these strategies are rejected though, and the value rule is accepted—which is reasonable, given that QM-probabilities do not seem to refer directly to limits of frequencies, the prism-type-models are usually quite artificial, and the assumption that systems always disguise their relevant properties automatically (over arbitrarily short time intervals) seems quite conspiratorial—, then even (FUNC*) can be demonstrated to be incompatible with upholding jointly the following two principles (cf. Redhead 1987, pp. 139–141):

Value Separability (VSEP) Let $\mathcal{H}_1$ and $\mathcal{H}_2$ be Hilbert spaces for two spatially separated systems S_1 and S_2, $\hat{A} \otimes \mathbb{1}$ a (degenerate)[56] operator on $\mathcal{H}_1 \otimes \mathcal{H}_2$ (representing a local observable A_1) where $\hat{A}$ is non-degenerate on $\mathcal{H}_1$, and $\hat{B}, \hat{C}$ two non-degenerate operators on $\mathcal{H}_1 \otimes \mathcal{H}_2$, with $[\hat{B}, \hat{C}] \neq 0$. Then for the observables represented by the operators in question, it holds that $v(A_{1,B}) = v(A_{1,C})$.

Value Locality (VLOC) Let S_1 and S_2 be two spatially separated systems, Q_1 an observable for S_1, and O and R non-degenerate observables for the joint system S_1S_2. Then if the difference between an apparatus set to measure O and one set to measure R is only in the setting of that part of it located at S_2, it holds that $v(Q_{1,O})_O = v(Q_{1,O})_R$, where $v(Q_{1,O})_X$ denotes the value of observable $Q_{1,O}$ if the measuring device is set to measure observable X.

(VSEP) is called *ontological locality* by Redhead, with the remark that it may equally be considered as a separability assumption (as we have chosen to do here), (VLOC) he calls *environmental locality*. (VSEP) tells us that the value of any observable which is non-degenerate for one of the (spatially) separated systems is independent of the values of incompatible observables which pertain to the system *as a whole*; (VLOC) tells us that choosing to measure something else only on *one* of the spatially separated systems should not change the value the *other* system has for any observable $Q_{1,O}$. As Redhead (1987, p. 141) puts it: "the value possessed by a local observable cannot be changed by altering the arrangement of a remote piece of apparatus which forms part of the measurement context for the *combined* system." (my emphasis—FB)

Both principles invoke the contextualization of an observable which is degenerate for the joint system, but non-degenerate for one of the subsystems. The crucial point is that considerations of *contextuality alone* give rise to questions of separability and locality, because the relevant 'context' may be spacelike separated. Arguably, this is 'the real bite' of the KS theorem, because it is hard to see how a ψ-epistemic model should cope with this implication while saving the core intuitions underlying Conjecture 1. Upon accepting the value rule (which, we have argued, is

[56] $\hat{A} \otimes \mathbb{1}$ is obviously degenerate since any state $|a_j\rangle \otimes |\phi\rangle$ with $\hat{A}\,|a_j\rangle = a_j\,|a_j\rangle$ will give a_j for $\hat{A} \otimes \mathbb{1}$, regardless of $|\phi\rangle$.

a reasonable thing to do) (FUNC*) implies that either (VSEP) or (VLOC) must go. And (FUNC*) still relies on (iii), i.e. that there is *something* in reality that accounts for the operationally well-defined numbers associated with self-adjoint operators; although that 'something' is now allowed to be influenced by the general measurement context, as we urged it should in sophisticated ψ-epistemic models.

Redhead, in fact, calls (iii) a 'reality principle'. But more clearly, it constitutes an instance of *abductive inference*. Distinguished by Peirce (1878)[57] from deduction (truth preserving reasoning as used in mathematics and logic) and induction (inferences from a sample to unobserved future cases, a general or a statistical regularity),[58] abduction may be viewed as a genuine kind of inference scheme or a family thereof (cf. Schurz 2008, for an extensive classification). Abduction, just as induction and in contrast to deduction, is *ampliative* ('content-adding') and uncertain.

The pattern originally described by Peirce (1878, p. 194) was that of "a fact quite different from anything observed, from which, according to known laws, something observed would necessarily result." One may formalize this, using '$\mid\!\sim$' to represent 'therefore it is conjectured that' and notation from first order predicate logic otherwise, as $\forall x[\mathrm{F}(x) \rightarrow \mathrm{G}(x)], \mathrm{G}(a) \mid\!\sim \mathrm{F}(a)$, or more generally $\forall x \forall y[\mathrm{R}(y, x) \rightarrow \mathrm{G}(x)], \mathrm{G}(a) \mid\!\sim \exists y \mathrm{R}(y, a)$ (where $\forall x \forall y[\mathrm{R}(y, x) \rightarrow \mathrm{G}(x)]$ is easily seen to be logically equivalent to $\forall x[\exists y \mathrm{R}(y, x) \rightarrow \mathrm{G}(x)]$; cf. also Schurz 2008, p. 208).

In a sense, this is a deductive inference 'upside down', because exchanging $\mathrm{F}(a)$ (or $\exists y \mathrm{R}(y, a)$ respectively) and $G(a)$ would lead to a deductively valid inference. The first premise may be construed as a *known law* (or lawlike connection), the second a *given datum* (something that is being observed). An often used example (of the second scheme, actually) is the inference from footprints in the sand ($\mathrm{G}(a)$) and the background knowledge that if somebody walks in the sand, they leave footprints ($\forall x \forall y[\mathrm{R}(y, x) \rightarrow \mathrm{G}(x)]$), to the belief in or conjecture of somebody having walked there in the sand ($\exists y \mathrm{R}(y, a)$). But arguably, a lot of inferences can be understood as abductions, in the broader sense of non-inductive ampliative inferences that strive for an *explanation* of a given observation. Abduction can hence be generally characterized by its *goal* of "inferring something about the *unobserved causes* or *explanatory reasons* of the observed events[...]." (Schurz 2008, p. 202; emphasis in original)

Importantly for us, a relevant subset of all abductive inferences can hence be understood as an inference to a *hidden cause*. The above 'reality principle' (iii) should be conceived of as an abductive inference to a hidden cause of a given observation, namely the (hidden) 'something' that exists mind-independently and causes the result of a measurement (the well-defined number of the self-adjoint

[57]In 1878, Peirce used the name 'hypothesis' instead of abduction. In his lectures on pragmatism, he later introduced the now-common name 'abduction' (cf. Buchler 1955, p. 150 ff.).

[58]E.g. Schurz (2014, p. 49) for some details on induction.

operator). In fact, the example of inferring that someone has walked along the beach from footprints *is* just such an inference, since the connection is certainly causal in this case.

But equally, the conjecture of *common* causes for observed correlations falls under this general scheme of inference. The PCC (or PCC*) is an instance of abductive reasoning (cf. also Schurz 2008, p. 221 ff.). And, as a matter of fact, so is EPR's use of the 'reality criterion': The known law is $p_1 + p_2 = 0$ (momentum conservation), the given datum is $p_1 = p$, and the explanatory reason is the decay event *together with* $p_2 = -p$, i.e. the *conjectured* momentum of the other particle. These factors would jointly explain the observed momentum, so appeal to the EPR reality criterion amounts to an abductive inference to the existence of the unobserved value $p_2 = -p$ *before* the execution of the measurement. In all three cases, EPR's, the KS- and Bell's theorem, (causal) abductive inferences are *crucially* involved—an issue that we should keep in mind.

4.4 Discussion (i): How Much Evidence and Evidence for What?

How reasonable is it to believe in the potential success of a prospective ψ-epistemic model, in spite of the numerous restrictions provided by the four theorems discussed above and by others like them?

Even though these theorems do not conclusively *rule out* ψ-epistemic models, their appeal is certainly drastically lessened. Many reasonable assumptions cannot be maintained simultaneously with the assumptions of (hidden) true states and ψ-epistemicity, as defined in Sect. 4.2.1. The strongest restriction for ψ-epistemic models is certainly posed by the experimental violations of Bell-type inequalities, which, as long as hidden variables-models do not explicitly include nonlocal interactions or 'inseparable' true states, cannot be reproduced if we accept the research results that claim to simultaneously close detection and locality loopholes.

Moreover, the corresponding theorems do not even specifically concern the OM-approach; Bell's and the KS theorem were in place long before the work of Harrigan and Spekkens, and they hence target a quite broad range of models with hidden variables. It is just that even the most modern versions of hidden variables-theories fall prey to these two theorems, as the discussion has clearly shown. So despite the fact that *some* of the apparent randomness and 'weirdness' of QM could be accounted for in terms of lacking knowledge about underlying true states, the extent to which QM phenomena *in general* can be explained in this way seems quite limited.

Some key features of QM, however, such as superposition and interference, were shown to have *prima facie* analogues in Spekkens' toy model, and so was even entanglement in special cases. So what are we to make of the fact that Spekkens' toy-analogues of superposition, interference, *and* certain entangled states appear to suggest that these features, even the observed correlations between remote measurement outcomes, might be a consequences of a previous lack of knowledge about the true states of the systems involved after all?

In fact, the model of Bartlett et al. (2012) provides something quite similar to Spekkens' treatment of entanglement but for systems with infinite degrees of freedom. In particular, Bartlet et al. model "maximal bipartite entanglement [...] by an epistemic state that describes perfect correlations between the pair of systems." (p. 8) This is done by appeal to a probability density $\mu_{AB}^{\text{corr}}(q_A, p_A, q_B, p_B) \propto \delta(q_A - q_B)\delta(p_A + p_B)$ for two systems A and B for which it is known that $q_A - q_B = 0$ and $p_A + p_B = 0$, i.e. which satisfies the exact conditions of the original EPR thought experiment (with $x_0 = 0$; cf. Sect. 4.3.3). But marginalizing for one of the coordinate pairs of one of the two systems leads to a uniform distribution, so that nothing is known about the true states of the single systems, and only relational properties are known for the joint system (the total values for position and momentum named above).

Just as the epistemic state in Spekkens' qubit-like toy model mirrors features of measurements on an entangled state of two qubits does the density in the restricted version of Liouville mechanics mirror the features of measurements on the actual EPR state.[59] In essence, we here get the same kind of informational update in consequence of a measurement on the total system. Because of her prior knowledge of the value of the total momentum of the two systems, say, Alice can determine the momentum value for Bob's system at once after measuring momentum on her system; and analogously for position.

Thus, as Bartlett et al. (2012, p. 12) put it:

> All that changes as a result of this measurement is how the observer refines her knowledge of the ontic state of particle B. She either refines her knowledge of its position or she refines her knowledge of its momentum. No 'spooky action at a distance' is required to understand the EPR experiment if one adopts the interpretation offered by [restricted Liouville] mechanics.

However, there are, *admittedly*, restrictions to how far one can take this view of entanglement, since neither of the models can reproduce violations of Bell-type inequalities or the like:

> The toy theory is, by construction, a local and noncontextual hidden variable theory. Thus, it cannot possibly capture all of quantum theory. In the face of these no-go theorems, a proponent of the epistemic view is forced to accept alternative possibilities for the nature

---

[59] There are, however, a few difficulties with the actual preparation and measurement of EPR states in the sense of the original paper: the state is not time dependent, and the descriptions used to set up the argument for incompleteness would only be valid at $t = 0$, whereas time evolution makes it unstable; and since a plane wave representation is used, there would be a non-vanishing probability of the two particles being basically anywhere in space, so that the assumption of spatial separatedness is actually unwarranted (cf. Home and Selleri 1991, p. 13). However, Praxmeyer et al. (2005) have constructed a scheme in which the EPR state appears as the limit of a two-mode squeezed state, and observables on it are considered which can be used to violate a Bell-type inequality.

of the ontic states to which our knowledge pertains in quantum theory. (Spekkens 2007, pp. 24–25)

> We emphasize that we are not arguing that a ψ-epistemic local hidden variable model could explain all quantum correlations, only that the particular correlations described in the EPR experiment can be so explained (in precisely the way that EPR suggested they should). This is not at odds with Bell's theorem because the correlations in the EPR experiment do not violate a Bell inequality.[60] Of course, because it is locally causal by construction, [restricted Liouville] mechanics cannot hope to reproduce Bell-inequality violations. Such violations are one of the quantum phenomena that [restricted Liouville] mechanics emphatically cannot reproduce, not even qualitatively. (Bartlett et al. 2012, pp. 24–25)

But this means that, effectively, both examples, that of Spekkens and that of Bartlett et al., are suggestive of something *false*. It is *not* that quantum non-locality or contextuality can be explained in this fashion in general. It is only by *selectively* choosing *particular* states which can be mirrored by ordinary probability densities and particular measurements on these that the *illusion* appears that one could interpret the correlations in harmony with Conjecture 1. One cannot *generally* explain the correlations observed on entangled systems by mere appeal to an 'information update'. This is exactly the gist of Bell's theorem, as Bell stressed in his 1981a *Bertlmann's socks and the nature of reality*. None of the ψ-epistemic models existing so far can help this fact.

What about the other achievements of Spekkens' model? The model is supposed to serve as 'evidence' for an epistemic view of quantum states (cf. Spekkens 2007). More precisely, we take it that this paper and that of Bartlett et al. (2012) are supposed to provide an *argument* for a ψ-epistemic view with the following general structure: (i) If one can provide evidence that QM seems to be about (a lack of) knowledge of the hidden, true states of (typically microscopic) physical systems, then it is reasonable to interpret QM in this way. (ii) The toy model (and spin offs) provide(s) such evidence. (iii) Therefore, by *modus ponens*, it is reasonable to interpret QM in this way. Let us call this the *argument from actual models*. (i) seems fairly uncontroversial and (iii) is just a logically valid step; so it is premise (ii) that we must put under scrutiny here.

The reproduction of interference was one of the core achievements of Spekkens' model, which was then explained, in the subsequent debate, in terms of "vacuum ontic states" (Leifer's phrase); states which carry the phase information and hence alter the behavior of the true states traveling the other path or, more generally speaking, the measurement statistics, but are otherwise undetectable. Clearly, this has quite an ad hoc character, but this need not be considered as so much of a flaw yet, given that these are 'just models' (cf. the remarks on the use of this term in Sect. 4.2.1).

However, now, after the discussion of Bell's and the KS theorem, we can show that the appeal to vacuum states which, to recall, was used in particular to avoid

---

[60] Depending on the specific setup used to implement the states appealed to in the EPR paper, this becomes a debatable claim; cf. Footnote 59.

admitting an immediate causal influence, i.e. nonlocal causation, may be a futile move, and strongly impacts the plausibility and evidential status of the model.

Namely, in relativistic AQFT, there is the so called *Reeh-Schlieder theorem*, originally proven by Reeh and Schlieder (1961), which says that for an open bounded region $\mathcal{O} \subset \mathbb{R}^4$ of spacetime and $\hat{\mathrm{A}}(\mathcal{O})$ an element of the algebra generated from all possible combinations of adjoints, sums, and products of operators $\hat{\phi}(f) := \int \mathrm{d}^4x\, f(x)\hat{\phi}(x),\ f \in C_0^\infty(\mathcal{O})$, the set of vectors $\hat{\mathrm{A}}(\mathcal{O})\,|\Omega\rangle$ is dense in the space $\mathcal{H}$ of state vectors ($|\Omega\rangle$ the vacuum state). This means that one can approximate (arbitrarily well) *any* state $|\psi\rangle$ by operations *local* to $\mathcal{O}$, even if $|\psi\rangle$ has implications for regions $\mathcal{O}'$ at a spacelike distance to $\mathcal{O}$ (e.g. Fleming 2000, p. S497 ff.). Most importantly, the theorem thus "demonstrates that the vacuum, and all other states of bounded energy, have long-distance correlations built into them. It is therefore not surprising to find that Bell inequalities are violated in these states—a standard sign of non-locality." (Dieks 2002, p. 216)

Here Dieks certainly refers to the works of Werner and Summers, who in the 1980s found "that already the vacuum fluctuations assure a maximal violation of Bell's inequalities for the appropriate detectors." (Summers and Werner 1985, pp. 258–259) Thus, if 'vacuum ontic states' are nothing but the quantum vacuum in disguise, then even the interferometer examples fail to work out in a local fashion—because the element of QFT appealed to in order to restore locality is itself a decisive expression of 'quantum non-locality'.

There are two foreseeable rebuttals, contingent on one another, so let us discuss them one by one. First, one may object that these implications follow only from the highly theoretical algebraic version of QFT, and that in practice, the canonical quantization approach is all that is needed and all that can be used. This worry gains support by Wallace's (2006, p. 33) observation that "no examples are known of AQFT-compatible interacting field theories, and in particular the standard model cannot at present be made AQFT-compatible." (Cf. also Sect. 2.2.4 on this point.) Hence one may suspect that these consequences of the Reeh-Schlieder theorem have no bearing on experimental practice and do not have to be taken seriously in virtue of a lack of empirical accessibility. But similar arguments were originally advanced w.r.t. the strong non-local correlations predicted by ordinary QM (most notably by Schrödinger 1935b, p. 166), and if there is anything we can learn from this example, it is that one is better off not to dismiss the implications of the quantum formalism easily.

The situation is arguably more subtle in the case of testing vacuum entanglement though, since, as Summers and Werner (1985, p. 259) note, "there would be experimental difficulties [...][because] the violation of Bell's inequality must vanish exponentially with the spatial separation of [two separate spacetime regions] on the length scale determined by the Compton wavelength of the lightest particle of the theory."

There are suggestions for *other* kinds of experiments, however, in which vacuum states enter crucially into entangled states, namely states entangled with those of a single photon. Take, as a simple example, the state prepared by the beam splitter in Eq. (4.6). Strictly speaking, the fact that in each term there is a photon in one path leaves a vaccum in the other; so in the occupation number representation this should be written rather as something like $|\psi\rangle = \frac{1}{\sqrt{2}}(|1_{\nearrow}\rangle\,|0_{\searrow}\rangle + i\,|0_{\nearrow}\rangle\,|1_{\searrow}\rangle)$ (e.g. Hardy 1994, p. 2280). Examples of experimental protocols that use states of this kind to demonstrate their nonlocal features are discussed, for instance, in Tan et al. (1991) or Hardy (1994). These protocols are typically understood as demonstrating that even a single particle is 'nonlocal'—in the same sense as already noticed by Einstein in examples discussed at the 1927 Solvay conference (cf. Jammer 1974, pp. 115–116) or by Reichenbach (1944, pp. 29): that in a double slit experiment, say, detection at the one slit seems to influence *immediately* what happens in the other slit (cf. also interlude I on this point). But since the states used to describe the single photon are states entangled with the vacuum, they demonstrate, at the same time, that the quantum vacuum is 'nonlocal' in just that sense.

Despite some original controversy (cf. Dunningham and Vedral 2007, p. 2 ff. for discussion) today there is a broad consensus that particular experiments can be used to test exactly for this 'single particle nonlocality', and the experiments that have been performed, e.g. by Hessmo et al. (2004) or Takeda et al. (2015), are reported to provide affirmations of the predictions.

Long story short: *Provisios* about the interpretation of the cited experiments aside, there are good theoretical *and* empirical reasons to suspect that quantum vacuum states are just the kinds of states which involve the problematic nonlocal correlations.

This forces a defender of the interpretation of interference in terms of 'vacuum ontic states' into the following dilemma: if one appeals to vacuum states in any sense sufficiently close to QFT, then one has *not* provided a local explanation of interference phenomena at all; if, on the other hand, one assumes an entirely new kind of nontrivial vacuum one has merely *shifted the burden* from explaining interference to explaining this new kind of state.

Let us assume, for the sake of argument, that the bullet is being bitten by taking the second horn of the dilemma, i.e. by postulating a radically new kind of vacuum state, called a 'vacuum ontic state'. This is the second of the aforementioned rebuttals, contingent on our elimination of the first. But this immediately provokes the question: what *defines* these 'vacuum ontic states'? Recall that we have argued, in Sect. 2.2.3, that 'vaccum state' is a theoretical concept of QFT; it cannot be understood in virtue of a single definition, but only by joint appeal to the foundational parameters ('axioms') of QFT. Models like that of Spekkens (2007) or Bartlett et al. (2012) operate in a (pardon the pun) theoretical vacuum, i.e. *without* a (specific) supporting background theory. The 'vaccum ontic state' (pardon the pun again) is *vacuous, a theoretical term without a theory*.

So assume now that one finds something worthy of the name 'theory' in the vein of ψ-epistemic OMs with a less provisional character. Then the new concept of 'vacuum ontic states' therein must preserve all the successful empirical predictions of the old one in QM—*including the nonlocal correlations*. And this means that quantum interference has *still* not been reproduced *purely* in terms of local interactions and information updates. We are back to square one.

There is an immediate impact on the evidential status of Spekkens' model and other models like it. If the *decisive* quantum phenomena such as nonlocal correlations *and* interference *cannot* be reproduced by a ψ-epistemic model—at least not in any *meaningful* way, i.e. other than by merely replacing kets by probability vectors and unitary operators by permutation matrices—, then these models do *not* provide any evidence for an epistemic view of quantum states, at least not in the sense envisioned in Conjecture 1, or in Spekkens' appeal to explanatory grounds for the quantum statistics. Premise (ii) of the argument from actual models, in other words, arguably does not hold.

If, hence, such remarkable and remarkably counterintuitive results which can be *derived* from quantum theory simply need to be presupposed and appealed to, instead of being satisfactorily explained by ψ-epistemic models, what good are these model then? Bartlett et al. (2012, p. 3) delineate their aims in finding epistemic models for QM as follows: "it is only by describing a broad landscape of possible theories that we can specify the sense in which quantum theory is special." This may be an honorable task and another recent investigation of the debate (Jennings and Leifer 2015) is more benevolent in this respect than we have been here. But we stress that if this is the *only* purpose of ψ-epistemic OMs, then they are hardly of any help in resolving the conceptual difficulties arising from QM. They rather emphasize them.

Einstein, in particular, viewed the QM of his time as "no useful point of departure for future development." (Einstein 1949a, p. 87) And the point of departure for Spekkens and others must be Einsteinian worries; a significant part of Harrigan and Spekkens' (2010) paper is dedicated, after all, to Einstein's work on the subject (cf. especially p. 144 ff. therein). In part this focus on Einstein is due to the fact that he suggested the route of showing QM to be incomplete by finding two (or more) distinct quantum states which should correspond to the very same true states of a system (cf. Harrigan and Spekkens 2010, p. 148; Howard 1985, p. 180; and our discussion of the EPR-paper above). But the very reason why Einstein was searching for ways to show that QM is incomplete was his major dissatisfaction with the theory's implications, *especially* its non-local character (cf. also Maudlin 2014b on this point). Thus, if the only purpose of OMs is to show how QM is special (as Bartlett et al. claim), then this must certainly be seen as somewhat of a 'surrender' to QM. From a point of view such as Einstein's, the aim must rather be to search for serious alternatives because QM is 'too special'.

Still, these models do serve an important purpose after all: Recall how, in a standard reception, Carnap's *Aufbau* is often hailed as an attempt to actually execute what others had merely claimed 'was possible', but that at the same time it signaled, to many, the demise of logical empiricism; because despite his remarkable efforts, Carnap was not able to reduce theoretical terms used in the sciences to an observational language (e.g. Friedman 1999, pp. 89–99). While these views about the *Aufbau* can be challenged on various grounds (cf. Friedman 1999, p. 90 ff.; Leitgeb 2011; Chalmers 2012) and while the models of Spekkens (2007), Bartlett et al. (2012), and others are, of course, much more limited in their scope than the *Aufbau*, we take it that their failure to reproduce core features of QM ψ-

epistemically signals the demise of the underlying project: to find an interpretation of QM in terms of incomplete knowledge about hidden states λ.

4.5 Intermediate Conclusions (i)

What the above discussion has shown at least is that meaningful OMs are *bound* to look like QM itself in many important respects. The move to such curiosities as 'vacuum ontic states' which nonetheless have to share certain peculiar features with vacuum states from QFT bears as witness. All the creativity and formal ingenuity used in the general approach and in particular models apparently cannot bring us past this point. In a similar vein, Timpson (2013, pp. 146–147) argues that opting for hidden-variables

> is unlikely to be attractive to anyone who is trying to appeal to information as a way of avoiding the problems caused by the seemingly odd behaviour of the quantum state. The aim, roughly speaking, was to circumvent the problems associated with collapse or nonlocality by arguments of the form: there's not really any physical collapse, just a change in our knowledge; there's not really any nonlocality, it's only Alice's knowledge of (information about) Bob's system that changes when she performs a measurement on her half of an EPR pair. But we all know that if we are to have hidden variables lurking around then these are going to be very badly behaved indeed in quantum mechanics (nonlocality, contextuality). (emphasis omitted)

We have given reasons to believe that contextuality by itself constitutes less of a hurdle than the confirmed violations of Bell-type inequalities, since it only requires a minimal revision of plausible assumptions for a ψ-epistemic hidden variable approach. However, contextuality constraints alone turned out to raise questions of locality and separability for observables on composite systems. Thus the central difficulty remains the 'grossly non-local character' that QM implies for any hidden-variable model. And this nonlocal character manifests itself in apparent reproductions of interference examples from 'purely' epistemic restrictions as well.

Moreover, we once more emphasize that not only are there many restrictions on ψ-epistemic models, but that conceptually *meaningful* models which reproduce QM to a significant degree are still *missing* to date, i.e. models which allow for an actual *ontology* of the underlying physical states λ in a less technical and more philosophical sense. This, however, was basically the original aim, to find "an explanation of the success of an operational theory" in terms of "physical systems that are [...] presumed to have attributes regardless of whether they are being subjected to experimental test, and regardless of what anyone knows about them", where these "attributes describe the real state of affairs of the system." (Spekkens 2005, p. 2) Given the modest success in pursuing this goal 'ψ-epistemically', it is understandable why more recently even Spekkens (2014, p. 7) has conceded that

> the investigation of [epistemically restricted] theories is best considered as a first step in a larger research program wherein the framework of ontological models [...] is ultimately rejected, but where one holds fast to the notion that a quantum state is epistemic.

Maybe the latter is indeed an interesting option, and we will give further thought to it in Chap. 7.

All in all it seems that the interpretation of QM cannot be *just* a matter of incomplete knowledge. If the conceptual problems associated with QM are a matter of knowledge *in some sense*, then this knowledge still is knowledge about something rather peculiar. We sum up the worries at this point with a mutilated version of one of Einstein's own comments on Schrödinger's wave mechanics:

> The successes of [Harrigan and Spekkens' – FB] theory make a great impression, and yet [we – FB] do not know whether it is question [*sic*] of anything more than the old quantum rules [. . .]. Has one really come closer to a solution of the riddle? (after Einstein 1926; as cited in Howard 1990, pp. 83–84)

To look for an ontology in the 'more philosophical sense' appealed to above should be our guiding principle for the next chapter. We may generally summarize it as "endeavouring to make sense of issues we should otherwise find perplexing" (Heil 2003, p. 3), or more specifically as a "careful analysis" of one of our "best scientific theories [. . .] with the goal of determining" what it implies "about the constitution of the physical world." (Maudlin 2007, p. 104) We will hence, after a second interlude, give some thought to interpretations of QM which allow for such sense-making—for obtaining a picture of what *the world according to QM* may be like.

Chapter 5
Philosophical Interlude II: Locality, Causality, Reality (Again)

How do the issues of 'realism', 'locality' and 'causality' raised in Chap. 4 connect? While we have already said something about the first two issues, it seems that we should now ask what we really *mean* when we talk of 'causation', as in the case of the common cause principle discussed in Sect. 4.3.3. That is all but a simple question, as was the case with 'probability'.[1] Following Hüttemann (2013), however, we can make the following general observations:

First of all note that, when we ask for causes, we typically have at least one of four desiderata in mind: (i) we want to be able to *explain* and (ii) *predict* events and processes, (iii) *manipulate* them, and (iv) attribute *moral responsibilities* (cf. Hüttemann 2013, p. 5). E.g.: (i) Why did a certain building collapse? (ii) Which building could collapse next, for the same reason? (iii) How can we keep that from happening? (iv) Whose actions lead to the building's collapse, who should pay?

Different accounts of causation typically focus more strongly on different ones of the desiderata (i–iv). However, one can equally easily list some *core intuitions* that *any* account of cause-effect relations should possibly respect (cf. Hüttemann 2013, p. 7): (1) cause and effect are *spatiotemporally located*, (2) causes *precede* their effects, (3) a cause '*produces*' or 'enforces' its effect, and (4) the same does *not* hold the other way around (effects do not produce their causes).

Prima facie the common cause principle may seem to violate (3); the statement is *merely probabilistic*, and it is questionable whether (though not excluded that) a probabilistic connection may count as 'production'. But recall that the PCC (or the PCC*) treats of *events types*, not of singular (event) *tokens*. Thus there is an important distinction between *type*- and *token-causation*, and an analysis might provide an account of either one or both. Incidentally, one could ask whether token-causation can reasonably be probabilistic; typically, we would assume that only

[1]The reader capable of the German language may be referred to Hüttemann's concise (2013) introductory exposition for details. Otherwise Psillos (2002, pp. 19–133) is a good starting point.

F. J. Boge, *Quantum Mechanics Between Ontology and Epistemology*, European Studies in Philosophy of Science 10,
https://doi.org/10.1007/978-3-319-95765-4_5

on the level of event types do we have to resort to statements of probabilities and predictions of correlations, in virtue of condition (3). But it is, again, not clear that this *must* be so, and Cartwright (1989, p. 105 ff.) defends an account of causation in which also singular causes can operate probabilistically (cf. also below).

Notable non-probabilistic theories capable of providing token-level accounts are e.g. Lewis' (1974) *counterfactual* account or *process theories* of causation such as that of Salmon (1998). Besides individual difficulties (e.g. Hüttemann 2013, p. 173, for discussion) both these approaches certainly latch onto out intuitions (1–4), to some extent and in different respects. The approaches that arguably have the greatest success regarding the *practical concerns* (ii) and (iii) do not merely recapture our intuitions though, but provide *formal frameworks* (e.g. Spirtes et al. 2000; Woodward 2003; Pearl 2009), in which specific conditions are stated for *finding* causal connections and exploiting these to manipulate the course of events. However, finding such connections and manipulating them presupposes that they occur *on a regular basis*; once a *particular* event has occurred, *it is gone* (cf. also Hüttemann 2013, p. 200). These theories are hence all—or at least primarily concern the—*type-level*.

Type-level theories will typically not be (fully) deterministic; quantum considerations aside, we still live in an 'imperfect world', where even the most cleverly contrived laboratory-experiments require averaging and statistical analysis. These considerations, moreover, are not merely *pragmatic* but also *epistemic*: When do we really have *justification* to *believe* in the fact that person X's smoking *caused* her cancer? Only after reference to sufficiently many known cases of smoking-cancer and non-smoking-no-cancer could we possibly convince a skeptic that she had better stop smoking.[2] And an even stronger case can be made in the case of drinking alcohol and getting headaches, where an outright *intervention* on the drinking is easily possible.

This notion of an 'intervention' is central to Woodward's (2003) framework for studying causation, and he introduces the notion intuitively (which will mostly suffice for our purposes) in the following terms: "an intervention on $\underline{X}$ with respect to $\underline{Y}$ changes the value of $\underline{X}$ in such a way that if any change occurs in $\underline{Y}$, it occurs only as a result of the change in the value of $\underline{X}$ and not from some other source." (p. 14; notation adapted) This requires, of course, to hold (most) other variables *fixed* (cf. ibid.)

Having somewhat justified a focus on (type-level) probabilistic approaches to causality, we must ask what we should conclude from the fact that even the PCC* implies Bell-type inequalities. Does QM straightforwardly imply a *failure of causation?* Is it *impossible* to account for Bell-type correlations in a causal way? The answer here cannot be an unqualified 'yes', even if van Fraassen (1982b, p. 28) basically thought so, as he believed EPRB experiments to constitute "a conceivable phenomenon [...] in which there is a correlation for which there *can* exist no

[2]Of course a strong *inductive* skeptic might still not be convinced (a little more on this in Chap. 7).

common cause." (my emphasis FB) This assertion is certainly too strong, as shall become apparent below.

First, we note, as a remark on the scope and meaning of (PCC) or (PCC*), that not every 'screener' or 'screener system' in the sense of Sect. 4.3.3 should count as a common cause (system); Wroński (2014, p. 36) gives two simple examples in which there is a screener system, but none of the cells in the partition (values of the variable) is *positively statistically relevant* for the correlated events (A and B), where he spells out positive statistical relevance as the conditional probability of A and of B given the various $\lambda_\ell \in \{\lambda_\ell\}_{\ell\in L}$ being greater than given some $\lambda_k \notin \{\lambda_\ell\}_{\ell\in L}$, i.e. $p(A|\lambda_\ell) > p(A|\lambda_k)$, $p(B|\lambda_\ell) > p(B|\lambda_k), \forall \ell \in L, k \notin L$ (Wroński 2014, p. 5).[3]

Arguably, statistical relevance is a necessary condition for a screener (system) to count as a common cause (system), and this requirement was also present in Reichenbach's original formulation (cf. Footnote 48). But obviously, it is not a problem that *not every* screener is a common cause: We have *inferred* the existence of a (hidden) common cause from the correlated behavior, (CEL), and (SEP), and now merely expect to *find* a screener, on account of the PCC$^{(*)}$. Our question is to the converse, whether *all common causes* have to be *screeners*, *viz.* have to satisfy (PCC) or (PCC*), and hence imply (FACT).

Supposedly the most prominent opponent to this view is Nancy Cartwright. Cartwright (1988, p. 184; 1989, p. 234) considers an atom which collides with a particle and subsequently emits two new particles. These particles are then emitted with probability 1/2 either at fixed angle $\tilde{\theta}$ or at fixed angle $\tilde{\theta}'$. Since momentum is conserved, if particle one is emitted at angle $\theta_1 = \tilde{\theta}$, then particle two must be emitted at $\theta_2 = -\tilde{\theta}$, and analogously for $\tilde{\theta}'$. But now suppose also that there is a (probabilistic) common cause λ such that when λ is present, the atom decays at $\pm\tilde{\theta}$ with probability r, and if λ is not present it decays at $\pm\tilde{\theta}'$. From this setup we immediately get that, since λ either occurs or does not, and since emission takes place at the respective angels 50% of the time, $p(\theta_1 = \tilde{\theta}, \theta_2 = -\tilde{\theta}, \lambda) + \underbrace{p(\theta_1 = \tilde{\theta}, \theta_2 = -\tilde{\theta}, \neg\lambda)}_{=0} = 1/2$, and since $p(\theta_1 = \tilde{\theta}|\lambda) = r$, we get from the definition of conditional probability that $p(\lambda) = \frac{1}{2r}$. But then $p(\theta_1 = \tilde{\theta}, \theta_2 = -\tilde{\theta}|\lambda) = \frac{1/2}{1/(2r)} = r \neq r^2 = p(\theta_1 = \tilde{\theta}|\lambda)p(\theta_2 = -\tilde{\theta}|\lambda)$. Hence the common cause λ does not screen off.

As the argument goes, it may be the case that the EPRB-situations are of similar kind and that a non-screening common cause may be found. More explicitly Cartwright (1989, p. 243) suggests, that "the quantum state consequent on the interaction operates, in conjunction with the separated apparatuses, as a joint cause of the results in each wing, with no direct causal connection between one wing and the other."

[3]One can similarly spell out negative statistical relevance for talking about negative correlation.

But there are several reasons to reject Cartwright's reasoning. Näger (2013a, p. 34) first of all shows how to redescribe the situation in terms of a (deterministic, hidden) two valued variable $\underline{\lambda}$ with values λ_1, λ_2, which determines exactly whether the particles will fly off at $\tilde{\theta}$ or not, and which *does* screen off.[4] But of course it need not be assumed (as Näger (2013a, p. 34) equally points out), that this is actually the case; it just makes the example less impressive since it *could* simply be a misguided description of a *deterministic* common-cause-scenario.

What is worse for Cartwright's case, however, is Näger's aforementioned observation that some dependence of an outcome on the remote setting needs to be assumed to avoid the derivability of a Bell-type inequality. Hence a non-screening off common cause "with no direct causal connection between one wing and the other" (as Cartwright has it) does not suffice to establish an adequate assessment of the situation.

Cartwright (1989), moreover, offers positive proposals for non-screening-off common cause models of EPRB scenarios, based on the formal theory of *linear causal models*, which include the "significant innovation" of a "built-in distinction between a cause being *present* and a cause's *action* ('firing') to bring about its effect." (Cachro and Placek 2003, p. 215; my emphasis—FB). But one of Cartwright's (1989, pp. 238–239) models has been found to suffer from mathematical deficits (imply probabilities >1; cf. Cachro and Placek 2003, p. 218) and the other one (Cartwright 1989, pp. 242–243), which avoids this difficulty, has been identified as being of a *conspirational* nature, in the same sense as violating (AUT) (cf. Cachro and Placek 2003, p. 219 ff.), whence it does not really suggest a *relevant* innovation after all.

Finally, it is not clear what the general plausibility of the non-EPRB models is, e.g. the appeal to *atomic decay* in Cartwright's plausibilizing example of causation without a screener (or screener system): scattering and atomic decay are processes for which a *quantum* treatment is indicated, and it is as doubtworthy that a causal assessment of the situation is apt *in this case* as it is in the EPRB case. Now suppose, for the sake of argument, that one was a strong 'quantum skeptic' or would accept the model as a statistical model of some *possible, non-quantum* world. Then one could *still* always take it that in this model there is a *statistical parameter* (λ) that simply *does not qualify as a cause*.

In fact, van Fraassen (1982a, p. 198) has provided similar *non-quantum* examples of (perfect) positive correlation in which a factor x exists that is probability-increasing but does not screen off. *His* conclusion, however, is exactly that the factor is *not* a common cause (e.g. his p. 200), i.e. that *no* common cause exists for the correlation. There hence seem to be two principled options for dealing with the problems raised the EPRB correlations, as regards the PCC*: (i) accept (PCC*)

[4] A somewhat similar but much more general result is proved by Hofer-Szabó et al. (1999). Here the authors show that any classical probability space containing pairs of correlated events but no screeners for some of these can be extended in such a way that the extension preserves the old probability measure but contains screeners for all the correlations (cf. also Wroński 2014, p. 70 ff. on this point).

as a suitable formalization of causal intuitions and accept that there are (unintuitive) cases where no cause can be found, or (ii) reject (PCC*) on the grounds of implying a failure of causation in situations which should be explained causally. For obvious reasons (i) has been called *van Fraassen's horn*, (ii) *Cartwright's horn* (cf. Näger 2015).

A good *prima facie* reason to accept Cartwright's horn is Salmon's discovery of *interactive forks*[5] (e.g. Salmon 1998, p. 133 ff.) which he (ibid.) introduced by way of a Compton scattering-example that we reconstruct (in a slightly idealized fashion) as follows. Imagine that we have a photon (γ) incident with energy E_0 on an electron (e^-) and, as a consequence of Compton scattering, a resulting e^- and γ with energies $E_\gamma + E_{e^-} = E_0$ due to energy conservation (neglecting the electron's rest energy). Now let A be the event that $E_\gamma = E_1$, B the event that $E_{e^-} = E_2$, and C the event that the scattering happens. Due to the conservation of energy, we should always have $p(A|B, C) = 1 = p(B|A, C)$, i.e. the energies of e^-, γ *will* assume the respective values, as soon as the scattering takes place and the scattering-partner assumes the other respective value. This, however, means that

$$p(A, C) = p(B, C) = p(A, B, C), \tag{5.1}$$

in virtue of the definition of conditional probability. Now assume that $p(A, B|C) \leq p(A|C)p(B|C)$. Then we have

$$\begin{aligned} \frac{p(A, B, C)}{p(C)} &\leq \frac{p(A,C)}{p(C)}\frac{p(B,C)}{p(C)} \overset{(5.1)}{=} \frac{p(A,B,C)}{p(C)}\frac{p(A,B,C)}{p(C)} \\ &\Leftrightarrow p(A, B|C) \leq p^2(A, B|C), \end{aligned} \tag{5.2}$$

which is false for all values < 1, and true for $p(A, B|C) = 1$, i.e. in the (empirically false) case that C will *always* effect $E_\gamma = E_1$ and $E_{e^-} = E_2$. This implies that we must have $p(A, B|C) > p(A|C)p(B|C)$—Reichenbach's screening off condition does not hold!

Notably, the example is again a *quantum* one, as was the case with Cartwright's atomic decay. Why does Salmon choose this example and not rather, say, billiard balls colliding? "[B]ecause there is good reason to believe that events of that type [Compton scattering—FB] are *irreducibly statistical*." (Salmon 1998, p. 134; my emphasis—FB) The point is, we take it, that the case for $p(A, B|C) = 1$ could easily be made in the billiard ball example if one resorts to a kind of

[5]Author's note: I was made aware of the debate on interactive forks by Paul Weingartner (private communication). Note, however, that in contrast to e.g. Suárez (2004, p. 289), Salmon himself does not even invoke the interactive fork against anti-causal appeals to EPRB correlations but rather accepts van Fraassen's arguments as "cogent" (cf. Salmon 1984, pp. 251–254). van Fraassen (1982a, p. 206), moreover, suggests that interactive forks lead to why-questions that only terminate when one ultimately reaches a Reichenbachian, non-interactive ('conjunctive') fork. So the appeal to interactive forks may be somewhat of a red herring in the first place.

graining-argument: as soon as C is sufficiently fine-grained, the case is perfectly deterministic, and $p(A, B|C) = 1$ (specific energies, given *specific* scatterings) is entirely reasonable. Only certain views on quantum theory make it plausible that such graining will not do any work.

But herein ('certain views') lies the problem with appeals to such examples. We can hold, without a problem, that the scattering event in Compton scattering is *not* the common cause of the resulting energies because there *is no* cause of these very energies. The scattering does not 'produce' the respective energies; *nothing* does. They *simply occur*. In a sense this is even somewhat of a *traditionalist* view on the situation of causality in QM (e.g. Bohr 1948, p. 313), although in early debates the distinction between 'causality' and 'determinism' was not sufficiently clarified. Assuming that there *must* be a common cause and then demonstrating that this common cause must be *non-screening* is, in other words, *question begging*.

The problem of the two horns has been reformulated in the framework of *causal graph theory*.[6] Causal graph theory models causal connections in virtue of *directed acyclic graphs* (DAGs) over a set of variables $\mathcal{V} = \left\{\underline{X}_j\right\}_{j\in J}$[7] *relative to which* cause-effect relations can be identified. The graphs, $\mathcal{G}$, consist of variables, possibly put into squares and circles ($\mathcal{G}$'s *vertices*), where certain of these variables will be connected by arrows (then called *directed edges*). Acyclicity means what it should: there are no direct paths from a vertex to itself. Uncaused vertices may be called *exogenous*, all other ones *endogenous* (cf. Wood and Spekkens 2015, p. 4).

Now the DAGs of interest in the present case are all *probabilistic*, meaning that the structure of a given graph must be inferred from probabilistic relations (in turn inferred deductively from theory or inductively from experiment), or that, conversely, only statistical correlations could be predicted from a given graph. In deterministic models, in contrast, only exogenous variables may be associated with a probability distribution, and all other relations between vertices are specified in terms of functional dependencies among variables (cf. Wood and Spekkens 2015, p. 4).

[6] Cf. Spirtes et al. (2000) and Pearl (2009) for detailed expositions. We here restrict our attention to the relevant points in the brief expositions given by Wood and Spekkens (2015), Näger (2016), and Schurz and Gebharter (2016) respectively.

[7] In this interlude we generically use curvy letters to denote *sets* of variables. We here again allow for a mixed notation wherein sets and variables may both appear as arguments of probability functions. As a side remark, note what it means that cause-effect relations are identified at the variable level: A causal relation could, e.g., be 'color (causally) influences visibility', where the *values* of the variable 'color' could be all the specific shades (or suitable classes thereof) perceivable by the relevant set of perceivers, and the values of the variable 'visibility' could e.g. be 'good' and 'bad' (or a suitable fine-graining of these, and possibly averaged over different perceptual conditions). Only a subset of all value attributions to the variable 'color' would then cause the value 'good' for 'visibility'. Since the assumption of a *value* by some variable describes an *event-type*, the theory of causal DAGs is even *above* type level, in relating whole variables to one another (cf. also Schurz and Gebharter 2016, p. 1075).

In this theory of causal DAGs, one can define two sorts of dependencies between sets of variables which are subsets of the variables in some graph: *probabilistic* conditional dependence, which we denote by $D_P(\mathcal{X}, \mathcal{Y}|\mathcal{Z})$, means that there are $\underline{X} \in \mathcal{X}, \underline{Y} \in \mathcal{Y}$ and values x, y such that it holds that $p(\underline{X} = x|\underline{Y} = y, \mathcal{Z}) \neq p(\underline{X} = x|\mathcal{Z})$, where, of course, one or multiple (though not all) variables in $\mathcal{Z}$ and/or some of their values could be redundant (cf. Schurz and Gebharter 2016, p. 1076). This dependence, moreover, allows for two subcases: $p(\underline{X}|\underline{Y}, \mathcal{Z}) > p(\underline{X}|\mathcal{Z})$ and $p(\underline{X}|\underline{Y}, \mathcal{Z}) < p(\underline{X}|\mathcal{Z})$ (value ascriptions omitted), the former of which we call *positive* conditional probabilistic dependence and denote by $D_P^+(\mathcal{X}, \mathcal{Y}|\mathcal{Z})$, and the latter of which we call *negative* conditional probabilistic dependence and denote by $D_P^-(\mathcal{X}, \mathcal{Y}|\mathcal{Z})$. $D_P(\mathcal{X}, \mathcal{Y}|\mathcal{Z})$ is the case *iff* either of the two holds, and its negation $\neg D_P(\mathcal{X}, \mathcal{Y}|\mathcal{Z})$ is easily seen to be equivalent to the screening-off condition $p(\underline{X}, \underline{Y}|\mathcal{Z}) = p(\underline{X}|\mathcal{Z})p(\underline{Y}|\mathcal{Z})$ of the PCC*.[8]

Following Schurz and Gebharter (2016, p. 1084) and Näger (2016, p. 1132), we can now understand the *causal* dependencies between sets $\mathcal{X}, \mathcal{Y}$, *relative* to some (non-overlapping) set $\mathcal{Z}$, as (exactly) the condition that there is at least one *path* π (a set of directed edges and vertices) from an $\underline{X} \in \mathcal{X}$ to a $\underline{Y} \in \mathcal{Y}$ such that no *intermediate* or *common cause* on π is in $\mathcal{Z}$ and every *common effect* on π is in $\mathcal{Z}$ or has effects in $\mathcal{Z}$. The intuition of the first part is clear: there is a causal influence between $\mathcal{X}$ and $\mathcal{Y}$ but not 'mediated by' 'or jointly received from' $\mathcal{Z}$. The intuition of the second part becomes clear when one interprets things in terms of *independence*: we would then, should $\mathcal{X}$ and $\mathcal{Y}$ be causally independent relative to $\mathcal{Z}$, expect to have only joint effects *not* in $\mathcal{Z}$. We will denote (relative) causal dependence by $D_C(\mathcal{X}, \mathcal{Y}|\mathcal{Z})$, and if $\mathcal{Z} = \varnothing$, one may speak of *unconditional* dependence in both cases (causal/probabilistic).

The generalization of the PCC to causal graph theory can then be stated as[9]:

Causal Markov Condition (CMC) In any causal DAG $\mathcal{G}$, if a set of variables $\mathcal{X}$ is *probabilistically* dependent on a set $\mathcal{Y}$ of variables conditional on some set of variables $\mathcal{Z}$, then it is also *causally* dependent on $\mathcal{Y}$ given $\mathcal{Z}$. Formally ('$\rightarrow$' denoting logical implication):

$$D_P(\mathcal{X}, \mathcal{Y}|\mathcal{Z}) \rightarrow D_C(\mathcal{X}, \mathcal{Y}|\mathcal{Z}).$$

The CMC is a generalized version of the PCC since it tells us that, should we discover that $D_P(\mathcal{X}, \mathcal{Y}|\mathcal{Z})$, i.e. the elements of $\mathcal{X}$ and $\mathcal{Y}$ are not screened off by the elements of $\mathcal{Z}$, $\mathcal{X}$ and $\mathcal{Y}$ are causally dependent relative to $\mathcal{Z}$ ($D_C(\mathcal{X}, \mathcal{Y}|\mathcal{Z})$). So we might want to look for causal connections between elements in $\mathcal{X}$ and $\mathcal{Y}$, other than being effects of some common cause.

[8]We take it as understood that $p(\mathcal{Z}) \neq 0$ if one conditionalizes on $\mathcal{Z}$.

[9]Note that this is not the standard formulation in the literature, but cf. Schurz and Gebharter (2016, p. 1085) for references on equivalence-proofs. These proofs, however, presuppose that the set of all variables, $\mathcal{V}$, be *finite* (cf. ibid.).

Now there is an additional principle in the theory of causal DAGs that states the *converse* of (CMC):

Faithfulness (F) In any causal DAG $\mathscr{G}$, if a set of variables $\mathcal{X}$ is *causally* dependent on a set $\mathcal{Y}$ of variables conditional on some set of variables $\mathcal{Z}$, then it is also *probabilistically* dependent on $\mathcal{Y}$ given $\mathcal{Z}$. Formally:

$$D_C(\mathcal{X}, \mathcal{Y}|\mathcal{Z}) \to D_P(\mathcal{X}, \mathcal{Y}|\mathcal{Z}).$$

Whatever our interpretation of 'probability', we will typically appeal to *frequency data* to justify the acceptance of probabilistic statements; even if we combinatorially infer $p(\text{heads}) = 1/2$ in a coin toss, we might want to perform a series of trials to check whether the coin is fair.[10] Hence, if we interpret both principles as *prescriptions* for *finding* causal structures, we can loosely understand (CMC) as telling us to 'draw arrows wherever the (frequency) data suggests to', while (F), stated contrapositively, loosely tells us to '*not* draw arrows where the (frequency) data *does not* suggest to'. For instance, if we have a correlation between $\underline{X}$ and $\underline{Y}$, but conditional on $\underline{Z}$ it vanishes (screening off), then we should not draw an arrow directly between $\underline{X}$ and $\underline{Y}$ (by *modus tollens* on (F)). On the other hand, if we do have the correlation and the statistical dependency of $\underline{X}$ and $\underline{Y}$ on $\underline{Z}$, then we should draw appropriate arrows from $\underline{Z}$ to $\underline{X}$ and $\underline{Y}$ (by *modus ponens* on (CMC)).

The dilemma between Cartwright's and van Fraassen's horn can now, in the framework of causal DAGs, be formulated as the dilemma between giving up either the condition (F) or the condition (CMC), and lately, the case for Cartwright's horn in this sense has been made by Wood and Spekkens (2015) and Näger (2016). Note that this is completely independent of any *locality* concerns: These two principles *by themselves* (assuming also a suitable independence of the preparation method and acyclcity) lead to a conflict with the empirically confirmed violations of Bell-type inequalities. Näger (2016, p. 1139) refers to this as the *causal problem of entanglement*, whereas considerations involving locality assumptions are referred to as the *spatiotemporal problem of entanglement* (Näger 2016, p. 1131). We shall have to say something about the connection between these two a little below.

A motivating reason for this strategy is that (F) is obviously somewhat restrictive: there could be cause-effect relations that *do not show up in the data*. One such example would be *canceling causal paths*. E.g., we could have a common effect $\underline{Z}$ of variables $\underline{X}$ and $\underline{Y}$ (where we again omit assumed values) such that $D_P^-(\underline{Y}, \underline{X})$, and $\neg D_P(\underline{X}, \underline{Y}|\underline{Z})$, because both $\underline{X}$ and $\underline{Y}$ have a positive causal influence on $\underline{Z}$, but $\underline{X}$ has a negative (preventive) causal influence on $\underline{Y}$ such that these influences always cancel. In a graph, this would be depicted by the vertex-directed edge combinations $\underline{X} \overset{+}{\to} \underline{Z} \overset{+}{\leftarrow} \underline{Y}$ and $\underline{X} \overset{-}{\to} \underline{Y}$, where $\pm$ indicate positive and negative causal influences respectively (e.g. Schurz and Gebharter 2016, p. 1082). Why is this a

[10] This also justifies loose interchange of 'probabilistic' and 'statistical' in parts of the discussion; we do not intend to claim thereby that the *meaning* of 'probability' is 'relative frequency' though.

violation of (F)? Because $\underline{Z}$ is a common *effect*, not a common cause, and there are no common or intermediate causes between $\{\underline{X}\}$ and $\{\underline{Y}\}$ contained in $\{\underline{Z}\}$. Hence $D_C(\underline{X}, \underline{Y}|\underline{Z})$, even though $\neg D_P(\underline{X}, \underline{Y}|\underline{Z})$.

The central suggestion of Wood and Spekkens (2015, p. 7) exactly is that there could be a violation of (F), in virtue of some *fine-tuning* of the *causal-statistical parameters*. The causal-statistical parameters here are the probabilistic dependencies of a vertex $\underline{X}$ on its direct causes or 'parents' $\text{Pa}(\underline{X})$, i.e. the probabilities $p(\underline{X}|\text{Pa}(\underline{X}))$ (e.g. Schurz and Gebharter 2016, p. 1086). Fine-tuning now means that there could be conditional probabilistic independencies as "a result of the causal-statistical parameters taking some *particular* set of values." (Wood and Spekkens 2015, p. 7; my emphasis—FB) However, any such fine-tuning will not be *robust* under small changes of the causal-statistical parameters. Change the conditional probabilities by a small amount, and the structure cannot hold true. This is exactly why it is a 'fine-tuning'; it depends on the precise probabilities assigned.[11]

How plausible is it, though, that there is some such fine-tuning involved in the EPRB-case? Näger (2016, p. 1148) suggests that what he calls "*unfaithfulness by internal cancelling paths* [...] is the way how the quantum mechanical formalism secures the unfaithful independences." (emphasis in original) How does this sort of unfaithfulness work? While in the canceling paths-scenario above the causal dependence is on the variable level, in Näger's internal canceling paths it occurs on the *value* level.[12] This means that for any pair of values x_j, y_j of variables $\underline{X}, \underline{Y}$ respectively, one specifies causal-statistical dependencies. It is then possible that one has (positive and negative) dependencies among the values of three variables $\underline{X}, \underline{Y}, \underline{Z}$ such that $D_P(\underline{Y}, \underline{X})$, $D_P(\underline{Z}, \underline{Y})$, but $\neg D_P(\underline{Z}, \underline{X})$, i.e. that causation becomes (statistically) *intransitive* (cf. Näger 2016, p. 1149, for illustrations and examples).

Now using unfaithfulness due to internal canceling paths, Näger (2016, p. 1151) ultimately comes up with the following causal structure for the EPRB case: The quantum state itself causes a later (time evolved) quantum state that will be causally influenced by both distant settings (rotations of the DuBois magnets, say), and then collapses into two independent quantum states that in turn cause the respective results in the measurements. The internal canceling paths occur, in this causal model, between setting, collapsed state, and outcome, thereby ensuring that no signaling is possible (there is no probabilistic information about the distant setting in the local outcome alone, due to the intransitivity). And, moreover, due to the

[11] Take the three variables $\underline{X}, \underline{Y}, \underline{Z}$ of the above canceling paths-structure. Then statistical dependencies implied by that would be $p(\underline{Y}|\underline{X}) < p(\underline{Y})$ and $p(\underline{Y}, \underline{X}|\underline{Z}) = p(\underline{Y}|\underline{Z})p(\underline{X}|\underline{Z})$. Then, using the definition of conditional probability multiple times, we get that $p(\underline{X}, \underline{Z})p(\underline{Y}, \underline{Z}) < p(\underline{X})p(\underline{Y})p(\underline{Z})$, which would easily be violated for, say, $p(\underline{X}, \underline{Z}) = p(\underline{Y}, \underline{Z}) = 1/2$, $p(\underline{X}) = p(\underline{Y}) = p(\underline{Z}) = 1/3$, since this would imply $1/4 < 1/9$ (↯).

[12] Assume that in our above canceling paths-example, $\underline{X}, \underline{Y}, \underline{Z}$ are binary, taking on values $\pm$. Then the proclaimed statistical (in)dependencies could mean that e.g. $p(\underline{Y} = \pm|\underline{X} = \pm) < p(\underline{Y} = \pm)$ and $p(\underline{Y} = \pm, \underline{X} = \pm|\underline{Z} = \pm) = p(\underline{Y} = \pm|\underline{Z} = \pm)p(\underline{X} = \pm|\underline{Z} = \pm)$ and the example otherwise repeats.

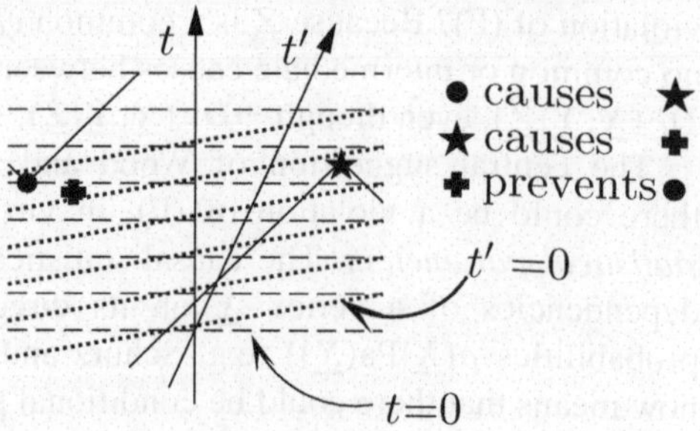

Fig. 5.1 Three events (●, ★, ✚) with causal relations as specified and variable temporal order according to primed and unprimed reference frame. The solid lines indicate the future and past light cones of ● and ★ respectively

unitarity of all transformations but measurement-interactions, Näger (2016, p. 1152) claims to be able to prove the stability of this particular sort of unfaithfulness (the proof is deferred to another paper, unpublished at the time of writing of this document).

While this is an inventive result and clearly refutes van Fraassen's impossibility claim for a causal explanation of the phenomenon, the causal structure suggested has a bunch of quite undesirable features: (i) It is inevitably *nonlocal*: the quantum state previous to collapse is instantaneously influenced by both remote measuring devices. (ii) It requires the *particular* sort of unfaithfulness suggested by Näger; and unfaithful causal models in general otherwise "rarely occur in contemporary practice, and when they do, the fact that they have properties that are consequences of unfaithfulness is taken as an objection to them." (Spirtes et al. 2000, p. 29) In this sense it is 'doubly ad hoc'. (iii) By the failure of (F), the common cause, the time-evolved quantum state, is hence exactly a common cause that *does not screen off*, and we had argued above that stipulating such causes is predicated on question begging arguments. (iv) The entire model presupposes a collapse-interpretation and hence becomes *interpretation-relative*.[13]

We had claimed, in the previous chapter, that nonlocality provokes a conflict with relativity, but we had not elaborated on this in detail. To see how conflicts can arise, consider Fig. 5.1. The special theory of relativity tells us that both time-parameters, t and t', are related by $t' = \gamma(t - vx/c^2)$ (e.g. Rindler 2006, p. 45), assuming v is uniform and the 'primed frame' (t' and doted hyperplanes of simultaneity) is associated with an observer moving in positive x-direction only (with v). This implies that if $t = 0$, t' is at $-\gamma vx/c^2$ ($\gamma = 1/\sqrt{1-(v/c)^2}$), which is generally $\neq 0$. Hence the lack of overarching simultaneity for both frames. Since the speed of light is c in both frames, the axes must, in appropriate units,[14] still be such that c

[13]Näger is, in fact, an adherent of GRW-like objective collapse interpretations (private communication), the likes of which we will discuss in Chap. 6. Possibly Näger's model could be adapted to a Bohmian interpretation with decohering wave packets though (cf. also Chap. 6). This would lessen the impact of (iv).

[14]If we set $c = 1$ then $t' = ct'$ and hence $x = ct' = t'$ for photons; so a photon's trajectory must also separate equal sectors in the primed frame.

separates two sectors in each frame by the same amount. Hence the tilting of both, the t' axis and the $t' = 0$ hyperplanes. Now if supeluminal causation is possible *in principle*, then nothing forbids a situation in which event ● causes event ★ which in the primed frame precedes event ✚ and therefore may (superluminally) cause ✚, which in turn prevents ● in the first place (↯).

But, as we already known, there is a no-signalling constraint enforced by the *probabilistic independence* $\neg D_P(A^x_j, i)$ (for all relevant A, x, j, i), where, to recall, i denotes the distant setting and A^x_j is the event that on one side (A) a particular value (x) is measured in a given local setting (j). The upshot is this: Assuming a causal model such as Näger's, nature *has* these causal connections that would in principle allow us to produce 'live contradictions' (branching spacetimes etc. aside), but she *prevents us* from *using* them by cleverly contriving the statistics of our interventions to be such that we simply cannot 'see' the causal connections. So we have a nonlocal common cause-structure without screening off, i.e. where the common cause cannot be inferred from the statistics in the usual way, and where the nonlocal character of the causal connection cannot be used to create causal paradoxes only in virtue of a lack of *access* or *information*. If anything is 'conspiratiorial', this certainly is.

Above we had also claimed that the spatiotemporal and causal problems of entanglement are not detached but connected. One sort of connection is the fact that the inference to a hidden common cause λ by the PCC$^{(*)}$ is driven by the observation that the local measurement events should not influence each other causally in virtue of spacelike separation. The entire *search* for common causes hence has to do with the spatiotemporal setup and the relativity theories. But the dawning danger of causal paradoxes even in common cause-scenarios makes for another connection: the same feature of the relativity theories (no spacelike causation) that suggests that there should be a common cause in the first place can be seen to suggest that there should *not* be one after all—because that common cause would *still* transmit causal influences superluminally. There is, in other words, not a spatiotemporal and a causal problem, but *two* causal problems; one *motivated* by spatiotemporal concerns and their causal *implications*, the other directly by concerns of causal modeling.

These are no rigorous arguments to strictly *refute* the causal account(s) suggested above, but one may certainly feel encouraged to look for alternatives at this point. While Näger (cf. 2016, p. 1142) thinks that accepting a failure of the CMC, i.e. the existence of uncaused correlations other than maybe those, say, between heads up and tails down in a coin toss, is basically 'giving up on science',[15] Glymour (2006, p. 124) reminds us that "[i]t is not a truth of logic that all experimental associations have a causal explanation [...]." And Adrian Wüthrich (2014, p. 603), a recent advocate of an *acausal* interpretation of EPRB-correlations, sets van Fraassen's impossibility claims of causal explanations aright: it constitutes "*an empirical hypothesis*" that "causal completeness, or 'closedeness', *may fail*." (my emphasis—FB)

[15]This makes him an "essentialist" in Fine's (1989, p. 182) sense.

Recall, moreover, that we can 'beat' EPR's, Bell's, *and* the KS theorem all at once if we reject causal reasoning in certain places: in Sect. 4.3.3, we identified EPR's reality criterion as invoking an abductive inference from an observed measurement result to a hidden cause; and in Sect. 4.3.4, we identified Redhead's 'reality principle' as an instance of abductive inference to a hidden cause as well. So if we reject that this sort of inference is *generally* valid (since reality is not 'causally closed'), we do not have to infer that QM is incomplete. And neither are we then motivated to accept (FUNC) or (FUNC*), so that no KS-contradiction arises. These are positive reasons to reject a causal interpretation of EPRB correlations and other aspects of QM.

We have seen how causality considerations are intimately connected to locality-considerations. But how, precisely, does this supposed lack of 'causal closedness' relate to issues of realism, as we had also suggested above? d'Espagnat (1997, p. 79) has it that "cause-effect relationships linking some phenomena with one another [are] considered by some upholders of conventional realism[16] as constituting a strong argument in favor of their conception." And van Fraassen (1980, pp. 23–28) similarly finds that

> arguments for realism [...] point to explanatory power as a criterion for theory choice. [...] The regularities in the observable phenomena must be explained in terms of deeper structure, for otherwise we are left with a belief in lucky accidents and coincidences on a cosmic scale. [...] The principle of the common cause [...] may be regarded as a formulation of the conviction that lies behind such arguments [...] requiring the elimination of 'cosmic coincidence' by science.

We already outlined, in Sect. 4.3.3, that giving up on the quantum state χ as a cause and rejecting additional causes λ leads to a partial rejection of the semantic condition of scientific realism: We cannot, if we accept the present course, always *interpolate* what we observe by (the behavior of) further, unobserved entities, and therefore not *speak about* or *refer to* everything using scientific methods. Parts of reality that require a quantum treatment would indeed become "unspeakable", as Bell (1984b, p. 171) would have it.

However, we have so far not even given serious thought to *how* the quantum state might *describe* reality after all, and how plausible causal accounts that take *it* to be a relevant (if peculiar) cause really are. The next chapter will give some thorough thought to the three currently most prominent *ontological* interpretations of QM, i.e. interpretations according to which the quantum state should figure in one's ontology.

[16]By "conventional realism", d'Espagnat means the conviction "that science is able to reach at mind-independent reality" (d'Espagnat 2011, pp. 1703–1704), which we understand as full endorsement of *scientific* realism.

Chapter 6
ψ-Ontology, or, Making Sense of Quantum Mechanics

Go to any meeting, and it is like being in a holy city in great tumult. You will find all the religions with all their priests pitted in holy war [...]. They all declare to see the light, the ultimate light. Each tells us that if we will accept their solution as our savior, then we too will see the light.

—C. Fuchs (2002, p. 1)

6.1 Bohmian Mechanics: Taking Wave-Particle Duality Seriously

6.1.1 General Outline and Connections to Orthodoxy

In Chap. 2, we located the importance of de Broglie's research for the development of QM in his speculating about matter waves, and hence in his indirect contribution to Schrödinger's discovery of the SE. But de Broglie's contributions to the early development of QM of course exceeded this point. In particular, he also proposed a so called *pilot wave theory*, in which there would be waves *and* particles, and which he hoped to be a precursor to a future (fully developed) '*theory of the double solution*'. In the latter there would be additional singular solutions to the SE or the KGE, highly peaked field amplitudes with a phase coinciding with that of the regular solutions, replacing genuine, additional particles (cf. Bacciagaluppi and Valentini 2009, p. 60; Jammer 1966, pp. 292 and 357; Mehra and Rechenberg 1987, p. 1209). Pilot wave theory and the theory of the double solution were certainly both motivated by the duality of wave-like and particle-like aspects exhibited in experiments and discussed at some length in Chap. 2. In contrast to the naïve (collapse) approach discussed therein, however, de Broglie's pilot wave theory would have particles be present at all times and only be *guided* or *piloted* by a

The term 'ψ-ontology', having a peircing phonetic ring to it (read: 'siontology'), has been traced back to Chris Granade by M. Leifer (2011).

F. J. Boge, *Quantum Mechanics Between Ontology and Epistemology*, European Studies in Philosophy of Science 10,
https://doi.org/10.1007/978-3-319-95765-4_6

simultaneously occurring wave-phenomenon. The theory of the double solution, he hoped, would then explain everything in terms of waves (fields) alone (cf. Dürr et al. 2012, p. 7; Mehra and Rechenberg 1987, p. 1209).

In a paper published in 1927, de Broglie developed his ideas by considering the KGE for a function $u(\boldsymbol{x}, t) = f(\boldsymbol{x}, t)e^{\frac{i}{\hbar}\varphi(\boldsymbol{x},t)}$ in addition to a solution of the form $\psi(\boldsymbol{x}, t) = Ae^{\frac{i}{\hbar}\varphi(\boldsymbol{x},t)}$. From this he derived the condition

$$\boldsymbol{v} = \nabla\frac{\varphi}{m}, \tag{6.1}$$

describing the velocity of a particle, or rather, the singularity-solution u in which f would be of a divergent form at some point (the singularity, mimicing a particle) and fall off quickly everywhere else. Equation (6.1) is analogous to the formula $\boldsymbol{p} = \nabla S$ in the Hamilton-Jacobi formulation of classical mechanics, where S is the action of a classical particle (e.g. Bacciagaluppi and Valentini 2009, p. 55 ff.; Jammer 1974, pp. 47–49). At the 1927 Solvay conference, de Broglie avoided the appeal to the function u and directly presented his approach as applying to particles (cf. Bacciagaluppi and Valentini 2009, p. 69). Equation (6.1) is usually called a *guidance equation*. It describes the localized particles as moving with definite velocities at all times; not in a Newtonian fashion, but rather guided by the ψ-function *via* the gradient of its phase φ, i.e. as being 'pushed' by the wavefronts.

But the ψ-function, of course, had to fulfill a dual role to connect to the probabilistic content of QM: as a guiding wave and as a 'probability wave'. This was acknowledged by de Broglie, but he still failed to demonstrate how the orthodox description of measurement processes could be recaptured within his formalism. According to Jammer (1966, p. 357) and others (e.g. Dürr et al. 2012), the harsh criticism of Pauli about this and related problems in particular lead de Broglie to abandon his work on the pilot wave theory, and as Bohm (1952a, p. 167) and Bacciagaluppi and Valentini (2009, p. 229) note, he discovered additional difficulties himself shortly after.

In 1952, however, David Bohm developed, in two subsequent papers (Bohm 1952a,b), an account similar to, or in many respects identical with de Broglie's pilot wave theory, without being aware of de Broglie's work at first. His starting point was to consider the time dependent SE on a wave function $\psi(\boldsymbol{x}, t) = R(\boldsymbol{x}, t) \cdot e^{\frac{i}{\hbar}S(\boldsymbol{x},t)}$ with S and R real valued. Plugged into the TDSE with potential V, this yields

$$ie^{\frac{i}{\hbar}S}\hbar\frac{\partial R}{\partial t} - Re^{\frac{i}{\hbar}S}\frac{\partial S}{\partial t} = -\frac{\hbar^2}{2m}e^{\frac{i}{\hbar}S}\left[\Delta R + 2\frac{i}{\hbar}\nabla R\cdot\nabla S + \frac{i}{\hbar}R\Delta S - \frac{1}{\hbar^2}R(\nabla S)^2\right]$$

[1]You can easily verify this by restricting attention to one of the coordinates, differentiating ψ twice using product- and chain rule, and realizing that $\nabla R\cdot\nabla S = \frac{\partial R}{\partial x}\cdot\frac{\partial S}{\partial x} + \frac{\partial R}{\partial y}\cdot\frac{\partial S}{\partial y} + \frac{\partial R}{\partial z}\cdot\frac{\partial S}{\partial z}$ (and so forth).

$$+ V \cdot Re^{\frac{i}{\hbar}S}.^{1} \tag{6.2}$$

Up to the exponential (which cannot generally vanish; cf. the Euler formula) and sorted for real and imaginary part, this in turn reads

$$i\left(\hbar\frac{\partial R}{\partial t} + \frac{\hbar^2}{2m}\left[2\frac{1}{\hbar}\nabla R\cdot\nabla S + \frac{1}{\hbar}R\Delta S\right]\right)$$

$$-\left(R\frac{\partial S}{\partial t} - \frac{\hbar^2}{2m}\left[\Delta R - \frac{1}{\hbar^2}R(\nabla S)^2\right] + V\cdot R\right) = 0 \tag{6.3}$$

(6.3) nicely decouples into the two equations

$$\frac{\partial R}{\partial t} = -\frac{1}{m}\left[\nabla R\cdot\nabla S + \frac{1}{2}R\Delta S\right] \tag{6.4}$$

$$\frac{\partial S}{\partial t} = -\left[\frac{1}{2m}(\nabla S)^2 + V - \frac{\hbar^2}{2m}\frac{\Delta R}{R}\right], \tag{6.5}$$

since both real and imaginary part must vanish individually for the entire expression on the LHS of (6.3) to be zero. Remarkably, one obtains the Hamilton-Jacobi equation from (6.5) if $\hbar$ is treated as zero, i.e. as a kind of '$\hbar \to 0$-limit'. Stated differently, (6.5) can be read as describing the motion of a classical particle with momentum $\boldsymbol{p} = \nabla S$ (as in de Broglie's approach), subject to a potential V and an additional '*quantum-potential*'

$$Q = -\frac{\hbar^2}{2m}\frac{\Delta R}{R}. \tag{6.6}$$

The expression $-\nabla(V + Q)$ could then be used by Bohm as an expression of a force, exactly as in classical mechanics. Using also that R^2 defines the probability density $\varrho = |\psi|^2$, Eq. (6.4) in turn implies the continuity equation $\frac{\partial\varrho}{\partial t} + \nabla(\varrho\cdot\boldsymbol{v}) = 0$, where $\boldsymbol{v} = \nabla\frac{S}{m}$, and $\varrho\cdot\boldsymbol{v}$ defines a probability current $\boldsymbol{j}$.

Based on this setup, Bohm provided treatments of stationary states, scattering problems, angular momentum, and may other aspects of QM, thereby exceeding de Broglie's previous treatment and curing it of certain problems. In virtue of the importance of Bohm's contributions to this version of QM, it is sometimes referred to as *Bohmian mechanics* (BM) (e.g. Dürr and Teufel 2009; Dürr et al. 2012), sometimes, to honor de Broglie's early contributions, the *de Broglie-Bohm theory* (e.g. Holland 1995; Friebe et al. 2015). For convenience, and convenience alone, we will use the acronym 'BM' throughout.

The central conjecture to be investigated here now is the following *dual ontology*:

Conjecture 2 (The dual ontology) Quantum mechanical systems are tiny objects which always possess a true, definite state, with precise values of position and momentum at all times. They move on usually non-Newtonian trajectories, guided, in some way, by a wave function defined on configuration space.

Above we have referred to BM as a 'version' of QM. Indeed, the physical interpretation is so much at variance with orthodox QM that BM has sometimes been thought of as an *alternative theory* (e.g. Dürr et al. 2012, p. 24; Ivanova 2014, p. 211 ff.). However, the same could be said about any 'non-orthodx' interpretation of QM, and we have seen that the essential ingredients of BM are already present in the QM formalism. Thus, BM is actually *formally conservative*. What is true though is that BM is *conceptually* highly revisionary: it assumes the existence of *particles* – not particles – at all times, and the wave function merely plays the role of a guiding or piloting field.

There is a sense in which this dual ontology is very appealing: we now can picture the situation such that a tiny lump of matter (a 'point particle') is being pushed through space by a wave (symbolized by ψ), and the influence of the wave *explains* the funny statistics that characterize the particle's behavior. And in fact, single-particle single- and double slit experiments can be emulated with oil droplets in vibrated baths of the same fluid, wherein the (bouncing) droplet creates a wavepacket around itself and the joint dynamics statistically reproduce characteristic single-particle interference patterns (cf. Couder and Fort 2006). Apparently our dual ontology hence comes quite close to fulfilling the 'dream' of Chap. 4, to give a clear, ontologically meaningful *explanation* of the success of QM when the latter is construed in a merely *operational* or *instrumental* fashion.

But we know already that the ψ-function is not a simple field on $3D$-space or $4D$-spacetime, but instead an often-times inseparable function on configuration space—a fact that we have taken into account in the formulation of our present conjecture. We have seen how far-reaching the consequences of this are in the discussion of Bell's theorem and associated issues. It is hence of utmost importance to understand the role of the ψ-function in BM more clearly, whence we will turn to a detailed analysis thereof in the next section.

Let us now first look at the central connections of BM to orthodox QM in a little more detail though. How is the 'dual role' of the wave function as guiding- *and* 'probability-wave' established? To some extent, its probabilistic content, which Dürr et al. (2012) refer to as the *quantum equilibrium hypothesis*, is always *postulated*[2]: Bohm (1952a, p. 171) explicitly introduced this as a fundamental assumption; Dürr et al. (2012, p. 44 ff.) go through some means to demonstrate

[2]Cf. Passon (2004, p. 7) for an overview of attempts to get around this feature.

that, if, possibly as a matter of incomplete knowledge about initial conditions, one assumes the actual configuration of the entire universe at a supposed initial time to be distributed according to the squared modulus of a *universal* wave function, then the configurations of all subsystems at later times will be distributed according to the squared modulus of a conditional wave function describing these systems alone.[3]

From this (motivated) assumption of ψ's probabilistic content follow basically all the further connections to orthodox QM and its predictions. The significance of self-adjoint operators as observables is taken to flow from their spectral decomposability $\hat{A} = \sum_j a_j \hat{P}_j$ and the statistical content of expressions like $\langle\psi|\hat{P}_j|\psi\rangle = |\psi_j|^2$. But Dürr et al. (2012) emphasize that this means that self-adjoint operators only play a role in the statistical description of certain *experiments*, and do not generally represent the statistics for obtainment of a value in a *genuine measurement*, i.e. "the ascertaining of the value of a quantity" (Dürr et al. 2012, p. 81):

> the notion of operator-as-observable in no way implies that anything is genuinely measured in the experiment, and certainly not the operator itself! In a general experiment no system property is being measured, even if the experiment happens to be measurement-like. (Dürr et al. 2012, p. 97)

By a "measurement-like" experiment, they mean "one which is reproducible in the sense that it will yield the same outcome as originally obtained if it is immediately repeated." (Dürr et al. 2012, p. 95) Given this view of self-adjoint operators as statistical devices, it is then a small step to introduce POVMs as describing the statistics of a more general set of (not necessarily reproducible) experiments (cf. Dürr et al. 2012, p 115 ff. for details).

The de-valuation of self-adjoint operators in particular has important implications for how the consequences of the KS theorem are viewed. Here Bohm and Hiley (1993, p 120; notation adapted; my emphasis – FB) write:

> In our interpretation we do not assign values such as $v(A)$, $v(B)$, $v(C)$, ... to the operators. For these operators do not correspond to *beables* in our approach. Rather the beables are the overall wave function together with the coordinates of the particles, both of the observed system, $\boldsymbol{x}$, and the [*sic*] of the observing apparatus, $\boldsymbol{y}$. These beables determine the results $R(A)$, $R(B)$, $R(C)$, ... of each individual measurement operation. But these results are not present before the measurement operation has been completed. [T]here is no pre-existing quantity that is actually revealed in this process.

This notion of a 'beable' (read: 'be-able')—or rather: *local* beable—stems from Bell (1976), and Bell introduced the term to denote something that *is* and is *there*, in contrast to QM's observables. Essentially, Bohm and Hiley here deny the very foundation of the KS theorem: the value definiteness rule (VD). Moreover, they seem to *embrace* contextuality w.r.t. QM's observables, and trace it back to the total interaction of system and apparatus (both being constituted by particles). Similar but slightly more specific remarks are made by Dürr et al. (2012, p. 149; notation adapted; my emphasis – FB):

[3] We will introduce the notion of a conditional wave function thoroughly below.

in Bohmian mechanics the random variables $Z_{\mathscr{E}}$ giving the results of experiments $\mathscr{E}$ depend, of course, on the experiment, and there is no reason that this should not be the case when the experiments under consideration happen to be associated with the same operator. Thus with any self-adjoint operator $\hat{A}$, Bohmian mechanics naturally may associate *many different random variables* $Z_{\mathscr{E}}$, one for each different experiment $\mathscr{E} \mapsto \hat{A}$ associated with $\hat{A}$.

The point, again, is that no self-adjoint operator truly represents a dynamical quantity of a system that can take on definite values, revealed by a suitable experiment, but rather represents a set of different experiments (or specific features thereof), each of which may be contextual in the sense that it depends on the total experimental setup. This is somewhat stronger than what we have presupposed for our minimal revision in Sect. 4.3.4; it is not that only *degenerate* operators are associated with multiple *observables*, but rather *all* (self-adjoint) operators are associated with multiple *experiments*. Talk of 'observables' is viewed as misleading here, since 'observables' are typically features of experiments, and the only 'real observable' is *position* (e.g. Dürr et al. 2012, p. 82). This special role for position has some intuitive appeal when one considers that usually the positions of pointers, pixels, or flashing light bulbs are ultimately observed in experiments.[4]

So far so good. But we noted in Sect. 4.3.4 that, when considered w.r.t. composite systems, the KS theorem has the implication that certain assumptions about separability and/or locality cannot be maintained (precisely: (VSEP) and (VLOC)). As Dürr et al. (2012, p. 154) put it: "whenever the relevant context is distant, contextuality implies nonlocality." Im BM, this feature of non-locality is built into the wave function for (certain) many-body systems. This leads us back to the necessity of confronting the ontological significance of the wave function in BM.

6.1.2 *Non-locality and the Ontological Status of the ψ-Function*

As we noted above, the ψ-function also has, besides its probabilistic content, the more directly ontological meaning of something that guides or pilots particles through space. Since it is defined on configuration space though, it is in need of explanation *how* ψ can act on the particles, situated in spacetime.

In particular, we need to turn to the Bohmian treatment of many-body systems to shed more light on this issue. Formally, there is not much difference to the single-body treatment considered above; since ψ is a complex-valued function, one can write it in the polar form $\psi = R(\boldsymbol{x}_1, \ldots, \boldsymbol{x}_N)e^{\frac{i}{\hbar}S(\boldsymbol{x}_1,\ldots,\boldsymbol{x}_N)}$ for N particles all the same. The Hamiltonian of the SE will then be of the form

[4]A 'measurement of the position operator' need not be a measurement of the actual position of a particle though; cf. the example in Dürr et al. (2012, pp. 142–143).

$$\hat{H} = -\frac{\hbar^2}{2}\sum_{j=1}^{N}\frac{\Delta_j}{m_j} + V(\boldsymbol{x}_1, \ldots, \boldsymbol{x}_N), \tag{6.7}$$

where Δ_j is the Laplacian w.r.t. $\boldsymbol{x}_j$, and the quantum potential will be of the form

$$Q = -\frac{\hbar^2}{2R}\sum_{j=1}^{N}\frac{\Delta_j}{m_j}R(\boldsymbol{x}_1, \ldots, \boldsymbol{x}_N) \tag{6.8}$$

(cf. Bohm 1952a, p. 175). Moreover, the expression

$$\boldsymbol{v}_j = \frac{\nabla_j S}{m} = \frac{\hbar}{m_j}\mathrm{Im}\frac{\psi^*\nabla_j\psi}{\psi^*\psi} \tag{6.9}$$

then defines a velocity for the j-th particle, where this form of the guidance equation is introduced, in particular, by Dürr et al. (2012). It can be easily seen that this is equivalent to (6.1) in the simple case of a spinless particle, since $\nabla_j\psi = \nabla_j R \cdot \frac{\psi}{R} + \frac{i}{\hbar}\nabla_j S \cdot \psi$ by product and chain rule. It is also easy to confirm that this is equal to $\frac{\boldsymbol{j}_j}{\varrho}$, with $\boldsymbol{j}_j$ defined in analogy to (2.56), and $\varrho = |\psi|^2$. In the case of a (Pauli-) spinor $\boldsymbol{\Psi} \doteq \begin{pmatrix}\psi_+(x)\\ \psi_-(x)\end{pmatrix}$, enumerator and denominator in (6.8) are simply interpreted in terms of inner products on $\mathbb{C}^2$ (cf. Bell 1971, p. 33; Dürr et al. 2012, p. 89). Indeed, we here notice another crucial feature of BM, namely that *spin* does not appear as a property of the *particles*, but of the *wave function* only.

To this extent, that BM introduces particles, characterized by definite velocities (momenta) and positions, and reduces 'decidedly nonclassical concepts' such as spin to the particles' *behavior*, or to a feature of that which governs them, it constitutes somewhat of a return to classical mechanics. But of course this is only possible in an empirically adequate fashion *via* the acceptance of the somewhat mysterious 'piloting' wave function and its statistical content.

Dürr et al. (2012, p. 30), moreover, explicitly introduce *actual* positions $\boldsymbol{X}_j$ of the particles in contrast to the 'generic' configuration space variables $\boldsymbol{x}_j$, and these actual positions together with the ψ-function are then construed as a complete description of the system in question. These actual coordinates also determine actual *velocities*, or rather, can be computed by solving $\frac{\mathrm{d}\boldsymbol{X}_j}{\mathrm{d}t} = \frac{\hbar}{m_j}\mathrm{Im}\frac{\psi^*\nabla_j\psi}{\psi^*\psi}(\boldsymbol{X}_1, \ldots \boldsymbol{X}_N)$.

It was noticed early on by Bohm (1952a, p. 168) that this assumption of existing definite positions and momenta in his interpretation would make it count as a *hidden variables-approach*, the hidden variables being the definite particle positions and momenta. But did we not see hidden variables approaches suffer from so many incapabilities in the last chapter? BM is *not* a ψ*-epistemic* hidden variables interpretation of QM, in the sense of Chap. 4. In fact, it is not even clear that it can be made to fit into the OM approach, for formal reasons (cf. Feintzeig 2014). And even if it could, it would (at least *prima facie*) rather qualify as ψ*-supplemented* in

Harrigan and Spekkens' terminology; ψ appears to play *some* role other than a mere representation of the knowledge of an observer. Whether this view of BM is correct, however, strongly depends on what ψ's role really *is*, as the subsequent discussion will show.

It also seems worth emphasizing that BM does not suffer from the objections we advanced against the historical ensemble interpretations in Sect. 4.1. BM can reproduce double-slit interference with single particles and quantum tunneling in virtue of the quantum potential Q, and even examples of (or analogous to) Mach-Zehnder interferrometry (for details e.g. Holland 1995, p. 200 ff.). The essence of the interference examples is that the measuring device will influence the quantum potential, and that particles will travel in curved trajectories that depend on the total setup, thus accounting for the observed patterns and frequencies. The essence of the tunneling example is that the quantum potential constitutes additional energy, whence a few particles will always escape the binding potential.

Notably though, since ψ depends on *all* the coordinates $\boldsymbol{x}_1, \ldots, \boldsymbol{x}_N$, (6.8) implies that the same goes for $\boldsymbol{v}_j$: the velocity of each particle in an N-particle configuration depends irreducibly on the coordinates of all the other $N-1$ particles, and (*via* S) on the ψ-function.[5] Bell (1966, p. 11) noticed this to imply that BM had "in general a grossly *non-local* character" (my emphasis – FB).[6] It hence seems to embrace the consequences of Bell's theorem in the most direct sense possible; but there is a sense in which it is nonetheless *not* in conflict with SR, while it is clear that the correlatedness and pre-measurement definiteness of positions and momenta raises worries about the possibility of superluminal signaling – and hence causal paradoxes. We will turn to this issue below, when we confront BM's relation to relativity in more detail.

So far we have seen that the wavefunction seems to play, among other things, a *physical* role, and that it somehow influences the particles, thus correlating them 'non-locally'. Given that ψ is defined on configuration space though, this still does not answer how it manages to transmit its 'non-local influence' to the particles. A first thing to note is that "in a universe governed by Bohmian mechanics there is a priori only one wave function, namely that of the universe, and there is a priori only one system governed by Bohmian mechanics, namely the universe itself." (Dürr et al. 2012, p. 85) To make sense of practice however, wherein the wave function of the universe never occurs, Dürr et al. introduce the notions of *conditional* and *effective wave functions*, used to describe the behavior of subsystems.

Consider the wave function of the entire universe, call it '$\Psi(x, y)$', here conveniently split for (generic) configuration space coordinates $x = (\boldsymbol{x}_1, \ldots \boldsymbol{x}_N)$ and $y = (\boldsymbol{y}_1, \ldots \boldsymbol{y}_M)$, where the first ones ($x$) correspond to the coordinates of a system of interest and the second ones (y) to its environment ('the rest'). Now the distinction of actual from generic coordinates defines a *conditional* wave function

[5] This only holds, of course, if the wavefunction is not factorizable, because otherwise all the factors in (6.8) that do not depend on $\boldsymbol{x}_j$ can be 'divided off'.

[6] However, cf. Norsen (2010) for an interesting first step towards a fully local view of BM.

for the x-system as the function $\psi(x,t) = \Psi(x, Y(t), t)$, i.e., one singled out by the actual configuration Y of the rest of the universe. However, the conditional wave function will not always evolve in accord with the SE. So it is useful to also introduce an *effective* wavefunction (at any given time; we suppress the t below) for the x-system, which is the function $\psi(x)$ for $\Psi(x, y)$ being of the form $\psi(x)\Phi(y) + \Psi^{\perp}(x, y)$, with the actual coordinates Y in the support of Φ, and where Φ and $\Psi^{\perp}$ are supposed to have 'macroscopically distinct' supports in the y-coordinates (cf. Dürr et al. 2012, p. 85–86). The effective wave function will always obey the SE, and thus corresponds to the wave functions used in practice. Note however that the other terms of the (highly entangled) wavefunction of the universe, $\Psi^{\perp}(x, y)$, are thus *empty*.

The use of the conditional wave function also leads to an 'apparent collapse'. Take an entangled state $|\Psi\rangle = \sum_i c_i |\psi_i\rangle |\phi_i\rangle$ of some system and apparatus, as results from their joint unitary evolution (cf. Chap. 2). Since only one part of $|\Psi\rangle$, projected onto configuration space variables (x, y), will contain the actual coordinates (e.g. pointer position) Y of the apparatus, only one of the possible outcomes will be measured, and we can thus assign the conditional wave function $\psi_i(x) = \Psi(x, Y)$ to the system after the measurement. This 'collapse', which is then merely an "act of convenience" (Dürr and Teufel 2009, p. 180), comes at the price of neglecting something that may or may not 'still be there' – the empty, non-effective wave functions – depending on one's view of Ψ. But this alleged price, Dürr and Teufel (2009, ibid.) have it, "amounts to nothing", because interference with them will quickly become very hard to establish due to environmental interactions.[7]

Again, so far so good. But we *still* do not know what the wave function 'really is', and how it 'interacts' with particles located in space(time). In fact, the accounts of what exactly the wavefunction represents differ grossly among 'Bohmians'. Bohm must have expressed his major dissatisfaction with Dürr et al.'s views being termed 'Bohmian mechanics' (cf. Hiley 1999, p. 117), broadly speaking because his account of BM is ontologically much 'richer'. It may hence not even be fair to discuss both Bohm and Hiley's and Dürr et al.'s views, as well as de Broglie's original attempts at a double solution, all under the same heading. But we have focused here mostly on commonalities, and some differences should become obvious from the discussion and the quotes therein. We stress that these are not *all* the differences though (e.g. Passon (2004) and Friebe et al. (2015, pp. 194–196) for a deeper discussion), and one crucial, further difference is e.g. the abandonment of the *quantum potential* by Dürr et al. (2012, p. 10) and others (e.g. Valentini 1996, p. 47).

Now Bohm (1952a, p. 170) himself thought of the wavefunction as

> an objectively real field [that] exerts a force on the particle in a way that is analogous to, but not identical with, the way in which an electromagnetic field exerts a force on a charge, and a meson field exerts a force on a nucleon.

[7] This will become clear after the discussion of decoherence in Sect. 6.3.2.

And Bell (1981b, p. 128) equally believed that:

> No one can understand this theory until he is willing to think of ψ as a real objective field rather than just a 'probability amplitude'. Even though it propagates not in 3-space but in $3N$-space. (emphasis omitted)

But there is a bunch of troubles with this view. The wave function behaves, for instance, unlike classical fields in that "there's no back action, no effect in the other direction, of the configuration upon the wave function[...]." (Dürr et al. 2012, p. 266) The more pressing problem is of course that it is far from clear *how* the wave function, residing in configuration space, effects the changes in spacetime. Bell (1981b, p. 128) merely referred to the determination of velocities by the wave function as "rather original"; and Bohm and Hiley (1993) helped themselves to an understanding by introducing a (somewhat elusive) notion of "active information" (p. 36), which was "ordered in the configuration space" (p. 60) and therefore implied that certain systems are "wholes guided by a pool of common information" (p. 61).

More concretely, Bohm and Hiley (1993, pp. 31–32) likened the situation of particle and guiding wave to "a ship on automatic pilot being guided by radio waves." Here the radio waves would equally not exert a mechanical force on the ship to effect the guided motion. But of course this analogy easily breaks down since, unlike the quantum wave function, the 'information' the ship acquires can be analyzed into *classical* physical processes 'ordered' in spacetime, effecting, in conjunction with the physics of the ship's motor, its total behavior. It is hence hard to see how it helps us to a proper understanding to replace a wave propagating in a configuration space by 'active information ordered in configuration space', both of which would have to somehow influence the goings on in spacetime. Not to mention the fact that the former concept (wave), in contrast to the latter (active information), is at least well understood *on* spacetime.

BM as presented by Dürr et al., in contrast, could be cashed out as essentially just a theory of strangely behaving particles. The particles are referred to by them as the *primitive ontology* of their theory, that which *primarily exists*. This concept of a primitive ontology is certainly heavily inspired by Bell's (1976) notion of local beables, and we will use both notions somewhat interchangeably.[8] About the wave function, they suggest that "one should think about [...] the possibility that it's nomological, nomic—that it's really more in the nature of a law than a concrete physical reality." (Dürr et al. 2012, p. 266) That this is plausible they motivate (pp. 267–268) by comparing the wavefunction to the Hamiltonian function, which is equally defined over configuration space, but about whose interpretation no one wonders (or at any rate by far not to the same extent).

But Dürr et al. (2012, p. 268 ff.) also concede that we should neither be able to *alter* the laws of nature, which we do with wave functions all the time, nor should they be *dynamical*, i.e., subject to (automatic) change.[9] To solve this difficulty, they rely on their belief that *only* the wave function of the *universe*, Ψ, is 'fundamental',

---

[8]It should be noted though that at least S. Goldstein (private communication) thinks that there are subtle but important conceptual differences.

[9]Cf. however Maudlin (2007, pp. 11–12) for some dissenting views.

and *this* wave function, they argue, is *neither* controllable *nor* dynamical. That ψ may not be dynamical they gather from the timeless *Wheeler-DeWitt* equation, $\hat{\mathscr{H}}\psi = 0$, where $\hat{\mathscr{H}}$ is a (kind of) general-relativistic Hamiltonian, and where on the right hand side the familiar $i\hbar\frac{\partial}{\partial t}$ from the TDSE is missing.[10]

This is an interesting suggestion, to be sure, and it reduces our dual ontology to a '*unal*' one; as Callender (2015, p. 3157) puts it: "no dualistic ontology [...] therefore no interaction problem." But it is still not sufficiently clear what these claims of 'nomicity' amount to. Roughly, the claim is that there are particles (the primitive ontology) that behave in a particular manner which, at times, is rather strongly and surprisingly correlated. That they do so is a law of nature. This law is or is encoded in/dependent on ψ. But *what notion* of a law of nature is being invoked here? For there is, of course, not one unified account of the laws of nature, but instead (as was the case e.g. with probability) a plethora of grossly differing accounts.

Consider the following two basic options: either the term 'laws of nature' signifies something in the real world, its 'modally robust structure(s)' if you will, or it does not. In the latter case, the appearance of law-like behavior, i.e. our so-far success in using inductive practices, typically summarized in terms of mathematical formulae, seems worthy of further explanation; what replaces the 'laws *qua* real entities', so as to account for the seeming uniformity of nature? A natural response, typically traced back to Hume's skeptical investigation of induction and necessary connections, summarized nicely in his remark that "[a]ll events seem entirely loose and separate" (Hume 1748, §26), is to state that laws are just that: *apparent regularities*, summarizable by a statement or mathematical formula. For obvious reasons such a view is often referred to as *Humean* (e.g. Psillos 2002, p. 5). But merely stating that laws are expressions of regularities is obviously insufficient, since there are all kinds of regularities that supposedly should not qualify as 'laws of nature'.

The most sophisticated version of a regularity view is the account precisified by David Lewis (1994), and advocated independently by Mill and Ramsey before him (cf. Psillos 2002, pp. 9 and 139 ff.). This is the so called *best system analysis* (BSA). Lewis (1994, p. 478) put the BSA's foundational idea as follows:

> Take all deductive systems whose theorems are true. Some are simpler, better systematized than others. Some are stronger, more informative, than others. These virtues compete: an uninformative system can be very simple, an unsystematized compendium of miscellaneous information can be very informative. The best system is the one that strikes as good a balance as truth will allow between simplicity and strength. How good a balance that is will depend on how kind nature is. A regularity is a law iff it is a theorem of the best system.

Now of course the BSA is not free of complications (what measures simplicity? what strength? how can laws of nature depend on our systematization?) but it nonetheless constitutes a serious contestant in an active field of research (the philos-

---

[10] ψ would here in fact be a functional on a space of three-metrics and $\hat{\mathscr{H}}$ contains a functional derivative (cf. Kiefer 2007, p. 141 ff. for details).

ophy of the laws of nature), in which (once more) none of the contestant accounts is free of complications.[11] Of more interest to us should be the consequences of a Humean view for BM. If the laws of nature are merely theorems (or axioms) of a best system, then this still means that 'things *just happen to be* the way the theorem/axiom says'. Period. If Ψ hence figures in a law of nature, describing the behavior of particles as correlated over large distances, *then they just happen to behave in this astonishingly correlated way. Period.* This is not to say that such an account cannot be maintained,[12] but it is certainly somewhat discomforting that particles should 'just' behave in such a peculiar manner.

We must ask, however, whether there is really much of an asymmetry between quantum and classical cases here. In particular, the 'brute factivity' of correlations could be viewed as *no more* mysterious in the quantum case than in the classical case.[13] Is it not equally 'spooky' that material particles scatter off each other in the way they do, i.e. that there are *these* particular correlations that we regularly observe after 'action by contact', once we accept a Humean view? Indeed, one could flesh this out into a somewhat 'therapeutic'[14] stance towards the laws of nature and our intuitions about them, and the apparent objection might be turned into a mere feature. Nevertheless, the *explanatory value* of ψ or Ψ and the guidance equation(s) is thereby grossly reduced[15]; and the nice feature outlined in Sect. 6.1.1 – that we now have strangely behaving particles whose strange behavior we can *understand* in virtue of a guiding field – *vanishes*. Again, this is a feature that Humeanism arguably bestows upon *any* kind of law (or lawlike entity): "on Humeanism, the laws of fundamental physics do not have any explanatory function. They sum up, at the end of the universe, what has happened in the universe, but they do not answer the question concerning why what has happened did in fact happen, given certain initial conditions." (Esfeld et al. 2014, p. 783)

There is an interesting additional feature that ψ has on a nomological view in general and a Humean one in particular. On any nomological account, ψ becomes somewhat *epistemic* again, although its 'epistemicity' is not formally explicated in the way it was in Chap. 4. The possibility of introducing an epistemic interpretation

[11]Of course it may also seem quite counter-intuitive that, on the BSA, it appears to depend on the existence of *minds*, being the carriers of descriptive systems, whether there *are* laws of nature or not. Even Lewis (1994, p. 479; emphasis in original) admitted that "*if* nature were unkind, and *if* disagreeing rival systems were running neck-and-neck, then lawhood might be a psychological matter [. . .]." We should appreciate, though, that at least the entities regularly exhibiting the same behavior *do* reside in the outside world and they *do* behave so mind-independently, even if they do not *have to*; and *given* certain standards, the *best* system may also be precisely (objectively) fixed. So there is certainly no radical subjectivism here (cf. also Psillos 2002, pp. 153–154).

[12]Cf. in particular Callender (2015) for a detailed treatment of some problems and potential solutions.

[13]Author's note: I owe this objection essentially to Sheldon Goldstein and Christian Loew independently (private communication in both cases).

[14]Cf. also Friederich's (2015) book-length investigation of a Wittgensteinian-therapeutic approach to QM, in this connection.

[15]Author's note: I owe thanks to Andreas Hüttemann for making me aware of this issue.

in this alternative fashion is acknowledged, in particular, by Harrigan and Spekkens (2010, p. 153); the distinction between the wave function Ψ of the universe and the effective wave function ψ of a given system is taken to indicate that at least the *latter* codifies knowledge in some sense. Accepting, however, that Ψ might be 'nomological', they believe that "it is presumably a category mistake to try to characterize the universal wave function as ontic or epistemic[...]." (ibid.)

According to our present analysis, this opinion is clearly flawed. Callender (2015, p. 3158) provides a similar analysis of laws into either 'its or bits', where the 'bits' are understood as "an aspect of our knowledge." (Callender 2015, p. 3154) Moreover, "for the Humean", Callender (2015, p. 3159) has it, the laws "are a special kind of *Bit*." (my emphasis – FB) They hence represent an aspect of our knowledge about something else (the particles, in BM). That this should be so is motivated by one particular way to spell out the BSA. Here is how Hall (as paraphrased by Callender 2015, p. 3160) puts things:

> (roughly) a proposition is a law iff an *ideal observer*, someone who is rational and has full information about what is being systematized and embraces our sciences' standards (which include simplicity and comprehensiveness), declares the proposition a law. (my emphasis – FB)

And similarly Schrenk (2014, p. 1788): "suppose you *knew* everything and you organised it as simply as possible in various competing deductive systems[...]." (my emphasis – FB)

On a Humean view, Ψ may thus come out just as epistemic as ψ, referring, to some degree, to the mental states of some observer and how she relates to the 'real goings on'. But if this is so, it appears *prima facie* problematic that we do not seem to *know* Ψ—so it codifies basically and aspect of *no one's* knowledge. Note, however, that in the above quotes, the BSA is cashed out in *counterfactual terms* and refers to an *ideal* (or future) scientist. The 'best system' hence plays a merely *definitional* role on this reading, and some solution Ψ to the fundamental equation of a future theory of quantum gravity could still be interpreted epistemically in the sense indicated.

Lewis (1994, p. 479), moreover, believed, or at least *hoped* (cf. ibid.) that "the best system will be *robustly* best—so far ahead of its rivals that it will come out first under any standards of simplicity and strength and balance." (emphasis in original) So Ψ might still be 'bit' rather than 'it', but an *objectively preferred* bit: it does not *directly* signify anything *in reality*, but the total behavior (or the totality of worldlines) of all particles in the universe makes some specific law (generalized guidance equation) in which Ψ occurs *inevitably true*. This makes Ψ no less epistemic; it aids in summarizing our knowledge of the behavior of something else (the entities contained in the 'primitive ontology' or the 'beables'). But this summary now comes in an objectively preferred way.

Still, not everybody is convinced by Humeanism, especially in the light of having to accept 'spooky' correlations as ontological bedrock. What alternatives do we hence have if we want to hold fast to a nomological reading of Ψ but do not want to swallow that particles *just* coordinate their behavior over large distances such as to give rise to the peculiar EPRB-correlations? A prominent alternative that adds a 'real something' to escape the apparent arbitrariness or contingency is the ADT account of laws (after Armstrong 1978, Dretske 1977, and Tooley 1977). This account is nicely summarized by Psillos (2002, p. 163) as follows:

> It is a law that all Fs are Gs if and only if there is a relation of *nomic necessitation* $N(F, G)$ between the properties (universals) F-ness and G-ness such that all Fs are Gs. (my emphasis – FB)

A first drawback of the ADT account is that physical laws are typically expressed as differential equations, and there is a general difficulty of how to relate these to statements of the form "all Fs are Gs" (cf. Smart 1993, p. 154; Maudlin 2007, pp. 11–12). Take, for instance, Faraday's Law of induction $\nabla \times \boldsymbol{E} = -\frac{\partial \boldsymbol{B}}{\partial t}$, one of Maxwell's equations. It *identifies* the rotation of an electric field $\boldsymbol{E}$ with the temporal change in a magnetic field $\boldsymbol{B}$. It is not of a conditional form, but could rather be translated into *two* conditional statements ('all rotating $\boldsymbol{E}$-fields are[16] changing $\boldsymbol{B}$-fields *and vice versa*'). What 'nomically necessitates' what?

Moreover, assume that we have found an 'order' for these terms, i.e. that we can identify one side of the equation as privileged over the other, e.g. by taking *practical* considerations into sight (i.e. what is usually determined in virtue of what) that tell us how to *read* such equations. For instance, we would probably first find a universal wavefunction Ψ (from its dynamical law) and then determine particle trajectories from a corresponding guidance equation, so we might speculate that something to do with Ψ necessitates the particle trajectories.

So assume for the sake of argument[17] that we can make sense of some such universal, generalized guidance equation of the form $\boldsymbol{v}_j = f(\frac{\Psi^* \nabla_j \Psi}{\Psi^* \Psi})$, where f would be any suitable kind of mapping, Ψ may depend non-trivially on all the coordinates of all particles in the universe, may be atemporal, may be of any desired mathematical complexity (a functional, a spinor-, tensor-, operator-valued function), and ∇ and $*$ may be replaced by whatever necessary generalization (cf. then next section for hints). Then the difficulty of transferring the treatment of the (classical) electromagnetic fields to the treatment of velocities in BM is that the field Ψ still 'lives' on configuration space, and that the latter sort of law hence does not treat (exclusively) of properties instantiated *in spacetime*.

"But", the alert reader may interject, "did we not liken the wavefunction to the classical Hamiltonian which equally 'lives' on configuration space?" This is certainly correct, but on the ADT account, we are looking for relations between universals, and the candidate universals on the RHS of our supposed universal guidance equation are gradients, ratios, proportionality constants... *and* Ψ. Now a classical Hamiltonian, once written out, does not occur as an additional universal in classical mechanical equations, and the terms $V_{jk}(\boldsymbol{x}_j - \boldsymbol{x}_k)$ in some such

[16] It may be tempting to say 'are associated with'; but in SR a simple coordinate transformation can turn $\boldsymbol{E}$-field components into $\boldsymbol{B}$-field components and *vice versa*, whence they are often viewed as two sides to the same phenomenon (cf. Griffiths 1999, p. 529 ff.), as we already noted in Chap. 2.

[17] Cf. the next section for difficulties of generalizing guidance equations.

Hamiltonian $H = \sum_j \frac{p_j^2}{2m} + \frac{1}{2}\sum_{j,k} V_{jk}(\boldsymbol{x}_j - \boldsymbol{x}_k)$, relating the behaviors of multiple particles at positions $\boldsymbol{x}_j$ (e.g. Schwabl 2006, p. 55 ff.), represent interactions as mediated by potentials V_{jk} that fall off at some power of the distance $|\boldsymbol{x}_j - \boldsymbol{x}_k|$ between the particles (i.e. decrease in strength with distance). So the candidate universals (relevant for comparison) that are being lawfully related in classical mechanics are fields and particles that *are* instantiated in spacetime.

The general point is that the situation of the crucial functions in classical mechanics (such as the Hamilton function) is notably different from that of Ψ (or ψ) in QM. This becomes most obvious when one tries to cast both theories in the same formal mold. This possibility has been explored in some detail by Holland (1995, p. 55 ff.) and Callender (2015, p. 3164 ff.), following him, makes us aware of the fact that classical mechanics can be rewritten in terms of a "classical wavefunction" $\psi_{cl.} = Re^{\frac{i}{\hbar}S}$, satisfying a "classical Schrödinger equation" that includes Bohm's quantum potential Q as an additional term, which then does *not* occur in the corresponding Hamilton-Jacobi equation. Here $\hbar$ merely occurs as a scaling for the appropriate units and $i = \sqrt{-1}$ as a mere convenience to express two equations at once. It is, in other words, possible to make both formalisms, those of classical and of quantum mechanics, look *remarkably* similar.

But there is then still "a precise sense in which the wavefunction is *forced upon* us in the quantum case but not classical case." (Callender 2015, p. 3169; my emphasis – FB) The difference lies in the fact that the phase-function S is not *needed* to determine the exact trajectories in classical mechanics whereas it *is* in BM (cf. also Holland 1995, p. 55 ff. for examples). This makes the instantaneous dependence of magnitudes describing particles on those describing distant ones *inevitable* in BM, and Callender (2015, p. 3171) thinks it also is "the reason why most Bohmians have agreed with Bell that it [the wavefunction – FB] must be treated ontologically."

Now if we accept this argument, we are back to square one. Wavefunctions are still fields on high-dimensional configuration spaces, determining the trajectories of particles in spacetime, even if there is a necessitation relation between the universals 'wavefunction' and 'particle trajectory'. However, even on the ADT account, where the laws are something over and above the behavior of the beables and may be said to *govern* the latter, our treatment need not be so naïve as to think of Ψ as field instantiated on configuration space. Callender (2015, p. 3159) e.g. urges us to not "mistake the mathematical representation of [...] governors with their physical reality." Thus Ψ in the supposed universal guidance equation could symbolize a property *distributed* across spacetime or *simultaneously* insantiated in multiple locations,[18] depending on the details of that supposed equation.

But even on such a sophisticated reading of the nomological understanding of Ψ in the ADT sense, not much is gained. This is due to a threefold complex of prob-

[18]In fact, any universal is simultaneously instantiated at many points in space, or cast in relativistic terms, instantiated on multiple points of the same spacelike hypersurface. The key point is that in BM the dependence of one quantity (velocity) is on the *multiple* instantiations (particle positions) on that hypersurface, not on only one of them (the particle's own one).

lems: van Fraassen's (1989, p. 96) "obviously related" problems of *identification* and *inference*, and the general problems with *instantiation* in the metaphysics of universals. Let us look at the three separately and then find their commonalities.

The identification problem is the problem that we may well wonder what kind of relation the relation N between two universals is, and the inference problem is how N 'does its job', i.e. *entails* that all Fs are Gs. The two are connected since solving the first problem must include solving the second (that is what N was introduced for). But neither task is easily fulfilled, if solvable at all; van Fraassen (1989, p. 97) considers, as a first approach to identifying N, the relation of *extensional inclusion*, namely: "A is extensionally included in B exactly if all instances of A are instances of B." If N were extensional inclusion, this would solve the inference problem, but: "if this qualifies as a necessitation relationship, then all ordinary universal regularities become matters of law." (ibid.) A strengthening of N (extensional inclusion + X) does not remove this trivialization, but any alternative seems to eschew the inference problem. There is a crucial gap between universals and their instances, which makes the problem appear unsolvable: $N(A, B)$ "is a singular statement about universals A and B. The conclusion to be drawn from it is about another sort of things, the particulars which are instances of A and B." (ibid.)

On the other hand, we have the basic metaphysical problems of *instantiation*. Instantiation is a cross-categorical relation between universals and particulars (the instances), of which it has been suspected, in virtue of philosophical argument, that it "cannot be explicated by any analysis, definition, or metaphor." (Armstrong 1989, p. 108) This alone may tip us off as being a problem, given that we were facing an interaction problem between high-dimensional configuration spaces and lower dimensional spacetimes before, and were attempting to remove it by reinterpreting the role of the Ψ-function. But now we have a no more illuminating 'interaction' between universals and their instances on our hands, a primitive unanalyzable relation of instantiation. Worse: instantiation itself appears to be a universal; every case of instantiation of a universal by a particular instantiates the universal of instantiation. This immediately leads to a "vicious or at least viciously uneconomical" infinite regress (Armstrong 1989, ibid.). Since both, the inference (and associated identification) problem and the problems of instantiation concern the 'gap' between universals and their instances, we may think of this (somewhat metaphorically) as another kind of 'interaction problem'—between entities in *platonic heaven* and spacetime, if you will.

The problems here encountered constitute a dilemma,[19] which we may call *Dürr et al.'s dilemma* for obvious reasons. If we think of Ψ as nomological, we apparently reduce the interaction problem, that Ψ resides in configuration space whereas the particles do not, to the problem of explicating what 'nomological' means. But if one accepts a Humean view, the explanatory value of Ψ and of guidance equations is greatly reduced (and with it the advantages of BM over orthodoxy). And if, on

[19]Author's note: Again thanks to Andreas Hüttemann are in order for the observation that the situation constitutes a dilemma.

the other hand, one accepts a realist (or non-reductionist) account of laws such as the ADT account, one either has to interpret Ψ as a real field—on configuration space—after all, thus falling back onto the interaction problem, or one simply faces a different 'interaction problem' between universals and their instances.

There are two further options that are seriously discussed in the literature on the laws of nature, but we can see quite easily why they do not resolve the dilemma in a satisfying way either. To wit, one might (a) accept laws as metaphysical entities *sui generis* (e.g. Maudlin 2007, p. 157 ff.),[20] or alternatively (b) replace them by other metaphysical entities such as (causal) powers or dispositions (e.g. Cartwright 1983, 1989, 1999; Mumford 2004). In either case, one would of course have to tell a compelling story about the occurrence of the strong Bell-type correlations, and how the laws (being 'primitive entities') effect them, or how the causal powers account for their occurrence.

Now (a) seems hardly more attractive than simply accepting the correlations as primitive, i.e. accepting the Humean view, or hardly more illuminating than introducing Bohm and Hiley's elusive concept of 'active information'. It is now a 'law' that somehow mysteriously guides and correlates the distant particles. So upon accepting (a), we face yet another 'interaction problem', though this time rather between a 'nomological realm' and the 'realm of (local) beables'.

The same can in fact be said about some versions of (b). Esfeld et al. (2014, p. 791) resort to introducing "a disposition of motion as a holistic property of the totality of the particles in the universe as primitve." But what this means beyond 'all particles in the universe move the way they do' is rather unclear. If a fanciful metaphysical story about the actual and possible is invoked to elucidate the 'holistic disposition of motion', on the other hand, wherein the disposition, a catalog of coordinated possibilities if you will, transitions into the actual behavior of the particles, then one faces the next interaction (or maybe transition) problem: how can the set of merely possible configurations *effect* the actual ones? How, in other words, is the disposition's transition from potential to actual to be understood? What would such an analysis add to a Humean account, other than a 'soft (metaphysical) pillow to rest one's head on'?

If one turns to 'non-holistic' dispositions instead, (b) appears as an endeavor not easily pursued if possible at all. We clearly saw the difficulties with establishing a causal interpretation of Bell type inequalities in the second interlude, and relying on 'non-holistic' dispositions or causal powers would imply that the *local* disposition of a distant particle would manifest *instantaneously* in virtue of what was done to its distant, EPRB-correlated partner—thus creating room for causal paradoxes again.

The bottom line is this: The move from interpreting Ψ as a field on configuration space to treating it as a 'nomolocial entity' certainly has the advantage of reducing two problems to one, as has been pointed out by Callender (2015, p. 3159). But unless any of the aforementioned accounts has been worked out in quite some detail

---

[20] In fact, Maudlin (2013, p. 151) basically suggests the same about the quantum state – that it may be an entity *sui generis*.

and is freed of its problems, one faces Dürr et al.'s dilemma—that one either, on the one horn, has to accept even spookier correlations than before and is rid of ψs basic explanatory value or, on the other horn, retains the interaction problem between configuration space and spacetime or merely replaces it by an 'interaction problem' between nomological and beable realms, the actual and the possible, or between spacetime and platonic heaven.[21]

6.1.3 *Relativity, Fields, and Possible Limitations of Bohmian Mechanics*

The last section introduced a philosophical difficulty of BM that may not be insurmountable, but at least demonstrates that there are essential ambiguities at its basis. In this section we will point to some physical difficulties that may account for why not most physicists accept a Bohmian interpretation.

Let us first take stock of our findings, however, evaluated against the agenda set out in Sect. 2.3. BM accepts, on a variety of readings, a dual ontology in which both, wave function *and* particles with well-defined trajectories play a crucial role. We already identified BM as formally conservative above, since the guidance equations 'pop out' by a mere rewriting of the TDSE. But at the same time BM is conceptually revisionary, since it embraces proper particles and their positions as additional variables. It should also be classified as non-collapse, since any 'collapse' at best occurs as a finding out of a trajectory.[22]

Now is BM ontological or non-ontological w.r.t. the wave function? This question cannot be unambiguously answered, as the previous discussion has demonstrated. We claimed that "on a variety of readings", the wave function *does* signify something in reality, and this is certainly the case in non-nomological and non-Humean nomological versions of BM, whence these should all be classified as ontological. A Humean version may, in contrast, be classified as *non*-ontological, since the wavefunction does not (directly) *describe* anything; it does not *itself* form part of the ontology. We will acknowledge this need for classificational refinements in Sect. 6.4, where we will compare multiple interpretations of QM directly.

How does BM fare w.r.t. the MAC and the DOC? The *dynamical task* of the MAC is tackled by modifying the dynamics – again: by a mere rewriting – in accepting, in addition to the TDSE, the guidance equation, taken to describe the movement of actual particles. The *kinematical task* is taken on by accepting that the wave function (of the universe) is either a somewhat mysterious entity ('active information', a 'real law', an entity *sui generis*, or a 'wave on configuration

[21] Cf. however Dorato (2015) for a quite different and somewhat more benevolent discussion on this context.

[22] One might thus be tempted to think of BM as a subjective collapse interpretation; but that would surely be misleading, since collapse plays no *substantial* role therein.

space') guiding the particles, or just our most convenient way of describing their behavior. No outcome problem really arises. Particles travel on definite—though classically unexpected and sometimes surprisingly correlated—trajectories with definite velocities. Their positions may be generally unknown, but they reveal themselves in suitable experiments. QM's observables are, in general, taken to represent features of experiments, not systems. Thus the sudden appearance of a value for some observable that seems like a 'collapse' of the quantum state is in a sense an illusion, accounted for by the funny behavior of material particles.

All of this has an immediate impact on the DOC. If one accepts a Humean-nomological view, BM is very clear on all three fronts: (i) 'classical' macroscopic objects are straightforwardly constituted by tiny particles (ii) whose strongly correlated (and 'non-classical') behavior can only be demonstrated when they are sufficiently isolated from one another, and (iii) the guidance equation and the TDSE in concert describe the behavior of these particles (the wavefunction becomes a mere 'calculation device' for the correct statistics). This is ontologically pretty clear, even if one has to swallow the weird correlations (which will be demonstrated to be even weirder a little below). This is not so on all other versions, where one faces interaction problems, i.e. where point (iii) of the DOC cannot generally be viewed as satisfied, as the discussion should have made clear. On grounds of ontological clarity, the Humean-nomological view seems preferable.

However, all our considerations so far were based on *non-relativistic* QM, which arguably provides the most 'natural setting' for BM. Herein lies the major crux: BM does not generalize so 'neatly' to even a *special* relativistic setting. But let us proceed step by step. First of all, we noted that the non-local character of the wave function and the implied immediate dependence of velocities on distant coordinates in BM raise worries about its compatibility wit SR. But we also claimed that there is *some* sense in which BM is *not* in conflict with relativity. This is the same sense in which the QM formalism generally is not in conflict with relativity, namely, in the sense that it does not allow for the transmission of superluminal *signals*. The reason is that, in virtue of the quantum equilibrium hypothesis, the very same arguments apply as in orthodox QM (cf. Dürr and Teufel 2009, p. 208): If signaling is to be established, it has to be in virtue of statistical changes, since we cannot control hidden parameters (in this case: the unmeasured and often times unmeasurable positions of particles). But in virtue of the quantum formalism, these changes cannot be detected, as was demonstrated in Sect. 4.2.4.

We must ask, though, whether this is a satisfying sort of 'compatibility'. Recall that the problem with superluminal signaling was that, if we accept it at least as a *sufficient* condition for superluminal *causation*, this potentially leads to causal paradoxes, because in virtue of relativistic geometry and the lack of an overarching simultaneity an agent could change her (causal) past with the aid of superluminal signals, thereby somehow stopping herself from doing so ($\lightning$). The point of the quantum equilibrium hypothesis and no-signaling constraints is that *we*, users of QM or BM, cannot signal into the past. But Bell (1990a, p. 111) once piercingly asked: "Who do we think *we* are?" (emphasis in original) His point (ibid.) was that "the 'no signalling…' notion rests on concepts which are desperately vague, or vaguely applicable", and that it was unclear whether "*we* include[d] chemists, or only physicists, plants, or only animals, pocket calculators, or only mainframe computers" (emphasis in original). The general worry that transpires is that there *could*, in fact, be entities (maybe future scientists) who knew how to *get around* the (no signaling-) constraints set up by the quantum equilibrium hypothesis, by gaining control over particle positions and velocities. Recall that the quantum equilibrium hypothesis only holds in virtue of assuming it to hold on the universal level at an initial time; it is thus a contingent hypothesis, or rather, only a *conditional* necessity. But even if it *were* a straightforward *theorem* of BM, then the very *existence* of mutual dependence at a distance would still raise worries about the possibility of *better* physics being found in the future, in which the quantum equilibrium hypothesis would be obsolete and one could signal at superluminal speeds.

At the very least, it seems quite 'odd' that BM and relativity should be compatible only in virtue of such a possibly contingent feature, a human shortcoming if you will. But is there not, since BM and orthodox QM *share* the prediction of violations of Bell-type inequalities and the no signaling theorems, a *general* such conflict between QM and relativity? The answer must be 'that depends!' It depends, that is, on whether one believes that the connection between the distant, correlated particles (nay, particles) is *causal* in orthodox QM, as it appears to be in BM. And in particular, BM assumes a piece of ontology that is not present in QM, namely definite particle positions at all times, from which the entire trouble transpires. It is these considerations that have prompted (Egg and Esfeld 2014, p. 190) to state that "Bohmian mechanics is [...] not committed to superluminal causation in an operational sense, but it is so committed in a metaphysical sense: given any initial particle configuration, the theory supports counterfactual claims of the type: 'If Alice had chosen a different setting, Bob would have obtained a different outcome'."

Whether Egg and Esfeld's analysis is correct is open to debate, since one can make a case that no *causal* counterfactuals are implied or even sanctioned by QM, and one could equaly make a case that the same holds for BM.[23] In the present context, however, the most obvious way to avoid the conclusion of a metaphysical commitment to causation is a Humean view of laws, where we have found the correlations to constitute 'ontological bedrock', and wherein the (*faulty*) impression of causation in EPRB-scenarios arises from the observed regularities.

Be that as it may, there are of course also *formal* difficulties in reconciling BM with relativity, in virtue of the non-local character of guidance equations. Recall (from Sect. 2.2.1 and the second interlude) that in both, SR and GR, spacetime is

[23] Here is how (in brief; Chap. 7 presents the argument for QM in more detail): In Healey (2012a, p. 22 ff.) and Friederich (2015, p. 132) it is argued that there can be no *interventions* $\underline{I}_A, \underline{I}_B$ for two agents (Alice and Bob) in remote places, sharing among them a pair of electrons in the singlet state, such that both Alice and Bob could perform their respective intervention to fix one of the possible values: "manipulability by the distant outcome always undermines the local control required for a genuine intervention." (Boge 2016a, p. 4) This is why *causal* counterfactuals are not 'sanctioned' by QM (the argument goes), if one accepts interventionism as definitive of causation. Since dependence is *mutual* in BM, this argument seems to transfer seamlessly.

conceived of as a unified whole (a Lorentzian manifold) without a preferred slicing into space- and time-components. What acts as the time parameter is dependent on one's own state of motion and the distribution of masses in the manifold. Here is where the troubles originate.

Consider the N particle treatment of the DE in the context of BM, first championed by Bohm and Hiley (1993, p. 274). Interpreting ψ as taking values in the space $(\mathbb{C}^4)^{\otimes N}$ of four component spinors for N particles,[24] and including the possibility of external electric and magnetic fields, one ends up (in units in which $\hbar = c = 1$) with

$$i\frac{\partial\psi}{\partial t} = \sum_{j=1}^{N}\left[-i\hat{\boldsymbol{\alpha}}^{(j)}\nabla_j - e\boldsymbol{\alpha}^{(j)}\boldsymbol{A} + e\Phi + \hat{\beta}^{(j)}m_j\right]\psi, \tag{6.10}$$

where $\hat{\boldsymbol{\alpha}}^{(j)} = (\hat{\alpha}_1^{(j)}, \hat{\alpha}_2^{(j)}, \hat{\alpha}_3^{(j)})^T$ is the vector of Dirac-α-matrices for the j-th particle, $\hat{\alpha}_\ell^{(j)} = \mathbb{1}\otimes\ldots\otimes\mathbb{1}\otimes\hat{\alpha}_\ell\otimes\mathbb{1}\otimes\ldots\otimes\mathbb{1}$ (with $\ell \in \{1, 2, 3\}$, and analogously for $\hat{\beta}^{(j)}$), ∇_j is the gradient w.r.t. $\boldsymbol{x}_j$ (the j-th particle's configuration space coordinates), and where Φ and $\boldsymbol{A}$ are electric and magnetic potentials respectively. From (6.10), the guidance equation

$$\boldsymbol{v}_k = \frac{\psi^\dagger\hat{\boldsymbol{\alpha}}^{(k)}\psi}{\psi^\dagger\psi} \tag{6.11}$$

quite naturally follows, where $\dagger$ indicates transposition and complex conjugation, and the term $\psi^\dagger\hat{\boldsymbol{\alpha}}_{(k)}\psi$ is to be understood as computing a three entry column vector of products $\psi^\dagger\hat{\alpha}_\ell^{(k)}\psi$ ($\ell \in \{1, 2, 3\}$).

Now the DE itself is Lorentz invariant (e.g. Bohm and Hiley 1993, pp. 276–278; Peskin and Schroeder 1995, p. 42 ff.), but the guidance equation of course still has an obviously nonlocal character as soon as ψ is entangled—the velocity of each particle depends, as was the case in the non-relativistic treatment, on the positions of all other particles whose generic coordinates appear in ψ—and it is far from clear how this 'squares with the spirit' of the relativity theories. Here is how Bohm and Hiley (1993, p. 285) phrase the problem:

> the entire calculation of the particle velocity will be *ambiguous* until we *specify the frame* in which the nonlocal connections are instantaneous. The *concept* of a particle guided in a nonlocal way will, in general, not be Lorentz invariant. [O]ne has therefore to assume some definite frame in which the connections are to be described as instantaneous, while in other frames they are described as working either backwards or forwards in time [...]. (my emphasis – FB)

[24]Note that ψ may be 'multi-time', i.e. depend on N 4-tuples of spacetime coordinates where the time coordinates do not necessarily coincide (e.g. Dürr et al. 2014, p. 227; Galvan 2015, p. 4).

More generally (and technically) speaking, this is the problem of a *preferred foliation* of spacetime $\tilde{\mathcal{M}}$ into 'leaves' Σ that can be 'stacked up' to give back $\tilde{\mathcal{M}}$ and provide, intuitively, a representation of 'space at different times'. Notably, Dürr et al. (2012, p. 227 ff.) provide such a foliation of the Minkowski spacetime $\mathcal{M}$ into $3D$ (spacelike) hypersurfaces, i.e. possibly curved submanifolds that can be ordered along a (then preferred) time-parameter. This foliation, $\mathscr{F}$, they define in terms of a smooth function $f : \mathcal{M} \rightarrow \mathbb{R}$, so that the sets $\mathcal{M}^{(3)}(\tau) := \{p \in \mathcal{M} | f(p) = \tau, \tau \in \mathbb{R}\}$, the *level sets* of f (e.g. Frankel 2004, p. 46), provide the leaves, i.e. intuitively the spaces at given times (τ). The foliation $\mathscr{F}$ thus given also uniquely provides a (future directed) vectorfield $\boldsymbol{n}(p)$ which is normal, at any $p \in \mathcal{M}$, to the hypersurface Σ through p and is defined as the normalization of the (generalized) gradient ∂f.[25]

Now using the γ-matrices in the DE instead of $\hat{\boldsymbol{\alpha}}^{(k)}$ and $\hat{\beta}^{(k)}$, one can ultimately rewrite the enumerator in (6.11), which provides a (probability) current for the k-th particle in an N particle system (recall that $\boldsymbol{v} = \boldsymbol{j}/\varrho$), as $\boldsymbol{j}^{(k)} = \bar{\psi}(\boldsymbol{\gamma}^{(1)}\boldsymbol{n}(p_1)) \ldots \boldsymbol{\gamma}^{(k)} \ldots (\boldsymbol{\gamma}^{(N)}\boldsymbol{n}(p_N))\psi$, where $\bar{\psi} := \psi^{\dagger}\gamma^{0}$,[26] $\boldsymbol{\gamma}^{(k)}$ is the (four component) vector of γ-matrices for the k-th particle, and $\boldsymbol{n}(p_k)$ is the (four component) normal vector to a leaf in the foliation (a hypersurface of simultaneity) at point p_k where the kth particle's spacetime path (its *worldline*) intersects it. The velocity for the kth particle is then ultimately given by

$$\frac{\mathrm{d}X^{(k)}}{\mathrm{d}s} = \frac{\boldsymbol{j}^{(k)}}{\partial f \cdot \boldsymbol{j}^{(k)}}, \tag{6.12}$$

where $s = f(X^{(k)})$ and $X^{(k)} = X^{(k)}(s)$ defines a parametrized path through spacetime thus foliated.

All of this is certainly an improvement; it liberates the dependence of one particle's state from the slicing of spacetime into different hyper*planes* of simultaneity and makes different foliations into curved spaces possible, thereby making a step into a 'more relativistic' or at any rate more general direction. But it does *not* impair the fact that a (preferred) foliation $\mathscr{F}$ is needed to specify the dependence of one particle on the configuration of all other particles *at some particular instant*, or on some 'hypersurface of simultaneity'.

How, in fact, should one expect to arrive at a preferred such foliation? One hope expressed by Bohmians is that it be "not simply posited as a novel piece

---

[25] ∂f is a generalized gradient for calculus on manifolds which can be locally written in a suitable coordinate representation (cf. Footnote 73 of Chap. 2) as $\partial f = \sum_{\mu,\nu} g^{\mu\nu} \frac{\partial f}{\partial x^{\nu}} \frac{\partial}{\partial x^{\mu}}$, with $g^{\mu\nu}$ the manifold's metric and where the $\frac{\partial}{\partial x^{\mu}}$ are conceived of as vectors in a 'tangent space' at a given point in the manifold for which the local coordinates x^{μ} are defined (e.g. Frankel 2004, p. 45 ff.; Nakahara 2003). Frankel (2004, p. 47) uses the notation '∇f' instead; we here follow that of Dürr et al. (2012, p. 227).

[26] This is the *adjoint spinor* that makes for a Lorentz-invariant scalar product $\bar{\psi}\psi$ (cf. Griffiths 2008, p. 236).

of absolute space-time structure, but is instead regarded as a dynamical object, itself obeying a Lorentz invariant law." (Dürr et al. 2014, p. 5) It is then hoped for that this law be somehow determined by Ψ itself; but there is not yet any worked out suggestion as to how this should come about, and even basic examples of how the wavefunction determines structures in spacetime (more precisely: tensorial quantities *via* operators in QFT) exhibit technical difficulties such as the non-well-definedness of a possibly resulting foliation in certain cases of interest (cf. Dürr et al. 2014, pp. 5–6).

Alternatively, one could deny that there even is the *need* for *one* preferred or dynamically selected foliation. This is what Galvan (2015) has recently attempted, by appeal to *typicality* conditions for spacetime trajectories of all existing particles, from which the statistical (and hence: empirical) content of BM follows but no preferred foliation; the typicality is rather evaluated over all possible foliations. Typicality is a probabilistic notion though, and it is hard to see in what sense it is meaningful to talk about the probability of a "trajectory of the universe" being "chosen at random" (Galvan 2015, p. 7). And it is equally hard to see how the conceptual difficulties are resolved by these considerations on an *ontological* level, arising from the explicitly non-local dependencies among particles.

We introduced relativistic notions in Chap. 4 in connection with QFT, and devising a Bohmian QFT is certainly a second major hurdle. Without going into too much detail, we here recapture a few features and restrictions that Bohmian QFTs face.[27] The general 'quest' for Bohmians is to find the 'beables' that any given QFT prescribes. Bohm (1952b, p. 189 ff.) first set out to find a Bohmian version of QED, by considering actual (though coarse grained, i.e. discontinuous) sets of field configurations $\phi(\boldsymbol{x}, t)$ as beables, over which "an objectively real superfield" (ibid.) Ψ would be defined, which is then mathematically a functional $\Psi[\phi_1(\boldsymbol{x}, t), \ldots, \phi_N(\boldsymbol{x}, t)]$, or more abstractly (avoiding reference to a countable index): $\Psi[\ldots\phi(\boldsymbol{x}, t)\ldots]$. This is the functional Schrödinger approach to QFT mentioned in Footnote 80, where, to recall, $\phi(\boldsymbol{x}, t)$ is the value of a multiplication operator $\hat{\phi}(\boldsymbol{x}, t)$, evaluated on (field-)states $|\phi\rangle$. In Bohm and Hiley (1993, p. 238 ff.), this treatment would be extended by letting $\Psi[\ldots\phi(\boldsymbol{x}, t)\ldots] = R[\ldots\phi(\boldsymbol{x}, t)\ldots]e^{\frac{i}{\hbar}S[\ldots\phi(\boldsymbol{x},t)\ldots]}$, making generalizations of the guidance equation to $\frac{\partial\phi}{\partial t} = \frac{\delta S}{\delta\phi}$[28] and of the quantum potential to the "super-quantum potential" $Q = -\frac{\hbar^2}{2R}\int \mathrm{d}^3\boldsymbol{x}\, \frac{\delta^2 R[\ldots\phi(\boldsymbol{x},t)\ldots]}{\delta\phi(\boldsymbol{x},t)^2}$ possible (Bohm and Hiley 1993, p. 240 ff.; cf. also Holland 1995, p. 520).

However, electromagentic fields are *bosonic*, and Bohm and Hiley (1993, p. 276) noticed an asymmetry between fermionic and bosonic QFTs, since "fermionic field

[27]The most important reference for further details is Struyve (2010).

[28]The *functional derivative* $\frac{\delta F[f]}{\delta f}$ of a functional $F[f]$ obtains a quite 'natural' understanding in close analogy to derivatives in ordinary calculus as $\lim_{\epsilon\to 0}\frac{1}{\epsilon}(F[f(x)+\epsilon\delta(x-x')] - F[f(x)])$, with $\delta(x-x')$ a Dirac-δ (e.g. Greiner and Reinhardt 1993, p. 37; Lancaster and Blundell 2014, p. 12), i.e. where one lets f vary with tiny 'strengths' (ϵ).

operators have only two possible states and cannot be put in correspondence with field beables that would change continuously", whereas "bosonic operators with their infinity of states can be represented in terms of continuous fields." To some this may suggest that a fermionic QFT treats of particles whereas a bosonic QFT treats of fields. This is true, for instance, in the model discussed by Dürr et al. (2012, p. 239 ff.), which they call a "Bell type QFT" (p. 240), due to Bell's (1984a) early contributions to the general agenda. Here the time dependent configuration $Q(t)$ of all particles in the universe behaves continuously for certain intervals and then jumps between different configuration space-sectors of definite particle number when particles are created or annihilated. The dynamics for these two types of process are then described by continuous and stochastic equations respectively (cf. Dürr et al. 2012, pp. 242–243), and field operators are taken to obtain their meaning only *via* determining PVMs (for probabilistic predictions) in terms of sets of number operators (cf. their p. 245).

It is not so clear, however, whether all of the successful predictions of modern QFT can be recaptured in this fashion (cf. in particular the comments in Wallace 2008, p. 84); and not all models generated to introduce Bohmian dynamics for fermionic field theories are alike. Struyve and Westman (2007), for instance, introduce a 'minimalist model' for QED, in which the fermionic degrees of freedom do not correspond to beables at all, but rather only appear as an index of the wavefunction (integrated out or summed over). And from basically two assumptions or features of the model, namely "the equilibrium distribution for the beables, together with the fact that wave functions representing macroscopically distinct systems have negligible overlap" (Struyve and Westman 2007, p. 3124), they claim to be able to generally reproduce QED's predictions.

The upshot (cf. Struyve and Westman 2007, p. 3125) is that, in a measurement involving a needle on some scale, there will be "particle positions representing the needle, so that the orientation of the needle [...] will be recorded and displayed in the particles' positions." When a field interacts with the needle, "on the level of the quantum state, the direction of the macroscopic needle will be correlated with the radiation that is scattered off [...] the needle." This leads to the field beable carrying "an image of a macroscopic needle, in a similar way to that in classical mechanics [...]."

This informal explanation of measurements, however, has the undesirable feature of presupposing significant chunks of *non-relativistic* (and non-field) BM, the *particle positions*. And even though the authors "see no problem in principle to construct a similar model [...] in the context of the standard model" (their p. 3124), it remains to be seen whether such a model can indeed be found and whether it reproduces the standard model in a convincing way. So far these are merely suggestions.

Valentini (1996) proposed yet another model, in which fermions correspond to "an objective field of Grassmann numbers evolving in time, guided by a wave functional Ψ." (Valentini 1996, p. 55) But not only are there technical difficulties with Valentini's approach (cf. Struyve 2010, pp. 26–27), it is also very hard to understand what kinds of 'beables' Grassman numbers[29] at spacetime points are supposed to be, and how they get us any closer to an understanding of QFT than did the operator-valued fields in Sect. 2.2.2.

Without going into any details about the remaining models (cf. Struyve 2010, for further reference), we here summarize that (a) in many cases – especially for fermionic fields – it becomes hard to see what the *beables* should be in Bohmian QFTs, that (b) it is in many cases still unclear how the project should be executed *formally*, that (c) it is not clear that *all* QFTs (or all successful predictions thereof) can be reproduced (compellingly), and that (d) the concerns about compatibility with *relativity* of course carry over from particle BM, whence models are often judged to be "not [...] Lorentz covariant at the *fundamental* level", whereas "Lorentz covariance is regained at the statistical level."[30] (Struyve and Westman 2007, p. 3116; my emphasis – FB). All of this leaves a foul taste to the advances in Bohmain QFT and relativity; but of course it does not mean that the project is impossible to execute or should be abandoned altogether (we will return to this issue in the discussion later).

Nevertheless, we have hence highlighted some of the current difficulties of BM and given reasons to be skeptical about its general aptness as an interpretation of QM. And we had claimed above that the 'most natural' setting is a non-relativistic particle setting, in which the only real difficulty is the interpretation of the wavefunction. Moreover, it seems that at least non-relativistic BM, especially in its nomological-Humean reading, comes quite far in recapturing the intuitions of our natural response discussed in Chap. 4, and that BM in general also provides a clear ontological *basis* for a theory of material objects in terms of tiny particles with definite states at all times that are, ultimately, 'simply unknown'.

There are, however, some further features of BM that make it hard to swallow already at the non-relativistic level. For one, we noted in Sect. 6.1.1 that position plays kind of a special role as an observable; and we equally noted that particles always possess definite velocities, as a consequence of the guidance equation(s). But these velocities are *immeasurable*: due to the measurement dynamics (i.e. the entangling unitary dynamics of system and apparatus), it becomes impossible to measure any quantity that "has a possible value (one with non-vanishing probability or probability density) when the wave function of the system is $\psi_1 + \psi_2$ that is neither a possible value when the wave function is ψ_1 nor a possible value when the wave function is ψ_2." (Dürr et al. 2012, p. 139; emphasis omitted)

Recall that the unitary dynamics will couple the terms (ψ_1 and ψ_2 respectively) in the system's wave function to suitable terms in the wave function of the apparatus.

---

[29]E.g. Nakahara (2003, p. 40 ff.) for an introduction to Grassmann numbers.

[30]'Covariance', strictly speaking, is not the same as invariance; it rather means that "a [...] quantity 'changes in the same way'." (Cheng 2005, p. 14) However, if all the quantities in an equation transform covariantly, the entire equation retains the same form (cf. ibid.), which is why the terms 'Lorentz invariant' and 'Lorentz covariant' are sometimes used interchangably in the literature, as regards equations and theories. Cf. also Friedman (1983, p. 45) for a deeper discussion and some subtleties in the transition from SR to GR.

But then obviously none of the apparatus-terms, one of which will contain the resulting particle configuration (the 'pointer position'), will be indicative of the true value of a quantity that is *not* possible in any single one of the system-terms (has vanishing probability therein). Now take any given wave function ψ and rewrite it, recalling that it spits out a complex number, as $\mathrm{Re}(\psi) + i\mathrm{Im}(\psi)$ (so that $\psi_1 := \mathrm{Re}(\psi)$ and $\psi_2 := i\mathrm{Im}(\psi)$). The velocity of a particle would then have to be either of the quantities $\boldsymbol{v}_{1/2} = \frac{\hbar}{m}\mathrm{Im}\frac{\psi^*_{1/2}\nabla\psi_{1/2}}{\psi^*_{1/2}\psi_{1/2}}$, both of which are always *zero* because the fractions in both cases are purely *real* (so their imaginary part is zero).

Since this can be done with any given wavefunction, particles would always have to be motionless if measurements of their Bohmian velocities were correct. But BM takes the unitary dynamics of QM for granted for (effective) wavefunctions (from which the remarkable success of QM is regained) and velocities are obviously *not* always zero. So the conclusion that velocities, together with other interesting quantities (cf. Dürr et al. 2012, p. 140), are immeasurable seems inevitable.

Another such 'feature' is the existence of what have been called "surrealistic" trajectories (Englert et al. 1992). To elaborate, consider a two slit experiment with atoms and with wave functions ψ_{up} and ψ_{down} representing those associated with upper and lower slit respectively. Due to the symmetry of the arrangement, ψ_{down} may be viewed as ψ_{up} reflected along the z-axis (chosen to be the middle axis between the two slits), i.e. $\psi_{\mathrm{down}}(x, y, z; t) = \psi_{\mathrm{up}}(x, y, -z; t)$. Then v_z, the z component of the particle velocity will be an odd function w.r.t. the z-axis as well, which implies that $v_z = 0$ on the $z = 0$-plane. This means that *particles do not cross the mid plane through the double slit* (cf. also Bell 1980, p. 113 or Dürr and Teufel 2009, p. 156). If one now places subtle, qubit-like detectors in front of the slits however, which make a transition between states upon passage of an atom, and where the "transition happens with virtual certainty and [...] the atom's center-of-mass wave function is not altered noticeably in the process" (Englert et al. 1992, p. 1178), the wavefunction will (approximately) become $\boldsymbol{\Psi}(\boldsymbol{x}, t) = \psi_{\mathrm{up}}(\boldsymbol{x}, t)\,|^{\mathrm{yes}}_{\mathrm{no}}\rangle + \psi_{\mathrm{down}}(\boldsymbol{x}, t)\,|^{\mathrm{no}}_{\mathrm{yes}}\rangle$. Since the detectors behave like two-state systems, this is effectively a spinor and the velocity is then given by the spinor-version of the guidance equation. The symmetry of the velocity vector is thereby preserved (no crossing of the $z = 0$-plane); but the probability density is now only $|\psi_{\mathrm{up}}|^2 + |\psi_{\mathrm{down}}|^2$, i.e. interference terms vanish (as should have been expected from the discussion in Sect. 2.1.1), which is easily seen since $\left\{|^{\mathrm{yes}}_{\mathrm{no}}\rangle, |^{\mathrm{no}}_{\mathrm{yes}}\rangle\right\}$ may be taken to form an ONB of $\mathbb{C}^2$.

But there is now the unpleasant consequence that, since $|\psi_{\mathrm{up}}|^2$, say, does not vanish below the $z = 0$-plane, the particle may be detected in the upper slit and still end up below $z = 0$—so that, in virtue of the guidance equation, it will be detected at a slit *which it never passed* (cf. Englert et al. 1992, p. 1178). The crucial fact is that both wavefunctions are almost unperturbed by the detectors, so they are both still relevant to the guidance equation.

To see this feature as a flaw can, of course, always be regarded as a failure to acknowledge the general non-local character of BM though. As Passon (2004, p. 9) puts it:

> The arrangement which has been considered by Englert et al. can be viewed as a special case in which 'empty waves' [...] show an effect if they are still coherent. In fact, the non-locality of the de Broglie-Bohm theory makes it possible to explain how the which-way detector can be excited even without any trajectory passing through it [...].

The wavefunction itself, which may, to recall, be mostly a representation of correlations between particles—in this case between the atom in the double slit arrangement and those constituting the detector—is what makes the detector 'go off'. What has (merely?) been demonstrated, to quote Passon (2004, ibid.) again, is that "the trajectories behave completely unclassical and that the de Broglie-Bohm theory is as unintuitive as the usual quantum theory."[31] As long as matters are not settled on the ontological status of wavefunctions, we do not seem to come *that* close to fulfilling the 'dream' of Chap. 4 after all.

6.2 Spontaneous and Induced Localizations: Taking Collapse Seriously

6.2.1 GRW's and Pearle's Formal Modifications and Two Ontologies

Our naïve view of Sect. 2.1.1 introduced the 'duality' between waves and particles differently than does BM, namely by means of a 'collapse' of wave-like extended entities into tiny lumps that could then, upon collapse, be thought of as particles. The advantages of this collapse are retained in the projection postulate of the orthodox interpretation in the sense that one finds, upon suitable measurement, particles in definite positions. But the projection postulate is devoid of meaning as a physical postulate; it does not provide a dynamics for the collapse, it does not 'tell a story' as to how the quantum state 'collapses'—it does not suggest a non-minimal interpretation.

So here is an alternative suggestion: Interpret ψ as an objectively real field, regardless of all its peculiar properties, and embrace some sort of more sophisticated collapse-dynamics to make sense of the particle-like findings in practice and the occurrence of non-quantum objects. Thus, we formulate another conjecture, which

[31] A lot more could be—and has been; cf. Passon (2004, p. 9) for references—said on this problem of 'surreal' trajectories, but for our present purposes the discussion seems fully sufficient. We briefly also mention the recent *experimental* work by Mahler et al. (2016), who show, by advanced experimental methods, that "the trajectories seem surreal only if one ignores their manifest nonlocality." (p. 1) Interpreting the results as being concerned with particle trajectories at all, however, obviously *presupposes* a Bohmianm understanding of QM.

we call, with reference to the *prima facie* similarity to the view discussed in Chap. 2, the *informed view*:

Conjecture 3 (The informed view) The quantum wavefunction ψ represents an objectively real field. The appearance of tiny particles in spacetime is accounted for by a sophisticated collapse-dynamics.

The first serious such proposal was the 'unified dynamics' of Ghirardi, Rimini, and Weber (1986) (short: GRW). To represent the dynamics, GRW used the density matrix formalism and the vNE with additional, stochastic collapse terms. More precisely, GRW (1986, p. 34) introduced a *superoperator* $\hat{\mathcal{T}}[\hat{\rho}] = (\frac{\alpha}{\pi})^{3/2} \int \mathrm{d}^3\tilde{\boldsymbol{x}}\, e^{-\frac{\alpha(\hat{\boldsymbol{x}}-\tilde{\boldsymbol{x}})^2}{2}}\, \hat{\rho} e^{-\frac{\alpha(\hat{\boldsymbol{x}}-\tilde{\boldsymbol{x}})^2}{2}}$, acting on density operators $\hat{\rho}$. Here α is an unspecified new constant (where $1/\sqrt{\alpha}$ is a length, representing the "sharpness of the localization"; Ghirardi et al. 1988, p. 386) and $\hat{\boldsymbol{x}}$ is simply the position operator. If one computes the matrix elements of $\hat{\rho}$ in the position basis, this yields, for a single particle pure state density operator $\hat{\rho} = |\psi\rangle\langle\psi|$,[32]

$$
\begin{aligned}
\langle \boldsymbol{x}|\hat{\mathcal{T}}[\hat{\rho}]|\boldsymbol{x}'\rangle &= \left(\frac{\alpha}{\pi}\right)^{3/2} \int \mathrm{d}^3\tilde{\boldsymbol{x}}\, \langle \boldsymbol{x}| e^{-\frac{\alpha(\hat{\boldsymbol{x}}-\tilde{\boldsymbol{x}})^2}{2}}\, |\psi\rangle\langle\psi|\, e^{-\frac{\alpha(\hat{\boldsymbol{x}}-\tilde{\boldsymbol{x}})^2}{2}} |\boldsymbol{x}'\rangle = \\
&= \left(\frac{\alpha}{\pi}\right)^{3/2} \int \mathrm{d}^3\tilde{\boldsymbol{x}}\, \psi(\boldsymbol{x})\psi^*(\boldsymbol{x}') e^{-\frac{\alpha(\boldsymbol{x}-\tilde{\boldsymbol{x}})^2+\alpha(\boldsymbol{x}'-\tilde{\boldsymbol{x}})^2}{2}} = \\
&= \psi(\boldsymbol{x})\psi^*(\boldsymbol{x}') e^{-\frac{\alpha(\boldsymbol{x}-\boldsymbol{x}')^2}{4}} \left(\frac{\alpha}{\pi}\right)^{3/2} \int \mathrm{d}^3\tilde{\boldsymbol{x}}\, e^{-\alpha(\tilde{\boldsymbol{x}}+(\boldsymbol{x}+\boldsymbol{x}')/2)^2} \\
&= \underbrace{\psi(\boldsymbol{x})\psi^*(\boldsymbol{x}')}_{\rho(\boldsymbol{x},\boldsymbol{x}')}\, e^{-\frac{\alpha(\boldsymbol{x}-\boldsymbol{x}')^2}{4}}, \qquad (6.13)
\end{aligned}
$$

so that the diagonal terms $\rho(\boldsymbol{x}, \boldsymbol{x})$ of the density matrix in position basis remain untouched ($e^0 = 1$) and off-diagonal terms will be damped away exponentially, depending on distance.[33] In other words: superpositions of states of different *localization* vanish quickly for larger distances.

To find a generalized evolution for the density operator, one can now appeal to the intuition that $\hat{\rho}$ evolves over short times ϵ according to the vNE with probability $(1-\lambda\epsilon)$ and with the remaining probability $\lambda\epsilon$ in accord with the evolution described

[32] The generalization to non-pure density matrices and multiple (distinguishable) particles is straightforward in virtue of the properties of tensor products and the linearity of sums.

[33] That the third line follows from the second can be verified by comparing the exponents; the fourth line may be derived by substituting a variable $\xi := -\sqrt{\alpha}(x - x')$ in all three spatial dimensions, so that the measure is rescaled by $(\sqrt{\alpha})^{-1}$ and the exponent becomes just $-\xi^2$ in all three dimensions. The integral can then be computed using Gauß's 'trick'.

by the superoperator $\hat{\mathcal{T}}[\hat{\rho}(t)]$ (cf. Bassi and Ghirardi 2003, p. 300). Writing out the differential operator $\frac{\partial}{\partial t}$ as a limit for small ϵ and neglecting the limit, the vNE takes on the form $\hat{\rho}(t+\epsilon) = \hat{\rho}(t) - \frac{i}{\hbar}[\hat{H}, \hat{\rho}(t)]\,\epsilon$. Thus one arrives, according to the above considerations, at the following 'master equation' (Ghirardi et al. 1986, p. 473):

$$\hat{\rho}(t+\epsilon) = (1-\lambda\epsilon)\left[\hat{\rho}(t) - \frac{i}{\hbar}[\hat{H}, \hat{\rho}(t)]\,\epsilon\right] + \lambda\epsilon\hat{\mathcal{T}}[\hat{\rho}(t)]$$

$$\Leftrightarrow \quad \frac{\hat{\rho}(t+\epsilon) - \hat{\rho}(t)}{\epsilon} = -\frac{i}{\hbar}[\hat{H}, \hat{\rho}(t)] + \frac{i}{\hbar}[\hat{H}, \hat{\rho}(t)]\,\lambda\epsilon - \lambda(\hat{\rho}(t) - \hat{\mathcal{T}}[\hat{\rho}(t)])$$

$$\xrightarrow{\epsilon\to 0} \quad \frac{\partial}{\partial t}\hat{\rho}(t) = -\frac{i}{\hbar}[\hat{H}, \hat{\rho}(t)] - \lambda(\hat{\rho}(t) - \hat{\mathcal{T}}[\hat{\rho}(t)]), \tag{6.14}$$

with λ (of dimension 1/time) an average *collapse frequency* (i.e. $\lambda\,\mathrm{d}t$ a probability for the damping of off-diagonal terms in the tiny time interval $\mathrm{d}t$). The additional terms in (6.14) have a position representation $\lambda(1-e^{-\frac{\alpha(x-x')^2}{4}})\rho(\boldsymbol{x}, \boldsymbol{x}'; t)$, so that the dynamics of diagonal elements is again unperturbed and obeys the vNE $(1-e^0 = 0)$.

Defining

$$\hat{\Lambda}(\tilde{\boldsymbol{x}}) := \left(\frac{\alpha}{\pi}\right)^{3/2} \exp\left(-\alpha(\hat{\boldsymbol{x}} - \tilde{\boldsymbol{x}})^2\right) \tag{6.15}$$

as the *collapse rate operator*, which is essentially a smeared out position operator (cf. Tumulka 2006b, p. 1899), the superoperator then becomes $\hat{\mathcal{T}}[\hat{\rho}] = \int \mathrm{d}^3\tilde{\boldsymbol{x}}\, \hat{\Lambda}^{1/2}(\tilde{\boldsymbol{x}})\hat{\rho}\hat{\Lambda}^{1/2}(\tilde{\boldsymbol{x}})$. The treatment for N particles can then be generalized by multiplying the $\lambda\hat{\rho}(t)$-term in (6.14) by N and summing over N superoperators $\hat{\mathcal{T}}_j[\hat{\rho}]$ (cf. Goldstein et al. 2012, p. 144). In principle one could allow also for different collapse rates λ_j depending, e.g., on the mass of the respective 'particle sort' (cf. Pearle and Squires 1994, p. 3), or allow for different localization accuracies $1/\sqrt{\alpha_j}$ (e.g. Bassi and Ghirardi 2003, p. 305).

The *prima facie* appeal of this model[34] is that we can go back, to some extent, to our very basic intuitions about wavefunctions and collapse, as nurtured in Chap. 2. The wavefunction somehow describes a real physical entity that has a tendency to collapse, repeatedly, so as to give rise to the impression of well localized 'particles' in spacetime, which in turn explains our experience of well-localized macroscopic objects. But of course, given everything that we know about the quantum wavefunction, there are many subtleties involved that we must elaborate on.

[34]Talk of 'models' here should be understood along the same lines as in Chap. 4: as highlighting the somewhat provisionary character. Ultimately all such 'models' here aim to provide an interpretation of the QM formalism, an explanation of our empirical success in using it. Of course the same considerations as in Sect. 6.1.1 hence come to mind; considerations of such collapse interpretations really constituting *alternative theories*.

First note, however, that the above proposal faces the "aesthetic drawback" that it "is not expressed in terms of a compact mathematical equation for the statevector [...]", and the "physical problem [...] that the dynamics does not preserve the symmetry character of wavefunctions describing systems of identical particles." (Bassi and Ghirardi 2003, p. 312)[35]

This difficulty was overcome by so called *continuous spontaneous localization models* (CSL models), originating with the work of Pearle (1989) and developed further in joint work with Ghirardi and Rimini (Ghirardi et al. 1990b). Essentially, the state vector evolves in CSL according to a modified *stochastic* TDSE like the following one[36]:

$$\frac{\mathrm{d}\,|\psi(t)\rangle}{\mathrm{d}t} = \left[-\frac{i}{\hbar}\hat{H}_0 + \sum_k \int \mathrm{d}^3\tilde{x}\, \hat{N}_k(\tilde{x}) w_k(\tilde{x}, t) - \gamma \sum_k \int \mathrm{d}^3\tilde{x}\, \hat{N}_k^2(\tilde{x}) \right] |\psi(t)\rangle\,. \tag{6.16}$$

Here $\hat{H}_0$ is a suitable 'traditional' Hamiltonian, $\hat{N}_k(\tilde{x}) = \sum_\sigma \int \mathrm{d}^3\boldsymbol{x}\, \Lambda^{1/2}(\tilde{x}) \hat{\phi}^\dagger_{k,\sigma}(\boldsymbol{x}) \hat{\phi}_{k,\sigma}(\boldsymbol{x})$ is the number operator for the k-th 'particle type' involved, where $\hat{\phi}^\dagger_{k,\sigma}(\boldsymbol{x})$ and $\hat{\phi}_{k,\sigma}(\boldsymbol{x})$ are creation- and annihilation operators for points $\boldsymbol{x}$ in space and particle types k respectively (σ is a spin-index that could also simply range from 1 to 1, for a spinless system), $\Lambda^{1/2}(\tilde{x})$ is the position representation of the collapse rate operator, and γ is typically defined as a suitable function of the frequency λ (e.g. Bassi and Ghirardi 2003, p. 323). The $w_k(\boldsymbol{x}, t)$ are a family (one for each particle type, k) of real valued functions describing a "white noise" (Bassi and Ghirardi 2003, p. 322) or a "universal fluctuating classical field" (Collett and Pearle 2003, p. 1495; cf. also Bassi et al. 2013, p. 478) interacting with the wavefunction *via* the particle number $\hat{N}_k(\tilde{x})$ such as to give rise to the collapsing behavior. So the collapse is *induced* in CSL models, and due to the coupling with the number operator, the 'more particles in a volume', the more frequent the collapse.

Certainly, the dynamical content of CSL is an improvement over GRW. But one can still say something about the behavior of the wavefunction in the original GRW model as well: Over certain time intervals τ, it will simply evolve unitarily and is then spontaneously affected, at random times, by the process described by $\hat{\Lambda}^{1/2}(\boldsymbol{x})$, so that it undergoes a 'spontaneous collapse'. For a general wavefunction $\Psi(\boldsymbol{x}_1, \ldots, \boldsymbol{x}_N; t_0) = \langle \boldsymbol{x}_1, \ldots, \boldsymbol{x}_N | \Psi(t_0) \rangle$ on configuration space at some initial time t_0, this means that the state vector at a later time t_f will be given by

[35]Ghirardi et al. (cf. 1988, p. 386) proposed a model for systems of 'indistinguishable particles', using a symmetrization of a then joint superoperator, so that individual positions of localization would not matter. The model however has the unsatisfying feature that it prescribes *simultaneous* localizations of all N particles, thereby leaving "no hope for a Lorentz-invariant version." (Tumulka 2006b, p. 1906)

[36]Cf. Ghirardi et al. (1995, pp. 8–11), Ghirardi and Pearle (1990a,b, pp. 30 and 35), or Bassi and Ghirardi (2003, p. 322 ff.).

$$|\Psi(t_f)\rangle = \frac{\hat{\Lambda}_{i_k}^{1/2}(\tilde{\boldsymbol{x}}_k)\hat{U}(\tau_k)\ldots\hat{\Lambda}_{i_2}^{1/2}(\tilde{\boldsymbol{x}}_2)\hat{U}(\tau_2)\hat{\Lambda}_{i_1}^{1/2}(\tilde{\boldsymbol{x}}_1)\hat{U}(\tau_1)\,|\Psi(t_0)\rangle}{\|\hat{\Lambda}_{i_k}^{1/2}(\tilde{\boldsymbol{x}}_k)\hat{U}(\tau_k)\ldots\hat{\Lambda}_{i_2}^{1/2}(\tilde{\boldsymbol{x}}_2)\hat{U}(\tau_2)\hat{\Lambda}_{i_1}^{1/2}(\tilde{\boldsymbol{x}}_1)\hat{U}(\tau_1)\,|\Psi(t_0)\rangle\,\|}, \tag{6.17}$$

for k collapse events in the time-interval $t_f - t_0$, and where $i_j \in \{1, \ldots, N\}, \forall 1 \leq j \leq k$. I.e.: the wavefunction will again and again 'collapse according to its i_j-th coordinate', around a random point $\tilde{\boldsymbol{x}}_j$, where that point is (randomly) chosen with probability $\Pr^{\Psi_-}_{\boldsymbol{x}_{i_j}}(\tilde{\boldsymbol{x}}_j \in d^3\boldsymbol{x}) = \langle\Psi_-|\hat{\Lambda}_{i_j}(\boldsymbol{x})|\Psi_-\rangle\, d^3\boldsymbol{x}$, with i_j and the times equally chosen at random with rate $\lambda\,\langle\Psi_-|\hat{\Lambda}_{i_j}(\boldsymbol{x})|\Psi_-\rangle$ (cf. Allori et al. 2008, p. 357–358; Goldstein et al. 2012, pp. 149–150). The subscript $\boldsymbol{x}_{i_j}$ in the probability function means that the i_j-th coordinate of Ψ_- is affected, where Ψ_- is the wavefunction right after τ_j, i.e. up until the collapse event. While in GRW this behavior occurs without any particular reason, in CSL the 'universal fluctuating fields' $w_k(\boldsymbol{x}, t)$ are to be blamed for the occurrence of a continuous process to a quite similar effect.

Both, GRW and CSL, share a bunch of appealing features; this should in fact be so as CSL preserves GRWs conditions on the density operator as expressed in (6.14) (cf. Bassi and Ghirardi 2003, p. 323). According to Bassi and Ghirardi (2003, pp. 297–298), GRW were driven, in the development of their original model, by the two following desiderata:

1. The 'preferred basis'—the basis on which reductions take place—must be chosen in such a way to guarantee a definite position in space to macroscopic objects.
2. The modified dynamics must have little impact on microscopic objects, but at the same time must reduce the superposition of different macroscopic states of macro-systems. There must then be an 'amplification' mechanism when moving from the micro to the macro level.

Both these desiderata are satisfied in both models. Now we know already how desideratum 1 is fulfilled: in GRW by spontaneous localization processes, and in CSL by the interaction with the randomly fluctuating field, both having the effect that off-diagonal terms in the density matrix in position representation are suppressed, i.e. the terms that signify interference behavior. Desideratum 2, in contrast, is fulfilled in GRW in virtue of the fact that on single systems the non-Hamiltonian part of the density matrix evolution (6.14) has little effect, but "when a large number of 'particles' interact with each other in appropriate ways, they end up being always extremely well localized in space, leading in this way to a situation which is perfectly adequate for characterizing what we call a 'macroscopic object'." (Bassi and Ghirardi 2003, p. 299) And in CSL, the coupling of the classical field to the number density ensures the fulfillment of desideratum 2, since the interaction strength now depends on the density of 'particles' in some volume (cf. Bassi and Ghirardi 2003, p. 321 ff.; and cf. ibid., p. 304 ff. for technical details in general). Both of these modifications are *stochastic* and *nonlinear*, where the former condition is required to reproduce the quantum statistics and disallow superluminal signaling (cf. Bassi et al. 2013, p. 482), and the latter one to ensure the approximate occurrence of definite properties such as localization (cf. Ghirardi 2016, pp. 6–7).

In the terminology introduced in Sect. 2.3, the class of all actual or possible models that satisfy similar parameters as GRW or CSL obviously constitute, or at least pave the way for, *ontological* collapse interpretations of QM. Due to the (necessarily) modificatory nature of the formal models, all such interpretations are clearly formally *revisionary*: they interpret by way of formal modifications to the TDSE or at least the vNE. Since these formal revisions pertain to the dynamics, it is also immediately obvious how the dynamical task of the MAC is fulfilled: the introduction of stochastic and nonlinear terms implies a quick suppression of the validity of the *dynamical* superposition principle (whereas under suitable, low-mass conditions it is at least still approximately valid) which in turn can be understood as a dynamical suppression of superpositions in the *kinematical* sense. Both models are also *conceptually* revisionary insofar as a spontaneous or continuous 'random collapse mechanism' is invoked to account for the non-observability of superpositions at a macroscopic level, and it is *not*, generally speaking, the *measurement* process alone that 'reduces the state vector'. Collapses are being considered as a *real process*, as they were in our first naïve approach, although now in an *informed* way. The kinematical task is a more subtle matter—essentially because one can, again, go either way, deny or allow the wavefunction a status independent of events in spacetime—and we will hence tackle it a little below.

Our discussion of the models so far, however, inevitably raises two questions: (a) How do the orthodox QM-measurement statistics arise, and (b) what about the experimentally confirmed *mesoscopic superpositions* as they occur in SQUIDs (cf. appendix B)? Question (a) has been adressed in most detail by Goldstein et al. (2012), and the reader is referred there for general reference. The upshot is that POVMs can here, as was the case with BM, be introduced to characterize statistical features of *experiments*. However, the POVMs in collapse models will not generally be the same (they arise by taking the collapse-inducing operators into account), so there will be small deviations in prediction from the orthodox quantum formalism (cf. Goldstein et al. 2012, p. 169 ff.).

Such deviations make it possible *in principle* to *test* for the aptness of GRW and similar collapse models, once all parameters such as the average collapse rate λ and width $\alpha^{-1/2}$ are fixed or reduced to known constants (cf. the next section). This possibility is discussed e.g. by Ghirardi and Pearle (1990a, p. 23), where they conclude that, regarding some exemplary neutron-interferometry experiment, "all that is needed is an improvement by a factor of 10^{15}" to "determine whether reduction dynamics really takes place." This is, of course, an *immense* improvement, and one might hence wonder whether empirical (dis-)confirmation is really ever possible in practice.

Question (b) has also been addressed in the literature (cf. Bassi and Ghirardi 2003, pp. 419–420 and references therein) with the upshot that the localization mechanism for individual electrons would only "break one of the Cooper pairs, which would result in the supercurrent being reduced by about one part in 10^{20}" (ibid., p. 419), i.e. that the overall influence of localizations is negligible in this case, which goes even more so if the *reformation* of Cooper pairs is taken into account (cf. ibid., p. 420). Given these considerations, Rae (2004, p. 106) has suggested that "it is possible that the model only really applies to an object like a pointer that is in a superposition of states that are separated in *space*." (my emphasis – FB)

So the formal models we have discussed so far are compatible – up to as yet unachieved measurement accuracies – with everything that has been said about QM before. Nevertheless, both models, GRW and CSL, also face inherent ambiguities. The wavefunction *still* resides in configuration space, and it is not entirely clear from the formulae how it 'manifests itself' in a particle-like fashion in spacetime or interacts (in CSL) with the classical fields $w_k(\boldsymbol{x}, t)$ located therein. Here is how Bell (1987, pp. 204–205) put things w.r.t. GRW:

> It is in the wavefunction that we must find an image of the physical world, and in particular of the arrangement of things in ordinary three-dimensional space. But the wavefunction as a whole lives in a much bigger space, of $3N$-dimensions. It makes no sense to ask for the amplitude or phase or whatever of the wavefunction at a point in ordinary space. It has neither amplitude nor phase nor anything else until a multitude of points in ordinary three-space are specified.

We know, of course, *something* about the joint dynamics in the CSL case in virtue of equations like (6.16); but it is one thing to write down a differential equation that describes interactions and another thing to interpret these interactions.

There are two basic ontologies that have been proposed to make sense of the GRW formalism, only one of which can be straightforwardly transferred to CSL (for reasons expounded on below). Let us hence first focus on GRW and then turn to CSL again. The two ontologies in question are the so called *flash-ontlogy* (GRWf) and the *mass-density ontology* (GRWm), the former originating with Bell's (1987) investigation of GRW's original model, the latter going back to Ghirardi et al. (1995).

The upshot of GRWf is that "matter consists of millions of so called *flashes*, physical events that are mathematically represented by space-time points." (Tumulka 2006b, p. 1898; emphasis in original) Phrased differently, "histories of matter are not made of world lines but of world points." (Goldstein et al. 2012, p. 151) It is these 'events', the collapses of the (high-dimensional) wavefunction, that represent, nay *replace* the point-particles of classical physics; they (or collections of them) should account for the satisfaction of all requirements of our phenomenological particle-concept from Sect. 2.1.3. And these flashes, accounting for the impression of particles, will occur in random locations distributed according to the squared modulus of the wavefunction that results from the collapse processes described in (6.17) at the respective times of collapse (cf. Goldstein et al. 2012, p. 150; Tumulka 2006b, pp. 1899–1900).

In GRWm one defines, in contrast, a *matter field*

$$
\begin{aligned}
m(\boldsymbol{x}, t) &:= \sum_{j=1}^{N} m_j \int \mathrm{d}^3\boldsymbol{x}_1 \ldots \mathrm{d}^3\boldsymbol{x}_N\, \delta(\boldsymbol{x}_j - \boldsymbol{x}) |\Psi(\boldsymbol{x}_1, \ldots \boldsymbol{x}_N; t)|^2 \\
&= \sum_{j=1}^{N} m_j \varrho^{(j)}(\boldsymbol{x}_j, t)\Big|_{\boldsymbol{x}_j = \boldsymbol{x}}
\end{aligned}
\tag{6.18}
$$

describing the *mass-* or *matter density*[37] throughout space and time, where the m_j are the masses of the respective 'particles' going into the wavefunction, and where by $\varrho^{(j)}(\boldsymbol{x}_j, t)\Big|_{\boldsymbol{x}_j=\boldsymbol{x}}$ we mean the quantity $|\Psi(\boldsymbol{x}_1, \ldots \boldsymbol{x}_N; t)|^2$ marginalized for all but the j-th coordinates and evaluated at $\boldsymbol{x}$. This means that the distribution of matter across space and time is determined, at any point in space at any given time, by adding up the contributions of all the respective 'particles' to the squared amplitude of the wavefunction (at that very point at that very time) multiplied by the respective 'particle mass', where the wavefunction is again determined by the evolution law (6.17). Of course 'particles' here become a mere metaphor, a manner of speaking for taking track of the structure of the wavefunction or its contributions to the matter field.

This treatment is somewhat similar to what Schrödinger (1926b) originally had in mind (cf. Sect. 2.1.3), as has been noted by Bacciagaluppi (2010, p. 16 ff.). But of course, due to the collapse-law, the situation is remarkably better in the GRW case, since the collapse dynamics avoids, for instance, Schrödinger-cat-like situations: Allowing for the metaphorical particle-way-of-speaking, a superposition $\alpha\,|$$\rangle + \beta\,|$$\rangle$ of dead and alive cat inside a box will represent a superposition of different (spatial) 'particle configurations', and if a single particle in the cat's heart will be 'hit' by the GRW mechanism, the entire superposed cat will quickly evolve into either $|$$\rangle$ or $|$$\rangle$, "*since the positions of the rest of the particles in the cat's heart are entangled with the position of this particle*[...]." (Maudlin 2011, p. 228; emphasis in original)

It seems worth noting, at this point, that GRWf and GRWm have been argued to be entirely *empircially equivalent*, i.e. that "there is no experiment we could possibly perform that would tell us whether we are in a GRWm world or in a GRWf world, assuming we are in one of the two." (Allori et al. 2008, p. 362) The simple reason is that, whatever the micro-ontology, the perceivable macroscopic matter will ultimately display the very same behavior at the end of any conceivable experiment since GRWm and GRWf *share* the GRW collapse law, the equation determining the *observable* behavior of matter density and flashes (cf. ibid.). This is not so w.r.t. other (ontological) interpretations such as BM, say, since the collapse mechanism introduces tiny deviations in the statistics (cf. the discussion above).

Now as we mentioned before, for CSL matters of ontology are a bit more subtle. First note that an equation like (6.16) (a generalization thereof) can be derived (cf. Nicrosini and Rimini 1990, p. 1320 ff.) as a special case of the infinite frequency limit of GRW processes, i.e. where the system is *constantly* 'being hit' by the collapse mechanism (cf. also Bacciagaluppi 2010, p. 13 on this point). However, as soon as one identifies the 'flashes' as the 'beables', i.e. that which exists in space and time, one encounters problems. The well-definedness of the limit in the above

[37] Allori et al. (2014, p. 330 ff.) have argued that it is contentious to use mass as the defining property, since one could set up a quite similar distribution using *charges* instead of masses, as was originally attempted by Schrödinger. It is hence preferable to use the more neutral term 'matter density'.

procedure requires that one sets $\lambda \cdot \alpha = \text{cst.}$, but this means that $\alpha \to 0$ as $\lambda \to \infty$. It can then be shown, in virtue of connections found by Diósi (1988, cf. p. 421), that the variance of the flash-positions becomes *divergent* (so the flashes typically occur *anywhere*). This has tipped off (Bacciagaluppi 2010, p. 14) to judge that "this picture of CSL as a beable theory turns out not to be viable."

The matter density ontology of GRWm, on the other hand, was in fact *first* postulated as a way to make sense of the CLS dynamics (cf. Ghirardi et al. 1995). Here the operators $\hat{N}_k(\boldsymbol{x})$ from equation (6.16) were multiplied by respective masses m_k for respective 'particle types', and γ was rescaled by m_{ref}^{-2} (m_{ref} an unspecified reference mass). From the resulting operators $\hat{M}(\boldsymbol{x}) = \sum_k m_k \hat{N}_k(\boldsymbol{x})$, a mass density function would then be defined as $m(\boldsymbol{x}, t) = \langle \Psi(t) | \hat{M}(\boldsymbol{x}) | \Psi(t) \rangle$ for a suitable (normalized) $|\Psi(t)\rangle$ (cf. Ghirardi et al. 1995, p. 16).

To sum up: in both, GRW and CSL, we have a wavefunction defined on configuration space, subject to either (GRW) *spontaneous* 'collapse events', i.e. suppressions of interference terms, each (*random*) time w.r.t. one (*random*) coordinate triple, or (CSL) to *continuous dynamics*, *inducing* the suppression of superpositions in the position basis *via* coupling to a randomly fluctuating function, representing a field in spacetime or a family of such, where the coupling is mediated by the particle number density. Then we have (at least) two ontologies to make sense of these (mostly) formal models: The flash ontology (GRWf) in which the spontaneous collapse events are interpreted as the occurrence of 'flashes', i.e. spontaneously occurring 'events' in spacetime, so that any macroscopic material object is then "a galaxy of such events." (Bell 1987, p. 205) And the matter density ontologies (GRWm, CSLm) according to which there is a matter density in spacetime whose dynamics is determined by the wavefunction, which is itself subject to either the spontaneous or the continuous collapse dynamics.

A bunch of obvious questions remain when these models and ontologies are taken under close scrutiny though. Let us talk ontology first. The question that obviously comes to mind is what *role* wavefunction and configuration space really play in these ontologies, as was the case in BM in the last section. Is the wavefunction an objectively existing field? Do spacetime and flashes or matter density 'emerge' from the dynamics of the wavefunction as the 'less fundamental reality'? Or are the (local) 'beables', matter density or flashes respectively, 'more real', and the wavefunction has a different, maybe merely derivative or possibly nomological status?

Prima facie virtually the same set of strategies for dealing with these questions (and hence with the kinematical task of the MAC) is available here as was the case in the discussion of BM and the role of the wavefunction therein (Sect. 6.1.2): One could take the wavefunction for real and would then end up, in this case, with something more similar maybe to de Broglie's 'double solution', i.e. something where flashes or matter densities emerge out of the wavefunction itself and are not postulated independently; but one would still face the problem of understanding the interaction between the wavefunction in configuration space and the points in spacetime wherein the flashes or matter density amplitudes occur. Or, on the other hand, one might try to relegate the wavefunction to the nomological again, opening up the possibility of turning it into a 'mere mathematical device' for a most effective (ideally: *the preferred*) description of stuff in spacetime; but one would, of course, have to tell a story about the laws of nature that is suitable *for* GRWf, GRWm, or CSLm. This means that we could equally refine our classification in this case (nomological, nomological-Humean, non-nomological) and repeat virtually the same arguments regarding the DOC as in Sect. 6.1.

The situation, in fact, involves some further subtleties in the case of the collapse ontologies though. First of all note that a Humean view of laws seems to fit quite well with GRW*f*, wherein the flashes simply occur in a regular statistical pattern described jointly by statistical and dynamical laws involving a suitable Ψ (maybe ultimately Ψ), then construed as a mere 'calculation device' or 'convenience'. Notably, Dowker and Herbauts (2005) demonstrate, for a simple discretized field model wherein binary field values (0,1) on lattice points play the role of the flashes in GRWf (cf. their pp. 503–504), that the dependence of predictions on an initial state $|\psi_0\rangle$ "dies away as time goes on until all we need to know to make predictions [...] is the field configuration back to a certain depth in time." (their p. 505) This prompts them to think of the wavefunction as "a convenient way of keeping the probability distribution up to date, given past events." (ibid.) If the laws (theorems, axioms) of GRWf involving Ψ turn out to be the *most* convenient (simplicity-fit-and-strength optimized) way, this fits perfectly well with a Humean understanding of these laws. The situation is comparable to that in BM.

On the other hand, Egg and Esfeld (2015, p. 3240) have argued that there is

> reason to be less attracted to Humeanism in GRW*m* than in BM. While Humeanism holds that the quantum state supervenes on the complete history of the primitive ontology, the mathematical structure of GRWm seems to imply just the opposite[...]. (my emphasis – FB)

To recall, *Humean supervenience* is the hypothesis "that every contingent property instantiation at our world holds *in virtue of* the instantiation of Humean properties", where a property is called 'Humean' "if its instantiation requires no more than a spatiotemporal point and its instantiation at that point has no *metaphysical* implications concerning the instantiations of fundamental properties elsewhere and eslewhen." (Loewer 1996, p. 102; emphasis in original) Of course the BSA can be construed as an implementation of Humean supervenience in the case of laws: whichever laws are true of our world are true *in virtue of* what happens *here and now* for varying heres and nows.

Now the matter density $m(\boldsymbol{x},t)$ itself specifies a local value for matter at every spacetime point and may hence be seen as a continuous collection of Humean properties. But $m(\boldsymbol{x},t)$ is defined in equation (6.18) in virtue of $|\Psi(\boldsymbol{x}_1,\ldots,\boldsymbol{x}_N;t)|^2$, and there is no obvious way to *infer* Ψ from $m(\boldsymbol{x},t)$ in virtue of (6.18)—Egg and Esfeld's argument has a *prima facie* plausibility. But it is (a) important not to confuse *deductive inference* with *metaphysical dependence*: Just because we infer the distribution of matter over space and time from the Ψ function does not force us to believe that either supervenes on the other.[38] And (b), more importantly, a regularity account of *laws* such as the BSA should not be confused with the *metaphysical* hypothesis of Humean supervenience. Cohen and Callender (2009, p. 3) e.g. have it that "the doctrine of Humean supervenience [...] is logically distinct from [the BSA]. Indeed, many versions of [the BSA] [...] are at odds with Humean supervenience." Nothing about the BSA commits us to believe in the supervenience of everything else on Humean properties, and since GRWf and GRWm are empirically equivalent, why should we not equally write down wavefunctions and laws such as the TDSE or (6.17) in virtue of the observed behavior of the *matter density*?

In any case, Humeanism about laws leaves us with the same discomforting feeling about correlations in spatially widely separated local values of the matter density or the local occurrences of flashes as it did about the particle velocities in the case of BM. Equally, the explanatory value of Ψ (ψ) is thereby significantly reduced: Flashes *just* occur, matter densities *just* fuzz around somewhat randomly, all the while exhibiting EPRB-correlations in both cases. According to a flash ontology, moreover, there is typically *nothing at all* in between source and screen in a (sufficiently evacuated) double slit experiment, because a flash only occurs in the presence of *other* masses (cf. Maudlin 2011, p. 237 for illustrations).

If, on the other hand, one would, for instance, take the arguments from Humean supervenience as crucial against the BSA and/or take the mathematics seriously and the direction of inference as indicative of metaphysical dependence, then one would end up with some kind of interaction problem between configuration space and spacetime (or nomological and beable realms or...) again since the wavefunction, being taken seriously, would *still* not describe anything residing in spacetime, whereas flashes and matter densities do. Ultimately, the impact of granting Ψ (ψ) a nomological status seems to leave us in no better situation in objective collapse interpretations than it did in Bohmian ones.

We have mostly left CSLm out of the discussion, since it plays kind of a special role when it comes to ontology. A reference to GRW instead of CSL is in fact widespread in the philosophical literature on QM, as has been noticed by Egg and Esfeld (cf. 2015, p. 3236). They, however, have it that

> if we remember that both theories were primarily introduced to explain the localization of macroscopic objects, then the difference between them does not seem so significant anymore: in systems consisting of a large number of constituents, the localization process is so fast that the difference between an instantaneous and a gradual process becomes negligible. (ibid.)

Again the argument is *prima facie* plausible, but ultimately not: CSLm contains the universal fluctuating field(s) $w_k(\boldsymbol{x}, t)$, whose ontological status raises questions

[38]This is acknowledged by Egg and Esfeld (2015, p. 3240), who recognize a "considerable flexibility in implementing the thesis of Humean supervenience of the quantum state on the primitive ontology." But they believe the "non-Humean options" to be "clearly [...] less revisionary" (p. 3241), which is reason enough for them to prefer those options.

beyond those occurring in GRWm. Bassi et al. (2013, p. 492) e.g. find it "tempting to suggest that such a field has a cosmological nature," but they concede that "at this stage this is only a speculation."

This, in fact, leads us back to the general properties of the *formal models* that we claimed raise obvious questions as well. The general worry is that there is a lot of *arbitrariness* in these models. Why the form of the operators $\hat{\Lambda}(\boldsymbol{x})$ (which, in fact, *need* not be that of equation (6.15); cf. Bassi and Ghirardi 2003, pp. 298–299)? Whence the average collapse rate λ and localization width $\alpha^{-1/2}$? Whence the universal fluctuating fields? Why should we accept all this non-Occamism? Before we flesh this out into a concise criticism though, let us first look at a class of related proposals that attempt to remove some of the arbitrariness in a more concrete fashion than by mere speculations about cosmological fluctuations.

6.2.2 *Can Gravity Account for Collapse?*

Certainly the most interesting proposal for a collapse model with a randomly fluctuating field just as in CSL, but where the nature of the field is *specified*, is that first investigated by Diósi and Lukács (1987) and Diósi (1987). To understand the basic proposal of these papers, first recall that the familiar formula for Newtonian force, $\boldsymbol{F} = m\boldsymbol{a}$, is $\boldsymbol{F}_G = m\boldsymbol{g}(\boldsymbol{x}, t)$ when applied to gravitation, where $\boldsymbol{g}(\boldsymbol{x}, t)$ is the local gravitational acceleration (with familiar mean value $\overline{|\boldsymbol{g}|} \approx 9,81$ m/s^2 on earth's surface). Moreover, Newton found that, when considered as a force exerted by a mass M on some other mass m, the force law reads $\boldsymbol{F}_G = -G\frac{mM}{r^2}\boldsymbol{n}_r$, where $r = |\boldsymbol{x} - \boldsymbol{x}'|$ is the distance between masses m and M at points $\boldsymbol{x}$ and $\boldsymbol{x}'$ respectively, G is the gravitational constant, and $\boldsymbol{n}_r$ is a unit vector pointing along a (Euclidean) straight line in the direction of m from M. So we can see that $\boldsymbol{g}(\boldsymbol{x}, t) = -G\frac{M}{r^2}\boldsymbol{n}_r$ (the t comes from the time-dependence of $\boldsymbol{x}$ and $\boldsymbol{x}'$).

Now if one writes $\Phi(\boldsymbol{x}) = -\frac{GM}{|\boldsymbol{x}-\boldsymbol{x}'|}$, one easily finds that

$$-\nabla\Phi(\boldsymbol{x}) = -GM\frac{\boldsymbol{x} - \boldsymbol{x}'}{|\boldsymbol{x} - \boldsymbol{x}'|^3} = -G\frac{M}{r^2}\boldsymbol{n}_r, \tag{6.19}$$

where $\boldsymbol{n}_r := \frac{\boldsymbol{x}-\boldsymbol{x}'}{|\boldsymbol{x}-\boldsymbol{x}'|}$. So the simple force law becomes $\boldsymbol{F}_G = -m\nabla\Phi(\boldsymbol{x}, t)$, and we can identify $\Phi(\boldsymbol{x}, t)$ as a *gravitational potential* and $m\Phi(\boldsymbol{x}, t)$ as a gravitational potential energy (e.g. Hartle 2003, p. 38 ff.), in virtue of the connection $\boldsymbol{F} = -\nabla V(\boldsymbol{x}, t)$ for V some potential energy and $\boldsymbol{F}$ conservative.

The original suggestion of Diósi and Lukács (1987, pp. 491 ff.) and Diósi (1987, pp. 378 ff.) now was to introduce a stochastically fluctuating gravitational potential Φ into the TDSE which satisfies $\langle\Phi\rangle_{\text{St.}} = \Phi_N$ for $\langle\cdot\rangle_{\text{St.}}$ some suitable stochastic average and Φ_N the Newtonian potential. The resulting TDSE is then of the form

$$i\hbar\frac{\partial}{\partial t}|\psi\rangle = \left[\hat{H}_0 + \int \mathrm{d}^3\boldsymbol{x}\,\Phi(\boldsymbol{x}, t)\hat{f}(\boldsymbol{x})\right]|\psi\rangle, \tag{6.20}$$

where $\hat{f}(\boldsymbol{x})$ is a local mass density operator and $\hat{H}_0$ the non-gravitational part of the Hamiltonian.

Using standard techniques for deriving master equations (as can be gathered e.g. from Le Bellac 2006, p. 526 ff. or Schlosshauer 2007, p. 153 ff.) and imposing also that Φ at different spacetime points correlates as $\langle \Phi(\boldsymbol{x}, t)\Phi(\boldsymbol{x}', t')\rangle_{\text{St.}} = \frac{\hbar G}{|\boldsymbol{x}-\boldsymbol{x}'|}\delta(t - t')$,[39] Diósi (1987, pp. 379) derived from the form of the total Hamiltonian in (6.20) the following master equation:

$$\frac{\partial}{\partial t}\hat{\rho}(t) = -\frac{i}{\hbar}[\hat{H}_0, \hat{\rho}(t)] - \frac{G}{2\hbar}\int\int \frac{\mathrm{d}^3\boldsymbol{x}\,\mathrm{d}^3\boldsymbol{x}'}{|\boldsymbol{x}-\boldsymbol{x}'|}[\hat{f}(\boldsymbol{x}), [\hat{f}(\boldsymbol{x}'), \hat{\rho}(t)]]. \tag{6.21}$$

Crucially, in a suitable coordinate representation $|\boldsymbol{X}^{(N)}\rangle := |\boldsymbol{x}_1, \ldots, \boldsymbol{x}_N\rangle$, one again obtains a damping of all off-diagonal terms $\langle \boldsymbol{X}^{(N)}|\hat{\rho}|\boldsymbol{X}'^{(N)}\rangle$ by a factor $1/\tau_d(\boldsymbol{X}^{(N)}, \boldsymbol{X}'^{(N)})$, with $\tau_d(\boldsymbol{X}^{(N)}, \boldsymbol{X}'^{(N)})$ some damping time defined in terms of the mass densities $f(\boldsymbol{x}|\boldsymbol{X}^{(N)})$, $f(\boldsymbol{x}'|\boldsymbol{X}'^{(N)})$ at points $\boldsymbol{x}$ and $\boldsymbol{x}'$ given configurations $\boldsymbol{X}^{(N)}$ and $\boldsymbol{X}'^{(N)}$ respectively (cf. Diósi 1987, p. 379; Bassi et al. 2013, p. 507). So based on gravitational considerations, spatial superpositions for larger systems will be quickly suppressed in this model as well, just as they should.

Again we have a fulfillment of the two main desiderata (well localized macroscopic objects and little effect on microscopic dynamics) that drove GRW, since the damping also depends on the mass density. In 1989, Diósi additionally "took the inevitable step of casting the master equation in the equivalent language of a *stochastic* Schrödinger equation" (Bassi et al. 2013, p. 508; my emphasis – FB), which made his model straightforwardly comparable to CSL (cf. Diósi 1989, for details).

But the resulting model clearly has an advantage over CSL in that the collapse-inducing field occurring therein is non-arbitrary—it is the gravitational potential. There also are no free parameters like λ and α from GRW: λ is replaced by the inverse of τ_d which depends only on natural constants and the mass density operators, which in turn do not introduce a new (average) localization length.[40]

However, the mass density operators themselves still constitute an element of *choice*, and the choice of operators $\hat{f}(\boldsymbol{x})$ in actual examples computed by Diósi (1989, p. 1171) turned out to be quite unfortunate, since it implied an unreasonable *increase in energy over time*. This is, in fact, a general worry about collapse models: due to the modification of dynamical equations, there is a *non-conservation of energy* that is already present in the original GRW model.

---

[39]The scaling can be motivated from considerations of the effect of the uncertainty relations on the gravitational acceleration, when it is assumed that Φ_N vanishes (cf. Diósi and Lukács 1987, p. 489 ff.; Bassi et al. 2013, p. 507).

[40]In fact, the ratio $\lambda\alpha$ still occurs in the paper, but it is treated as a constant and "assume[d] its order is of unity." (Diósi 1989, p. 1169) Subsequent discussions drop the factor altogether (e.g. Ghirardi et al. 1990a, p. 1059).

Solving their equation (6.14) in position representation, Ghirardi et al. (1986, p. 474) found that the square of the momentum operator satisfies $\langle \hat{p}^2 \rangle = \langle \hat{p}^2 \rangle_0 + \frac{\alpha \lambda \hbar^2}{2} t$ when averaged w.r.t. the resulting density matrix, where $\langle \hat{p}^2 \rangle_0$ is the average that would result for the usual Schrödinger dynamics. Since $\langle E \rangle = \frac{\langle \hat{p}^2 \rangle}{2m}$, this implies a linear increase in energy over time. Now while in GRW this increase was estimated to be minute for 'realistic' values of α and λ (cf. Ghirardi et al. 1986, p. 481), Ghirardi et al. (1990a, pp. 1061 and 1064) also estimated the average energy increase in Diósi's (1989) model, which, for a system of nucleons in the order of 10^{23} constituents and based on the stochastic Hamiltonian, turns out to be about 0,1 mW—that is close to the power range of modern smartphones in standby.

But Ghirardi et al. (1990a, pp. 1058 and 1061) also demonstrated that one can avoid these difficulties if one uses the Gaußian-smeared operators $\hat{M}_k(\boldsymbol{x}) = m_k \sum_\sigma \int \mathrm{d}^3\boldsymbol{x}\, \Lambda^{1/2}(\tilde{\boldsymbol{x}}) \hat{\phi}^\dagger_{k,\sigma}(\boldsymbol{x}) \hat{\phi}_{k,\sigma}(\boldsymbol{x})$ that we already mentioned in the discussion of the CSL model. We can see that the ideas of CSL and Diósi's gravitational model fit together quite nicely, and indeed, Pearle and Squires (1996, p. 292) have also investigated the possibility of interpreting $w(\boldsymbol{x}, t)$ as a mass density, related to the gravitational potential *via* the Newton-Poisson equation $w(\boldsymbol{x}, t) = \frac{1}{4\pi G} \Delta \Phi(\boldsymbol{x}, t)$, which reappears in GR as a limit for a static weak-field metric (cf. Hartle 2003, pp. 38 ff. and 485 ff. for details).

These models raise interesting possibilities, but they still all suffer from an obvious deficit: They are *semi-classical*, and there are numerous good reasons to be skeptical about the aptness of a semi-cassical treatment of gravity, among them empirically false predictions about the behavior of a cavendish balance in a suitably contrived experiment (cf. Kiefer 2007, p. 20 and p. 15 ff. and references therein for further discussion). In agreement with these objections, Penrose (1996) has offered a quite different treatment for gravitational collapse models. He starts from considering "a rigid lump of material" (his p. 584) put into a superposition $|\psi\rangle = \alpha |\xi\rangle + \beta |\varphi\rangle$ of spatially non-overlapping localizations by some clever mechanism, where $|\xi\rangle, |\varphi\rangle$ are eigenstates of some suitable Hamiltonian with the same energy E_0. This means, of course, that the total state $|\psi\rangle = \alpha |\xi\rangle + \beta |\varphi\rangle$ for given coefficients α, β would also be of energy E_0, and hence *stationary* w.r.t. the action of the Hamiltonian, whence it "must [...] persist unchanged for all time" (Penrose 1996, p. 585) unless affected by a different process.

Penrose then considers the (stationary) gravitational fields created by the two 'lumps' and equips them with quantum states $|G_\xi\rangle$ and $|G_\varphi\rangle$ which should incorporate "whatever is to be meant by the quantum state of a stationary gravitational field – including all the internal degrees of freedom of the field[...]." (Penrose 1996, p. 588) The resulting state would hence be the entangled state $|\Psi\rangle = \alpha |\xi\rangle\, |G_\xi\rangle + \beta |\varphi\rangle\, |G_\varphi\rangle$; but the crucial question now is whether $|\Psi\rangle$ is *still* stable (*qua* being a stationary state), since the states $|\xi\rangle\, |G_\xi\rangle$ and $|\varphi\rangle\, |G_\varphi\rangle$ each "must involve a reasonably well-defined (stationary) space-time geometry, where these two space-time geometries differ significantly from each other." (Penrose 1996, ibid.)

Crucially, the 'stationarity' of *quantum states* becomes a subtle matter in two superposed spacetimes: *for each spacetime*, there will be a *Killing vector* that genreralizes the operator $\frac{\partial}{\partial t}$, i.e. a vector that 'lies in the direction' of a translation $t \mapsto t+\epsilon$ along which the metric does not change (cf. Hartle 2003, p. 176; Nakahara 2003, p. 279), as long as that spacetime is itself stationary,[41] so that a notion of 'stationary state' is well defined therein. However, here *two spacetimes* are being superposed, and one must wonder whether an operator can be found that represents 'time translations' for *superposed* spactimes. Penrose's (1996, p. 589; emphasis in original) conclusion is that "the notion of time-translation operator is *essentially* ill defined" in this context, but that one can give "a clear-cut measure of the *degree* of this ill-definedness for such a superposed state", and that the associated "fuzziness" in the concept of energy for such a state "is consistent with the view that such a superposed state is *unstable*[...]." A "lifetime" (ibid.), τ, of the state can then be quantified in virtue of a measure that exploits an energy-time uncertainty, i.e. $\tau \propto \hbar/\Delta E$.[42]

This is an equally interesting and remarkably different proposal, compared to that of Diósi, although to a similar effect (fast collapse of large-scale spatial superpositions). Given that there are independent reasons for subjecting gravitational fields to a quantization procedure (e.g. Kiefer 2007, p. 3 ff.) and reasons to be skeptical about semi-classical approaches to gravity, a 'fully quantum' treatment of gravitational fields is certainly an improvement. But Penrose's considerations do not constitute a fully worked-out proposal,[43] and to the present author's knowledge, none exists to date.

In summary, gravitational considerations help to remove some of the arbitrariness, but in semi-classical ones it is not clear whether they can be made compatible with a prospective future theory of quantum gravity. The approach of Penrose (1996) lays interesting ground work for quantum-gravitational concerns, but no fully worked out model is presented, and any prospective one may (or may not) face difficulties such as the fact that "if the evolution is deterministic and nonlinear, the possibility of superluminal propagation appears to be present", or that "[i]t is not clear [...] how the Born rule will be recovered dynamically." (Bassi et al. 2013, pp. 508 and 509) These details obviously depend on the eventual model though.

---

[41] For details on the stationarity of spacetimes e.g. Ruetsche (2011, p. 207).

[42] A first estimate for ΔE is also computed by Penrose from squared differences in gravitational accelerations for different spacetimes in a Newtonian limit (cf. Penrose 1996, p. 594 ff. for details).

[43] In particular, that would require a "precise measure of uncertainty that is to be assigned to the 'superposed Killing vector' and to the corresponding notion of 'stationarity' for the superposed space-time." (Penrose 1996, p. 596)

6.2.3 Ad Hocness and Loose Ends

As we have seen in the last section, gravitational models still face a bunch of inherent difficulties or are not (yet) developed in sufficient detail. This leaves us with the objection raised at the end of Sect. 6.2.1, that there is a lot of arbitrariness in CSL and GRW. In fact, the whole introduction of stochastic collapses, be they dynamical or spontaneous, parametrized by new constants (λ, α) or induced by unidentified fields ($w_k(\boldsymbol{x}, t)$), has a very unpleasant ad hoc character to it.

Based on the desire to fulfill the desiderata 1 and 2 of Sect. 6.2.1 (localization of macroscopic objects and little effect on microscopic ones), moreover, Ghirardi et al. (1986, p. 480) have provided estimates for λ and α. In fact, setting $\lambda_{\text{micro}} = \lambda_{\text{macro}}/N$, where N is the number of particles involved, they gave the values $\lambda_{\text{micro}} \approx 10^{-16}$/sec, $\lambda_{\text{macro}} \approx 10^{7}$/sec and $\alpha^{-1/2} \approx 10^{-7}$ m, which would mean that tiny particles are almost never well localized and that macroscopic objects almost never in (spatial) superposition states. But this only means that the model is being fitted (in a somewhat ad hoc fashion) to data known beforehand. Fitting of parameters is of course a widespread scientific practice, but it would still be *preferable* to remove the ad hocness, and to account for the occurrence of the (apparent) macroscopic world of chairs, tables, and houses in a *less arbitrary* fashion.

The same worry of arbitrariness and ad hocness of course carries over to CSL; the universal fluctuating field implies a new parameter, nay, a continuous collection thereof. And the same also goes for Pearle and Squires' (1996) aforementioned gravitational modification of CSL, where they would investigate models for the spontaneous localization of massive monopoles and dipoles, but had to concede that they simply "have no good argument for choosing" (their p. 301) certain values for parameters such as a localization probability and mass density to ultimately determine a localization length. This is a pathological problem for the entire endeavor of defining collapse models.

But the occurrence of *parameters* to fit is, of course, not the only ad hocness concern. Here is a maybe philosophically more serious worry. The introduced stochastic Schrödinger equations or master equations *formally* do *not* have the effect that a mass density suddenly occurs *in spacetime* or that well defined flashes occur at spacetime points, i.e. that the multi-coordinate wavefunction is replaced by some suitable function of three spatial coordinates and one temporal one. The two ontologies have to be *imposed upon* the formalism and do not 'arise naturally' from it.

This is quite in contrast to BM where at least in the non-relativistic particle case the guidance equation, describing the motion of a particle in space, is derived straightforwardly from the TDSE. In GRW, the modified (density matrix) dynamics has the implication that "linear superposition is consequently transformed into a *statistical mixture of states*" (Ghirardi et al. 1986, p. 478; my emphasis – FB), *not* that a multi-coordinate function is replaced by a unique set of single-coordinate functions, individually describing pieces of matter in space(time). The mass density formula (6.18) or the interpretation of the 'collapse events' as 'flashes' have to be imposed *in addition*. Thus w.r.t. ontological matters, collapse models arguably fare slightly worse than does BM, which is equally Maudlin's (2010a, p. 126) judgement: "it is not at all obvious how the GRW collapses, by themselves, can do any of the work that makes the Bohmian account comprehensible." In other words: while part (i) of our DOC is somewhat satisfied in virtue of the collapse mechanisms, (ii) is not, and neither is (iii) really: the relation between objects in spacetime and the collapsing wavefunction requires an additional narrative, and the ontological significance of the wavefuntion depends strongly on that narrative.

There are also some further loose ends in this connection, such as so called *tails problems* (e.g. Bassi and Ghirardi 2003, p. 357 ff.; Ghirardi and Pearle 1990b, p. 37 ff.; Wallace 2008, p. 56 ff.). Wallace (2008, p. 56) distinguishes two different such problems, namely the problem of *bare tails* and that of *structured* ones. The bare tails problem concerns the fact that the localization in collapse models is typically Gaußian or similarly 'smeared', which implies that it has vanishing support *nowhere* (or equally: has infinitely long 'tails'). There is lots of things to say, however, that make it appear that the problem "has little or nothing to do with the GRW theory." (Wallace 2008, p. 57). The main reason is that such 'tails' occur, of course, already in *orthodox QM*; we discussed the *spreading* of the wavepacket in Sect. 2.1.3, but we also (consciously) merely claimed that wavepackets 'raise hopes' for suitably describing well-localized particles in Sect. 2.1.1—which, strictly speaking, they do *not* describe. Replacing, in some collapse model, the Gaußian operators by ones with compact support would be of little help, since whenever "the evolution equation contains the kinetic energy term, any function, even if it has compact support at a given time, will instantaneously spread acquiring a tail extending over the whole of space." (Ghirardi 2016, p. 21)

In a mass density ontology, this is not so much of a problem though (some subtleties about, say, mind-brain supervenience aside), as Monton (2004, p. 418) has argued: "macroscopic objects appear highly localized" therein, because "most all [*sic*] of their mass is concentrated in a small region of space, the region where the object appears to be localized."

The occurrence of *structured* tails is different and poses a more serious problem, even for collapse ontologies with mass density. Recall that we claimed that Schrödinger-cat situations would not arise because the collapse dynamics would quickly drive a state like α |  $\rangle + \beta$ |  $\rangle$ into either of the two superposed states. But this is not quite right; rather the collapse mechanism will yield something like $\sqrt{1-\epsilon^2}$ |  $\rangle + \epsilon$ |  $\rangle$, which still has a dead *and* a live cat in it, no matter how small ϵ becomes. This problem is more serious; there should also be a 'low density dead cat', if the mass density formula is taken seriously. Suggested solutions to this problem appeal to unjustified talk of different 'worlds' (e.g. Egg and Esfeld 2015, p. 3235) or to introducing operators with compact support after all (e.g. Wallace 2014, p. 4), thereby 'fully eliminating' the dead cat—by essentially *another ad hoc move*.

Of course worries about relativistic generalizations also come to mind for collapse models as they did in BM. But it appears that matters are considerably better for collapse models than for BM: Work by Bedingham (2009, 2011) on a relativistic CSL-like model avoids earlier worries about the possibility of superluminal signaling in collapse models (cf. Squires 1992), problems of divergent increases in energy in previous relativistic versions of CSL (cf. Bassi and Ghirardi 2003, p. 331 ff.), limitations of previous relativistic and field theoretic models as regards interactions (cf. Tumulka 2006a,b),[44] and can be supplemented with a relativistic law for the matter density function (cf. Bedingham et al. 2014). Bedingham's relativistic model also requires some *foliation* for its formulation, but Bedingham (2011, p. 692) thinks that this foliation "has no physical consequences since given a fixed initial state and a complete realized set of stochastic information [...], for any two foliations which share a common leaf, the assigned state on that leaf is unique." This does not impair the ad hocness worries raised before, however; Bedingham introduces a relativistic replacement of the CSL universal field that simply constitutes a "*nonlocal hidden variable*[...]." (Bedingham 2011, p. 693; my emphasis – FB)

6.3 Many Worlds: Taking Superposition Seriously

6.3.1 *Everett's Brave Proposal*

So far we have seen moderately successful attempts at interpreting QM by either rewriting and reinterpreting the dynamical equations or by explicitly modifying them so as to remove the conflict between projection postulate and unitary evolution. Here is an (apparently) entirely different suggestion: Just *drop the projection postulate*. How could a serious proposal along these lines work? One would have to specify a mechanism that creates the *illusion* of the occurrence of definite outcomes as are predicted by quantum postulate (VII) and observed in experiment. The first to champion such an approach was Hugh Everett III in his PhD thesis, first published in 1957 in a version "cut down to a quarter of its size on Wheeler's [his supervisor's – FB] insistence" (Saunders 2010, p. 6), and only later (1973) published in its 'full glory' and with a more overt statement of the remarkable philosophical consequences.

The basic proposal goes as follows (cf. Everett, 1973, p. 64 ff.).[45] Recall (once more) from (2.33) that the unitary evolution on a system S interacting with a measurement apparatus A will have the effect $\hat{U}_{S,A} \sum_j \alpha_j |S_j\rangle |\mathcal{A}_0\rangle = \sum_j \alpha_j |S_j\rangle |\mathcal{A}_j\rangle$, if S is in a suitable superposition state and A in some 'ready state' $|\mathcal{A}_0\rangle$, and if $\hat{U}_{S,A}$ effects a suitable joint evolution.[46] One of Everett's (1973)

[44]Cf. also Maudlin (2011, p. 243 ff.) for an accessible informal analysis of at least *Tumulka's* relativistic model.

[45]Cf. also Saunders (2010) for a general introduction to the subject.

[46]For obvious reasons, we will here generically use $|S_j\rangle$ to refer to system states, $|\mathcal{A}_j\rangle$ to apparatus states... and so on.

suggestions now was to think of an *observer* O as a *physical system* with a memory register [...], where the dots indicate arbitrary memories of past interactions with an environment up to a certain point. This would allow to equip the observer with a quantum state $|\mathcal{O}_0[\ldots]\rangle$ before she interacts with the apparatus A to read out an outcome, i.e. in between stages (i) and (ii) of the measurement process as we had analyzed it in Sect. 2.1.4. Taking QM seriously, there should also be a suitable interaction Hamiltonian describing the physical interaction, (ii), between observer O and apparatus A. The result would then (neglecting once more state perturbations due to the interaction) be of the form

$$|\Psi_{SAO}\rangle := \sum_j \alpha_j\, |\mathcal{S}_j\rangle\, |\mathcal{A}_j\rangle\, |\mathcal{O}_0[\ldots]\rangle \xmapsto{\hat{U}_{SA,O}} \sum_j \alpha_j\, |\mathcal{S}_j\rangle\, |\mathcal{A}_j\rangle\, |\mathcal{O}_j[\ldots, j]\rangle$$
$$=: |\Psi'_{SAO}\rangle\,, \qquad (6.22)$$

where we have used $\hat{U}_{SA,O}$ to denote the unitary operator that describes the state change, the index SAO on the quantum state refers to the fact that the state is the joint state of S, A, and O, and where the j indices in O's final state indicate that O *has a* state 'coupled' to the j-th state of the joint system SA *in which* she has *registered* the j-th outcome on A. In other words: There should now be *several* 'relative states' (cf. Everett, 1957, p. 456) of O of *having registered all the outcomes respectively*.

Put even more straightforwardly, the unitary evolution here puts O into a certain superposition state as well, entangled with that of S and A. And that is all there is, a global entangled state vector, subject to unitary dynamics. In other words, we here investigate the following 'unal ontology':

Conjecture 4 (The unal ontology) The quantum mechanical state vector is a truthful representation of all the systems in the universe. It and the unitary dynamics that pertain to it are all there is.

But how can we even make sense of such a proposal, given that we ourselves do not know, from introspection, what it is *like* to be in a superposition of states? We will return to this issue in the next section, but now first review some reasons to even accept such a 'preposterous' suggestion in the first place.

An explicit concern of Everett (1973, pp. 8 and 118–119) was that QM be applicable *universally*, i.e. also in the form of some prospective final quantum theory of the gravitational field, which we mentioned in Sect. 6.2.2 to be indicated by empirical evidence and theoretical concerns. After much initial resistance (cf. Byrne 2010a), this and similar reasons ultimately served as a motivation for a whole range of other physicists to accept Everett's ideas; and today Everett's interpretation is, in fact, among the most widely accepted ones, as reflected in some recent and less recent polls (cf. Tegmark 1998; Schlosshauer et al. 2013). One of the first 'Everettians' was Bryce DeWitt, who found that "Everett's view of the world is a very natural one to adopt in the quantum theory of gravity, where one is accustomed to speak without embarassment of the 'wave function of the universe.'" (1967, p. 1141)

Another advantage is that at least *prima facie* this approach does not require additional interpretive or formal elements; just take the quantum state to represent what it *seems* to represent. Or as DeWitt (1970, p. 160; emphasis omitted) had it: "The mathematical formalism of the quantum theory is capable of yielding its own interpretation." And similarly Wallace (2012, p. 38): "The 'Everett interpretation of quantum mechanics' is just quantum mechanics itself, 'interpreted' the same way we have always interpreted scientific theories in the past: as modelling the world."

DeWitt (1971, p. 182), in fact, also coined the now-familiar talk of "many worlds", referring to the individual terms in a superposition state such as $|\Psi'_{SAO}\rangle$. So Everett's interpretation is a *many worlds interpretation* (MWI) of the quantum formalism. It tackles the MAC by leaving the (*unitary*) dynamics alone but taking them seriously (the dynamical task) and by simply interpreting the meaning of the quantum state to be an accurate description of reality at all scales—as an expression of *multiplicity*, not *indefiniteness* (cf. Wallace 2012, p. 37)—even if what is real thereby exists inevitably partly unperceived by 'us' (the kinematical task). The MWI, moreover, is *formally* quite conservative and avoids the difficulty of collapses: one simply has to remove one of the postulates from the list (I)–(VII), namely the contentious projection postulate, (VII). And, as in the two interpretations reviewed in the previous two sections, we (apparently) also remove the 'dawning danger' of Wigner's friend, that *consciousness* has anything to do with inducing the (physical) 'collapse' of the state vector, simply because there is no collapse.

But the MWI of course implies *conceptual* revisions w.r.t orthodoxy, and clarifications are certainly in order in many places. First of all note that the sort of evolution needed to lead to superpositions of apparatus, observers, laboratories... and so forth, is *ubiquitous*. Here is DeWitt (1970, p. 161) again:

> This universe is constantly splitting into a stupendous number of branches, all resulting from the measurementlike interactions between its myriads of components.

So basically any old kind of interaction between any old pair of systems 'splits' the universe into a multitude of 'different branches'. This does not explain, however, how *conscious observers* are located or 'dynamically created' in these branches, and how it comes that they not observe *each other*. How are the 'branches' even well defined in the first place? And *how many of them* are there? Let us start shedding some light on these issues and their possible resolutions.

6.3.2 *Quantum Decoherence and Its Importance for the Many Worlds Interpretation*

There is something quite problematic about a state like $|\Psi'_{SAO}\rangle = \sum_j \alpha_j |S_j\rangle |\mathcal{A}_j\rangle |\mathcal{O}_j[\ldots, j]\rangle$, namely, that it could be rewritten in a *different basis*. To make our point with the aid of a simple example (analogously given in Albert and Loewer 1988, p. 202 ff. or Schlosshauer 2007, p. 53 ff.), imagine a simple spin measurement with a detector (the 'apparatus', *A*) that is itself a two-state system. Then we could have an evolution like

$$|\psi^i_{SA}\rangle = \frac{1}{\sqrt{2}}(|\uparrow_z\rangle + |\downarrow_z\rangle)\,|\mathcal{A}_0\rangle \mapsto |\psi^f_{SA}\rangle = \frac{1}{\sqrt{2}}\left(|\uparrow_z\rangle\,|\mathcal{A}_{\uparrow_z}\rangle + |\downarrow_z\rangle\,|\mathcal{A}_{\downarrow_z}\rangle\right), \tag{6.23}$$

where the superscripts i and f indicate initial and final states of the interaction respectively, and $\mathcal{A}_x$ are labels for suitable states of the detector (of having registered nothing, up, or down respectively). But $\left\{|\mathcal{A}_{\uparrow_z}\rangle, |\mathcal{A}_{\downarrow_z}\rangle\right\}$ is construed as a basis of $\mathbb{C}^2$, so with an eye on connections like (2.14), we can immediately rewrite $|\psi^f_{SA}\rangle$ in (6.23) as

$$\begin{aligned}|\psi^f_{SA}\rangle &= \frac{1}{\sqrt{2}}\frac{1}{2}\Big[(|\uparrow_x\rangle + |\downarrow_x\rangle)(|\mathcal{A}_{\uparrow_x}\rangle + |\mathcal{A}_{\downarrow_x}\rangle) + (|\uparrow_x\rangle - |\downarrow_x\rangle)(|\mathcal{A}_{\uparrow_x}\rangle - |\mathcal{A}_{\downarrow_x}\rangle)\Big]\\ &= \frac{1}{\sqrt{2}}\left(|\uparrow_x\rangle\,|\mathcal{A}_{\uparrow_x}\rangle + |\downarrow_x\rangle\,|\mathcal{A}_{\downarrow_x}\rangle\right).\end{aligned} \tag{6.24}$$

What did the detector detect? What, in fact, did *happen* at all? Quantum states, unless specified instrumentally for a specific purpose, have an *inherent ambiguity* in them in virtue of the possibility of a *basis change* in a vector space. What 'right' do we have to claim that O in our earlier state $|\Psi'_{SAO}\rangle$ is in all the observational $|\mathcal{O}_j[\ldots, j]\rangle$ states after the interaction, not in some entirely different set of states, expressed by viewing each of the $|\mathcal{O}_j[\ldots, j]\rangle$ in some alternate basis? It seems that the MWI "must add something if it is to support its position that the worlds split along a *preferred basis*." (Albert and Loewer 1988, p. 203; my emphasis – FB)

We do (or at least consider ourselves to) observe certain outcomes and not others. So maybe consciousness does at least single out a preferred basis (as the one being observed), even if it does not collapse quantum states. Albert and Loewer (1988, p. 206) have in fact suggested to equip the system O with an *infinity* of minds, and that only certain distinguished brain states would be capable of carrying such a (conscious) mind, thereby singling out a preferred basis. This 'many minds' interpretation was subsequently picked up and improved by Lockwood (1996), but in (any of) its original version(s) has never become the standard way to understand the MWI. This is mostly due to the fact that it adds to the Everettian formalism a collection of ad hoc postulates which [...] undercut the motivation for taking Everett seriously, namely that it purports to explain how to make sense of quantum theory without adding extra equations or interpretational postulates. (Kent 2010, p. 311)

The general situation for understanding the MWI *with* a preferred basis, and possibly also the locus of conscious observers therein, became arguably much better, however, with the advent of *decoherence theory*.[47] Discovered and outlined first by H. D. Zeh (1970), and later developed further by Zurek (1982), and with contributions by E. Joos, C. Kiefer, and others (cf. Joos et al. 2003), decoherence essentially describes the vanishing of interference terms, the selection of a suitable preferred basis, and the 'emergence of classicality' in virtue of interaction of a system with its *environment*.

Thus take, again, a system already coupled to an apparatus and consider in addition a larger environment E—which would be indicated for any 'realistic' description of almost all experimental or other situations *anyway*—and equip E with quantum states $|\mathcal{E}_j\rangle$.[48] After a suitable interaction, one would end up this time with a state $|\Psi_{SAE}\rangle = \sum_j \alpha_j |S_j\rangle |A_j\rangle |\mathcal{E}_j\rangle$, and the projector $|\Psi_{SAE}\rangle\langle\Psi_{SAE}|$ onto this state defines a pure state density matrix $\hat{\rho}_{SAE}$. Now we can compute a density matrix for S and A alone by executing a partial trace over E's degrees of freedom, which yields:

$$\begin{aligned}
\hat{\rho}_{SA} &= \mathrm{Tr}_E(\hat{\rho}_{SAE}) = \sum_{i,j} \alpha_i\alpha_j^* |S_i\rangle\langle S_j| \otimes |A_i\rangle\langle A_j|\, \mathrm{Tr}\big(|\mathcal{E}_i\rangle\langle\mathcal{E}_j|\big) = \\
&= \sum_{i,j} \alpha_i\alpha_j^* |S_i\rangle\langle S_j| \otimes |A_i\rangle\langle A_j| \sum_k \langle\phi_k|\Big(\sum_{\ell,\ell'} \beta_\ell^{(i)}\beta_{\ell'}^{(j)*} |\phi_\ell\rangle\langle\phi_{\ell'}|\Big)|\phi_k\rangle = \\
&= \sum_{i,j} \alpha_i\alpha_j^* |S_i\rangle\langle S_j| \otimes |A_i\rangle\langle A_j| \sum_k \beta_k^{(i)}\beta_k^{(j)*} \\
&= \sum_{i,j} \alpha_i\alpha_j^* |S_i\rangle\langle S_j| \otimes |A_i\rangle\langle A_j|\, \langle\mathcal{E}_j|\mathcal{E}_i\rangle\,, \qquad (6.25)
\end{aligned}$$

where we have assumed that the environment states can be written as $|\mathcal{E}_j\rangle = \sum_\ell \beta_\ell^{(j)} |\phi_\ell\rangle$ in some suitable (orthonormal) basis of the space of environmental states. If we assume also that the environment states $|\mathcal{E}_j\rangle$ are sufficiently 'dis-

[47] Some of the following is discussed in more detail in Boge (2016b, p. 12 ff.).

[48] Why the (larger) *environment*? Because the treatment given below would not lead to stable states if it were applied to the state of system and apparatus only; the preferred basis is selected as the basis that is stable w.r.t. the action of the environment on the apparatus, i.e. as the basis of eigenstates ('pointer states') of some observable ('pointer observable') that commutes with the interaction Hamiltonian (cf. Joos et al. 2003, p. 166; Schlosshauer 2007, p. 77). Why only the *interaction Hamiltonian*? Because it can usually reasonably be assumed that this part dominates the total Hamiltonian; this is called the "quantum measurement limit" (cf. Schlosshauer 2007, p. 77; emphasis omitted).

tinguishing' between different system-apparatus states, i.e. that $\langle\mathcal{E}_j|\mathcal{E}_i\rangle \approx 0$ if $i \neq j$, then we have $\hat{\rho}_{SA} \approx \sum_j |\alpha_j|^2 |\mathcal{S}_i\rangle\langle\mathcal{S}_j| \otimes |\mathcal{A}_i\rangle\langle\mathcal{A}_j|$, which very much looks like a statistical mixture in which the 'true' quantum state is simply unknown. Note, moreover, that the supposition that the environment states are orthogonal for different system-apparatus states is not an empty formal requirement. There are, for instance, concrete scattering models (cf. Boge 2016b, p. 13 ff.; Joos et al. 2003, p. 64 ff.; Schlosshauer 2007, p. 119 ff.) from which, given certain well-established physical considerations, a differential equation for the density matrix in position representation follows that has off-diagonal (interference) terms damped away exponentially over time and as a function of the distance of points in space (i.e., loosely speaking, $\langle\mathcal{E}(\boldsymbol{x})|\mathcal{E}(\boldsymbol{x}')\rangle \longrightarrow 0$ for $\boldsymbol{x} \neq \boldsymbol{x}'$ as $t \longrightarrow \infty$). The details of the damping depend on the respective scattering cross section.

This implies the proclaimed 'emergence of classicality' due to decoherence: Systems will usually be driven, by decoherence, into states that are quite definite in a *position* representation ('well localized'), but still sufficiently 'unsharp' therein so as to also allow for a quite definite ascription of *momentum*. Incidentally this means that the infamous spreading of the wave packet will be suppressed in any 'realistic' environment: what is (approximately) 'well localized', stays well localized in any such environment, due to the scattering with environmental particles. Moreover, decoherence implies the dynamical creation of an approximate or 'effective' *superselection rule* (mentioned briefly at the end of Sect. 2.1.4), because "the interaction with the environment forces the system to be in one of the eigenstates of the pointer observable, rather than in some arbitrary superposition of such eigenstates." (Zurek 1982, p. 1836)

However, it must be stressed that the resulting density matrix still represents a *collection* of multiple (well localized) states, not a single one. This is the reason why Bell (1990a, p. 24) found himself "quite puzzled" by claims that decoherence *alone* would solve the outcome problem: the reduced density matrix still appears to represent an 'and' rather than an 'or'. Moreover, the above approximate 'statistical mixture' due to decoherence comes about by *tracing out* the environment which means 'neglecting' its degrees of freedom—the mixture is an *improper* one. This implies that *in principle* there are observables that could be exploited to *distinguish* the resulting decohered state from one represented by a proper mixture, in virtue of different resulting expectation values (cf. d'Espagnat 1990, p. 1154 ff. for details).

Note that this is not all unlike the situation in GRW or other collapse models, where "linear superpositions of states separated by distances larger than 10^{-4} cm are transformed into *statistical mixtures*." (Ghirardi et al. 1986, p. 481; my emphasis – FB) But collapse models come, as they are discussed in the literature, as a 'package deal' with one of the two ontologies, flash- and matter density, and as long as one does not supply anything alike for the decoherence case, the situation is different. More importantly, the mixtures in decoherence are improper (they include Born rule weights) whereas those in GRW are approximate proper ones (they include stochastic weights). This should also make for a significant difference, given the above considerations on the in-principle distinguishability of proper and improper mixtures (cf. also Ghirardi et al. 1987, p. 3288, on this point).

Still, it is not entirely unreasonable to suppose that the influence of environmental degrees of freedom can be utilized to ultimately find some sort of *less arbitrary* collapse model in which the collapse is 'mutually induced', i.e. wherein systems enforce the damping of interference terms on each other *via* interaction, which process is then construed, with the aid of a suitable additional ontology, as the occurrence of well-localized objects in spacetime.

This is also the opinion of Schlosshauer (2004, p. 1296), who thinks that "the similarity of the governing equations [in decoherence theory and collapse models – FB] might enable one to choose the free parameters in collapse models on physical grounds rather than on the basis of empirical adequacy." Numerical models by Tegmark (1993) have shown how decoherence effects are typically much stronger than those effects brought about by the dynamics of the GRW model; so decoherence in any case *constrains* collapse models – which may not be the worst thing if one attempts to interpret decoherence *in terms of* collapse.[49]

Moreover, in its foundational ideas at least, decoherence is also quite compatible with BM: Since it induces the loss of coherence (the vanishing of interference terms), it can be taken to specify the conditions under which 'empty waves' do not influence particles associated with a certain conditional wave function anymore, i.e. under which conditions the 'surrealism' of the trajectories *vanishes* and certain correlations will not occur. In a sense, this leads to a reoccurrence of quasi-classical trajectories under suitable circumstances. However, "while the basic idea of employing decoherence-related processes to yield the correct classical limit of Bohmian trajectories seems reasonable," Schlosshauer (2004, p. 1298) informs us, "many details of this approach still need to be worked out", as classical trajectories can not always be recovered in BM by appeal to decoherence where they should. For instance, "even when coherence is fully lost, and thus interference is absent," it can be shown that in a double-slit experiment "nonlocal quantum correlations remain that influence the dynamics of the particles in the Bohm theory, demonstrating that in this example decoherence does not suffice to achieve the classical limit in Bohmian mechanics." (Schlosshauer 2004, p. 1298; cf. also the references therein for further problems and achievements)

So while decoherence does not straightforwardly *enforce* acceptance of the MWI, it certainly aids to rid it of fundamental ambiguities while many open questions as to decoherence's compatibility with BM and collapse models remain. Moreover, decoherence theory has produced various empirically successful predictions such as the precise conditions for the vanishing of discernible interference patterns in experiments with C_{70} molecules (cf. Hackermüller et al. 2004), or how correlations between states of Rydberg atoms (effective two-state systems) vanish as a function

[49]On the other hand, if one employs collapses *in addition* to decoherece, this "might actually allow for an experimental disproof of collapse theories" (Schlosshauer 2004, p. 1296), since then there could be experimentally realizable situations in which a respective collapse model predicts localizations where decoherence would *not*. But such experimental protocols would be extremely difficult to realize, due to the approximate 'omnipresence' of decoherence effects (cf. Schlosshauer 2004, ibid.).

of time when these atoms are successively sent through a cavity that contains a field (cf. Brune et al. 1996; cf. also Schlosshauer 2007, p. 246 ff. for a detailed and accessible analysis of the experiment). Decoherence, in other words, is 'here to stay'; it has an *empirical impact* that has to be recognized by *any* interpretation of QM.

So we now have reasons to believe in a preferred basis due to interactions between quantum systems in the MWI. But how, exactly, does the "splitting" into "branches" occur that DeWitt (1970) was talking about? This is often elucidated in the literature by appeal to the so called *decoherent histories*-formalism, first investigated by Gell-Mann and Hartle (1989). The upshot is the following: define, for a given time t_j, a *time-dependent* PVM $\{\hat{P}_k(t_j)\}_{k\in K}$, in the Heisenberg picture, i.e. where every element of the PVM is of the form

$$\hat{P}_k^j := \hat{P}_k(t_j) := e^{\frac{i}{\hbar}\hat{H}t_j}\hat{P}_k e^{-\frac{i}{\hbar}\hat{H}t_j}, \tag{6.26}$$

(ignoring the possible time-dependence of $\hat{H}$). Then a sequence $\hat{P}_k^j\hat{P}_m^{j-1}\ldots\hat{P}_\ell^0$ of such projectors from $j+1$ respective time-dependent PVMs (with times decreasing from left to right) defines a *history operator* $\hat{C}_\alpha$ for the history α of a system, described by some initial quantum state $|\psi\rangle$ at $t = 0$. Applying this operator to $|\psi\rangle$ then leads to

$$\hat{C}_\alpha|\psi\rangle = e^{\frac{i}{\hbar}\hat{H}t_j}\hat{P}_k e^{\frac{i}{\hbar}\hat{H}(t_{j-1}-t_j)}\hat{P}_m e^{\frac{i}{\hbar}\hat{H}(t_{j-2}-t_{j-1})}\ldots e^{\frac{i}{\hbar}\hat{H}(t_0-t_1)}\hat{P}_\ell e^{-\frac{i}{\hbar}\hat{H}t_0}|\psi\rangle. \tag{6.27}$$

In the orthodox interpretation, this would be interpreted as the system being successively projected (by suitable measurements) onto states $|\phi_k\rangle, |\phi_m\rangle\ldots$ and evolving unitarily for time intervals $\Delta t_n = t_n - t_{n-1}$ in between. For the MWI, this formalism is used to define *branching histories* of the universe instead, i.e. the aforementioned branches.[50] Two histories α, β will be said to be *branching* in case for any given $|\psi\rangle$ and time t_j, $\hat{P}_k^j\hat{P}_m^{j-1}|\psi\rangle \neq 0$ and $\hat{P}_k^j\hat{P}_\ell^{j-1}|\psi\rangle \neq 0$ implies that $\hat{P}_m^{j-1} = \hat{P}_\ell^{j-1}$, or in words: "there is a *unique* way to connect projectors at later times to projectors at earlier times [...]." (Wallace 2012, p. 88; my emphasis – FB)

Sated differently this means that two histories α, β described by $\hat{C}_\alpha, \hat{C}_\beta$ will *agree* on the past up to some time t_j, and thereafter possibly diverge from each other in content. Moreover, the expression $D(\alpha, \beta) = \langle\psi|\hat{C}_\alpha^\dagger\hat{C}_\beta|\psi\rangle$ now defines a *decoherence functional*, and a set H of histories can be said to be *decoherent*

[50] We critically remark at this point, however, that the *ontological significance* of the projectors is somewhat opaque in the MWI; Saunders' (2010, p. 42) appeal to von Neumann's (1932, p. 409) construal of projectors as the "elementary building blocks of the macroscopic description of the world", for instance, does not really help this fact. Maybe we can think of them, in virtue of the branching-decoherence theorem mentioned below, as representing 'emergent structures' due to decoherence in some sense. But it would certainly be desirable to find more clarity on this issue in the literature.

(relative to $|\psi\rangle$) if any two (classically) incompatible histories $\gamma, \delta \in H$ satisfy $D(\gamma, \delta) = 0$.[51] Crucially now, there is the so called *branching-decoherence* theorem (cf. Wallace 2012, p. 93 and references therein) which, by appeal to these quantities, establishes "that branching entails decoherence and (up to possible coarse-grainings) vice versa [...]." (Wallace 2012, ibid.)

Using this formalism, we can understand DeWitt's earlier quote much better: when we take all the mutual interactions between the myriads of systems in the universe into account, it will look as if a multitude of histories is created, in virtue of these interactions, that pairwise agree up to a certain point in time and then 'branch of' to represent different 'worlds'. This, basically, constitutes the now-standard way of thinking about the MWI. Importantly though, decoherence never *strictly* implies disconnected worlds. Off-diagonal elements in a density matrix will be zero at temporal or spatial *infinity* only, so strictly speaking *never*.

What about the locus of consciousness in the MWI, that we had claimed above, might obtain a clearer status in a decoherence based MWI? Wallace (2012, p. 3) here has it that

> Everettian quantum mechanics really is both a many-worlds and a many-minds theory, in the sense that it entails that there are a great many versions of myself, living in surroundings much like my own and interacting with other versions of yourself, elsewhere in physical reality.

Zeh (2000, p. 226), in contrast, finds the entire talk of 'many worlds' "misleading":

> The quantum world (described by a wave function) would correspond to *one* superposition of myriads of components representing *classically* different worlds. They are all dynamically coupled (hence 'actual'), and they may in principle (re)combine as well as branch. It is not the real world (described by a wave funtion [*sic*]) that branches in this picture, but consciousness (or rather the state of its physical carrier), and with it the *observed* (apparent) 'world' [...]. Once we have accepted the formal part of quantum theory, only our experience teaches us that consciousness is physically determined by (factor) wave functions in certain *components* of the total wave function. (emphasis in original)

So some factors in certain components of the highly entangled but decohered wavefunction can be identified as carriers of consciousnesses according to Zeh, while Wallace straightforwardly embraces many worlds *and* minds. But both views ultimately boil down to the same thing[52]: Wallace equally relies on decoherence which, strictly speaking, leaves the one 'undivided' quantum universe (or multiverse) intact. And Wallace's sense in which the many worlds exist is *in relation to* 'selves', 'living' in environments, components of the state vector of the universe.

---

[51] More precisely, one can distinguish different *degrees* of decoherence in virtue of $D(\alpha, \beta)$ being *complex valued*: one can make a difference between the whole functional vanishing or only the real or imaginary part. The details do not matter for the present context though, so we refer the interested reader to Joos et al. (2003, p. 241 ff.).

[52] There may be important differences between the two authors' views on the role of the *wavefunction* in the MWI though that will become apparent in the subsequent discussion.

Notably, there is, then, a *decisive* role for *consciousness* in this MWI after all, according to both Wallace and Zeh: *only* in relation to the experience of conscious beings can these (classical) 'worlds' really be said to *exist*, because decoherence preserves *one* (highly entangled) state vector and *never* fully eliminates the off-diagonal terms in a density matrix, while 'we' (the somehow dynamically created conscious beings) do not perceive ourselves as simultaneously 'partly' in this and partly in that 'branch'.

6.3.3 The Problem of Probabilities

As we noted initially, the MWI has a bunch of appealing features. It does not raise specific concerns as regards compatibility with SR, since there are fully Lorentz covariant QFTs, telling us how interactions between systems proceed in a Lorentz covariant way. And a key motivation for endorsing it is, as we have seen, that one can talk, in a very straightforward sense, about the wavefunction of the universe, which is conducive to such projects as quantum cosmology and gravity.

Moreover, Wallace (2012, p. 310) thinks that "in Everettian quantum mechanics, violations of Bell's inequality are relatively uninteresting." The reason is that there are no objectively *unique* 'outcomes' anymore – without collapse, the state vector contains all possible outcomes – but only subjectively perceived ones. Now the objectively evolving state vector predetermines which observers will find their results coinciding (or 'anticoinciding') with those in the other arm of an EPRB experiment *whenever they meet and compare*. So a certain divide between interactions and states accounts for the *observed* correlations, because, as Wallace (2012, ibid.) puts it, "in Everettian quantum mechanics interactions are local but states are nonlocal." This includes, of course, the interactions between carriers of consciousness. It was (supposedly) first discovered by Albert and Loewer (1988, p. 210) in their original 'many minds'-proposal, that if two observers "measured the same spin component they will end up (after communication) believing that they obtained opposite values", where the communication is, of course, a *local* interaction, and 'in reality', there are now multiple versions of both observer having observed different possible outcomes. This feature remains true within a decoherence-informed MWI.

However, the *precise* situation of violations of Bell-type inequalities becomes more subtle if one does not exactly follow Albert and Loewer's approach, because Albert and Loewer (1988, p. 208) explain the *probabilities* involved in these violations in terms of fractions of observers (or suitable limits of such) that observe any one outcome. It is not clear whether this treatment can be carried over to a decoherence based MWI, wherein one does not *stipulate* that there are these desired fractions. More generally speaking: even if the MWI comes with all the aforementioned desirable features, and is, as we must stress against the background of the previous discussion, certainly no 'stranger' than any of its alternatives, it still equally faces some serious problems. The central one of which is the indicated *problem of probabilities*.

Recall that orthodox QM embraces the Born rule, and that the statistical averages QM predicts of course equally depend on the probabilistic content of the theory. This probabilistic content is hence intimately connected to the theory's empirical success: whenever we predict and confirm an interference pattern, this will constitute a statistical distribution predicted by Born's rule; whenever we predict and confirm the *general* value of a certain magnitude, this prediction will imply averaging some operator w.r.t. a quantum state. Now given that, apparently, everything that *can* happen according to the state vector *does* happen in the MWI, what sense is there to predicting a probability for its happening? All events potentially subject to Born's rule (the observation of certain outcomes by observers) *will* happen—with probability 1. This difficulty of even making sense of the word 'probability' in the MWI has been coined the *incoherence problem* by Wallace (2008, p. 47). But a second problem arises even if one assumes that there be some sense to 'probability' in the MWI after all, namely: why would these probabilities conform to Born's rule? This Wallace (2008, p. 49) calls the *quantitative problem*.

Historically, Everett (1973, p. 71 ff.) believed to have proven that the Born rule would hold at least subjectively, i.e. that the squared modulus of the amplitudes would quantify the statistics of the observations of a typical observer. His treatment, however, has been characterized as wanting "on the grounds of insufficient motivation" (Graham 1973, p. 236), as "circular and question-begging" (Wallace 2012, p. 127), or at any rate conceptually confused (cf. Byrne 2010b, pp. 260–261). Similar accusations of circularity were made against Graham's (1973) attempt of an improvement in terms of the *relative frequencies* of outcomes in the branching Everettian universe (cf. Kent 1990, p. 1752), and it seems that comparable criticisms can be advanced against all other historical proposals (cf. Schlosshauer and Fine 2005, p. 198, for references).

In Boge (2016b, p. 21 ff.), two more recent approaches were compared and discussed individually at some length and we will here merely summarize the central results instead of going into the proofs again. Some more emphasis will be laid here, in contrast, on the associated philosophical issues. The upshot, however, is that both approaches ultimately cannot solve the *quantitative problem* because they appeal to premises that suffer from the *incoherence problem*—premises connected to meanings of 'probability' that do not make sense or are hard to entertain once one takes the MWI seriously.

The first kind of approach is the decision-theoretic program first proposed by Deutsch (1999). The underlying idea is to define a *utility function*

$$u(a) := \sum_{x \in \mathcal{S}_M} p(x|M)\pi(x), \tag{6.28}$$

where $a = (M, \pi)$ is an *act*, consisting of a *chance setup* M, a situation where any one out of a given set of events might occur but it is impossible to tell which (as with the rolling of dice), and a *payoff function* π that maps from a set S_M of states compatible with M (the 'possible results in M') to their *consequences*.[53] $p(x|M)$ is a *decision weight* for the states compatible with M, given that M is the case. Intuitively, this 'weight' of course denotes the *probability* of result x occurring given that chance setup M obtains, and the formula thus tells us that we (should) evaluate how useful some act is to us by summing up possible rewards π in some 'chance setup' (not necessarily an actual game), weighted by the probability that the state associated with the respective reward occurs, conditional on participating in the setup in the first place.

The strategy to prove the Born rule by appeal to u then is to demonstrate that, given that u and π satisfy some plausible constraints and $M = (|\psi\rangle, \hat{O})$ is a quantum measurement where one receives payoffs for the eigenvalues of $\hat{O}$, $p(x|M)$ must be the Born probability $\text{Pr}^{\psi}_{O}(o) = |\langle o|\psi\rangle|^2$ (for details cf. Deutsch 1999; Wallace 2002, 2003, pp. 418–431; Boge 2016b, pp. 22–26).

In Wallace (2012, p. 172 ff.) a much-advanced treatment is given: a whole range of specifications on 'quantum decision problems' are made, a large list of axioms is presented, subdivided into 'richness' and 'rationality' axioms, and the theorem is phrased in terms of the (Born-rule) utility function *representing* the preference order of a rational agent. But the general agenda is exactly the same.

This Deutsch-Wallace approach has, however, been met with a range of serious criticisms. The first problematic feature is the mere occurrence of a decision weight. What does it quantify? We had intuitively related it to the probability of some outcome (state) occurring in the chance setup. But how does this talk of 'chance setups' and probabilities even make sense in the present context? This is, basically, a restatement of the incoherence problem, so we should maybe make this point more explicit.

Let us first ask: What does a probability assertion express in general? As interlude I should have made clear, there is no unified answer to this question. But both kinds of probability conceptions, the objective and epistemic ones, may be said to express or rely on some sort of *uncertainty* or *indeterminacy*. Take a propensity account of probability; single case or long run, no matter. What does the assertion of a propensity of some setup/system to exhibit a certain outcome/behavior with probability $p < 1$ express? It expresses an *objective indeterminacy* of the setup/system *before* an appropriate stimulus condition is realized, and quantifies the strength of a tendency that in turn determines the frequency with which a given outcome will be realized in a longer run of similar experiments on similar systems. Of course this also implies a subjective uncertainty as to which outcome/behavior will be displayed; a firm belief in the occurrence some *particular* outcome/behavior would simply be mistaken, before the stimulus conditions are realized. What, on the other hand, does probability as a degree of belief or confirmation quantify? Of

[53] We here take it that π is real valued for convenience, but one can of course introduce an additional 'intermediate' map that assigns real values to consequences such as 'I am being handed a sandwich' (cf. Wallace 2002, p. 6).

course $p < 1$ *directly* expresses the *subjective uncertainty* of a real or ideal agent or observer, even in situations where the outcome or observed behavior might actually be predetermined.

Relative frequency accounts may appear to play kind of a special role, but they need not be viewed so: Assume that a deterministic mechanism were known that exactly specifies which outcome will occur under which precise conditions, in some experimental setup. Then counting the frequency of a given outcome in a number of runs with varied initial conditions (and correspondingly varying outcomes) and abstracting from it a 'probability' would not *formally* be objectionable; but it would certainly create confusion due to a violation of Grice's (1975, p. 46) *relevance maxim*: why even *state* a probability when it is perfectly known which outcome would occur when and in virtue of what? So the assertion of probabilities, we here take it, is generally only *meaningful* in contexts of *subjective uncertainty* or *objective indeterminacy*. Neither of these seems to fit the MWI.

Not so, has argued Saunders (1998, pp. 383–384). He considers a scenario wherein an observer, let us call her O_0, performs a spin measurement upon the conclusion of which there will be two observers, $O_\uparrow$ and $O_\downarrow$, with obvious associated observations. Now Saunders asks who O_0 should anticipate to *be* after the measurement, (i) *none* of the two, (ii) *both* of the two, or (iii) only *one* of the two. (i) is dismissed by Saunders on the grounds that the only reason to expect none of the options would be that O_0 does not exist anymore after the measurement; but any state *at some time* equally does not exist anymore at *later* times in (deterministic) classical physics, which does *not* lead us to "expect nothing at all" (p. 383). (ii), on the other hand, is dismissed as "straightforwardly inconsistent", because $O_\uparrow$ and $O_\downarrow$ "do not speak in unison; they do not share a single mind; they witness different events." (ibid.) So the conclusion seems to be that O_0 has to expect to become either $O_\uparrow$ or $O_\downarrow$. This has been coined the *subjective uncertainty view* (e.g. Greaves 2007, p. 116).

This view, however, is hard to defend, once one takes the MWI seriously. Who is the 'rightful heir' of O_0's memories, $O_\uparrow$ or $O_\downarrow$? Which one *has* O_0 become? Moreover, Saunders' (1998) arguments against the other two options are not fully compelling. (i) could e.g. be defended as follows. Note, first of all, that questions of expectations about the future in the light of ceasing to exist *only* become relevant when applied to *conscious beings*: We cannot meaningfully ask a 'classical particle', say, for *its* expectations about its future *experience*. Depending on one's views on diachronic identity in the philosophy of mind, one could of course argue that O_0 *does* cease to exist, in a relevant sense, when 'giving birth' to $O_\uparrow$ and $O_\downarrow$, and that O_0 *herself* should hence not expect anything about *her* future experiences at all. In contrast, an observer in a classical scenario will have *one unique* successor so that the question as to one's expectations about the future unambiguously refers to *that successor's* experiences.

And (ii) is defensible on similar grounds, namely, if one believes in the *supervenience* of (conscious) mental states on brain states,[54] then after the branching there will be two different (approximately non-overlapping) brain states on which $O_{\uparrow}$'s and $O_{\downarrow}$'s consciousnesses supervene, and these *each* bear a sufficient *continuity* to the brain states of O_0 to say that O_0 'has become *both*', in a relevant sense.[55] This is essentially the view of Greaves (2004, p. 441), who, building on Parfit (1984, p. 215 ff.), defends the view that O_0 is sufficiently psychologically connected and continuous to both $O_{\uparrow}$ and $O_{\downarrow}$ to be certain about her future experience, once she accepts the MWI. O_0 will indeed become *both*.

Greaves, however, also thinks that the Deutsch-Wallace approach is defensible nevertheless, if one interprets $p(x|M)$ in formula (6.28) as a "*caring measure*" (Greaves 2004, p. 430; my emphasis – FB), i.e. as a quantification of how much an agent should care about her future selves, given that they will receive different rewards for their actions. This is an interesting approach, to be sure, and Greaves (2004, ibid.) argues that such a measure can satisfy all the requirements of a probability measure as a consequence of so called 'Dutch book coherence'.[56] But it has been objected that this hardly suffices to establish that agents *should* care about their future selves as quantified by the Born-rule caring measure in an Everettian universe. In particular, Albert (2010, p. 362) has suggested, as a *reductio*, to use a 'fatness measure' instead, whereby one cares more about the branches with 'more of oneself' on them (why should one not?).

Whether one can or cannot make sense of such caring measures may be irrelevant in the end though, because there are good reasons to suspect that the very *axioms* used by Deutsch (1999) and Wallace (2012) respectively are *unjustified* in the context of the MWI. The most concise criticism to this effect has been given by Maudlin (2014a, p. 803 ff.). Maudlin considers, following Wallace (2012, p. 193),[57] a scenario in which a student, maybe our O_0, really desires to study history *and* physics, but only has the (cognitive, financial...) resources to study one of the two, whence she settles for physics, say. She then meets an experimental physicist who

---

[54]Note that a decoherence-based MWI effectively rules out mind-brain *identity*, because (as was our earlier observation) the brain states will still be (very weakly) overlapping, even if decoherence has taken place, but the conscious states will not: O_0 will never, we take it, have the experience of observing both, ↑ and ↓, not even 'remotely'.

[55]Note briefly that there is also the possibility of interpreting the situation in terms of *divergence* (cf. Greaves 2007, p. 117): there could be multiple copies of oneself all along, coexisting together as O_0 up until the measurement and then just 'leaving off' into different branches. This would sanction the subjective uncertainty view, but it has the obvious deficit of introducing Albert-Loewer-like *many minds* after all, and hence undesirable ad hoc surplus structure.

[56]Cf. Chap. 7 for some details and cf. Boge (2016b, p. 27) for a brief discussion of the details of Greaves' proposal.

[57]...who in turn adapts the scenario from Adam Elgar (cf. Wallace 2012, p. 193). Note that Wallace (2012, ibid.) of course presents own arguments for why he thinks his strategy is immune to the criticism; but we here agree with Maudlin (2014a, p. 803), that "Wallace's response [...] fails to make contact with the case considered."

offers her to perform a spin measurement on a spin-$\frac{1}{2}$ system, depending on whose result O_0 may reconsider (stay with physics if she observes $\uparrow$, switch if $\downarrow$).

Now according to the MWI, there will be *two students*, $O_\uparrow$ and $O_\downarrow$, after the measurement, one of which will study physics and the other of which will study history. Since studying both subjects 'on one branch' was deemed impossible for O_0 and, as we have argued above, $O_\uparrow$ and $O_\downarrow$ both bear sufficient continuity to O_0, this *comes as close as it gets* to fulfilling O_0's dream (cf. Maudlin 2014a, p. 802 ff.).

The point is that the Deutsch-Wallace approach treats 'outcomes' of a 'chance setup' (or a 'how much to care'-setup, in Greaves' version), in which decisions are due, as *mutually exclusive*. But if the MWI is correct, there is a *common outcome* to all the 'observations of eigenvalues' in the different branches which is not even recognized in the very *formulation* of the decision problem, namely: *the branching itself*. Here is how Maudlin (2014a, p. 804; emphasis in original) puts it: "Wallace's decision theory has been set up *in its axioms* to rule out a rational agent acting in a way that takes Everettian quantum mechanics seriously."

Maudlin's reference is specifically to an axiom (cf. Wallace 2012, p. 175; Maudlin 2014a, p. 804) which states that outcomes are always mutually *orthogonal*—which $|\text{physics}\rangle$ and $|\text{history}\rangle + |\text{physics}\rangle$ would *not* be. But there are further, related axioms with similarly unconvincing implications. While in Deutsch (1999) this is only a tacit assumption (outcomes being identified with eigenvalues of a self-adjoint operator), Wallace (2012, p. 170) also states an axiom of *branching indifference*:

> An agent doesn't care about branching per se: if a certain operation leaves his future selves in N different macrostates [represented by subspaces of a separable Hilbert space – FB] but doesn't change any of their rewards, he is indifferent as to whether or not the operation is performed.

Somewhat in line with Maudlin (2014a), Dizadji-Bahmani (2013, p. 7) has argued that branching indifference "is highly counter-intuitive as an axiom of rationality", since it "is not saying that you *need not* care about the number of your future descendants", but rather that "it is *rationally required* that you *do not care*!" (my emphasis – FB) Why would Wallace believe such a thing? His central reason seems to be that "branching is uncontrollable and ever-present in an Everettian universe." (Wallace 2012, p. 170). All kinds of minute interactions between systems can trigger branching processes, whence we would care about something entirely out of our hands if we cared about branching (or so the argument might go). This would clearly make us irrational.

The latter appears to be a *non-sequitur* though, as can be seen when the situation is applied to *moral matters*. Maudlin (2014a, p. 805) considers an additional scenario in a nuclear power plant in which a catastrophic development is imminent and one is faced with the following choice: perform action A, which has a Born probability of 99,99% of saving everybody and 0,01% of killing 1000 people, or perform action B with the (*certain*) result of killing half a dozen but saving everyone else. On the MWI, option A would mean sending 1000 people to death on some 'low Born weight'-branch(es). It is far from clear that under these circumstances we would necessarily be irrational if we wanted to *avoid* the branching (i.e. were not indifferent about it).

However, the scenario above does not really target branching indifference (nor is it supposed to) because branching indifference presupposes that the *rewards* are *the same* on all resulting branches. But imagine the same imminent catastrophe without any of the two options to avoid it, and imagine also that someone would deliberately perform a spin measurement before the catastrophe happens, thereby effecting a twofold split and (at least) twice as many branches on which 1000 people died.

It does not matter whether there is a precise branch count delivered by decoherence (which is more than controversial, as we shall see below), because it is still safe to say that there will be well-distinguished conscious experiences, including horrible suffering due to the catastrophe, associated with the spin-up and the spin-down parts of the universal state vector. And it equally does not matter that all the *other* aforementioned minute interactions will multiply these branches *further*—the person performing the spin measurement would *still* have caused a multitude of *additional* horrible deaths due to his inconsiderate actions leading to branching. The point is this: Just because we cannot control *all* branching events does not mean that we cannot control *some* of them. That we are *required* to not care about branching per se to not count as irrational seems by far too strong a requirement.[58]

On top of these concerns it has been suggested, in particular by Baker (2007), that the entire approach suffers from a vicious *circularity*. Recall how we identified the branching structure by decoherence and how decoherence relies on the partial trace to define when branches are 'approximately separated'. But why do we even use the partial trace? Typically, the use of the partial trace is motivated *entirely* in virtue of its *statistical/probabilistic* properties, namely, that it is the "the *unique* operation which gives rise to the correct description of observable quantities for subsystems of a composite system" (Nielsen and Chuang 2010, p. 107; emphasis in original), in the sense that it preserves the (*probabilistic*) predictions generated for single-system observables with operators of the form $\mathbb{1} \otimes \hat{A}$ (cf. ibid.). And of course, the trace operation on one of the subspaces computes *Born probabilities*. Thus, Baker (2007, p. 164) argues, "the employment of decoherence to identify branches depends upon the *unlikeliness* of low-weight events, the framework of quantum games in which the theorem is formulated *presumes its conclusion*." (my emphasis – FB)

[58]More technically phrased, Wallace (2012, p. 163) represents an 'act' by a unitary $\hat{U}$ and one's quantum state $|\psi\rangle$ is assumed to be in a 'macrostate' $\mathcal{M}$ which is a subspace of the Hilbert space $\mathcal{H}$, as is the 'reward', $\mathcal{R}$, because not every detail about these quantum states matters. So to avoid the language of measurements, self-adjoint operators, and eigenvalues, Wallace (2012) appeals only to unitary operators and subspaces to model reward situations and actions. He then (p. 179) defines branching indifference such that if one's own quantum state $|\psi\rangle$ is in $\mathcal{M}$ and $\mathcal{M} \subset \mathcal{R}$, then one is indifferent at $|\psi\rangle$ about either performing $\hat{U}$ such that $\hat{U}\,|\psi\rangle \in \mathcal{R}$ or leaving $|\psi\rangle$ alone (performing $\mathbb{1}_{\mathcal{M}}$). This makes the problem quite obvious: in our scenario above, the macrostate would lie in the 'reward' space (imminent catastrophe), but we could still dislike a unitary with the effect of 'more catastrophes happening' due to branching, i.e. we could *rationally* want to avoid *multiplication* within $\mathcal{R}$.

Wallace (2012, p. 253), in contrast, urges us to "think of the significance of the Hilbert space metric [as occurs in the partial trace – FB] as telling us when some emergent structure really is robustly present, [...] a perfectly objective feature of the physics, prior to any considerations of probability." It is unclear what the force of this suggestion is though, since the standard reason to use the partial trace rather than some other operation is that it gives rise to the correct subsystem *statistics*.

An additional worry about the Deutsch-Wallace approach is that the targeted *explanandum* is a different one. What really startles us are our *observations* in experimental situations that conform to the Born rule (correlated distant spins, build-ups of interference patterns...). So in asking about the validity of the Born rule in the MWI, we were inquiring rather how and whether it predicts appropriate *frequencies*. *Even if* the argument that rational agents would let the Born rule guide their actions could succeed, this would hardly *explain* the (surprising) frequency data.

The second aforementioned approach to proving the Born rule, due to Zurek (2003, 2005), is far better off in this respect. But it requires a lot of philosophical clarification and ultimately also suffers from (different) problems.[59]

Zurek's (2003, 2005) approach rests on the following observations. Recall (from Sect. 2.1.3) that one can write any entangled state (in a separable Hilbert space) of system S and environment E in a bi-orthogonal (or 'Schmidt') form, $|\Psi_{SE}\rangle = \sum_j \alpha_j |\mathcal{S}_j\rangle |\mathcal{E}_j\rangle$, wherein $\{ |\mathcal{S}_j\rangle\}_{j\in J}$ and $\{ |\mathcal{E}_k\rangle\}_{k\in K}$ are ONBs of the system space $\mathcal{H}_S$ and the environment space $\mathcal{H}_E$ respectively. Since the α_j are complex numbers, one can always write them as $|\alpha_j|e^{i\varphi_j}$, where $|\alpha_j|, \varphi_j \in \mathbb{R}$ ($|\alpha_j|$ the modulus). Given these preliminaries, Zurek (2003, p. 1) points out that one can perform certain actions (represented, of course, by operators) on the system states alone and subsequently perform other actions on the environment states alone, and then end up *with the original state*, so that it is as if one had *not done anything to the system at all*. In other words: quantum states have a kind of 'environment-assisted invariance' or *envariance* (cf. Zurek 2003, ibid.).

To make things precise,[60] take the following definition (as gathered in Boge 2016b from Zurek's somewhat more casual writings):

Definition (***Envariance***) Let $|\Psi_{SE}\rangle \in \mathcal{H} = \mathcal{H}_S \otimes \mathcal{H}_E$ be a joint state of a system of interest S and its environment E, and let $\hat{U}_S = \hat{u}_S \otimes \mathbb{1}_E$ be a unitary operator acting non-trivially solely on the space $\mathcal{H}_S$. Then $|\Psi_{SE}\rangle$ is called *envariant* under $\hat{U}_S$ *iff* there is an $\hat{U}_E = \mathbb{1}_S \otimes \hat{u}_E$, acting non-trivially solely on $\mathcal{H}_E$, s.t.

$$\hat{U}_E \hat{U}_S |\Psi_{SE}\rangle = |\Psi_{SE}\rangle .$$

[59]Again, we will only give highlights; the reader interested in details of the proof may be referred either to the original papers or to Boge (2016b, p. 32–41) and references therein.

[60]In what follows, we focus on the details given in Zurek (2005) and recaptured in Boge (2016b, p. 31 ff.), since these are already informed by criticism (e.g. Barnum 2003; Caves 2004; Mohrhoff 2004; Schlosshauer and Fine 2005).

Operators may also be called envariant on a given set of states if all states in the set are envariant under them. A crucial thing to note is that Zurek considers *only unitary* transformations, and that he *has to do so* in order to derive certain lemmas, but that other, non-norm preserving operations are thinkable without problem (cf. Boge 2016b, p. 32) for which these lemmas would not hold. One thing at a time though.

To see envariance at work, take the simple example of two qubits S, E with joint Schmidt state $|\Phi_{SE}\rangle = |\alpha_1|e^{i\varphi_1}|\mathcal{S}_1\rangle\,|\mathcal{E}_1\rangle + |\alpha_2|e^{i\varphi_2}|\mathcal{S}_2\rangle\,|\mathcal{E}_2\rangle$. Now take the two operators

$$\hat{U}_S := \left(|\mathcal{S}_1\rangle\langle\mathcal{S}_1|\,e^{i(\varphi_2-\varphi_1)} + |\mathcal{S}_2\rangle\langle\mathcal{S}_2|\,e^{i(\varphi_1-\varphi_2)}\right)\otimes\mathbb{1}_E, \tag{6.29}$$

$$\hat{U}_E := \mathbb{1}_S\otimes\left(|\mathcal{E}_1\rangle\langle\mathcal{E}_1|\,e^{i(\varphi_1-\varphi_2)} + |\mathcal{E}_2\rangle\langle\mathcal{E}_2|\,e^{i(\varphi_2-\varphi_1)}\right) \tag{6.30}$$

These are clearly unitary (just complex conjugate the phases to see this), and acting with $\hat{U}_S$ on $|\Phi_{SE}\rangle$ has the effect of switching the phases,

$$\begin{aligned}\hat{U}_S\,|\Phi_{SE}\rangle &= |\alpha_1|e^{i(\varphi_1+\varphi_2-\varphi_1)}\underbrace{\langle\mathcal{S}_1|\mathcal{S}_1\rangle}_{=1}|\mathcal{S}_1\rangle\,|\mathcal{E}_1\rangle + |\alpha_1|e^{i(\varphi_1+\varphi_1-\varphi_2)}\underbrace{\langle\mathcal{S}_2|\mathcal{S}_1\rangle}_{=0}|\mathcal{S}_2\rangle\,|\mathcal{E}_1\rangle\\ &\quad+|\alpha_2|e^{i(\varphi_2+\varphi_1-\varphi_2)}\underbrace{\langle\mathcal{S}_2|\mathcal{S}_2\rangle}_{=1}|\mathcal{S}_2\rangle\,|\mathcal{E}_2\rangle + |\alpha_2|e^{i(\varphi_2+\varphi_2-\varphi_1)}\underbrace{\langle\mathcal{S}_1|\mathcal{S}_2\rangle}_{=0}|\mathcal{S}_2\rangle\,|\mathcal{E}_2\rangle\\ &= |\alpha_1|e^{i\varphi_2}|\mathcal{S}_1\rangle\,|\mathcal{E}_1\rangle + |\alpha_2|e^{i\varphi_1}|\mathcal{S}_2\rangle\,|\mathcal{E}_2\rangle,\end{aligned} \tag{6.31}$$

whereas $\hat{U}_E$ will basically do the same thing in reverse on $\mathcal{H}_E$, and we have $\hat{U}_S\hat{U}_E\,|\Phi_{SE}\rangle = |\Phi_{SE}\rangle$ as desired.

It can then be proven, as a lemma, that an operator $\hat{U}_S$ is envariant on some Schmidt state $|\Psi_{SE}\rangle = \sum_j |\alpha_j|e^{i\varphi_j}|\mathcal{S}_j\rangle\,|\mathcal{E}_j\rangle$ *iff* the system-part $\hat{u}_S$ of $\hat{U}_S$ can be represented as $\hat{u}_S = \sum_k e^{i\tilde{\varphi}_k}|\mathcal{S}_k\rangle\langle\mathcal{S}_k|$ ($\tilde{\varphi}_k \in \mathbb{R}$), and more importantly, that a state $|\Psi_{SE}\rangle = \sum_j |\alpha_j|e^{i\varphi_j}|\mathcal{S}_j\rangle\,|\mathcal{E}_j\rangle$ is envariant under a *swap* (we only give the system part here)

$$\hat{u}_S(j \leftrightharpoons k) = \left(e^{i\varphi_{j,k}}|\mathcal{S}_j\rangle\langle\mathcal{S}_k| + e^{-i\varphi_{j,k}}|\mathcal{S}_k\rangle\langle\mathcal{S}_j| + \sum_{j\neq\ell\neq k}|\mathcal{S}_\ell\rangle\langle\mathcal{S}_\ell|\right) \tag{6.32}$$

iff $|\alpha_j| = |\alpha_k|$. Swaps are *exchanges of states* (just contemplate their effect on a given state by a calculation comparable to that in (6.31)) with an additional arbitrary modification of the phase. Both of these lemmas require that one presupposes the transformations to be *unitary*. In the first case this is because the uniqueness of the representation would obviously fail (one would have the *modulus* $|\alpha_j|$ as another degree of freedom). And in the second case it is because swaps are undone by counterswaps on the environment,[61]

$$\hat{u}_{\mathcal{E}}(j \leftrightharpoons k) = e^{-j(\varphi_{j,k}+\varphi_j-\varphi_k)}\,|\mathcal{E}_j\rangle\langle\mathcal{E}_k| + e^{i(\varphi_{j,k}+\varphi_j-\varphi_k)}\,|\mathcal{E}_k\rangle\langle\mathcal{E}_j| + \sum_{j\neq\ell\neq k} |\mathcal{E}_\ell\rangle\langle\mathcal{E}_\ell|, \tag{6.33}$$

and in case one would allow for a *rescaling* due to the initial swap, i.e. a multiplicative argument that has the effect $|\alpha_j| \mapsto |\alpha_j| \cdot \frac{|\alpha_k|}{|\alpha_j|}$, then the counterswap would merely have to do the opposite, $|\alpha_k| \mapsto |\alpha_k| \cdot \frac{|\alpha_j|}{|\alpha_k|}$, and the original state would be restored *without the implication* $|\alpha_j| = |\alpha_k|$ (cf. also Boge 2016b, p. 32).[62]

Why are these two lemmas important anyway? Zurek (2005, p. 4) uses them to derive two intermediate results, namely, that probabilities cannot depend on the *phases*, basically because these can be exchanged without affecting the state of the system that is supposed to be associated with a respective probability, and that component states $|\mathcal{S}_j\rangle\,|\mathcal{E}_j\rangle$ and $|\mathcal{S}_k\rangle\,|\mathcal{E}_k\rangle$ in a global (Schmidt) state $|\Psi_{SE}\rangle$ are associated with equal probabilities if they are associated with equal moduli ($|\alpha_j| = |\alpha_k|$).

How can Zurek establish these things? To that end, he has to assume, first of all, the following three premises[63]:

Premise 1 To represent the alteration of the state of a system S by a unitary operator $\hat{u}$, $\hat{u}$ must act on the Hilbert space $\mathcal{H}_S$ of that system.

Premise 2 All measurable quantities pertaining to a system S *and their respective probabilities* are fully and exclusively specified by S's state.

Premise 3 The state of a subsystem S_j included in a larger system $S_{\text{tot}} = S_1 S_2 \ldots S_j \ldots S_{N-1} S_N$ is fully and exclusively specified by the state of S_{tot}.

Premise 1 seems fairly uncontroversial, given the structure of QM; Premise 3 is an equally uncontroversial (almost 'analytic') statement about the composition of systems.[64] Premise 2 may strike us as somewhat odd though: Did we not inquire where the probabilities come from in the first place? That observation is certainly correct, but Premise 2 should be understood more as a *bridging principle*; it would be unreasonable to derive probabilities from an inherently non-probabilistic theory, or, as Schlosshauer and Fine (2005, p. 211) put it: "we need to 'put probabilities in to get probabilities out.'" In other words: In embracing Premise 2 we merely assume that there may be probabilities *somehow*, and we take it that they are (exclusively)

[61] We omit the additional phase degree of freedom that is effectively 1 (cf. Zurek 2003, p. 2).

[62] Author's note: I am indebted to Rochus Klesse for raising my awareness on this point.

[63] We here use the paraphrases also given in Boge (2016b, pp. 32–33).

[64] Note that the converse would be much less uncontroversial, due to the considerations of 'holism' that have long pervaded the philosophy of QM (e.g. Teller 1986; Healey 2009).

associated with a system's state. What these probabilities *are* is then determined by the details of the proof (as we shall see).

To derive the result that equal moduli are associated with equal probabilities, and to subsequently derive the Born rule, Zurek (2005), however, has to assume one of three further premises. In Boge (2016b, p. 39) it was argued that one of these additional premises is inherently implausible (Premise (4b) on p. 33 therein), so we will here focus only on the remaining two:

Premise 4(a) If swaps of two orthonormal states in a joint Schmidt state $|\Psi_{SE}\rangle$ leave the state of S unchanged, the probabilities for the outcomes associated with these states must be equal.

Premise 4(b) The outcomes associated with states in a tensor term $|\mathcal{S}_j\rangle\ |\mathcal{E}_j\rangle$ of some Schmidt state $|\Psi_{SE}\rangle$ are perfectly correlated, i.e., if state $|\mathcal{S}_j\rangle$ is measured on $|\Psi_{SE}\rangle = \sum_j \alpha_j\ |\mathcal{S}_j\rangle\ |\mathcal{E}_j\rangle$, state $|\mathcal{E}_j\rangle$ will be measured with probability 1 as well (and *vice versa*).

The proof from premise 4(b) is fairly straightforward and requires in addition only premises 1–3 and the definition of conditional probability. But this proof does not convey any meaning to the word 'probability' in the context of the MWI (does not address the incoherence problem). We knew already that quantum states imply the observation of perfect correlations (after sufficient post-selection and general correction for noise etc.). How does that tell us *anything* about the occurrence of certain frequencies of observations of a given type on any one branch in an Everettian universe?

The proof from premise 4(a) is pretty much immediate, given the other premises and aforementioned lemmas. However, it is not really clear here why we should accept premise 4(a) either. Zurek (2005, p. 5) relates his approach to Laplace's (1814, p. 6) *principle of indifference* (only later named thus) and considers it as a kind of 'objectivization' thereof, as he believes to show the "'objective indifference' of the physical state of the system in question rather than the observer's subjective indifference based on his state of knowledge." (Zurek 2005, p. 5; emphasis omitted)

As it stands, this is hardly a meaningful statement. 'Indifference' is a state of conscious beings with preferences, at least if taken non-metaphorically. So it cannot be (straightforwardly) applied to an unconscious physical system. Laplace's (1814) original intention in devising the principle of indifference presumably was a *rationality constraint*, that one *should* not believe in one out of a bunch of equal-appearing options simply 'out of a hunch'. And it is treated this way in objective Bayesian approaches to probability and formal epistemology (cf. interlude I and the 'equivocation norm'), wherein the principle of indifference, which itself ultimately leads to paradoxes, is typically replaced (or 'approximated') by *entropy maximization* (cf. Williamson 2010, pp. 21 ff. and 25 for reference).

Now in order for his approach to even *address* the incoherence problem of the MWI, Zurek has to supply some suitable *meaning* to 'probability'. While this is not being done in early stages of the proof, more insight transpires when one consults the final proof of the *Born rule*,[65] including also cases of *unequal coefficients*, and Zurek's (2005, pp. 9–10) subsequent discussion.[66]

In this proof (that, to recall, presupposes the intermediate equal-modulus result) Zurek (2005, p. 7) first discusses the case in which the moduli are of the form $|\alpha_j| = \sqrt{\frac{m_j}{M}}$, for $m_j, M \in \mathbb{N}$, i.e. are all square roots of positive rational numbers. He then extracts from the environment in a Schmidt state $|\Psi_{SE}\rangle$ a "*counterweight-counter*" (ibid.; emphasis in original), which could be the measuring apparatus, and which is assumed to have a sufficiently high-dimensional Hilbert space to *count the multiplicities* m_j, i.e. to 'spawn off' m_j subsequent *branches* for a tensor term $|\mathcal{S}_j\rangle\, |\mathcal{E}_j\rangle$ that has coefficient $\alpha_j = \sqrt{\frac{m_j}{M}} e^{i\varphi_j}$. In other words: The probability of the occurrence of a state $|\mathcal{S}_j\rangle$ under environmental conditions $|\mathcal{E}_j\rangle$ can be interpreted as a *tendency* (propensity) to 'spawn off' m_j further branches, in which the state is still $|\mathcal{S}_j\rangle$. This leads to the prediction of a frequency m_j/M in longer runs of experiments on similarly prepared states and nicely explains our observations as 'observers' in these branches.[67]

In principle, this is a quite brilliant approach; but one smells trouble when one confronts the *possibility* of such a *branch counting*, as is employed in the proof at the stage where the 'counterweight counter' takes track of the multiplicities. There are two general reasons. The first is the reason multiply appealed to by Wallace (2012), that "the branching structure is given by decoherence, and decoherence does not deliver a structure with a well-defined notion of branch count." (his p. 120) His reason to believe so is that

> [v]ery small changes in how the *decoherence basis* is defined, or the *fineness of grain* that is chosen for that basis, will lead to wild swings in the branch count. Insofar as a particular mathematical formalism for decoherence does deliver something that looks like a branch count (and many do not), that something is a mathematical artefact of no physical significance. (ibid.; my emphasis – FB)

A similar opinion is expressed by Dawid and Thébault (2015, pp. 1561–1562), who write:

> Since there is no unique way to specify at which stage two branches have fully decoupled and therefore must be counted separately, it is impossible to specify one definitive branching structure for a quantum process. This in turn implies that no definitive probabilistic conclusions can be drawn from branch counting since the number of branches is inherently indeterminate.

---

[65]The proof is first only given for finite dimensional spaces. All other cases are treated in a quite straightforward and compelling manner by Zurek (2003, p. 3 and 2005, p. 27 ff.) though, once one accepts the proof in finite dimensions (cf. also Boge 2016b, pp. 36 and 37 ff.).

[66]It should be noted that Zurek does not commit to the MWI directly (e.g. Zurek 2009, p. 185), but his method of proof has been picked up in this context (cf. the discussion below), and ubiquitous talk of 'branches' in Zurek's writings (e.g. Zurek 2005, 2009) certainly invites for this.

[67]We have been careful to appeal to a propensity rather than directly to frequencies, so that cases with moduli that are not square roots of rational numbers and are treated by a limiting procedure (cf. Zurek 2003, p. 3) may be understood on equal terms.

Fig. 6.1 Branching structure with unequal distribution of an event over branches at different times

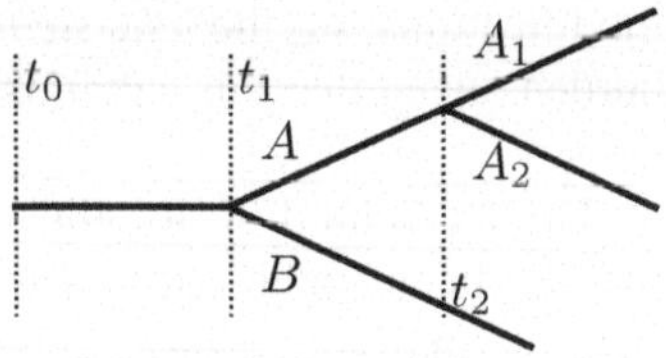

Whether these arguments are compelling is open to debate. We have argued multiple times that *consciousness* must 'cut off' in a way that the physical states do not, even after the decoherence process has taken place. No conscious observer ever *feels* like being only 'mostly' in a state of having observed 'spin up', say, and still 'a little bit' like having observed 'spin down'. Nor is it even clear what a *perceptual* state of that sort would be like. Since the probabilities only pose a pressing problem in relation to (conscious) observations, it is not entirely unreasonable to suggest that consciousness may induce 'states with compact support' which at least fully eliminate the overlap between Zeh's (2000, p. 226) observed or apparent worlds (cf. also Boge 2016b, pp. 42–43). But of course the appeal of the MWI as entirely providing an interpretation of QM 'from within' is thereby significantly reduced.

There is a second, independent reason to be suspicious of counting-based proofs of the Born probabilities in a branching structure like the MWI, which is also discussed by Wallace (2012, p. 120). Consider, following Wallace, a very simple branching toy-universe as depicted in Fig. 6.1, and assume that one gets handed a coin at t_1 in the A-world but not in the B-world and that having the coin persists throughout all A worlds, where A splits further into A_1 and A_2 at t_2. If one identifies the probability $p(\text{coin}_{t_1 \leq t < t_2})$ of having a coin for the specified times with the fraction of branches on which one has the coin, then this would clearly be $\frac{1}{2}$. However, in all A-worlds the having of the coin remains constant, so that for times $t \geq t_2$ this probability becomes $\frac{2}{3}$. But by the law of total probability, one would have

$$
\begin{aligned}
p(\text{coin}_{t_2 \leq t}) &= p(\text{coin}_{t_2 \leq t} | \text{coin}_{t_1 \leq t < t_2}) \cdot p(\text{coin}_{t_1 \leq t < t_2}) + \\
&+ p(\text{coin}_{t_2 \leq t} | \text{no coin}_{t_1 \leq t < t_2}) \cdot p(\text{no coin}_{t_1 \leq t < t_2}) = 1 \cdot \frac{1}{2} + 0 \cdot \frac{1}{2} \neq \frac{2}{3}. \quad \lightning
\end{aligned}
\tag{6.34}
$$

The very *laws of probability* would have to be changed if one were to make sense of probabilities in terms of branch counting, and it is not even straightforwardly foreseeable *how* this should be done: Assume, for instance, that the A worlds after the splitting would have respective modulus-weights of $\sqrt{\frac{1}{3}}$ and $\sqrt{\frac{2}{3}}$. Then according to a Zurek-style argument, there would be four worlds in total, one A_1 world, two A_2 worlds, and one B world. So $p(\text{coin}_{t_2 \leq t})$ would be $\frac{3}{4}$ which is still not $\frac{1}{2}$, as required by the law of total probability.

Fixing this problem would spare us modifications of the (unitary part of the) quantum formalism, but it would presumably imply, at the same time, massive modifications of the *classical probability calculus*—which seems like quite a burden, given that this calculus has been put to good use in the sciences long before the advent of QM.

There is a kind of 'middle way' in between the two approaches discussed so far that has been suggested by Carroll and Sebens (2014). Their proof is, in its central steps, virtually the same as that of Zurek (2005), in that they reduce, for instance, "the problem of two branches with unequal amplitudes to that of three branches with equal amplitudes." (Carroll and Sebens 2014, p. 166) However, extending a basic proposal by Vaidman (2012, p. 304 ff.), they interpret this probability they derive in terms of a *self-locating uncertainty*. Shortly after the physical measurement process, when O_0 *has already become* $O_\uparrow$ and $O_\downarrow$ (or died off to give life to them), and *immediately before* $O_\uparrow$ and $O_\downarrow$ have the conscious result ↑ or ↓ respectively (between stages (ii) and (iii) in our analysis of the measurement process; cf. Sect. 2.1.4), there will be 'room for' an uncertainty as to which observer one *is*:

> The timescale for decoherence for a macroscopic apparatus is extremely short, generally much less than 10^{20} sec. Even if we imagine an experimenter looking directly at a quantum system, the state of the experimenter's eyeballs would decohere that quickly. The timescale over which human perception occurs, however, is tens of milliseconds or longer. Even the most agile experimenter will experience some period of self-locating uncertainty in which they don't know which of several branches they are on, even if it is too brief for them to notice. (Carroll and Sebens 2014, p. 161)

Since the probabilities do hence not concern O_0's credences but rather "how the various future selves into which you will evolve should apportion their credences" (Carroll and Sebens 2014, p. 168), this immediately avoids the difficulties surrounding the subjective uncertainty view that we discussed in the Deutsch-Wallace approach.

But there are still serious difficulties: First of all, the very *meaning* of 'probability' now depends on somewhat uninteresting subjective uncertainties during tiny timescales of some "tens of milliseconds", which impairs on the question of the correct explanandum again (frequency data). One might come to terms with this 'insight', but keeping in mind that the proof is basically the same as Zurek's, *it still implies a branch counting strategy*. Even if the decoherence-induced problems with branch counting might be overcome by leaving some room for consciousness the MWI (*which has to be incorporated anyways*), the difficulties arising from the rules of the probability calculus seem much harder to tackle. And Carroll and Sebens (2014, p. 162), in fact, only address difficulties with branch counting by appeal to the *naïvest* kind of strategy, wherein each branch is simply associated with an *equal* weight or multiplicity.

In conclusion, there are many open questions, and there is still a general lack of clarity in all these approaches. As Kent (2015, pp. 215–216) puts it, in a piercing remark:

> Wherever one thinks of the scientific status of many worlds quantum theory, one cannot reasonably [...] think it is so obvious how to translate equations into statements about a many-worlds reality that arguments and explanations are redundant.

6.3.4 *Matters of Ontology*

Kent's observation at the end of the previous section, that translations of equations into statements about a many-worlds reality are typically not that obvious, also has an impact in the non-probabilisitic part of the MWI. Recall that Zeh (2000, p. 226) has the quantum world "described by a wave function" that corresponds to "*one* superposition of myriads of components" where the "branches" rather correspond to "the observed (apparent) 'world'".

If we take this to mean that for Zeh the world *really is* 'all wavefunction', the MWI presented in this fashion comes out as an instance of what Wallace and Timpson (2010) refer to as *wavefunction realism*. This general ontological stance in the foundations of QM was especially popularized by David Albert (1996, p. 277; my emphasis – FB) who describes it, in a different context, as follows:

> the space we live in, the space in which any realistic understanding of quantum mechanics is necessarily going to depict the history of the world as playing itself out (if space is the right name for it [...]) is *configuration-space*. And whatever impression we have to the contrary (whatever impression we have, say, of living in a three-dimensional space, or in a four-dimensional space-time) is somehow *flatly illusory*.

As formulated above, this makes the *position representation* of the state vector somewhat preferred, as representing a real (waving) *field* on a high-dimensional space (cf. also Wallace and Timpson 2010, p. 706, on this point). One finds similar passages in Everett (1973), namely that "the wave function itself is [...] the fundamental entity, obeying at all times a deterministic wave equation" (p. 115), and one also finds Everett struggle to relate "the existence of macroscopic objects, and [...] their ordinary (classical) behavior in the three dimensional world to the underlying wave mechanics in the higher dimensional space." (p. 86) Only in the modified version published first did he more modestly present his theory such that the wavefunction is "the basic physical entity with *no a priori interpretation*." (Everett, 1957, p. 455; emphasis in original).

Wavefunction realism seems to be deeply connected to the MWI. Yet there are many good reasons to object to wavefunction realism. Maudlin (2010a, p. 127; emphasis in original), for instance, observes that in many expositions of the MWI, macroscopic objects are still described as being "somehow [...] 'made up' of a very large number of *atoms* that all are located in a *common space* and therefore can have particular *configurations*. [H]ow, in a truly *monistic* theory, to get from a high-dimensional space to a configuration space is not even asked." The point is that the wavefunction cannot evolve as an object on 'configuration'-space if there are, strictly speaking, no *configurations* of lower-dimensional particles. But even if one were to make sense of these configurations as some sort of 'mere metaphor', the following basic question remains:

> if all there is the wavefunction, an extremely high-dimensional object evolving in some specified way, *how does that account for the low-dimensional world of localized objects that we start off believing in, whose apparent behavior constitutes the explanandum of physics in the first place*? (Maudlin 2010a, p. 132–133; emphasis in original)

How does Albert's 'flat illusion' really arise, and how does acceptance that it may be an illusion help us to *understand* the existence and behavior of the *apparent* lower-dimensional objects?

Monton (2002), moreover, argues against wavefunction-realism on the grounds that there is not even *supervenience* of possible three-dimensional spaces on the wavefunction in a $3N$-dimensional space. Among other things this is so because "nowhere in the $3N$-dimensional space is it specified which dimensions correspond to which particles" (p. 267), so that "given the state of the objects in $3N$-dimensional space, one cannot establish the state of the objects in three-dimensional space." (p. 268) This implies, *by definition* (e.g. McLaughlin and Bennett 2011, p. 1), that no supervenience holds—there *can* be changes in the 3-dimensional spaces *without* there being changes in the $3N$-dimensional space.

It is hard to understand how the apparent objects in spacetime *precisely* relate to the high dimensional space on which the wavefunction 'lives', and as Monton's arguments show, there are reasons to suspect that there is no clear-cut connection such as supervenience. This is related, of course, to the plain old interaction problem that was present in BM and collapse interpretations. As long as one appeals to wavefunction-realism, similar such concerns are present in the MWI.

Considerations like these have motivated Wallace and Timpson (2010, p. 709 ff.) to suggest an alternative which they call *spacetime state realism*. Basically, the suggestion here is to "associate a set of properties (represented by a density operator) to each region of spacetime." (Wallace and Timpson 2010, p. 712) How can this association be effected? Recall, from Sect. 2.2.4 and Appendix C, that in virtue of the GNS-construction, one can represent a C^*-algebra of local observables by operators in a Hilbert space, and that the folium of the representation will define local density operators. This is basically the route ultimately taken by Wallace and Timpson (2010, p. 711), although they also suggest that "one can even remain at the more abstract level, forego the representation theorems and just take the C^*-algebraic state itself as denoting the properties of a region." (their p. 712)

Applied to the MWI, the implication in both cases is that, for any given 'observable', there will be a *collection* of values associated to a spacetime region. Assume also that one has a state $\hat{\rho}_{\mathcal{O}\cup\mathcal{O}'}$ for two spacelike separated regions $\mathcal{O}$ and $\mathcal{O}'$. Then by tracing over the degrees of freedom for one of these regions, say $\mathcal{O}'$, one obtains a state $\hat{\rho}_{\mathcal{O}}$ for the respective other region. But of course the properties measurable in $\mathcal{O}$ can be correlated with the values measurable in $\mathcal{O}'$ in virtue of entanglement. Thus local spacetime states will not generally be *separable*, in the sense that respective information about $\mathcal{O}$ and $\mathcal{O}'$ will not suffice to determine the state of $\mathcal{O} \cup \mathcal{O}'$ (cf. Wallace and Timpson 2010, pp. 712–713) This is, of course, exactly the sort of separability at stake in our (Howard's) SEP from Sect. 4.3.3,[68] and what is being denied here is part (2) thereof.

Using spacetime state realism, 'quantum states' do not occur as ineffable fields on configuration space in the MWI, but rather as multivalued properties associated

[68] Wallace and Timpson refer back to Healey (1991, p. 406) instead.

with spacetime points, where the multiple values will rarely interfere with one another and will otherwise typically quickly cease to do so in virtue of decoherence. This is certainly a massive improvement over wavefunction realism: wavefunction realism does not properly satisfy our DOC, mostly because (ii) is not addressed (how do high-dimensional space and lower-dimensional spactimes relate?). (i) and (iii), in contrast, are more or less satisfied even on wavefunction realism: (i) 'classical' objects 'emerge' out of the wavefunction in virtue of decoherence, and (iii) the only thing with real ontological significance are the wavefunction and its dynamics. Spacetime state realism removes the difficulty associated with satisfying (ii); the multiplicity is found *in spacetime regions*, and (i) is satisfied in virtue of observers being 'trapped' inside decohered branches for most of their lives.

However, spacetime state realism only works out for the MWI and on point (iii) of the DOC once one is willing to accept the additional prose of density operators representing 'multivalued properties', as we have called them, because otherwise the meaning of density operators as representatives of properties becomes *elusive*. Not so, say Wallace and Timpson (2010, p. 701), as they think of their conception as *interpretation-neutral*. Moreover, they find the association of density operators to spacetime points or regions comparable to the association of vectors to points or regions *via* vector-valued fields in classical electromagnetism, since they "know of no rule of segregation which states that only those mathematical items to which one is introduced sufficiently early on in the schoolroom get to count as possible representatives of physical quantities [. . .]!" (Wallace and Timpson 2010, p. 710)

These considerations miss an important point, and the epistemically more cautious or more empirically-minded philosopher may respond to the polemic as follows: When one specifies, say, some particular classical magnetic field $\boldsymbol{B}$ over some region of points $\{\boldsymbol{x}\}_{\boldsymbol{x}\in\mathcal{O}\subseteq\mathbb{R}^3}$, generated e.g. by some electric current with density $\boldsymbol{j}$ and computed with the aid of the Biot-Savart law $\boldsymbol{B}(\boldsymbol{x}) = \frac{\mu_0}{4\pi}\int \mathrm{d}^3x' \, \frac{\boldsymbol{j}(\boldsymbol{x}')\times(\boldsymbol{x}-\boldsymbol{x}')}{|\boldsymbol{x}-\boldsymbol{x}'|^3}$ (cf. Jackson 1990, p. 178), one *knows exactly what to expect in a single case observation*. The vector valued field $\boldsymbol{B}(\boldsymbol{x})$ specifies the behavior of (strength and direction of the field's effect on) test-objects with magnetic properties for any such *singular* observation, and while deviations are possible and one resorts to multiple trials to *confirm* the underlying theory, the *prediction* is still single-case. Not so in the quantum case: *only* when we specify a long series of repeated trials do we know what to *expect*, as single, 'non-multivalued' observes, and we do so in virtue of the *Born-rule statistics* predicted by the local density operator $\hat{\rho}_{\mathcal{O}}$. There is all the difference in the world between the two associations, and it has nothing to do with early exposure to concepts in schoolrooms.

Our conclusion from this is the following: While spacetime state realism is a great improvement over wavefunction realism for the MWI, it can *only* help to clarify the ontology once the probability problem is solved—in both its dimensions, quantitative and incoherence-wise. The problem of probabilities is the central problem of the MWI. And since statistical predictions form the "empirical heart" of QM (Lewis 2007, p. 62), as long as the MWI "has not been shown to recover the astounding empirical success of the textbook quantum recipe" (the orthodox interpretation), it "cannot appeal to that success for empirical support." (Maudlin 2014a, p. 808)

6.4 Discussion (ii): No Clear Winner?

Let us take stock of our findings, first on 'dialectically neutral' or merely classificatory grounds, and then in the form of the promised discussion of advantages and disadvantages of respective interpretations. BM introduces particles with definite positions and velocities at all times as a primitive ontology, and explains quantum phenomena in terms of the statistics of these. The statistics, in turn, depend on an effective wavefunction ψ or ultimately on the wavefunction Ψ of the universe, and it becomes a matter of debate what ψ really represents: (a) a guiding field, residing in a different, higher-dimensional space, or active information ordered therein; or (b) an expression of a law-like connection.

In the latter case one can either understand Ψ or a respective law featuring it as an expression of something *real* and *independent* of the particles, or merely as an expression of their regular behavior as described by a 'best system'. Let us denote the 'pure interpretation(s)' that embrace (a), i.e. where both, particles *and* ψ (or Ψ) have an independent ontological significance as real, independent and non-nomological entities, simply by BM. Then the other two options may be called BM_O^N and BM_H^N respectively, where N stands for *nomological*, O for *ontological* (meaning not-merely-regularity-like), and H for *Humean*.

As we have seen, the same clarification is possible w.r.t. GRW, CSL, and gravitational collapse interpretations. Since all of these could respectively fall under the three respective (coarse) categories, we here refer to the 'pure interpretation(s)' (GRWm, GRWf, CSLm, interpreted gravitational models) wherein the wavefunction is 'the real stuff' as CI (short for 'collapse interpretations'), and to the two nomological versions as CI_O^N and CI_H^N respectively. So as promised in Sects. 6.1 and 6.2, we can use the introduced terminology to devise a more fine-grained classification (cf. Fig. 6.2).

As for the MWI, the distinction between spacetime state realism and wavefunction realism does not seem to validate a similar distinction between 'nomological' and 'non-nomological' versions. Moreover, we introduced our notion of ontological interpretations in Sect. 2.3 w.r.t. wavefunction *or* state vector, and when we used the term 'wavefunction', we did not generally and exclusively intend to refer to the *position representation* of the wavefunction, as is being done in wavefunction realism. Thus, since state vectors and density operators are still intimately connected, at the very least *via* algebraic states in the GNS-construction, we take it that there is no 'non-ontological' MWI in our sense, regardless of whether one endorses wavefunction- or spacetime state realism.

It is interesting to see which spaces are filled and empty respectively in Fig. 6.2 and why. Note first that, if the wavefunction is interpreted nomologically in CI

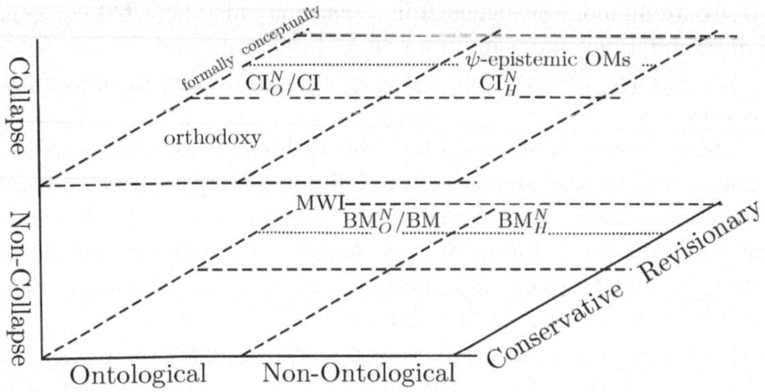

Fig. 6.2 Classification of different interpretations according to the scheme from Sect. 2.3. The nomenclature is explained in the text

or BM and one embraces a Humean account of laws (in the sense of BSA, not necessarily with Humean supervenience), then according to our terminological choices, these versions should be classified as non-ontological. Our choices for CI/CI$_O^N$, BM/BM$_O^N$, and the MWI have been justified before, so they need no further explanation. We have set the 'orhtodoxy', i.e. the minimal Dirac-von Neumann interpretation into the ontological non-revisionary collapse corner, since the quantum state is taken to represent the physical state of the system, and since all revisions were evaluated w.r.t. this very interpretation (so it cannot be revisionary, by definition).

ψ-epistemic models are set on the boundary of the formally and conceptually revisionary interpretations, since they suggest to revise both, the formalism and the orthodox interpretation (the quantum state here does not refer to the physical state of the system). Why are there no conservative non-collapse interpretations? Because if one removes the collapse, this is already a conceptual revision, regardless of whether one also deprives the wavefunction of its ontological significance or not. These options are forestalled *by default*, given our terminological choices. The same may be said about merely formally revisionary non-collapse interpretations; dropping the projection postulate *is* a conceptual revision. However, any *full* version of BM may be set on the boundary of the formal/conceptual dichotomy, since a Bohmian *QFT* or in general a relativistic BM seems to introduce the need for formal revisions as well, even if a Bohmian guidance equation in the non-relativistic domain is 'already in' the TDSE.

What options are still possible and underrepresented in our scheme? A possibility is to *reinterpret* the orthodox scheme, i.e. leave the formalism and the postulates intact as they are and interpret the quantum state as the physical state of the system, but assign a different physical meaning to other parts of the theory (upper rear left corner). Or one could keep the formalism and the postulates intact, but not interpret the quantum state as a *full description* of the physical state of the system (upper rear right corner). This is basically what the historical ensemble-interpretations attempted—unsuccessfully—and may also be what some advocates of modal interpretations have in mind (e.g. Lombardi and Dieks 2012, for an overview). We will also find another sort of interpretation that fits into this corner in the next section.

A final option seems to be to adopt the following strategy: accept that the quantum state is the physical state, as in the orthodox interpretation, and embrace all the postulates, but redefine what is *meant* by 'physical state' in such a way that the ontological commitment of the term is loosened. This would be an interpretation that fits into the front upper right corner, and we will ultimately suggest an interpretational strategy that is of this sort.

How do the interpretations discussed so far fare in comparison to one another and w.r.t. our two guiding principles, MAC and DOC? We have argued that both BM and CI suffer from a lack of ontological clarity as regards the connection between the realm of the wavefunction and spacetime if one interprets the wavefunction ontologically, regardless of whether it is construed as a 'real law' or not. Humean-nomological versions were argued to be ontolgically much clearer, on the other hand, but they come with a loss of explanatory value. The MWI can be made ontologically clear if one adopts spacetime state realism; but as long as the probability problem is not solved, it does not predict the observed statistics and hence has a problem with the constraint of empirical adequacy involved in the MAC.

Moreover, BM *in general* was said to (still) be in conflict with relativity, and all CIs were said to involve too much of an ad hoc character. And due to the non-conservation of energy in CIs, one may here equally arrive at an empirical inadequacy, should experimental accuracies improve significantly. Finally, we found ψ-epistemic OMs to be incapable of successfully reproducing the QM predictions entirely (thereby immediately not satisfying the MAC) and/or to be ontolgically unclear ('vacuum ontic states', gerrymandered formal models...), and we may also say that the *existing* models ultimately all involve ad hoc moves, either formally (cf. the model of Lewis et al. 2012, for comparison) or conceptually (cf. the model of Spekkens 2007, and our discussion on 'vacuum ontic states').

Table 6.1 summarizes our findings. '✓' symbolizes an advantage, '✗' a disadvantage of the respective interpretation or interpretational program/paradigm; if there is no (unambiguous) answer (yet), we write '*?*'. Some further important features,

Table 6.1 Comparison of different interpretations. '✓' symbolizes an advantage, '✗' a disadvantage, '*?*' that there is no unambiguous answer or no answer yet

	Empirical adequacy	Compatibility with SR	Ontological clarity	ad hocness
ψ-epistemic OMs	✗	*?*	✗	✗
BM_O^N/BM	✓	✗	✗	✓
BM_H^N	✓	✗	✓	✓
CI_O^N/CI	*?*	✓	✗	✗
CI_H^N	*?*	✓	✓	✗
MWI	✗	✓	✓	✓

aside from overall ontological clarity and empirical adequacy, have crystallized from the debate that need to be taken into account here as well: that an interpretation may have difficulties in compatibility with the relativity theories, and that it may introduce additional elements in an ad hoc fashion. As we can see, ψ-epistemic OMs fare worst, followed by CI_O^N/CI, according to our analysis. In the case of ψ-epistemic OMs, this may be at least partly due to the fact that we do not *have* any full ψ-epistemic interpretation yet, but only a collection of partially plausibilizing models. And in the case of CI_O^N or CI, evidence *for* an objective collapse (e.g. measurable increases of energy) would massively tip the balance towards these interpretations anyway, since the other points (ontological clarity and ad hocness) are 'merely philosophical': maybe the world happens to be hard to understand and can only be understood by appeal to ad hoc moves.

The interpretations that fare best are the MWI and BM_H^N, since they only have one '✗' respectively. However, as long as no contradicting evidence is (or even can be) found, CI_H^N fares equally well. In fact, from a purely 'philosophy of science'-point of view, that all CI make deviating predictions should count as a *virtue*, not a *vice*: They are not immune to falsification.

Notably, the collapse- and Bohmian interpretations that fare best according to our catalog also have to be amended with a suitably therapeutic attitude: one 'simply has to accept' certain kinds of phenomena and certain correlations, such as flashy events occurring *out of nowhere* and *for no reason* (CI_H^N), detectors indicating particles where there are none *on a regular basis* (BM_H^N), and correlations between particles or flashes over large distances, again *for no reason*.

The MWI avoids all these complications, but it does so at a remarkably high price: due to the persistence of the probability problem, one looses empirical adequacy. BM in general additionally suffers from difficulties of reconciliation with SR (not to mention GR), which is not the case with the MWI.

But none of these disadvantages *force* one to give up either of the associated research programs. Maybe one *can* find a way to reconcile decoherence, branch counting, standard probability calculus, and the Born rule in the MWI after all. Maybe one *can* argue for the possibility of a reconciliation of BM and relativity on the grounds of reinterpreting the content of the relativity theories, or modifying them to the extent that evidence allows. And so forth.

The latter sort of strategy is, in fact, basically suggested by Bohm and Hiley (1993, p. 288):

> We have to be careful [...] not to assume that a theory has an absolute and unlimited validity just because it has agreed with a very wide range of experiments and because its form is aesthetically pleasing. From these reasons it does not follow that the theory of relativity is an absolute truth.

This is certainly a valid point, and it leads them (ibid., p. 290) to speculate about an alternative to SR (and presumably GR) that

> would be reminiscent of the Lorentz-type ether theory within which there were large scale objects with structures undergoing processes that would change with velocity in such a way as to bring about Lorentz invariance in terms of frames defined through these structures.

However, no such theory exists to date and GR, containing the predictions of SR, continues to make successful predictions (e.g. Castelvecchi and Witze 2016, for a more recent, 'spectacular' one). Note, finally, that we went only briefly into the reasons to believe that CI fare better than BM on grounds of compatibility with SR, and of course these reasons could be disputed as well. This would then change things in *disfavor* of all CI and turn the advantages in this respect into disadvantages.

6.5 Intermediate Conclusions (ii)

The insight that crystallizes from the foregoing debate is that there is as yet no clear winner among the interpretations considered. Note that our choice in considering only BM, CI, and the MWI was not without reason: these are the currently most hotly debated (ontological) interpretations, since (arguably) all remaining interpretations suffer from even worse fundamental deficits, whence some of them have already largely been 'purged from the debate'.

Once more we hit a road block. None of the ontological interpretations discussed so far strikes us as preferable, none of the priests has shown us the light (cf. the quote by Fuchs at the beginning of this chapter), and we have given detailed reasons for why that is. Moreover, the natural response discussed in Chap. 4 and its execution in terms of ψ-epistemic OMs seem to be among the worst faring interpretational schemes. So QM neither promotes a unique and clear ontology, nor is it reducible to a 'mere lack of knowledge'.

There are, however, *still* good reasons to think that QM *does* have to do with knowledge *in some sense*. We had focused mostly on the deficits of Spekkens' toy model w.r.t. *explanations* of the 'classics' of QM (interference and entanglement) in terms of true states λ. But the treatment of non-commutativity and collapse as belief-update were somewhat convincing nonetheless, and we mentioned that there are further examples, specifically from QIT, that Spekkens reproduces in terms of knowledge. We also demonstrated in detail how the assumption of true states λ leads to all kinds of trouble, among other things to the derivability of Bell-type inequalities from the PCC. Fuchs (2014, p. 388) similarly judges that "the phenomena [that Spekkens' model reproduces – FB] arise in the *uncertainties*, never in the mechanical configurations. It is the states of uncertainty that mimic the formal apparatus of quantum theory, not the toys' so-called ontic states[. . .]." (my emphasis – FB) So maybe there is a way to make sense of QM 'epistemically' *but without the* λ*s*.

Moreover, there are reasons to suspect that QM is an *inherently probabilistic* theory, and depending on one's interpretation of probability (or the probabilities at play, at any rate) this could again be fleshed out to mean that knowledge or information or belief. . . are at stake *in some sense* in the interpretation of QM all along. Recall how we argued in Sect. 6.3.3 that crucial parts of Zurek's (2005, 2003) proof are impossible if one does not presuppose a restriction to *unitary* operators as representations of state-transformations. Why does one, in fact, use unitary operators to describe state changes in QM? Hughes (1989, p. 115 ff.) motivates the use and structure of unitaries as state transformations *entirely* from *probabilistic* concerns; and similarly (Mohrhoff 2004, p. 228), in analyzing Zurek's (2003) proof, asks why the operators encoding the evolution given by the TDSE are unitary, and answers: "Because *probability is conserved* whenever the system in question persists (is stable)." (my emphasis – FB)

This and the fact that a whole brand of interpretations of 'probability' make it an expression of epistemic *uncertainty*, together with the apparent failure of the ontological interpretations discussed, certainly suffice to motivate a return to broadly 'epistemic' considerations in the next chapter.

But even if we thereby give up on the attempt to say what the 'world *according to* QM' is like, we must acknowledge, as a result of the discussion in the previous two chapters, that QM does put *constraints* on what the 'world in spite of QM' *can* be like. By and large, this is due to *entanglement*, the feature of QM that "enforces its entire departure from classical lines of thought." (Schrödinger 1935a, p. 555) And all of the ontological interpretations acknowledge this fact: Either one is faced with unexplained correlations in the behavior of distant (Bohmian) particles, flashes, or the local values of a mass density, or one must embrace novel metaphysics of 'holistic dispositions of motion' or 'nomological entities *sui generis*' to account for the correlated observations, or, alternatively, accept that one is 'tricked' into experiencing the appearance of such correlations in virtue of the dynamics of multiple possibilities that are concurrently realized in spacetime regions and 'branch off' after suitable interactions. These are (basically) the possibilities laid out by the ontological interpretations discussed in this chapter for dealing with entanglement and its implications.

Chapter 7
Reconsidering Knowledge, or, Coming to Terms With Quantum Mechanics

There is no quantum world. There is only an abstract quantum mechanical description. It is wrong to think that the task of physics is to find out how Nature is. Physics concerns what we can say about Nature.

—Attributed to N. Bohr by Petersen (1963, p. 12)

7.1 Preliminaries: A Change in Perspective

In conclusion of the previous chapter, we argued that the strong involvement of probabilities in the formalism of QM and the fact that one is not *bound* to introducing (formally explicit) hidden variables λ justifies to reconsider knowledge. We also briefly mentioned that there are further reasons in QIT to consider quantum states as being concerned with knowledge. There is, for instance, a theorem[1] that any pure quantum state $|\psi\rangle$ cannot be reliably distinguished from any other, non-orthogonal one, $|\phi\rangle$, *by means of any measurement* representable as a POVM. Based on this, Caves et al. (2002a, p. 3) have argued that the only means of reliably identifying the pure state assigned to some system "requires consulting the assigner or the records he leaves behind", which they take as a strong indication for quantum state-assignments being expressions of the assigner's knowledge or information.

However, that this helps us understand QM only holds if one can *find* an 'epistemic' interpretation that works out without talk of hidden states λ, can accommodate the fact that we *have* to use the QM formalism for certain applications, that

Author's note: I owe the contrast between 'making sense' and a mere 'coming to terms' to Markus Schrenk (private communication).

[1]Cf. Nielsen and Chuang (2010, p. 87); the proof is quite straightforward.

F. J. Boge, *Quantum Mechanics Between Ontology and Epistemology*, European Studies in Philosophy of Science 10,
https://doi.org/10.1007/978-3-319-95765-4_7

we *do* find the correlations predicted in certain experiments, and that QM has lead us to a great many *practical applications* that we could not even have dreamed of before its advent. Arguably, the concepts surrounding the term 'Copenhagen interpretation' are roughly of this kind; but we argued already in Chap. 1 that there is no well-defined, indisputable core to this interpretation, and that many different (sometimes incompatible) ideas are typically subsumed under the label. A more unified contestant interpretation of the same sort, only advanced in more recent years, goes by the name *Quantum Bayesianism*.

Before going into details, let us make some general remarks. First of all, the interpretations presented in this chapter all aim at (re)solving the OP by largely embracing QM *as it is being put to use*, while reassessing the content or meaning of the state vector. This means, however, that grossly the same strategy for dealing with the kinematical part of the MAC is employed as in Chap. 4; the quantum state is deprived of its ontological significance, whence the PP, when applied, does not pose a problem anymore: we do not believe that the system 'literally' undergoes a 'sudden change' from a superposition of eigenstates of some observable to one of the eigenstates of that observable—because neither the superposition state *nor* the eigenstate represents the 'outside world conditions' of a physical system.

The crucial difference to the project discussed in Chap. 4 is that no formal revisions, or at least none on the 'ontological side', i.e. w.r.t. the states and dynamics standardly supposed to represent mind-independently evolving systems, are being suggested here. We hence formulate the following more cautious conjecture (which, for want of a better name, we call 'purely epistemic' for now):

Conjecture 5 (The purely epistemic view) The quantum mechanical state vector *somehow* has to do with the epistemic conditions of real or ideal observers.

An immediate worry might be this: 'Are we then, dropping formal revisions, not back to the old ensemble-interpretations?' The answer is 'no', because no *unmentioned*, '(quasi-)classical' hidden variables like hidden momenta etc. are assumed here either, and a 'position eigenstate' is not at all assumed to represent an ensemble of systems in a definite position but with varying momenta. In fact it remains to be seen if and what the quantum state *represents* at all.

How can any such proposal work? How can it satisfy the DOC, if talk of the mind-independent states of systems is altogether avoided? The answer is that it cannot, and that we here *explicitly* take a step back and re-evaluate what an interpretation of QM *should* provide, given that all ontological interpretations considered seemed to fail on some grounds. All the interpretations discussed in the following have to do with epistemology rather than ontology. So while we may appreciate ontological clarity to the extent that it *can* still be provided by such an approach (e.g. regarding the subject-object relation), it seems sensible to formulate an alternative to our DOC at this point:

Desideratum of Epistemological Clarity (DEC) Any non-minimal but also non-ontological interpretation of QM that does not introduce additional formal elements for alternative ontological considerations ('hidden variables') should be *epistemologically* as clear as possible. This means that it should

(i) explain the *use of words* like 'classical objects', and of definite property-*assertions*,
(ii) explain the switch from 'quantum' to 'classical' *concepts* as precisely as possible,
(iii) and specify the significance of all (formal *and* informal) ingredients as regards the *epistemic situation* of a user of QM.

Applying (DEC) rather than (DOC) implies a switch in perspective. QM is a physical theory that was developed during the course of the twentieth century. Undoubtedly, it has philosophical consequences because it confronts us with the task of accommodating new empirical evidence and new phenomena into our world view that do not fit well with world views established in the context of previous physics. But who ever said that the physics up to the nineteenth century are a suitable guide to philosophy in the first place, just because they fit somewhat nicely with most of our everyday convictions and life and common sense intuitions? We stand on an equally long history of *philosophical* investigation and argument, developed, to a considerable extent, *independently* of the specific contents of the sciences, and with an appeal in their own right. Why not rather try to understand QM against *this* background than trying to mold it into a 'more physically sounding' cast, as the interpretational programs presented in Chaps. 4 and 6 arguably do?

In fact, this may be seen as the general agenda underlying this entire chapter, and equally, of course, as a project that has its roots already in the writings of Bohr, Heisenberg, Pauli, Schrödinger,[2] and others of that era.[3] In the following,

[2]... who, however, was not so comfortable with this "Deliberate About-Face of the Epistemological Viewpoint" (cf. Schrödinger 1935b, p. 157 ff.).

[3]Zeh (2012, p. 19) points out that the 'founding fathers', especially those who contributed to the 'Copenhagen' tradition, were *driven*, in their interpretive efforts, by their respective *world views*, not necessarily even by well worked-out philosophies. To some extent this is certainly correct and to some extent it even emphasizes our point; that one might seek for answers consulting sources outside physics. But one should also caution against attributing a false systematic value to the insight. For one can equally localize the reasons for endorsing any of the ontological interpretations discussed in Chap. 6 in strong 'realistic' intuitions that may in turn be influenced by a naïvely realist 'world view'. To infer from this (which Zeh does not, at least not overtly) the inferiority of *any* of the respective views would simply mean to confound the context of their discovery with the context of their justification.

we investigate ways to take this program seriously, and ultimately show how to combine different strands therein into a proper 'epistemological' interpretation of QM, i.e. one that takes QM as an expression of how we *generate* knowledge out of experience.

Now we will successively see that the desiderata of the DEC are met by different such interpretations to varying degree and with varying presuppositions. And we will use this insight to develop a viewpoint from which some of them should count as formally and conceptually revisionary w.r.t. orthodoxy while others may be said to be conservative on all fronts, given an appropriate reading of the notions occurring in the quantum postulates.

7.2 The Quantum Bayesian Program and QBist Epistemology

7.2.1 The Formal Epistemological Background

The first interpretation of interest here is the aforementioned Quantum Bayesianism, and the heart of Quantum Bayesianism is a subjective Bayesian view of probability. The heart of the subjective Bayesian view of probability, in turn, are so called *Dutch book theorems*, as originally developed by Ramsey (1926) and De Finetti (1937).[4] Dutch book theorems presuppose that one can quantify one's degrees of belief in the occurrence of some event A in the form of some *belief-function*, which we suggestively call $p(A)$. This is not too far-out an assumption, as it is usually not difficult to at least *order* one's beliefs according to strength, as we also do in everyday life (cf. Hájek 2009a, p. 173 for some examples). So it is not too large a step to map them to suitable numbers either.

Dutch book theorems now provide a mark for *rational* belief in proving that if one's beliefs $p(A)$ do not conform to the Kolmogorov axioms (cf. Appendix A), a Dutch book can be made against one, i.e. a bet or a (finite) series thereof in which one is *certain* to loose. Non-Dutch-bookable behavior is often called *coherent* (e.g. Earman 1992). These theorems are usually supplemented by *converse* Dutch book theorems, that if one's beliefs $p(A)$ *do* conform to the Kolmogorov axioms, then a Dutch book *cannot* be made against one. And such converse theorems are clearly indicated to sustain an interpretation of probability in terms of rational belief, since Dutch book theorems alone provide only a *necessary* condition for coherence, not a sufficient one (e.g. Earman 1992, p. 39; Hájek 2009a, p. 177).

The above already indicates in what terms degrees of belief are assumed to be quantifiable, and we have basically also touched on this in the context of the MWI, namely in terms of *betting behavior*. Dutch book theorems hence proceed from

[4] The origin of the term is not unambiguously clear, but folklore has it that it stems from the bad reputation of seventeenth century dutch bookmakers (cf. Hájek 2009a, p. 174).

a betting situation in which a bookie offers a (potential) bettor some amount of $w = a€$ (w for 'win'), say, in case event A occurs, and in which the bettor bets $\ell = b€$ (ℓ for 'loose') on A's occurrence. Then the 'stakes' is $s = (w + \ell)$ and the 'betting odds' are w/ℓ (cf. Earman 1992, p. 38). One's willingness to bet $r \cdot (a+b)€$ in a bet where one receives $0€$ if A does not occur and $(a + b)€$ if it does (which means that the net win is $(1-r) \cdot (a+b)€$ if A occurs and the net loss is $r \cdot (a+b)€$ else) is then taken to reflect the fact that one's degree of belief in A is $p(A) = r$. Now assume that s may be positive or negative, i.e., that both parties to the bet use the same criterion for their willingness to bet on A. The first Kolmogorov axiom, $p(A) \geq 0$, then follows straightforwardly, because if one allows $p(A) = q < 0$, the betting party for whom $s < 0$ would be willing to indulge in bets where her net win is $(1 - q)s < 0$ if A occurs, and her net loss is $q \cdot s > 0$ else—a Dutch book. Disobeying the other Kolmogorov axioms leads to Dutch books in a similar way, and the converse theorems proceed by the same basic scheme (e.g. Williamson 2010, p. 35 ff.).

So far so good, but *subjective* Bayesianism has been confronted with a bunch of objections, most of them based on one's freedom to choose initial probability assignments therein, combined with the practice of belief-updating *via conditionalization*. Conditionalization proceeds from some *prior probability* (short: prior) $p(H)$ for a hypothesis H, which may be based on one's total beliefs and evidence $\mathscr{E}$ in some present state (whence we may write $p_{\mathscr{E}}(H)$). The dependence or non-dependence on $\mathscr{E}$ is not clear a priori though—one could not believe something *in spite of evidence*—and the allowance of a neglect of $\mathscr{E}$ in the formation of a prior can in fact be taken to mark off the *radically* subjectivist stance from what Williamson (2010, p. 15) calls *empirically based subjective Bayesianism* (cf. the first interlude). If $\mathscr{E}$ does not imply anything about H though, $p_{\mathscr{E}}(H)$ can be chosen *arbitrarily* even by the empirically based subjectivist, aside from being constrained by the Kolmogorov axioms.

After obtaining some new evidence E, however, one's degrees of belief must be changed to $p_{\mathscr{E} \cup \{E\}}(H) = p_{\mathscr{E}}(H|E)$ in evidence-based Bayesianisms, which is exactly the method of conditionalizing on evidence. In fact, *Bayes' theorem*, $p(H|E) = p(E|H)p(H)/p(E)$ (here given in its simplest form) provides a means for updating one's degrees of belief, based on the prior $p(H)$, the *likelihood* $p(E|H)$ of the evidence obtaining given that the hypothesis is true, and the degree $p(E)$ to which the evidence is believed to occur (cf. Howson and Urbach 2006, p. 21).

But apparent problems arise from this freedom to assign priors and the practice of conditionalization. Assume for instance (cf. Bacchus et al. 1990, p. 490) that some agent's beliefs are such that $p(A) = 1$, $p(B) = 0,2$, $p(C) = 0,8$, $p(D) = 0,2$, and $p(B, D) = p(C, D) = 0,1$, where A could mean that Peterson is Scandinavian, B that Peterson is a Swede, C that Peterson is a Norwegian, and D that 80% of all Scandinavians are Swedes. Upon learning that *D is the case*, the agent would have to update her belief that Peterson is a swede, the argument goes, to $p(B|D) = p(D|B)p(B)/p(D) = p(B, D)/p(D) = 1/2$. But this is unequal to 0, 8, which *should* be assumed, according to the belief in the hypothesis that 80% of all Scandinavians are Swedes.

Prima facie this means trouble for conditionalization and free choice of priors, but the argument is easily seen to be *question begging*: It is simply assumed here that the statistical statement provides the ‘correct’ probability assignment. Indeed, since the Scandinavian population is finite, D could (under very favorable circumstances) just express an *analytic* probability for ‘drawing’ one Swede from an ‘urn’ of Scandinavians. But it is still far from clear that this analytic probability provides the ‘correct’ degree of belief: maybe the agent remains more skeptical than 0,8 about Peterson being a Swede for good reasons contained in his background beliefs and evidence $\mathscr{E}$, even in the light of the new evidence; maybe she never assigns probability 1 to the statistical (or analytical) hypothesis in the first place, keeping in mind possible counting errors and imperfection of statistical methods. And, more importantly, if one takes the subjectivist stance seriously, *then there is just no such thing as a ‘correct probability assignment’ at all*. Indeed, de Finetti (1970, p. x) put it in bold letters: “PROBABILITY DOES NOT EXIST”. But of course even de Finetti accepted probabilities as existing *qua* expressions of belief of (ideal) agents; what he meant to deny certainly were propensity- or similar accounts (cf. Healey 2012d, p. 734), wherein probability occurs as “a property of something in the physical world independent of the epistemic state of anyone making judgments about it.” (ibid.)

There are many arguments similar to that of Bacchus et al., and they typically ultimately fail for the same reason: they are question begging. Williamson (2010) e.g. argues against the background of a “physical chance function” (p. 28), as does (e.g.) Schurz (2014, p. 160) somewhat more subtly. It is hence unclear that subjective accounts of probability are doomed to failure due to a miscorrespondence to objective chances and their acceptance of conditionalization and free choice for priors—since the existence of objective chance is disputed in the first place. If the world just happens to be such that one cannot make sense of objective ‘propensities’ or ‘chances’, and the most sense that can be made of the word ‘probability’ is that it quantifies one’s (informed) beliefs (against a background of subjective uncertainty), or rather, provides a normative guide to acknowledging certain rationality constraints (avoidance of a Dutch book), then appeals to a “chance function” or other objective notions are simply deluded.

For now, let us provisionally accept this line of reasoning, which provides grounds enough to take subjective Bayesianism seriously in the first place. Additionally we can, to superficially defend subjectivist accounts of probability further, appeal to *tu quoque* arguments: in interlude I we laid out how other conceptions of probability suffer from fundamental difficulties, and how none of them seems to be clearly preferred (somewhat ironically not all unlike interpretations of QM). These are no positive arguments for endorsing a subjectivist view though (empirically based or not) and we should later reevaluate if and to what extent it is *positively plausible* to accept any such view, specifically in the context of QM. Let us not settle this matter yet, but rather focus on how the subjective Bayesian account of probability relates to *quantum* states first.

7.2.2 *How to Carry the Program Over to Quantum Mechanics?*

In principle, the generalization of the subjective Bayesian framework for probabilities to the quantum case is quite straightforward: interpret *all* probabilities, even those appearing in QM, as quantifying degrees of belief of some agent. Since *quantum states* are associated with probabilities, this also means that they themselves should be taken as quantifying something to do with the agent, not the system. As already noted, the Quantum Bayesian approach only posits the epistemic role of the quantum state $|\psi\rangle$ but no hidden states λ are assumed. We will spell out the philosophical implications of this in more detail below, in particular what it means in regard to an agent's or observer's relation to (mind-independent) reality. But we should first focus on the details of the *formal* treatment of quantum states in Quantum Bayesianism.

Some central ideas have been developed in joint papers by Caves et al. (2002a,b). Here (2002a, p. 3 ff.) the authors identify a family of orthogonal, one-dimensional projectors $\left\{\hat{P}_j\right\}_{j\in J}$ with a set of answers to (experimental) *yes-no questions* that can be posed about a system. They then appeal to *Gleason's* famous (1957) theorem, which states that for any such family $\left\{\hat{P}_j\right\}_{j\in J}$ with $|J| \geq 3$ and any function μ that maps the $\hat{P}_j$ into [0, 1] and satisfies $\mu(\sum_{j=1}^{m}\hat{P}_j) = \sum_{j=1}^{m}\mu(\hat{P}_j)$ $(m \leq |J|)$ and $\mu(\mathbb{1}) = 1$, there exists a density operator $\hat{\rho}$ such that $\mu(\hat{P}_j) = \mathrm{Tr}(\hat{\rho}\hat{P}_j)$, which is of course a *Born probability* $\mathrm{Pr}_M^{\hat{\rho}}(j)$ (M the measurement represented by the $\hat{P}_j$ and j the j-th outcome).[5] The appeal for Quantum Bayesianism here is that the assumptions on μ are an implementation of the Kolmogorov axioms for projectors, whence μ can be interpreted as a coherent degree of belief in the Dutch book sense. I.e.: even subjective Bayesian probabilities must apparently satisfy the Born rule in the quantum context.

Next, Caves et al. (2002a, ibid.) consider a case of *maximal information* and demonstrate why they think one is urged by coherence to assign pure states in this case. They first consider the 'classical case', in which it is assumed that maximal information implies certainty as to an event A's occurrence, whence assigning $p(A) = r < 1$ would lead to a Dutch book as follows: in the betting-analysis of probability, $p(A) = r < 1$ would mean that one was willing to accept a bet in which one receives $r \cdot s$ up front (as bookie), and then pays s if A occurs (which it is assumed to do), leading to a sure net loss of $(1 - r) \cdot s > 0$.

It appears somewhat problematic though, that it is not specified in this analysis what the 'classical case' really is. One can easily imagine worlds where there are *fundamental* epistemic restrictions (as investigated in Chap. 4), in spite of these worlds being otherwise 'classical' (in then sense of not requiring a quantum treatment), whence A could *never* be assumed to 'occur with certainty'. In fact, if it

[5]For a generalization to elements of a POVM see Busch (2003). Note also the similarity to the discussion of folia in AQFT in Sect. 2.2.4.

were not for the empirical evidence forestalling any compelling ψ-epistemic model or interpretation, one might even have been inclined to believe that *we live in* exactly such a 'classical world'.

Let us grant, however, that there *could be* a classical world as required for the previous Dutch book argument. Then (the argument might go) an agent in *that* world who would use QM and would be certain that some outcome j will occur would have to assign, in accordance with Gleason's theorem, a pure quantum state $\hat{\rho} = |j\rangle\langle j|$ to the system to ensure that $p(j) = 1$, and to avoid the Dutch book. The point being that there is no other means *for expressing* certainty in QM than by pure states assignments (projectors). But even then, a quantum state will represent a situation of 'maximal information' *and* incomplete knowledge, since it will still not allow for answers to *all* possible experimental questions: think *non-commuting* observables and projectors.

Assuming that Dutch book coherence thus requires pure state assignments for cases of maximal information, one can also gather that probability assignments will coincide with long-run frequency predictions in suitable cases as follows. According to the foregoing (Dutch book/probabilistic certainty) argument, a joint quantum state assignment for N systems to which the exact same maximal information applies would be of the form[6]

$$\hat{\rho}^{(N)} = \underbrace{|\psi\rangle\langle\psi| \otimes \ldots \otimes |\psi\rangle\langle\psi|}_{N \text{ times}}, \tag{7.1}$$

and if a measurement of N projectors from a set $\left\{\hat{P}_{o_j}\right\}_{j=1}^{D}$ is performed on these systems (the $\hat{P}_{o_j}$ being single system projectors for D measurable values o_j), then the probability of finding a given sequence $o_1, \ldots, o_N$ is immediately given by

$$p(o_1, \ldots, o_N) = \mathrm{Tr}\left(\hat{\rho}^{(N)} \bigotimes_{i=1}^{N} \hat{P}_{o_{j_i}}\right), \tag{7.2}$$

which is *factorizing* ($p(o_1, \ldots, o_N) = p(o_1) \cdot \ldots \cdot (o_N)$), and where $1 \leq j_i \leq D, \forall 1 \leq i \leq N$. This corresponds to an assignment for *independent and identically distributed* outcomes of repeated measurements (cf. Caves et al. 2002a, p. 4). Crucially, the probability for measuring outcome o_j n_j times can, on account of this assignment, be pieced together (by mere considerations of independence and the possibility of multiple occurrences of each o_j; cf. de Finetti 1970, p. 182) as being given by a *multinomial distribution*

[6]Of course convictions of *independence* go into the very *formulation* of such a quantum state. Note also that the N systems could be investigated in temporal succession, i.e. that our concern here is not with a box of interacting particles or the like.

$$p(n_1, \ldots, n_D) = \frac{N!}{n_1! \ldots n_D!} p(o_1)^{n_1} \cdot \ldots \cdot p(o_D)^{n_D}, \tag{7.3}$$

with D here the Hilbert space dimension. This distribution can be shown to peak at $n_j \approx Np(o_j)$ for large N, so that it predicts *relative frequencies* n_j/N close to $p(o_j)$ (cf. Caves et al. 2002a, p. 4). This basically provides a law of large numbers for QM (cf. ibid.). Again, *given* the acceptance of the aforementioned Dutch book-argument, this is an impressive carry-over of a subjectivist attitude towards probability to QM.

Another strategy invoked by Caves et al. (2002b, p. 4546) is to argue that "if a density operator is even partially a reflection of one's state of knowledge, the multiplicity of ensemble decomposition means that a pure state must also be a state of knowledge", since the multiplicity of ensemble decompositions implies that "the distinction between subjective and objective becomes hopelessly blurred." (p. 4545) The problem can be illustrated as follows (cf. ibid.).

Consider how we expressed a general density operator in the Bloch sphere representation as $\hat{\rho} = \frac{1}{2}(1 + \hat{\boldsymbol{\sigma}} \cdot \boldsymbol{r})$, where $|\boldsymbol{r}| < 1$ for a mixed state. Now the Euclidean vector $\boldsymbol{r}$ can be expressed as a linear combination of three unit vectors $\boldsymbol{n}_j$ ($j \in \{1, 2, 3\}, |\boldsymbol{n}_j| = 1$), drawn from an arbitrary (uncountably infinite) number of triplets of such unit vectors. Interpreting unit-length pointers in the sphere as pure states, this means that there is an arbitrary number of probability-weighted sums $\boldsymbol{r} = \sum_{j=1}^{3} p_j \boldsymbol{n}_j$, or equally, an arbitrary number of decompositions $\hat{\rho} = \sum_{j=1}^{3} p_j \frac{1}{2}(1 + \hat{\boldsymbol{\sigma}} \cdot \boldsymbol{n}_j) = \sum_{j=1}^{3} p_j \, |\boldsymbol{n}_j\rangle\langle\boldsymbol{n}_j|$. For instance we have $\hat{\rho} = \frac{3}{4} \, |\boldsymbol{n}_z\rangle\langle\boldsymbol{n}_z| + \frac{1}{4} \, |-\boldsymbol{n}_z\rangle\langle-\boldsymbol{n}_z| = \frac{1}{2}(\, |\boldsymbol{n}_+\rangle\langle\boldsymbol{n}_+| + |\boldsymbol{n}_-\rangle\langle\boldsymbol{n}_-|)$, where $\boldsymbol{n}_z$ is the unit vector pointing in positive z-direction in a Cartesian coordinate frame, and $\boldsymbol{n}_\pm := \frac{1}{2}\boldsymbol{n}_z \pm \sqrt{\frac{3}{4}}\boldsymbol{n}_x$. But the former decomposition seems to be a biased expression of ignorance as to whether $\boldsymbol{n}_z$ or $-\boldsymbol{n}_z$ is really the case (representing e.g. $|\uparrow_z\rangle$ and $|\downarrow_z\rangle$) whereas the latter seems to be an unbiased expression of ignorance as to the pertaining of $\boldsymbol{n}_+$ or $\boldsymbol{n}_-$ respectively.

Now the probability of finding $\boldsymbol{n}_z$ (or $\uparrow_z$) on $\hat{\rho}$ is $\langle\boldsymbol{n}_z|\hat{\rho}|\boldsymbol{n}_z\rangle = \frac{3}{4}| \, \langle\boldsymbol{n}_z|\boldsymbol{n}_z\rangle \, |^2 = \frac{3}{4}$, or equally (from the second decomposition), $\langle\boldsymbol{n}_z|\hat{\rho}|\boldsymbol{n}_z\rangle = \frac{1}{2}| \, \langle\boldsymbol{n}_z|\boldsymbol{n}_+\rangle \, |^2 + \frac{1}{2}| \, \langle\boldsymbol{n}_z|\boldsymbol{n}_-\rangle \, |^2 = 2 \cdot \frac{1}{2} \cdot |\frac{\sqrt{3}}{2}|^2 = \frac{3}{4}$.[7] The problem with this is that if the Born probabilities are here taken to reflect something objective, then in the first decomposition it looks as though the z-value was objectively *determined*, but subjectively only expected with a certain bias, due to (e.g.) ambiguities in the

[7]To see the latter, just consider that both $\boldsymbol{n}_\pm$ lie in the x-z plane, so the azimuthal angle φ is zero; a trigonometric consideration then gives $\cos\left(\frac{\pi}{2} - \theta\right) = \pm\frac{\sqrt{3}}{2} \Leftrightarrow \theta = \pm\frac{\pi}{3}$. Using (2.19), one obtains two matrices $\hat{\sigma}_{\pm\frac{\pi}{3}} \doteq \begin{pmatrix} \frac{1}{2} & \pm\frac{\sqrt{3}}{2} \\ \pm\frac{\sqrt{3}}{2} & -\frac{1}{2} \end{pmatrix}$, each with (obvious) eigenvalues ± 1 respectively and $+1$-eigenvectors $|\boldsymbol{n}_\pm\rangle \doteq \begin{pmatrix} \pm\frac{\sqrt{3}}{2} \\ \frac{1}{2} \end{pmatrix}$ (in the standard z-spin basis). To avoid confusion, note that $|\boldsymbol{n}_-\rangle \neq |-\boldsymbol{n}_+\rangle$, i.e. the two are *not* related as spin-up-$\pi/3$ and spin-down-$\pi/3$.

preparation method; whereas in the second decomposition, the probability of $\uparrow_z$ in state $\hat{\rho}$ would be *objectively* 3/4. Objective and epistemic probabilities are indeed "hopelessly blurred" in this example.

To conclude from this, though, that "pure state[s] *must* also be [...] state[s] of knowledge" (my emphasis—FB) seems to be a *non sequitur*. All that the previous argument shows is that mixed states, assuming that one thinks of them as reflecting incomplete knowledge in the first place, are extremely *ambiguous* representations of incomplete knowledge. And one could argue that dependent on context, one or the other decomposition of a given $\hat{\rho}$ is *preferred*: if one toys with a preparation method that is believed to be certain to prepare $|\boldsymbol{n}_z\rangle\langle\boldsymbol{n}_z|$, and one uses a computer program to smuggle in, in a deterministic but unknown fashion, $|-\boldsymbol{n}_z\rangle\langle-\boldsymbol{n}_z|$-states with presumed long run occurrence of $1/4$, then one might prefer the first decomposition; if one does something similar with $|\boldsymbol{n}_+\rangle\langle\boldsymbol{n}_+|$, $|\boldsymbol{n}_-\rangle\langle\boldsymbol{n}_-|$, and occurrences of $1/2$, the second decomposition is supposedly preferred. All of this, of course, depends on how one understands probabilities in the first place—but it is surely enough to show that the 'must' in Caves et al.'s claim is too strong.

Still, the arguments do make it *plausible* to (re)consider the possibility that (*all*) quantum states are 'epistemic' in a sense; but in *what* sense? Hidden variables, as endorsed by the ψ-epistemic approach, are not present in the Quantum Bayesian program, as we pointed out above, nor is a straightforwardly operational reading of quantum states in which they are only P/M-states. Since a subjectivist Bayesian understanding of probabilities is assumed though, and since in a subjectivist understanding of probabilities these quantify *beliefs* (in hypotheses or the occurrence of events), the quantum state, being considered as an 'extension' or 'liberalization' of a classical probability assignment, should be called a *doxastic state*.[8] At the same time, the view of QM provided here should be called *epistemological* rather than epistemic, since it is concerned with how belief is quantified and assigned in a specific sort of uncertain context, and with the question of how we make inferences about the future or the unobserved, in the light of a particular sort of empirical evidence. In contrast, the 'epistemic' view discussed in Chap. 4 was rather concerned with how QM 'disguises' a lack of knowledge about an otherwise pretty domestic outside world.

Is this all there is, then, to the Quantum Bayesian program, a reduction of quantum states to states of *belief*? The answer here depends on what one understands as 'the Quantum Bayesian program', since this subjectivist approach to probabilities and quantum states has spawned off many philosophical considerations that exceed the merely *formal*-epistemological part by far, and are not necessarily shared 'in full glory' by all of the original contributors. These philosophical considerations are generally subsumed under the label '*QBism*', and QBism is what interests us here in the first place. We will consider the informal side of QBism in the following section, but to round things off, we should now first take a look at a 'hurdle' that Quantum Bayesianism (in the formal-epistemological sense) takes, in making sense

[8]This point is also significantly clarified in Mermin (2012, p. 8).

of the notion of an *unknown quantum state*; a notion endemic to 'quantum state tomography' (cf. Nielsen and Chuang 2010, p. 336).

An obvious way for making sense of an 'unknown quantum state' within the Quantum Bayesian program, given that we have identified quantum states as doxastic therein, would be the denial of the principle that one knows one's own beliefs,[9] in analogy to the rejection of the KK-principle—that if one knows some proposition p, then one knows that one knows p, as defended e.g. by Hintikka (1962, p. 103 ff.)—by externalists and externalism-affine philosophers (e.g. Williamson 2002, p. 135 ff.). But of course denial of such a principle being *generally* valid does not mean that *all* of one's beliefs are unknown to oneself. And the sorts of beliefs supposedly expressed by quantum states do not really qualify as suitable candidates of doxastic states that one unknowingly entertains—they rather constitute *explicit judgments*.

The Quantum Bayesians indeed take a completely different route to making sense of 'unknown quantum states', in proving a quantum analogue of a *representation theorem* that goes back to de Finetti (1930). Following Caves et al. (2002b, p. 4543 ff.), we can summarize it as follows: Take a probability distribution $p(x_1, \ldots, x_N)$, which is *symmetric* under any *permutation* π (i.e. $p(x_1, \ldots, x_N) = p(x_{\pi(1)}, \ldots, x_{\pi(N)}), \forall \pi \in S_N$, S_N the permutation group for N items), and which satisfies $p(x_1, \ldots, x_N) = \sum_{x_{N+1},\ldots,x_{N+M}} p_{N+M}(x_1, \ldots, x_N, x_{N+1}, \ldots, x_{N+M})$, for arbitrary $M \in \mathbb{N}$, where p_{N+M} is equally permutation symmetric, meaning that p can be extended indefinitely in a permutation symmetric fashion. Such a distribution is usually called *exchangeable* (cf. de Finetti 1970, p. 215; Caves et al. 2002b, p. 4543). The theorem then states that any exchangeable distribution can be written as

$$p(x_1, \ldots, x_N) = \int_{\mathcal{S}_k} \mathrm{d}\boldsymbol{p}\, \varrho(\boldsymbol{p}) p(x_1) \cdot \ldots \cdot p(x_N) = \int_{\mathcal{S}_k} \mathrm{d}\boldsymbol{p}\, \varrho(\boldsymbol{p}) p_1^{n_1} \cdot \ldots \cdot p_k^{n_k}, \tag{7.4}$$

where $\boldsymbol{p} = (p_1, \ldots, p_k), \mathcal{S}_k = \left\{\boldsymbol{p} | \forall 1 \leq j \leq k : p_j \geq 0, \sum_{j=1}^k p_j = 1\right\}$, and $\int_{\mathcal{S}_k} \mathrm{d}\boldsymbol{p}\, \varrho(\boldsymbol{p}) = 1, \varrho(\boldsymbol{p}) \geq 0, \forall \boldsymbol{p} \in \mathcal{S}_k$. The probability density $\varrho(\boldsymbol{p})$ is interpreted as quantifying belief about the 'unknown probabilities' p_j; $p(x_1, \ldots, x_N)$ is interpreted as a prior probability, quantifying belief about the occurrences of outcomes or events.

This innocent looking result has the forceful implication that the assumed 'unknown probabilities' p_j can be taken to merely *appear* objective; conditionalizing $p(x_1, \ldots, x_N)$—which is specified solely by exchangability considerations of the observed events—on observed frequencies, $\varrho(\boldsymbol{p})$ will peak around some particular value (cf. Caves et al. 2002b, p. 4555 ff.), otherwise almost regardless of the form of $p(x_1, \ldots, x_N)$ and $\varrho(\boldsymbol{p})$, up to the fact that $\varrho(\boldsymbol{p})$ must also be non-zero everywhere on $\mathcal{S}_k$, even if arbitrarily close to it (cf. Fuchs 2002, p. 46). The point is that different agents, assigning possibly different priors, will come to an

[9]This is sometimes assumed as an axiom in doxastic logics (e.g. Kraus and Lehmann 1988).

agreement on the 'unknown probabilities', solely on the basis of obtained frequency data, modest restrictions (exchangeability) on their conception of the experimental setup, and no *ad hoc* exclusion of possibilities ($\varrho(\boldsymbol{p}) > 0$ everywhere). As Timpson (2013, p. 199) puts it: "homing in is coming to agreement."

This result is carried over to the quantum case by Caves et al. (2002b, p. 4546 ff.) by requiring that the matrix elements of an N-system density operator $\hat{\rho}^{(N)}$ be permutation symmetric, and that $\hat{\rho}^{(N)}$ be (permutation symmetrically) extendable *via* $\hat{\rho}^{(N)} = \mathrm{Tr}_M \hat{\rho}^{(N+M)}$. To effect the Bayesian updating, Caves et al. also introduce a *quantum Bayes rule*

$$\varrho(\hat{\rho}|D_K) = \frac{\varrho(D_K|\hat{\rho})\varrho(\hat{\rho})}{\varrho(D_K)}, \tag{7.5}$$

where D_K is some measurement result (with POVM element $\hat{D}_K$) for K measured systems ($\hat{\rho} = \hat{\rho}^{\otimes K} := \underbrace{\hat{\rho} \otimes \ldots \otimes \hat{\rho}}_{K \text{ times}}$), and

$$\varrho(D_K) = \int \mathrm{d}\hat{\rho}\, \varrho(D_K|\hat{\rho})\varrho(\hat{\rho}), \tag{7.6}$$

which corresponds to a continuous version of the law of total probability applied to density operators. Here $\mathrm{d}\hat{\rho}$ is a suitable measure, $\varrho(\hat{\rho})$ a probability density, and the integral is taken over the space (convex set) of density operators.

Using these ingredients, it is then proven that

$$\hat{\rho}^{(N)} = \int \mathrm{d}\hat{\rho}\, \varrho(\hat{\rho})\hat{\rho}^{\otimes N}, \tag{7.7}$$

i.e., that the exchangeable $\hat{\rho}^{(N)}$, assigned by an agent, can be used to replace the 'unknown' quantum state prepared on N systems. The quantum Bayes rule is here used to generate a state-update prescription, which for large enough K again enforces an agreement

$$\int \mathrm{d}\hat{\rho}\, \varrho(\hat{\rho}|D_K)\hat{\rho}^{\otimes N} \mapsto \hat{\rho}_{D_K}^{\otimes N}. \tag{7.8}$$

This is an important step for the Quantum Bayesians, since, as we noted, state-tomographic methods in QIT suggest that there should be such a thing as an unknown quantum state that can be found out.

Once more a worry arises at this point though: is the condition of *exchangability* not an expression of *equivocation* between different alternatives? I.e., do Quantum Bayesians not retreat from their decidedly subjective (even though empirically based) Bayesian stance in assuming exchangable priors? In fact, Williamson (2010, p. 19) believes it to be a "hitch [...] that under de Finetti's strict subjectivism, there is no reason to suppose that degrees of belief will be exchangeable." But we need to be careful here, since the theorem is of a conditional form: *if* one assigns an exchangable prior (quantum state) and is sufficiently admissive ($\varrho(\boldsymbol{p}) > 0$), *then* there will be agreement with others that also satisfy these constraints, after a sufficiently long time and 'repeated trials'. What the theorem provides is not so much a guide to suitable priors, but only a replacement of the notion of an 'unknown quantum state' by long-run agreement between certain kinds of agents under suitable circumstances. To that extent, the objection misfires;[10] but it still demonstrates that the scope of the quantum de Finetti theorem's implications is limited—an issue that we will return to a little below.

7.2.3 *Quantum States Do Not Exist, Nor Do Hidden Variables... But Then What Does?*

Let us grant for the moment that subjectivism about probabilities is somewhat plausible in the light of Dutch book theorems and de Finetti's representation theorem. And let us also grant that the Quantum Bayesians make a somewhat convincing case for carrying over the subjective Bayesian program to QM. Then of course this *prima facie* success and the fact that QM is such a remarkably successful physical theory together imply the need for a cautious clarification of many philosophical issues.

As mentioned earlier, the philosophical views that have arisen from the quantum Bayesian project are usually conjoined under the name *QBism* by their proponents, where the meaning of this term has been loosened from a mere shorthand for 'Quantum Bayesianism' to denoting multiple possible alternatives (cf. Fuchs 2010; Mermin 2013). It is clear that not all contributors to the original project share all of the views discussed in the followings to an equal extent; but we will nevertheless use 'QBism' or 'QBist views' as umbrella terms to collectively refer to them.

Given that we have identified quantum states as doxastic on the QBist account, the question 'beliefs about what?' offers itself. In fact, there is a well-known objection of unclarity by Bell (1990a, p. 34) against information-based interpretations, which he phrased in terms of a similar question: "*Whose* information? Information about *what*?" (emphasis in original) Now the question of 'whose beliefs?' is easily answered on the doxastic-QBist view: the beliefs of the agent who makes the particular assignment (cf. Fuchs 2002; Fuchs et al. 2014; Fuchs and Schack 2014). To the second question, 'beliefs about what?', the QBists' answer is: "the potential consequences of our experimental interventions into nature" (Fuchs 2002, p. 991), where by 'our' any single agent/observer/intervenor is meant, and by the "potential consequences" they mean "the content of *her* subsequent experience." (Mermin 2012, p. 8; my emphasis) A *measurement*, according to the QBists, "is *any* action an agent takes to elicit a set of possible experiences. The measurement outcome is the

[10] Author's note: I am indebted to Chris Timpson for helping me sort this issue out.

particular experience of that agent elicited in this way." (Fuchs et al. 2014, p. 749; emphasis in original)

What we end up with is a view of QM in which, as Fuchs (2002, p. 41) has emphasized, "QUANTUM STATES DO NOT EXIST", in the same sense in which probabilities do not on de Finetti's account. But neither, to recall, do hidden variables or hidden true states λ. Here is how the QBists think of such λs in the context of EPRB-correlations and violations of Bell-type inequalities:

> What the parameter λ expresses is a classical intuition that correlations in the experiences of agents in widely separated regions ought to find their explanation in correlations in conditions prevailing in those regions. In particular when the local experiences are mediated by the arrival of particles originating at a common source, λ is supposed to represent common objective features of those particles imposed on them at that source. These features affect the outcomes Alice and Bob experience. It is an important fact, surprising to one's classical intuition that the correlations in Alice's and Bob's outcomes cannot be accounted for in this way. But this does not mean that anything in Alice's experience is influenced by Bob's choice of setting, or vice-versa. The variable λ is nothing more than a version of the discredited EPR elements of reality. For a QBist the nonexistence of such objective facts-on-the-ground as λ no more implies nonlocality than does the nonexistence of elements of reality in the original EPR argument. (Fuchs et al. 2014, pp. 752–753)

This is somewhat in agreement with many of our findings in Chap. 4 and interlude II, that one can avoid the consequences of EPR's, Bell's, and the KS theorem if one disallows certain (abductive) inferences to a (hidden, and sometimes common) cause for (correlated) measurement outcomes.

As we can see, Fuchs et al. (2014) also refer back to the original EPR paper here, and EPR's famous 'reasonable criterion for reality' that we briefly discussed in Sect. 4.3.3. This criterion of EPR, to recall, is of conditional from and introduces a prediction "with certainty (i.e., with probability equal to unity)" as a sufficient condition for the assumption of an "element of physical reality". But, the QBists argue, "probability-1 (or probability-0) judgments are still judgments, like any other probability assignments[...]." (Fuchs et al. 2014, p. 752)

To support their views on probability, Fuchs et al. refer to Hume's (1748, p. 112 ff.) *problem of induction*, anticipated already by the Pyrrhoneans in antiquity (cf. Sextus Empiricus, §204), the problem of how to *justify* the inductive inferences we constantly make. To date, no all-agreed and unproblematic solution to this problem exists,[11] and one might suspect that none can be found.

[11]Cf. Schurz (2014, p. 80 ff.) for a general overview of purported solutions and their problems. Schurz's own proposal does provide an optimality-based justification for induction on the object-level under certain epistemological background assumptions—which, however, means that object-induction might still do pretty baldy but that (even under the respective, favorable circumstances) we just have no better alternative. Arnold (2010), moreover, assesses some general limitations to Schurz's solution when applied to the meta-level of prediction methods, which is where cleverly adapted (meta-)inductive practices are proven, for a whole range of possible scenarios, to have the same long-run success rates as the best of all other epistemically accessible prediction methods (e.g. Schurz 2009).

What is compelling about this appeal is that induction of course lies at the heart of all reasoning from observed evidence to general or predictive claims invoking regularities, be they statistical or particulate (cf. Carnap 1950, p. 207 ff.) And induction *is* an *uncertain* inference with an as yet shaky status of epistemological justification, no matter how dearly we hold it and how much we rely on it. On this line of reasoning even probability-1 assertions—unless, maybe, devoid of content *qua* being about tautological or analytical claims—, as they are appealed to in the EPR argument, cannot be more than an expression of *subjective* certainty on the basis of previous evidence.

Now while inductive skepticism *in general* is in good support of the QBist's views, in the present context the appeal to *in*duction seems misguided. Inductive skepticism ultimately questions the *uniformity* of nature and perceivable events (in the future or as regards synchronously unobserved cases), or rather the justification for our beliefs in it. In the EPR scenario, uniformity or regularity are not at stake; the doubted inference is from something observed (the momentum of the particle on one side) and a 'known law' (momentum conservation) to something unobserved (the momentum of the particle on the other side)—an instance of *ab*duction rather than induction, as we had already pointed out in Chap. 4.

All of this connects to long-standing debates in the philosophy of QM. As a matter of historically well-established fact, Bohr was quite baffled by the EPR paper. His assistant Rosenfeld (1967, p. 142) famously described it as an "onslaught" that "came down upon [them—FB] as a bolt from the blue", and reported how Bohr would abandon all other work to concern himself immediately and exclusively with EPR's argument. How deep and what exactly the impact on Bohr's philosophical views ultimately was is a matter of considerable debate (e.g. Beller and Fine 1994 vs. Halvorson and Clifton 2002b; and cf. Whitaker 2004 for a nice overview and reassessment). But Bohr certainly retained a substantial amount of his previous beliefs, and he apparently even believed to have answered the incompleteness problem in a satisfying way.

His main line of criticism towards EPR, justifying this retention of beliefs, was that there was, in his opinion, "an essential ambiguity" in EPR's reality criterion, more specifically in the phrase "without in any way disturbing", since any measurement on one of the two entangled particles would of course non-negligibly (and 'uncontrollably') alter its (*quantum*) state, and thereby exert, in Bohr's opinion, "an influence on the very conditions which define the possible types of *predictions* regarding the *future behavior* of the system." (my emphasis—FB) These he believed to "constitute an inherent element of the description of any phenomenon to which the term 'physical reality' can be properly attached [...]." (Bohr 1935, pp. 696 and 700)

Bohr's elaborations are somewhat confusing, e.g. his appeal to the word "influence" while allowing "no question of a mechanical disturbance [...]." (p. 700) We here 'choose to' read the comments as supplying a stronger *necessary* condition for the assumption of 'elements of reality' in terms of future predictability, maybe in the sense of re-identifiability of 'the same thing'. While this may be an imposition on Bohr, we hence understand him here as disencouraging an abductive inference from observed phenomena to something 'real', and as seeing his necessary criterion for 'physically real phenomena' violated in the EPR scenario—thereby ultimately embracing some sort of view of reality that EPR had deemed 'unreasonable'.

The QBists' argument against EPR, in comparison, is much more straightforward but ultimately to the same effect: they hold it to be an "unwarranted assumption that probability-1 judgments are necessarily backed up by objective facts-on-the-ground—elements of physical reality[...]." (Fuchs et al. 2014, p. 752) We emphasize again that they too effectively argue against (the universal applicability of) *ab*duction here as well ("objective facts-on-the-ground"), not against induction. If our analysis is correct, then both Bohr and the QBists ultimately attack the same point as did we in Chap. 4, albeit each on slightly different grounds.

David Mermin (2014a, p. 422) has pointed out that *Schrödinger* once commented, in a letter to Sommerfeld, that "in Quantum Mechanics, statements about what 'really' is, statements about the *object*, are forbidden, they only treat of the *relation* object-subject—and that obviously in a much more decisive sense than is the case, after all, in *any* description of nature." (Schrödinger 1931, p. 490; emphasis in original; my translation—FB)[12] Intriguingly, when one consults the original letter, one also finds that Schrödinger, building on discussions he must have had in Berlin and which possibly included Einstein (cf. Schrödinger 1931, p. 490; von Meyenn 2011, pp. 442 ff.; Moore 1994, p. 175), reports a kind of crude EPR-example, basically (though not explicitly) involving entanglement with *macroscopic* degrees of freedom—an '*EPRS-Cat*', if you will: Bouncing a photon in a highly localized state off a mirror with precisely prepared momentum (zero), one can use the mirror—which will receive twice the original momentum of the photon by momentum conservation and hardly experience a change in position due to its comparatively large mass—to indirectly measure *either* the position *or* the momentum of the photon to accuracies jointly forbidden by the uncertainty relations.

Since one is at liberty to chose between these two measurements at will, even though one cannot perform both at the same time, both quantities should *already exist*—unless one were to allow that the photon would only assume a specific momentum or position in virtue of the measurement performed on the mirror (action at a distance). The latter option Schrödinger rejected out of hand, but the possibility of the simultaneous existence of position and momentum he also considered "too strict and paradoxical" (Schrödinger 1931, p. 490; my translation—FB) in the light of all that was already known by the time.[13] So he could only retreat, as

[12]German original: "in der Quantenmechanik sind Aussagen über das, was 'wirklich' ist, Aussagen über das Objekt, verboten, sie handeln nur von der Relation *Objekt-Subjekt*—und zwar offenbar in einem noch viel einschneidenderen Sinn, als dies schließlich von *jeder* Naturbeschreibung gilt."

[13]German original: "Man möchte darum schließen, daß das Lichtquant jederzeit einen ganz bestimmten Ort und einen ganz bestimmten Impuls besitzt—eine Auffassung, die wir doch eigentlich längst als zu hart und paradox verlassen haben."

an "emergency decree" (ibid.; my translation—FB),[14] to QM being most radically about the subject-object relation.

Now the QBists bite the same bullet. But for them, this is an *insight*, not an emergency decree:

> For the QBist, there is [...] a split [...] between the world in which an agent lives and her experience of that world. [...] Vagueness and ambiguity only arise if one fails to acknowledge that the splits reside not in the objective world, but at the boundaries between that world and the experiences of the various agents who use quantum mechanics. (Mermin 2012, p. 8)

We can see how QBism nicely connects to the debates that have been around since the advent of QM, and the outlined connections to Bohr put it in the proximity of the Copenhagen tradition. Certainly though, QBism is much clearer on many issues regarding the role and meaning of the quantum state than is the elusive Copenhagen interpretation. And indeed, it is quite helpful in dealing with some of the puzzles raised by QM. Consider, for instance, Wigner's friend again:

> in QBism Wigner's Friend is transformed from a paradox to a fundamental parable. Until Wigner manages to share in his friend's experience, it makes sense for him to assign her and her apparatus an entangled state in which her possible reports of her experiences (outcomes) are strictly correlated with the corresponding pointer readings (digital displays, etc.) of the apparatus. (Mermin 2014b, p. 8)

Still, there are many open ends to QBism on philosophical grounds. The quantum state, as we have outlined above, is doxastic in QBism. As such it quantifies what is believed by the assigner about the consequences of her interventions into nature, and *only* that. QBists believe this to include situations of '*maximal* knowledge', meaning that the quantum state may contain 'all there is to say' about an investigated system for a respective assigner. And they also reject the assumption of hidden variables λ as a mere appeal to unnecessary 'classical intuitions'. But *prima facie* this creates quite a dilemma: if the quantum state is not the true state of a given system, and if no hidden variable model delivers the true state of the system...and if there is no foreseeable way in which we can otherwise talk about the true state of the system...*then there is just not true state of the system!*

7.2.4 *What is the QBists' Epistemology?*

Let us try to make some sense of all that has been said so far. We took the fact that pure quantum states codify maximal but incomplete knowledge in QBism to mean that they may contain all there is to say about a system, from the perspective of some assigner. This cannot be quite right though, as there are of course many additional things to say about any kind of system in a *non-quantum* vocabulary. So let us take

[14] German: "Notverordnung"

a pure quantum state *plus* everything there is to say about a system *without* using QM, to specify a doxastic state of some agent, codifying 'maximal but incomplete knowledge'. Let us call this, in an abuse of terminology introduced by Healey (e.g. 2012d, p. 760), the '*quantum state assignment*$^{+}$' (short: QSA^{+}). Then the claim that there is no true state of the system in spite of everything contained in the QSA^{+} may be taken to mean that in spite of the QSA^{+}, we still lack any *description* of that part of mind-independent reality, but that there is also *nothing closer* to a description than the QSA^{+}.

Hence, if quantum states are non-descripitve of the underlying reality, there may be just *no description whatsoever*. That is: on an interpretation such as QBism, we would have to understand the remarkable success in *using* a theory like QM to imply a limit of *conceivability* of the external world, a limitation of our capacities to refer to and conceptualize it in certain respects. At the same time, QBists, in some of their writings, tend to think of themselves as 'realists' (e.g. Fuchs 2002, 2016) and endorse a notion of "the world" on which actions are performed (e.g. Fuchs et al. 2014; Fuchs and Schack 2014). But how can it fit with any reasonable kind of 'realism' that for a whole range of situations we lack any tools whatsoever to describe and conceptualize them? This clearly needs some spelling out.

A possible reply would be that the foregoing does not imply *metaphysical* anti-realism, when 'metaphysical realism' is read in the weak sense; one may still grant that there is such a thing as a mind-independent world. But it *does* commit one to a certain degree of *scientific* anti-realism, in that we are thereby denying the semantic condition, at the very least for the broad domain of applications requiring a 'quantum treatment'.

A lot about the QBists' views is also revealed in the two quotes by Fuchs et al. and Mermin on pages 306 and 309 in this document. First of all, we notice the primacy of the concept of 'experience', which is taken as a "primitive concept [...]." (Fuchs et al. 2014, p. 749) Experience is also used to explain how and when correlations come into play, namely, at the interface between two—*time-like* separated—experiences of a single agent. In particular, the correlation in the EPRB scenario comes about when Alice and Bob compare their 'measurement results', i.e. report to each other what they have 'observed' or 'experienced' (thereby inducing new experiences in one another). Since no single agent can move faster than light, all her experiences will be related to one another in a time-like fashion, so even the correlations involved in EPRB scenarios are not 'created nonlocally', the argument goes (cf. Fuchs et al. 2014, pp. 750–751).

A second point to note is the strong affinity to dispensing with 'classical intuitions', those which we identified in Chap. 4 to be the key motivation for designing ψ-epistemic models. QBists bite the bullet and accept that our usual explanatory practices of supplementing hidden causes to account for aspects of our experience are not ubiquitously applicable.

And thirdly, we notice the particular view of *science* indicated in the above considerations; correlations "*cannot* be accounted for" (Fuchs et al. 2014, p. 753; my emphasis – FB) by stipulating further elements of reality, λ, and the central concept of experience is even treated as "*fundamental* to an understanding of *science*." (Fuchs et al. 2014, p. 752; my emphasis – FB) According to QBism, scientific investigation is thus not about 'nature', 'mind-independent reality', 'the external world', or whatever you prefer to call it, at least not directly. Instead, the QBists explicitly side with Bohr to the extent[15] of believing that "in our description of nature the purpose is not to disclose the real essence of the phenomena but only to track down, so far as it is possible, relations between the manifold aspects of our experience." (Bohr 1934, p. 18)

Thus presented, there are decisively *empiricist* elements in this view of science. But anyone with a background in the philosophy of science knows how difficult it is to spell out and defend such an empiricist view. As is well known, *logical* empiricism was the dominating philosophy of science in the early twentieth century and the dominating version of empiricism deemed capable of providing an analysis of scientific practice. Logical empiricism departed from traditional empiricism in embracing the *falliblity* of *observation sentences* such as 'there is a table' (cf. Schurz 2014, p. 6); but such sentences were nevertheless believed to "have an epistemically favored role within the whole system of discovery [...]." (ibid.) Its most extreme form, logical *positivism*, endorsed a *verificationist* account of *meaning*, i.e. that "a synthetic proposition [...] should, in principle at least, be conclusively verifiable" (Ayer 1936, p. 147), where the (im)possibility of such a verification was viewed as a demarcation between "literal sense and nonsense[...]." (ibid.) This means, in other words, "that the meaning of a proposition was its mode of its verification[...]." (Kenny 2007, p. 58) Verificationism, however, was easily seen to be problematic, since the status of the verification principle *itself* as *meaningful* appeared doubtworthy by its own lights (cf. Kenny 2007, ibid.).

A more general characteristic trait of early logical empiricism was a form of *reductionism*, endorsing that terms of theories *not* referring directly to sense experience should be reducible, *via* chains of definitions, to terms directly referring to sense experience (e.g. Schurz 2014, p. 6). By the 1950s, however, the project of reductionism was largely abandoned *qua* lack of adequate executability (e.g. Carnap 1956, p. 48), and with it (arguably) the core of logical empiricism.

Van Fraassen's (1980) younger *constructive* empiricism, certainly the most widely discussed form of empiricism (or in general: anti-realism) in modern philosophy of science—hence worthy of a brief review, in the present context—is of a different flavor. First of all, van Frassen views the *aim* of science to be the point of contention between scientific realism and the position he has in mind: according to scientific realism, van Fraassen (1980, p. 8) believes, science (the entire enterprise as a whole) "aims to give us, in its theories, a literally true story of what the world is like; and acceptance of a scientific theory involves the belief that it is true." (emphasis omitted) This excludes ("literally") construals of theories according to which the terms have to be interpreted correctly first, as e.g. religious scripture does according to most modern theologians. And it is considerably weak ("aims to"),

[15]Cf. Mermin (2014b) and Faye (2016) for further discussion of differences and commonalities between QBism and Bohr's (or 'Copenhagen') views.

so as to target an otherwise broad class of 'realisms'. According to van Frassen's own constructive empiricsm, however, "[s]cience aims to give us theories which are *empirically adequate*; and acceptance of a theory involves as belief only that it is empirically adequate." (his p. 12; my emphasis—FB)

Empirical adequacy of a theory *T* is spelled out here as *T* being *true* w.r.t. all *observable* phenomena (ibid.), and as a "rough guide" to what is meant by 'observable', he proposes the following (his p. 16):

> *X* is observable if there are circumstances which are such that, if *X* is present to us under those circumstances, then we observe it.

Thus 'observable' is construed here as an *indexical* concept ("us", "we"), the index referring to a specific, possibly time-variant, *epistemic community* (his p. 18). And some emphasis is laid on the possibility of observing by the *unaided* senses, so bubble-chamber 'observations' do not count as genuine observations (of elementary particles) in van Fraassen's sense (cf. his p. 17).

In contradistinction to logical positivism and (other) kinds of semantic anti-realism (e.g. Dummett 1982), van Frassen hence holds that scientific theories are *capable* of being true, in *all* areas, even when endorsing concepts not referring to observable entities; it is just that we *cannot be certain* as to their truth in *those* domains, and we are not committed to such, in pursuing science. But while constructive empiricism thus manages, by its comparative moderateness, to get around a whole host of difficulties that logical empiricism did not (cf. Ladyman 2000, pp. 840–845 for a short survey), there still remains a somewhat non-empiricist—*qua modal*—flavor to the use of the notion 'observ*able*' and the general vocabulary crucially relied on by van Fraassen (cf. in particular Ladyman 2000, p. 849 ff. for detailed criticism). These difficulties have been answered to some extent (cf. Monton and Van Fraassen 2003), but on the pains of appealing to elements not 'endemic' to constructive empiricism (e.g. Stalnaker-like semantics for counterfactual conditionals; cf. Monton and Van Fraassen 2003, p. 410) and with several rebuttals by Ladyman (2004).

Our general point, again, is that an empiricist philosophy of science is not easily facilitated, even in spite of the caution exhibited in van Fraassen's 'constructive' version. The QBist view, however, does not seem to be a variant of either logical- *or* constructive empiricism, but rather borrows elements from both. To the extent that one important (mature, well-confirmed) scientific theory (QM) is *not* viewed as *capable* of being true (of an external reality) QBism is closer to *positivism* rather than to *constructive empiricism*; the quantum state is a means of quantifying beliefs, and *not descriptive*. Beyond that, however, QBism neither seems to promote verificationism nor a specifically reductionist agenda regarding scientific *vocabulary*. All that is being claimed is that 'experience' and (cor)relations therein are the subject matter of science—not the goings on in mind-independent reality. On the other hand, Fuchs (2010, p. 21) wishes to include "everything experience*able*" (my emphasis—FB) into the scope of science, which brings him 'dangerously' close to the problems encountered by van Fraassen.

Additionally, the QBists seem to be committed to a degree of *instrumentalism*, in viewing quantum states as non-descriptive but useful for making predictions (cf. also Bub 2016, p. 232, on this issue), despite their reluctance to the label (cf. in particular Fuchs 2016, pp. 2–3). QM, to wit, is called "a user's manual" (Fuchs 2010, p. 9), and "a single user theory" (Fuchs and Schack 2014, p. 3), which sound quite instrumentalist.[16]

Finally, we can identify another label that the QBists eschew even though their position seems to be characterized quite well by it, namely: *solipsism*. QM is, after all "a *single user* theory" on their account (cf. also Norsen 2016, p. 215 ff. on this point).

Possibly *ambiguity* is to blame for their reluctance to at least the label 'solipsism': Introductory texts on epistemology (e.g. Borst 2010) tell us that we should distinguish between *metaphysical*, *epistemological*, and *methodological* solipsism, only the first of these being the single-mind pendant to Berkeleyan idealism, i.e. the thesis that *all that exists* is *one mind*. Epistemological solipsism, in contrast, rather characterizes the position that only the own mind's existence is *certain*; and methodological solipsism is usually characterized as either the program of *constructing* 'the world' and 'other minds' out of the experiences of the inquiring self, a position quite often associated with Carnap (e.g. Carnap 1936, p. 423 ff.), or (more weakly) as a general method of inquiry, *starting* from the inquiring subject, as was the case at some stage in Descartes' (1642) famous meditations.

Now QBists have it that "any user's own experience constitutes all of the raw material out of which she constructs her world." (Fuchs et al. 2014, p. 753) Like it or not, this is at least a statement of methodological solipsism. But once more trouble transpires from philosophical conduct, as was the case with empiricism. Putnam (1982, p. 10) once noticed that "a methodological solipsist [...] kindly adds that you, dear reader, are the 'I' of this construction when you perform it [...]." (emphasis omitted) This he found "ludicrously incompatible" with the fact that from the methodological solipsist point of view, "your experiences [...] are a construction out of your bodily behavior, which [...] is a construction out of *my* experiences" (ibid.; emphasis in original) and that "if it's really true that the 'you' of the system is the only 'you' he can understand, then [that] the 'you' he addresses [...] cannot be the empirical 'you' of the system [...] is *unintelligible*." (p. 11; emphasis in original)

The force of Putnam's argument is unclear though, since it is equally unclear that the methodological solipsist must embrace an *unintelligibility* stance towards that which is not constructed out of her experience: She can certainly entertain the belief that what she has constructed (other minds) is 'paralleled by' something which has an existence independent of herself, and is not *only* a construction. This may commit her to a degree of (weak) metaphysical realism after all, at least about other minds—but this is why it is called *methodological* solipsism in the first place.

---

[16]Somewhat ironically, Fuchs (2010, p. 7) also happily refers back to Pierce's claims of theories as *instruments*, despite his strong reluctance to the label.

Still, a methodological solipsist is certainly much closer to the epistemological and metaphysical solipsist than is the full blown realist: Accepting, as a methodological step, an epistemological basis in one's own mind and experience, there may be a remainder of doubt about the existence and content of other minds and of the external world, even after one has dared the extra step of imagining that the construction is indicative of or paralleled by something external.

So the QBists are metaphysical realists to the extent that they believe in nature responding to our interventions. And even though they dislike most of these labels, they are committed to a degree of empiricism, as they place experience at the center of scientific inquiry; to a degree of scientific anti-realism, as they deny the semantic condition of scientific realism w.r.t. quantum states and view science in general not to be about the "real essence of the phenomena"; to a degree of instrumentalism, as they find quantum states to be useful even if non-descriptive; and to a kind of methodological solipsism, as they view construction out of the experience of a single 'user' to be at play, when we form a vision of 'the world' or 'other minds'.[17]

The crux of the matter is this: While QBism certainly exceeds the Copenhagen tradition in clarity in many respects, it presents, as it stands, not a well worked out interpretation either, but rather a collection of intriguing ideas—not entirely unlike the 'Copenhagen interpretation' itself. To highlight the problems more clearly, we should make use of our desiderata and adequacy criterion in this context as well.

QBism *does* satisfy the MAC 'by default', so to speak: QM is simply viewed here as a quantitative description of expectations, and quantum states may hence be spontaneously updated when new information is available. No outcome problem arises since the quantum state does not represent the conditions of a system, so there is no transition from multiple outcomes to just one. But QBism does *not* satisfy desiderata of ontological clarity to the extent that we had deemed this still possible: the subject-object relation is not being sufficiently specified. And, as we shall see a little below, the appeal to 'consequences of interventions' *as crucial* makes this relation all the more obscure.

But, more importantly, neither does QBism really satisfy the DEC: Point (ii) is mostly fine: there basically *is no* 'classical world', according to QBism, and QM is supposed to be applicable, in principle, to *any* situation (e.g. Fuchs et al. 2014, p. 750). So no switch from 'quantum' to 'classical' either, quantum states can be quite generally used to quantify expectations about consequences of interventions. Still, when and how do we come to apply a suitably 'classical' description anyways, in terms of which we describe these consequences of interventions? This bears on point (i) of the DEC: there is not too much of an explanation being offered, at least not in any systematic way, of how we can make sense of *objects* we *treat* 'classically' in many circumstances—i.e. when we do not assign quantum states *de facto*—, given the split between a user's experience and the world. And part (iii) of the DEC appears to be unfulfilled in QBism as well: whence, say, the role of unitary operators,

[17]Cf. however Fuchs (2010, p. 20) for a quite different characterization of the project (which is partially his).

appealed to in so many places e.g. in Fuchs (2002) or Caves et al. (2007)? And how should we understand decoherence?

7.3 How Much Subjectivity Is There and How Much Is Good?

7.3.1 *Objectivism About Probabilities and Objectivism About Quantum States*

We have identified many difficulties in the epistemological basis of QBism above, but there are also difficulties at a systematically prior stage. QBism relies on Quantum Bayseianism, Quantum Bayesianism relies on (some sort of) subjective Bayesianism. The plausibility of the latter has an impact on the former two. How plausible are the arguments for subjective (Quantum) Bayesianism, given in Sect. 7.2.2?

In fact, many of the plausibilizing arguments are ultimately rather unsatisfying. Recall that pure state-assignments were argued to be rationally compelling in the case of maximal information, in virtue of a loss of Dutch book-coherence otherwise. But the Dutch book argument proceeds from a $p(A) = 1$ assignment, in the case of 'certainty', and concludes that one would face a sure loss if one would choose a different probability assignment. What could enforce such an assignment? What could justify, rationally compellingly, the belief that one would otherwise face a sure loss? If a subjectivist account of probability is taken seriously, then the answer must be: *nothing!*

Probability assignments are understood as expressions of belief only, and one should be at liberty to *remain uncertain* ($p(A) < 1$). Judging that there could be a case where $p(A) < 1$ leads to a sure loss because it is *certain*—for everyone—that A will happen means to abandon the entire approach on the meta-level: There can be no such judgment, or at least not one that *everyone* is *committed* to. If the statement of the theorem is intended to be 'purely semantic', on the other hand, i.e. if it means that a pure state assignment encodes *subjective* certainty, then the most that the theorem would demonstrate is that 'I am certain that A is the case' should be translated into 'I fear that I will loose a bet if I do not bet on A occurring'. In any case, the Dutch book argument either cannot get off the ground or merely establishes something quite weak, as it otherwise appears as much question-begging as do the arguments of opponents to subjectiv Bayesianism(s) that appeal to 'objective chance functions' etc.

Much depends on this Dutch book argument though, as we have seen (subsequent arguments build on it), and so with it the overall appeal of the Quantum Bayesian project is put in question. And note that we already provided critical remarks about another plausibilizing argument, namely the inference from multiple decompositions of mixed states to the subjectivity of Born probabilities in Sect. 7.2.2.

What about the quantum de Finetti theorem, the result that different agents who assign initial states that satisfy certain (rather weak) structural constraints, and who have an 'open mind' ($\varrho(\hat{\rho}) > 0$ on the entire space of density matrices), will converge in their opinions, i.e. will agree on some 'unknown' quantum state, given a large range of observations? Now this result still depends on a notion of "*frequency data*" (Caves et al. 2002b, p. 4554; my emphasis – FB), and what precise state the different agents will agree upon will depend on those data.

While we acknowledged that the statement of the theorem is merely of the conditional form 'if you accept states of this and that sort, you will come to an agreement with others', a satisfying *explanation* of the practice of 'finding out' some particular 'unknown' quantum state, which the theorem is supposed to ensure, *does* depend on two quite non-subjective sources: (a) *intersubjective agreement* on what *sort* of quantum state to assign and on which states are at least possible, and (b) an *external source* that provides the appropriate frequency data. That makes the entire approach much less subjectivist than it appeared to be in the first place, since even Williamson (2010, p. 24) has it that his "apparently unparsimonious appeal to chance is in fact eliminable in favour of talk of indicators of chances such as *sample frequencies and symmetries*." (my emphasis—FB)

Why, however, would the theorem be of interest, other than for *sociological* reasons, if one was not interested in how we *can* come to an agreement? Why, in other words, would a mere conditional statement about the agreement of certain agents be interesting in the context of questions about the nature of physical reality? The worry here is similar to our concerns about a false explanandum in the Deutsch-Wallace approach to Born probabilities in the MWI: We are interested in *why those frequencies occur*, why long-run evidence *suggests* certain state assignments to us, not just under what conditions different agents may come to agree on how to *take track* of frequencies. The quantum-de Fintetti theorem targets only the latter aim.

Criticism towards the plausibility of certain similar arguments of the QBists is also uttered by Stairs (2011, p. 161 ff., in particular). Stairs argues that just because different agents may assign different quantum states for the same system, this does not imply any sort of (radical) subjectivism. For suppose, e.g., that Alice and Bob share an electron pair in the singlet state, and Alice measures spin up along the z-axis of some agreed coordinate frame, assigning $|z+\rangle$ to her electron in consequence. She would then supposedly also, in virtue of her measurement and conditional on the assumption that Bob also measures for z, assign the state $|z-\rangle$ to Bob's electron. But:

> suppose that Bob did not measure z-spin at all, but measured spin in direction d, skew to z, and got the result -1. Bob will assign the state $|d-\rangle$. Do he and Alice disagree? Not at all. They are *making use of different information*. (Stairs 2011, p. 164; my emphasis – FB)

Just because the 'best' or preferred quantum state assigned by different users of QM *to the very same system* may differ from situation to situation, this does not mean that there is no such best or preferred state *relative to that respective situation*. It merely means that each state is *subject relative*, in the sense of being sensitive to the totality of an individual agent's *epistemic conditions*.

Stairs (2011, p. 164–165), moreover, thinks that

> Alice's state assignment does, of course, go with a subjective probability—a willingness to bet conditionally, if you like. Should she find out that Bob actually measured z-spin, she will be certain that the result was -1. But Alice, objectivist that she is, will add that this credence reflects something about the world: the objective probability that Bob found result -1, conditional on Alice and Bob both measuring z-spin and Alice getting result $+1$, is one.

As much as the QBist view on QM lifts the 'paradoxical' character from Wigner's friend or Schrödinger's cat, it bestows a paradoxical character on EPRB-correlations. Why on earth *would* Alice and Bob (almost) always find opposite values, when they get together and compare their results on runs in which they happened to measure for the same direction? Why if not due to a 'rigidity' or 'recalcitrance' *in nature*? However, we have claimed multiple times that this recalcitrance is not well explained *causally* (by abducing a hidden common cause) and Stairs (2011, p. 165) basically agrees with this:

> when Alice makes her probability-one claim about Bob's qubit, she does not need to infer pre-existing properties nor attribute counterfactuals. On the contrary, if she wants to square her objectivism with causal locality, those are exactly the things she should not do.

But how *else* could we 'explain' the apparent recalcitrance? This will be another headache, deferred at this point to Sect. 7.4. At the same time, we shall then have to say something about the involvement of *counterfactuals* there.

The objectivism of Stairs is, in the first place, an objectivism about the *probabilities* and *correlations* predicted by QM. Related views are expressed by Friederich (2015, p. 79), who thinks that quantum probabilities are "*objective* inasmuch as they are fixed as soon as all relevant features of the epistemic conditions of the agents ascribing them are made *explicit*." (emphasis in original) His main reason to reject the radically subjectivist views at the heart of Quantum Bayesianism is that "quantum Bayesianism denies the existence of a determinate answer to the question of *which* observable is measured[...]." (Friederich 2015, p. 62; emphasis in original)

Assuming that one also accepts the Lüders rule as an 'objective feature' of the measurement, the argument to this conclusion is simple (cf. Fuchs 2002, p. 39): If we resolve an observable as $\hat{Q} = \sum_j q_j \hat{P}_j$ and perform a selective projective measurement on it, then the *objective* final state, according to Lüders' rule, is one of the $\hat{P}_j$, in contradiction to the very foundation of Quantum Bayesianism. So it cannot be an objective feature that Q was measured. This is reason enough for Friederich (2015, p. 66 ff.) to regard Quantum Bayesianism as (at least) an implausible account of *physical practice*.

Friederich (2015, p. 75 ff.) additionally struggles with the question "Probabilities of what?", and while this is only half the battle and not at all in contradiction to the QBists allusions to future experiences, it is certainly a significant point of clarification that the claims at stake in probability assertions provided by the

Born rule are *non-quantum magnitude claims* (NQMCs), as pointed out by Healey (2012d, p. 740) and embraced by Friederich (2015, ibid.).[18]

Recall how our probability expressions are of the form $\Pr^{\psi}_{O}(o_j) = p$, or more restrictively and realistically, $\Pr^{\psi}_{O}(O \in \Delta) = p$. The 'non-quantum' part is here the magnitude claim $O \in \Delta$ or $O = o_j$, which should be read, to recall, as $O(S) = o_j$ or $O(S) \in \Delta$, with S either variable or fixed, meaning that the value of variable (observable magnitude) O on some system S lies in range Δ or is o_j.

Now the *quantum* part to this is the fact that these probabilities are given by the Born rule and that "[t]here is no *joint* probability space for the statistical relations specified by the quantum algorithm for two *incompatible* magnitudes." (Bub 1974, p. 35; my emphasis – FB; cf. also Healey 2012c, p. 14) In other words: While QM *positively advises us* to assign particular degrees of belief to NQMCs in specific contexts, it *negatively advises us* to refrain from asserting certain kinds of ('incompatible') NQMCs at the same time.

To a large extent, this is the upshot of Healey's (2012d) views on QM, where

> quantum probabilities given by the Born rule do not describe any natural property of the system or systems to which they pertain, or of any other physical system or situation: nor is it their function to *describe* any actual agent's state of belief, knowledge, or information. Their function is to *offer advice* to any actual or hypothetical agent on the extent of its[19] commitment to [NQMCs][...]. (p. 735; my emphasis—FB)

The conditions under which a state may be assigned are called its *backing conditions*, those about which it offers advice its *advice conditions* (cf. Healey 2015, p. 1). Moreover, Healey, like Stairs and Friederich, embraces a form of *objectivism*, but his objectivism is quite strong and directly concerned with quantum states:

> Knowing a state's backing conditions, one is justified in assigning that state: but one would be *warranted* in assigning the state whether or not one knew these conditions, just as a test result may warrant a diagnosis whether or not the doctor knows about it. (Healey 2015, p. 4; emphasis in original)

We may read this as expressing the conviction that the advice a quantum state offers is 'dictated' by an external source, and it is hence not entirely up to the agent which quantum state to assign: Given that one is interested in having long-run future success, there is a *preferred* quantum state that is to be assigned in a given physical situation. Notably, Healey thus also captures the upshot of Stair's earlier mentioned comments on possible divergence in quantum state assignments by different agents: "Any application of a quantum model is *perspectival*—it is from the perspective of a hypothetical, physically situated, agent." (Healey 2015, p. 2)

---

[18] Why 'only half the battle'? Because immediately the question arises: 'The probability of... these NQMCs being *true*? *Appropriate* to future *experience*? *Assertible*...?'

[19] 'It' here refers to the *agent*. Healey wishes to include also non-conscious 'agents'.

7.3.2 *Healeyan Views and the Involvement of 'Meaning'*

Healey (2012d) calls his approach *pragmatist*, not least because he accepts pragmatist views on *semantic content*. His specific pragmatist stance towards meaning or semantic content has its roots in Carnap (1937, p. 42), where the *logical* content of some statement *s* or a class thereof is identified, within a specific formal language, as the "class of non-analytic sentences [...] which are *consequences* of [*s* or the class—FB] respectively [...]." (my emphasis—FB) Carnap (1937, p. 27) also had it that by so called "transformation rules, [...] we determine under what conditions a sentence is a *consequence* of another sentence or sentences (the *premisses*)." (emphasis in original) This should be evident for a formal first-order predicate language and a logical calculus defined thereover. But Carnap (1937, p. 180) also allowed for *extra-logical* transformation rules being operative in some suitable language, which he called "P-rules" ("P" for *physical*, construed broadly).

As an example for a P-valid inference, Carnap (1937, p. 185) names the inference from '*a* is made of iron' to '*a* cannot float on water'. From the point of view of the logician this is merely an *enthymeme*: the inference would be *logically* valid if one introduced the additional premise 'for all *x* it holds that if *x* is made of iron, it cannot float on water', so it appears as a mere *abbreviation*.

While Carnap (1937, p. 180) thought of the addition of P-rules to a language as "a matter of convention and hence, at most, a question of expedience", i.e. believed that they were completely *dispensable*, Sellars (1953, p. 320) has argued that they should be viewed as drawn from a class of inferences *sui generis*, which he calls *material inferences*. The class of these inferences is probably not precisely delineated, or at least it is hard to find precise specifications; Brandom (1994, p. 97) describes them as "[t]he kind of inference whose correctnesses essentially involve the conceptual contents of its premises and conclusions", whereas Healey (2012d, p. 746; my emphasis – FB) describes them as "inferences of the kind anyone with a normal understanding of [some sentence—FB] will *naturally* make[...]." Material inferences hence are clearly connected to pragmatism in the sense of *linguistics*, i.e. to Gricean maxims and implicatures (cf. Grice 1975), not merely in the sense of epistemology.

To substantiate his views, Sellars (1953, p. 323) let a "Metaphysicus" argue that

> we must interpret [certain—FB] *subjunctive conditionals* [...] as expressions of material rules of inference. 'If there were to be a flash of lightning, there would be thunder', giving expression to some such rule as 'There is thunder at time *t*-plus-*n* may be inferred from there is lightning at time *t*', and this rule is not in any obvious way a specification of a purely logical rule of inference. (my emphasis—FB)

Sellars' investigation was before Stalnaker (1968) and Lewis (1973), and in principle the quote cries out for an investigation in terms of the logic of (counterfactual) subjunctive conditionals.[20] But this is not the place, and we merely acknowledge

[20] Point of clarification: a subjunctive conditional has the (schematic) natural language-form 'if it were the case that *x*, then *y* would be the case as well'. A subjunctive counterfactual has a

that certain 'material rules of inference' are being thought of here as expressed by (typically counterfactual; we will use that term generically below) subjunctive conditionals.

The position of Sellars (1953) and others (e.g. Brandom 1994, 2000), called *inferentialist pragmatism*, now is that the material inferences that can be drawn from some sentence are *constitutive* of that sentence's *meaning*: "the meanings of linguistic expressions and the contents of intentional states [...] should be understood [...] in terms of playing a distinctive kind of role in *reasoning*." (Brandom 2000, p. 1; emphasis in original) Below, we will give reasons to doubt that this inferentialism can provide an appropriate, exhaustive theory of meaning; but we still acknowledge here that in everyday life we certainly do 'infer materially' (from lightning to thunder, say, or from being made of iron to not being able to float on water), whether we interpret this act of inference as performing an enthymeme in *some* logic (regardless of the involvement of subjunctive conditionals) or not. And of course the inferences we can draw from an expression are *connected* to its semantic content, even though the latter is probably not *exhausted* by them.

What does all this have to do with QM? Healey (2012d, p. 746) suggests that QM, construed as a normative calculus for quantifying expectations, limits our capacity to draw certain material inferences, and thereby, according to inferentialist pragmatism, the content of certain statements. His example is an interference experiment with molecules (fullerenes), and a disjunctive statement (which he calls 's_{or}') about the positions of the molecules on a suitable screen where the interference fringes occur: 'The position of molecule S is $x_1 \pm \epsilon$, or $x_2 \pm \epsilon$, or $x_3 \pm \epsilon$, ...' (The ϵs represent experimental errors.)

Now according to Healey, once one accepts QM, the definiteness of positions as predicted by decoherence does "license" (his p. 744) endorsement of s_{or} when the screen is investigated, but endorsement of s_{or}, in turn, does not license inferences to statements such as "It is possible reliably to observe through which slit each particle passed without altering the interference pattern", or "If this is not so, then that can only be because any physical mechanism that permitted reliable observation of which slit each particle passed through would inevitably disturb the particle while doing so." (Healey 2012d, p. 746)

On the basis of the (specific kind of) inferentialism accepted by Healey, this means that "the *content* of s_{or} must be understood very differently within a community that has accepted a quantum theoretic analysis of the situation[...]." (Healey 2012d, ibid.; notation adapted; my emphasis – FB) This conviction of Healey's certainly stands in some continuity to Bohr's (1935, p. 700) emphasis on "the very conditions which define the possible types of predictions regarding the future behavior of the system", and Healey (2012b, p. 3) in fact believes that

false antecedent (x is *not* the case), whereas subjunctive conditionals *in general* may also express epistemic uncertainty as to the truth of the antecedent. Note also that Lewis (1973, p. 4) cites a case where the expression is not (or at least not overtly) subjunctive but still a counterfactual conditional.

"inferentialist pragmatism about content promises a better treatment of meaning than that offered by Bohr and his followers." It seems safe to say that Healey's treatment is more up to date than Bohr's because where Bohr (1935, p. 696) insisted that "[t]he extent to which an unambiguous meaning can be attributed to such an expression as 'physical reality' [...] must be founded on a direct appeal to experiments and measurements", Healey (2012d, p. 744) has it that "*decoherence* licenses [...] any suitably physically situated agent, human, conscious, or neither, to make some [particular NQMC—FB]." (my emphasis—FB)

More precisely Healey (2012d, p. 747; notation adapted; my emphasis - FB) thinks that, applied to the fullerene-interference case,

> it becomes more and more appropriate to think and speak of the fullerenes as having a well-defined path through the interferometer as the degree of thermally-induced electromagnetic decoherence into their environment increases. But note that on the present inferentialist view of content, this progressive definition of content *has no natural limit* such that one could say that, when this limit is reached, a statement like s_{Or} *is simply true* because one has finally succeeded in establishing a kind of *natural language-world correspondence relation* in virtue of which the statement *correctly represents some radically mind- and language-independent state of affairs*.

Connections to Putnam's version of *internal realism* should come to mind at this point. Recall, from the first interlude, that Putnam was eager to leave the correspondence theory of truth behind on behalf of a "idealized rational acceptability"-theory of truth, where creatures with "a rational and sensible nature" were invoked to delineate what counts as rationally acceptable. Much in the same way, material inferences constitutive of semantic content are those drawn by "anyone with a normal understanding" for Healey. The connections are not accidental: Healey was a student of Putnam, and Putnam (1977, p. 485) also calls his realism "Peircean", in honor of the arch-pragmatist C. S. Peirce.[21]

And there is an obvious connection to another twentieth century philosopher here, namely to *Wittgenstein*: The pragmatist inferentialism Healey endorses is heavily influenced by the specific rule-based account of meaning put forward by the later Wittgenstein. Here is Brandom (1994, p. xii): "One of the overarching methodological commitments that orients this [Brandom's—FB] project is to explain the meanings of linguistic expressions in terms of their use—an endorsement of one dimension of Wittgenstein's pragmatism."

But despite the fact hat Healey's pragmatist approach, like QBism, lets certain riddles of QM disappear, these considerations on reference and meaning make it arguably 'even more radical' than the QBists' one, much in the same way as skepticism about meaning, such as that of Wittgenstein (1968), has been judged to be "more radical than epistemological scepticism" (Miller 2006, pp. 91–92) or even

[21] For historical details on Peirce's role in the development of pragmatism e.g. Kenny (2007, p. 34 ff.). Burch (2014, p. 8), in particular, argues that "even when Peirce calls himself a 'realist' or is called by others a 'realist,' it must be kept in mind that Peirce was always a realist of the Kantian 'empirical' sort and not a Kantian 'transcendental realist.'" We identified Kant as a specific kind of internal realist in interlude I, so the connections run quite deep.

"the most radical and original sceptical problem that philosophy has seen to date" (Kripke 1982, p. 60). It seems that on Healey's views, it is not only that QM implies a failure of *knowledge* and *certainty* about the truth of certain propositions or a failure to conceptualize certain *aspects* of the external world; it ultimately implies a failure of successful *reference* to mind-independent reality *at all*, at least in the sense of a mind-world correspondence. Healey's considerations have an impact on the very *meaningfulness* of *non-quantum* claims (his NQMCs), i.e. statements expressed in terms of everyday language.

But what *is* 'reference'? What 'meaning'? These are, of course, deep philosophical questions well beyond the scope of this book. Quine, for instance, has been accredited with the quite devastating view that "[n]o scientifically satisfactory sense can be made of the concept meaning, and not even [...] of reference[...]: the use of those concepts is inescapably intuitive, unpredictably interest-relative, and subject to radical indeterminacy and even paradox." (Kemp 2012, p. 2; emphasis omitted) We succumb to Quine's purported authority at this point, but relative to our particular interests, we should make some distinctions for the sake of clarity. In particular, we reserve the intuition that meaning has a component *purely internal* to the speaker's mind—which even the semantic externalist Putnam (1975a) allowed (cf. also Neander 2006, p. 377)—and that reference is specifically "a *relation* that obtains between certain sorts of representational tokens [...] and *objects*." (Reimer and Michaelson 2014, p. 1; my emphasis – FB) So reference 'crosses boundaries' whereas meaning need not; and when it comes to the merely possible, there are good reasons to sharply distinguish the two (cf. Quine 1939, pp. 702–703; Quine 1948, p. 26). This distinction allows us the freedom to largely agree with Healey when it comes to questions regarding the language-world relation while disagreeing with him about meaning.

What, however, are the reasons for disagreeing with Healey on the impact of QM on the meaning of specific statements? Now we had readily acknowledged that Healey offers an improvement on Bohr's appeals to meaning, in virtue of his inclusion of decoherence, and we had also admitted that the material inferences that may be drawn from a statement have *something* to do with its meaning. But it seems far fetched to embrace *full* inferentialism about meaning: Carnap's (1937, p. 42) inferentialism was carefully formulated w.r.t. *logical* content only, and taken to apply with certainty "*so long as nothing psychological or extra-logical is intended by it*." (my emphasis—FB) Fodor and Lepore (2001, pp. 468 and 473), moreover, argue that neither is Brandom (2000) clear on *which* inferences are constitutive to meaning—even among the 'material' ones, which we found rather vaguely defined in the first place—, nor can inferentialist pragmatism *in principle*, *ever* account for *compositionality*, the "uncontroversial" assumption "that, apart from idioms, the meaning of any complex expression-type (such as a sentence) depends on the meanings of its component words and on how those words have been combined with one another." (Horwich 2006, pp. 47–48)

This is also the point where Friederich (2015) branches off from Healey, whose conception is otherwise quite strongly influenced by the latter. Friederich (ibid., p. 79), however, *only* takes the scope of the *Born rule* to be at stake, instead of semantic content, i.e. the range of NQMCs about which it (the Born rule) licenses particular inferences. We can only partly agree. Certainly, the scope of the Born rule is at stake as well. But what Healey seeks to clarify is the Bohrian heritage which promotes the view that QM has an impact on language use and meaning and is still prevalent in physical conduct, as witnessed by the following textbook passages (my emphases—FB):

> In the example of the Stern-Gerlach experiment, sorting into the two categories $\mu_z = +\mu_0$ and $\mu_z = -\mu_0$ *loses all its meaning* if one attempts to sort the systems into subcategories $\mu_x = +\mu_0$ and $\mu_x = -\mu_0$. (Basdevant and Dalibard 2002, p. 166)

> Before quantum mechanics was born, the thermodynamic properties of an ideal gas [...] were obtained by summing over the phase-space locations of each molecule independently. [...] Quantum mechanics teaches that the state of the gas is completely specified by listing the three occupied states, $|1\rangle$, $|2\rangle$ and $|3\rangle$ for it is *meaningless to say* which molecule is in which state. (Binney and Skinner 2014, pp. 157–158)

> it is *not meaningful* to regard a quantum particle as possessing any intrinsic property, independent of the (classical) measuring apparatus used to observe it. This interpretation is remarkably useful, and is *used unthinkingly by thousands of physicists*. (Le Bellac 2006, p. 186)

Healey's considerations, we take it, establish is *in what sense* and *to what extent* our use of QM has an impact on the meaning or the content of certain statements (NQMCs) in certain contexts. It may be a *useless practice* to *state* that the path of a fullerene, electron, or photon in a suitable and suitably isolated interferometer is such and such—but we urge that this does not make the statement *meaningless altogether*: we still seem to *understand* it very well[22]; classical physics would otherwise not have been possible.

In other words: We appreciate that QM has an impact on the *inferential* content of certain NQMCs, where the inferences that are not being promoted need not be deductively valid ones. But we deny that a statement's meaning is exhausted by its inferential content, and hence, as Healey seems to think, that its capacity to mean anything can depend on e.g. the degree of thermally induced decoherence. The more restricted impact on *inferential* content *only* is probably also closer to what (most) physicists have in mind, or would subscribe to when pushed on this issue, when they talk about the 'meaningfulness' of this or that statement as in the three quotes above: From the quantum state $|\mu_x = +\mu_0\rangle$ we cannot infer anything about the states $|\mu_z = \pm\mu_0\rangle$; from the listing of three distinct states in occupation number representation, we cannot infer anything about the thermodynamic properties of the individual molecules; from the properties exhibited in measurement, we cannot infer anything about the properties outside *any* observational situation. This seems to be the intended meaning of 'meaning' here. But to re-emphasizes our conclusions from

[22]Davies (2006, p. 23) equally refers to the understanding of a sentence as an explication of what it is to know its meaning. Since knowledge implies truth on most conceptions, we may take it that a sentence has to *have* a meaning in order for a competent speaker to understand it. Here is Quine (1939, p. 703, emphasis in original): "The noun 'Pegasus' *is* meaningful. If asked its meaning, we could reply with a translation into other words: 'the winged horse captured by Bellerophon.'"

Chap. 4: in all three cases, it is a mistake to think that this is a result of a limitation in *epistemic access alone*, as the thermodynamic example, involving entanglement, shows most forcefully.

7.3.3 What to Make of Decoherence?

While Healey relies heavily on decoherence, "there is", much in contrast to this, "no foundational place for decoherence in the Quantum Bayesian program." (Fuchs and Schack 2012, p. 246) Fuchs and Schack aim to replace the decoherence mechanism by a story based on van Fraassen's (1984, p. 244) *reflection principle*, that our degree of belief in the occurrence in some event A at time t, given that we assign the degree of belief q to it at time $t+\delta$, should be q as well, i.e. $p_t(A|p_{t+\delta}(A)=q)=q$. They fist argue (their pp. 239–240) for the adoption of this principle from a diachronic Dutch book argument, and then, subsequently (their pp. 244–245), that the principle, translated to the quantum formalism, implies an update rule for measurement situations with two subsequent measurements according to which the updated state "has the form of a 'decohered' state," which then "*is* the agent's quantum state [...] *as far as the second measurement is concerned*." (Fuchs and Schack 2012, p. 245; emphasis in original)

This is hardly an apt replacement of the decoherence mechanism *with all its implications* though. What about decoherence times and confirming experimental evidence? What about concrete implementations such as scattering models or decoherence based on spin-couplings (cf. Schlosshauer 2007, p. 88 ff.)? It is doubtworthy that the argument from the reflection principle can capture the *full* content of decoherence theory, which, we have argued, 'is here to stay'.

A question that should be bugging us by now, however, is what the status of decoherence actually *is*, in Healey's pragmatist account or any conception relevantly similar to it. In Sect. 6.3.2, decoherence was treated more or less directly as a *physical process*, aiding the MWI to some extent, while also creating problems for reconciliation with the Born rule. We also made reference to experimental evidence supporting the predictions of decoherence theory, and should this evidence not commit us to a belief in there being a physical process that is described by the decoherence mechanism after all? And are we then not faced with Bell's problem of retrieving an 'or' from an 'and' again, i.e. of how to interpret QM "the same way we have always interpreted scientific theories in the past: as modelling the world", as Wallace (2012, p. 38) urged us to?

Now we could pose Bell's "Who do we think *we* are?"-question against, and even shed doubts on the historical accuracy of Wallace's comment (e.g. Mach 1910; or Cantor et al. 1990, p. 191 ff., for an overview of some 'anti-realist' contentions in the histroy of science). But compare, more illuminatingly, a typical situation in which decoherence becomes relevant to the following situation (cf. Schwabl 2006, p. 429 ff.) of (semi-)classical particles, subject to some stochastic dynamics driven by a force $-\frac{\partial V}{\partial x}$ with suitable damping Γ, and described by a (time dependent) probability density $P(x, t)$ over particle positions x. The temporal evolution of the probability density is given, in this context, by the *Smoluchowski equation*

$$\frac{\partial P}{\partial t} = \Gamma \frac{\partial}{\partial x}\left[\left(\beta^{-1}\frac{\partial}{\partial x} + \frac{\partial V}{\partial x}\right) P\right], \tag{7.9}$$

where $\beta = 1/(k_B T)$, k_B Boltzmann's constant and T the temperature. An ansatz for this equation is $P(x, t) = \rho(x, t)e^{-V\beta/2}$, where ρ can be separated for x and t and developed as $\rho(x, t) = \sum_{n=0}^{\infty} c_n \varphi_n(x) e^{-\frac{\Gamma E_n}{\beta}t}$, according to (discretized, and hence semi-quantum) energy contributions E_n. Using that P is normalized, one finds the zeroth contribution $c_0\varphi_0$ to be

$$c_0\varphi_0 = \frac{e^{-V\beta/2}}{\int \mathrm{d}x\, e^{-V\beta}}. \tag{7.10}$$

This factor is independent of time, and multiplying by the factor $e^{-V\beta/2}$ of the ansatz, one obtains

$$P(x, t) = \frac{e^{-V\beta}}{\int \mathrm{d}x\, e^{-V\beta}} + e^{-V\beta/2} \sum_{n=1}^{\infty} \varphi_n(x) e^{-\frac{\Gamma E_n}{\beta}t}. \tag{7.11}$$

As we can see, the distribution will approximately evolve into an *equilibrium distribution* $P_{\text{eq}} = \frac{e^{-V\beta}}{\int \mathrm{d}x e^{-V\beta}}$ for large enough t, since deviations are given by the rest of the series ($n > 0$) and vanish exponentially for $t \to \infty$.

We have put some physical considerations into this sketch, and it appears that over a short time, the 'ensemble' represented by P evolves into a familiar form—a process accessible to thermodynamic experiment. Does this, however, give *any* credibility to P representing anything physical at all? Does it mean that there is a real, physical process, described by the above steps, according to which a ('real, physical') *substance*, the 'probability density', evolves over time into a ('real, physical') equilibrium density? Clearly the answer is: *no!* It only means that insofar as P and P_{eq} have empirical significance (can be connected to experience), we should believe that there is *something* in reality that requires us to switch from *using* P to using P_{eq} after appropriate intervals of time τ, when the suitable initial conditions are given to use the Smoluchowski equation and the above ansatz.

Healey (2012b, p. 1538; emphasis in original) similarly has it that his

> pragmatist interpretation does not deny that environmental decoherence *involves* a physical process: but it does deny that the role of a system's quantum state is to describe or represent properties of systems involved in such a process.

Quite in contrast to this,

> [f]or a Quantum Bayesian, the only *physical* process in a quantum measurement is what was previously seen as 'the selection step'—i.e., the agent's action on the external world

> and its unpredictable consequence for her, the data that leads to a new state of belief about the system. (Fuchs and Schack 2012, p. 246; emphasis in original)

This is (too) much of a fallback onto Heisenberg's disturbance theory and onto the 'other side' of Bohr's thinking, with its emphasis on "sources of uncontrollable interaction between objects and measuring instruments." (Bohr 1948, p. 52) It is all but clear how these considerations help to understand the use of the decoherence formalism or elucidate the 'weirdnesses' of QM more generally (cf. in particular the comments on EPRB correlations above).

Surprisingly, our above treatment of the *semi-classical* particles and probabilities can, in fact, be described quite well in terms of Healey's aforementioned *backing* and *advice conditions*: *only* under suitable such backing conditions can we justify our *use* of the Smoluchowski equation and the appropriate ansatz to infer the 'evolution into equilibrium'—and we do so to thereby successfully form expectations about future observations in appropriate advice conditions, so these conditions even warrant the use. There seems to be much more continuity between classical physics and QM than expected, when viewed from this perspective.

All in all, our conclusion w.r.t. decoherence is the following: Once the assumption is dropped that quantum states are representational w.r.t. anything 'external', the thrust of arguments from decoherence to particular ontological interpretations such as the MWI or prospective decoherence-based collapse interpretations, or even to the loss of interference with 'empty waves' in BM, vanishes. Anyone familiar with the way in which physicists *use*, say, charts and diagrams to depict all kinds of connections *not* directly indicative of ('real, physical') processes should find some appeal in this argument. Emphatically, we are *not* claiming that quantum states 'are just' probability densities like those discussed above, which would be obviously *false*. The appeal to the dynamics of classical probability densities here serves merely as a *plausibilization* for viewing quantum states *and their dynamics* more *like* probability densities, on an appropriate reading of the latter: as *cognitive tools* for making predictions about future experience.

The OP, as we already noted, basically dissolves once one denies an 'ontic' status to quantum states. If one appreciates the Born rule as well, empirical adequacy is built into one's interpretation. And if one accepts the unitary time evolution as a preferred norm-preserving time evolution on (partly) empirical grounds, one can even appreciate that there is no need to quarrel over dynamical questions. Given these ingredients to a specific interpretation, the MAC is fulfilled quite trivially and, in contrast to QBism, even on dynamical grounds.

Note that Healey also believes decoherence to help solve what he calls "the residual measurement problem", which loosely coincides with Pitowsky's (2006, p. 233) or Bub's (2016, p. 208) 'small' measurement problem(s): "Given a superposed entangled state (such as that of quantum system and quantum detector), under what circumstances is it *legitimate to infer* that (at least) one of the entangled systems has some definite property, with probability given by the Born Rule?" (Healey 2012c, p. 8; my emphasis – FB)

How can decoherence do the trick? While in some places Healey (e.g. 2012c, p. 12) resorts to the standard realist vocabulary of environmental interactions 'dislocalizing the phase' of a system 'into the environment', which becomes somewhat elusive against the background of his inferentialist pragmatism, in his 2015 he is much clearer:

> The Born rule may be legitimately applied to [some system—FB] *only when* [...] decoherence has occurred, and then only to those privileged magnitude claims corresponding to projection operators onto subspaces in the relevant pointer basis. Such decoherence is never perfect, and nor is a 'pointer basis' precisely determined and perfectly constant. But the advice provided by the Born rule concerning only magnitude claims privileged by pointer bases in the same narrow neighborhood will be consistent and typically prove reliable: the corresponding advice conditions will typically obtain with relative frequencies closely corresponding to their Born rule probabilities. (p. 3; my emphasis—FB)

We take it that decoherence, in virtue of its (statistical) empirical adequacy, should be seen as just that: an important (if not QM's *unique*) *normative guide* to expectations about observing this or that value for a measurable magnitude, i.e. as 'legitimizing' expectations about certain NQMCs. This implies a sort of *relationalism* about quantum states (cf. Healey 2012d, p. 752 ff.): A suitable preparation on a photon may warrant the assignment of a superposition state in the basis in which a suitable detector measures, and the decohering measurement interaction may then, in turn, warrant one of the states from that basis, i.e.:

> the quantum state of the photon's polarization *is* a superposition of horizontal and vertical *relative* to the situation of an agent *prior* to the decohering interaction with the detector and its environment, but horizontal relative to the situation of an agent *after* that interaction. (Healey 2012d, p. 753; my emphasis – FB)

And, importantly, decoherence should be viewed as legitimizing expectations about particular NQMCs in a much broader context than merely Bell's (1990a, p. 34) "piddling laboratory operations":

> Much of what we know about the *solar system*, and almost everything we know about what lies outside it, is based on evidence provided by analyzing electromagnetic radiation, especially that emitted or absorbed by excited atoms and molecules. [...] No single, simple model of decoherence can be expected to encompass all such phenomena. But in many cases the atoms and molecules involved will be in an environment that decoheres their internal states in an energy basis [...]. It is such decoherence that justifies one in assuming that emission or absorption occurs between states of well-defined internal energy, and so applying the Born Rule to calculate absorption or emission probabilities. (Healey 2012b, p. 1552; my emphasis – FB)

Suggested correction to avoid the descriptivist touch: It is not *decoherence*, construed as a 'real', 'physical' process, that justifies these assumptions, but whatever 'dooms' us to *use* the decoherence description of state changes, or rather, legitimizes our preference for one basis over the other and advises us to quantify our expectations about NQMCs suggested by that basis in accord with the Born rule. This will always imply that we cannot think of the system of interest as isolated, but must rather regard it as embedded in an environment. And the most interesting part is that this advice comes *dynamically*: decoherence times, computed in virtue of scattering models etc., tell us over which *time scales* and in what ways our expectations should change, given that we know the backing conditions for states of both, system of interest and environment.

7.4 From an Epistemological Point of View: Piecing Together a Positive Proposal

7.4.1 *Material Inferences and Abduction*

A major complaint in Sect. 7.2.3 was that QBism does not really offer an epistemologically clear interpretation, and while Healey's views certainly help to resolve some issues, many questions still remain unanswered here as well.

Above, we have sided with Healey, Stairs, and Friederich and against QBism in that quantum states are not merely an expression of the beliefs of some agent. They provide a normative calculus that, when applied under typically well-understood 'backing conditions', gives advice about what to expect in specific future circumstances ('advice conditions'). The timescales under relevant (de-isolating) conditions that determine when to expect what and with what probability (frequency in repeated similar circumstances) are also provided by the decoherence mechanism. In particular this has the implication that the *correlations* implied by QM do *not* depend on the beliefs of agents; there is a sense in which they are '*objectively* there'.

We do, however, agree with the QBists on the *meta-level*: the belief in the *appropriateness* of the normative content of QM is just a belief: the hypothesis that QM will *continue* to enable our remarkable scientific success as it has done in the past is merely a hypothesis, *formed on the basis of induction*. It expresses a belief in a certain uniformity of nature. And we agree with the QBists also on the (object-level) issue that "there is [...] a split [...] between the world in which an agent lives and her experience of that world." (Mermin 2012, p. 8)

A split between mind and world is, however, also *tacitly* acknowledged by Healey (2012d, p. 747), when he denies that a statement 'licensed' on the basis of sufficient decoherence "correctly represents some radically mind- and language-independent state of affairs." But Healey (2015, p. 12; my emphasis – FB) commits an apparent crime against this attitude when he writes: "There are *real* patterns of statistical correlation *in the physical world*. Correctly assigned quantum states *reliably* track these through legitimate applications of the Born rule." How, if QM never perfectly allows for the establishment of a "natural language-world correspondence", is the term 'physical world' even to be *understood*? How can we 'rely' on QM or the NQMCs it licenses, if we cannot establish their *truth* in the sense of mind-world correspondence?

This is the point where we find Healey's pragmatism unsatisfactory and to equally not satisfy the DEC. (i) seems fine in Healey's account: 'classical objects' come out as those (we take it) that are well-described by those (strictly incompatible)

NQMCs which are simultaneously licensed by decoherence to a satisfactory degree. (ii) is fine as well: the quantum-classical transition is given by decoherence, understood as a guide to advice conditions wherein one can make use of the Born rule, and somehow 'backed up' by a physical process. The 'somehow' in the previous sentence however indicates where the trouble lies: What 'licenses' even the notion of a 'physical world', when the correspondence theory of truth is given up? More precisely: What is *meant* by that term? What is meant by the process 'involved in', but not described by, decoherence? Part (iii) of our DEC is not appropriately addressed.

Another, ultimately closely related point of contention with Healey's views was with the role of material inferences and their impact on meaning. While we found the class of material inferences to be rather ill-defined, we appreciated that standard examples are typically formulated as counterfacutals. This makes for an interesting connection to a reoccurring theme in this book: Rosenberg (1974, p. 76), a student of Sellars, suggests to "call the credibility attaching to [...] a material rule of inference or system of such rules by virtue of the fact that their acceptance or espousal enhances our explanatory competence '*abductive* credibility'." (my emphasis—FB) Recall from Sect. 4.3.4 that abductive inferences follow the general scheme from known laws and a given datum to the conjecture of an explanation. So if the material inferences that we draw (typically) express, in a more familiar albeit problematic parlance, 'lawlike' connections, as evidenced by previous examples like the inference from lightning to thunder, acceptance of them of course enhances our ability to infer, abductively, explanatory reasons for observed data.

Since QM disencourages material inferences like 'we measured $\mu_z = \pm\mu_0$ on S in a $|\mu_x = +\mu_0\rangle$-state because S already had this and that property/was disposed to do this and that w.r.t. μ_z in state $|\mu_x = +\mu_0\rangle$', it thereby disencourages abductive inferences to the existence of a definite property w.r.t. μ_z; or to the existence of a particular thermodynamic configuration in the number representations $|1\rangle$, $|2\rangle$, and $|3\rangle$; or to the existence of particular pre-existing properties in general. This is the connection to, nay restriction on, abduction that, as we have multiply emphasized in the course of this book, should be seen as playing a major role in the interpretation of QM, given the negative results from Chaps. 4 and 6. The general limitation implied by QM, we take it, is hence better phrased in terms of *abductive* than material inference: we cannot abductively infer that there 'really was a particle all along' at any given, particular position within an interferometer from the fact that we observe a spot on some screen or hear a click from some detector. So we cannot appeal to the particle's *causal history* as an explanation of our present observations in the same sense as this is possible according to pre-quantum intuitions. The situation is radically different from footsteps in the sand and the inference to some person having been walking there.

But *which* counterfactuals—and hence: which abductive inferences—are disencouraged and which ones are not is, in fact, a quite subtle matter. Stairs (2011, p. 164), considering the entangled electron pair of Alice and Bob, measured at a spacelike distance and with Alice measuring $+1$ for z-spin, points out that Alice should not assert that

> if Bob *were to* measure z-spin, he *would* find -1. Instead, she will say that if Bob *did* measure z-spin, he *did* find -1. But if he did not, she will say nothing about what he would have found if he had. (emphasis in original)

Why should Alice act this way? Since both events are at a spacelike distance, asserting that Bob *were to* measure -1 for z-spin must be understood either *nonlocally causal* or *superdeterministic*: if the assertion is not intended to mean that the course of events predetermines Bob's -1 measurement anyway (*and* Alice's $+1$ one, *and* her choice of axis...), it can only be taken to mean that Alice's $+1$ measurement *makes it so* that Bob measures -1, i.e. causes it, or *vice versa*—nonlocally, in each case. How else could it be understood?

Stairs nonetheless believes that there *is* a "lawful, counterfactual-supporting generalization" (his p. 165): Given that the probabilities are construed as objective and the correlations as objectively occurring, one can safely say that if Bob and Alice *were* to measure along the same axis, they *would* find opposing spin values. This basic opinion is also expressed by Healey (2012a, p. 24), but while Stairs (2011, p. 165) eschews discussion on *which* counterfactuals QM encourages and which ones it does not, and remains with the "vague remark" that it "seems plausible that whether we attribute counterfactuals has to do with what we actually interact with", Healey, using the Alice-Bob case, manages to narrow things down much better.

Recall from Chap. 6 (in particular Footnote 23) that we shed doubts on whether QM supports certain *causal* counterfactuals. Woodward (2003, p. 15), in particular, believes "that the sorts of counterfactuals that matter for purposes of causation [...] are just such counterfactuals that describe how the value of one variable would change under interventions that change the value of another." This is quite plausible: Woodward (2003, pp. 14–15) appeals, as a motivating example, to the correlation in barometer readings and storms occurring, and while 'if the barometer reading were to fall, a storm would occur' seems to be true (or is at least supported by evidence), 'if one were to intervene on barometer readings, a storm would/would not occur' is clearly false. So only the latter seems to capture the causal relation appropriately, in virtue of the intervention terms.

Now given this interventionist touch to causal counterfactuals, there will be no causal connection between Alice and Bob, because there *can be no* interventions $\underline{I}_A, \underline{I}_B$ such that Alice and Bob could perform $\underline{I}_A$ or $\underline{I}_B$ respectively to fix their system to one of the possible values. To see this, consider that for some $\underline{I}$ to count as a genuine intervention on a variable $\underline{X}$ w.r.t. variable $\underline{Y}$, it is required (among other things) that "certain values of $\underline{I}$ are such that when $\underline{I}$ attains those values, $\underline{X}$ ceases to depend on the values of other variables that cause $\underline{X}$ and instead depends only on the value taken by $\underline{I}$." (Woodward 2003, p. 98; notation adapted) Now assume that there could be an $\underline{I}_B$ which would allow Bob to fix his spin-value to $\uparrow_z$, say, which would (causally) fix Alice's outcome to $\downarrow_z$. Since the singlet state is perfectly symmetric, the same thing would apply to Alice and some $\underline{I}_A$, whence Bob's value *would still depend on Alice's distant value, in spite of* $\underline{I}_B$. This is in contradiction to Woodward's requirement of independence from all other variables; there can hence be no such interventions.

This argument is executed by Healey (2012a, pp. 23–24) and repeated by Friederich (2015, p. 132), and Healey (2012a, p. 2) concludes that "the only counterfactuals that hold in this case manifest epistemic rather than causal connections between distant events." What is being demonstrated, more precisely, is that QM does not encourage causal reasoning in *specific contexts* such as EPRB-experiments, in the sense of hidden *common* causes *or direct* causation. It is in *this* sense that "quantum mechanics as a means of ordering an immense amount of evidence" suggests a "departure from accustomed demands of causal explanation[...]" (Bohr 1963, p. 3), much more so than in the sense Bohr probably had in mind.

But these considerations still fall short of a general characterization of which counterfactuals are or are not being rendered moot. Stairs (2011, p. 165) refers to a counterfactual of the form 'if we were to prepare a single spin as $|\uparrow_z\rangle$ and measure it shortly after, we would measure $\uparrow_z$ again'. This counterfactual *is* supported by QM, since for short enough times and sufficient isolation the effect of decoherence may be regarded as negligible, and the free Schrödinger-evolution

$$i\hbar\frac{\partial}{\partial t}\boldsymbol{\psi} = -\frac{\hbar^2}{2m}\Delta \otimes \mathbb{1}\boldsymbol{\psi}$$

of a spinor $\boldsymbol{\psi}(\boldsymbol{x}, t) = \psi(\boldsymbol{x}, t)\,|\uparrow_z\rangle$ between two Du Bois magnets aligned along the z-axis will leave the spin-up character of the state, as well as the probabilistic predictions for spin up along z, entirely untouched. The spin up-state may hence be said to cause a later, time evolved spin up-state and the counterfactual *does* seem to be of the required causal sort.

This is closely related to the point emphasized by Born (1926, p. 804), that *probability itself* was governed by a causal law in QM. It is hence the probability-assertions implied (*via* the Born rule) by quantum states that are related to one another by a unitary evolution which can be summarized in terms of lawful counterfactuals, counterfactuals that could be viewed as 'causal' in even the strong sense of deterministic causality. In fact, we should provide a more encompassing analysis of the role of (unitary) *transformations* in the interpretation of QM, the lack of which we found wanting in QBism.

7.4.2 From Bohr and Einstein to the Constitution of a 'Quantum Reality'

In Sect. 7.3.2, we noted Healey's connections to Putnam. And we also noted, in the first interlude, that according to Brown's (1988) analysis, Putnam and Kant both "hold that the world we know and talk about is empirically real, but [...] also [...] mind-dependent", and that both also "consider [...] the view that there is an unknowable noumenal world behind the phenomena." Due to his denial of our representational access to "some radically mind- and language-independent state of affairs", when Healey talks about the 'physical world', this can only be in the 'phenomenal' sense, i.e. as referring to an 'empirical', 'mind-dependent' world—or put less philosophically bloated: *the world as it appears*. 'Physical reality', if understood this way, is dependent to some degree on our cognitive interior and perceptive inventory.

We cautioned against embracing Healey's pragmatist inferentialism in full, arguing that there is more to 'meaning' than the (material) inferences promoted by some statement. What we took QM to signal, however, is: (a) a failure of *reference* of formal and natural languages, at least whenever one *has to* resort to QM to 'reason about' a given situation and no 'classical description' is available; and (b) a failure of *causal interpolability*, because e.g. neither the endorsement of direct causation nor of a hidden common cause in EPRB-scenarios are promoted by the quantum calculus. As a first approximation, we can hence, if a separation between two 'worlds' is acknowledged, take both our conclusions, (a) and (b), to signify that empirical reality has 'gaps'.

Causal interpolability is here taken as a constitutive feature of an empirical world, as was somewhat justified by the realism-considerations in interlude II. And this, of course, is quite a *Kantian* point. Kant (1783, p. 70; my emphasis – FB) believed, in particular, to "have amply shown" that causality, understood as "the reference of the existence of one thing to the existence of another, which is *necessitated* by the former", could be "firmly established a priori" and had an "undoubted objective value, though only with regard to experience." Now 'necessitation' of effect by cause is not endorsed by everyone anymore, as we saw in interlude II, and Kant's (1781, pp. 176–177) 'proof' of the "the principle of sufficient reason" (where 'reason' is construed causally) is certainly wanting in many respects.

But we already noted in interlude II that the 'production' of effect by cause is at least a typical *intuition* associated with causation, and there is certainly also a strong intuitive reluctance to accepting an 'uncaused correlation' in a situation like that described by EPRB or actual experimental realizations thereof. Why else would almost 1/3 of the over 10,000 citations of Bell's (1964) famous paper contain the word 'explanation', as can be gathered using modern search engines? Thus while causal reasoning may fall short of an a priori principle with the objective strength claimed by Kant, it is still certainly well routed in our intuitive modes of thinking and conceptualizing, and a failure of causality might still count as signaling a 'gap' in reality as experienced.

More generally speaking, Kant's *synthetic a priori*, which lies at the heart of his 'necessity' claims, has been identified as a central deficit in his philosophy, especially in the light of developments in non-Euclidean geometry in the late nineteenth century (e.g. Friedman 1999, p. 6). And it has also been suggested to connect to inconsistencies, or at any rate incoherences, in Kant's own philosophy (e.g. Moore 2012, p. 134): Kant suggests that "[t]he concept of a noumenon is [...] merely *limitative*, and intended to keep the claims of sensibility within proper bounds, therefore of negative use only" (Kant 1781, p. 255; emphasis in original), while also declaring, in some places, that space and time are "pure forms of sensuous intuition" (Kant 1781, p. 22), a crucial step for promoting the principles of Euclidean geometry to a priori valid synthetic judgments (his p. 24). But it looks as though he is thereby immediately lead into claiming at least *negative knowledge of the nuomenon* as a-spatial and a-temporal.[23]

Moreover, Kant's (1781, p. 18) doctrine that "[a]ll thought [...] must, directly or indirectly, go back to intuitions (Anschauungen)", where the *form* of *empirical* intuitions would be determined by the synthetic, has been suspected to be at least partly responsible for Schrödinger's initial insistence on *visualizable* wave-concepts (e.g. d'Espagnat 1995, p. 314). And equally, Einstein's (1948, p. 321) aforementioned insistence on e.g. the separability principle, without which, he thought, "physical thinking in the sense familiar to us *would not be possible*" (my emphasis and translation—FB) may have been partly due to Kant's impact on his thinking.[24] Einstein, for instance, once argued (1944, p. 22) that while Kant's concepts of pure reason had "nothing of the [...] inherent *necessity*, which Kant had attributed to them" (my emphasis—FB), it still appeared to him "correct in Kant's statement of the problem" (of a basis for "assured knowledge") that "in thinking we use, with a certain 'right,' concepts to which there is no access from the materials of sensory experience[...]."

It is quite ironic that Einstein was so influenced by Kant in his thinking which furthered his *resistance* to QM, when at the same time *Bohr* has been characterized by Folse (1994, p. 121) as a "pragmatized Kantian". The irony is somewhat lifted though, when one realizes that Bohr and Einstein seem to have fixated on different *aspects* of Kantian philosophy. And this also helps to sort out what parts of Kant's philosophy we can (and cannot) make good use of, in interpreting QM 'epistemologically'.

First note that while Einstein (1944, p. 22) insisted that "the concepts which arise in our thought and in our linguistic expressions are all [...] the free creations of thought which cannot inductively be gained from sense experiences", Folse (1994, p. 121) describes the pragmatized Kantianism which he attributes to Bohr as characterized by the fact that "it is not Reason but Nature as that which is subsumed under the concepts provided by the organizing mind which grounds the applicability of the concepts." In other words: Both, Bohr and Einstein, give up on the *a priori validity* of certain synthetic judgements. But while Bohr also allows experience to *alter* our categories or their application quite *substantially*, Einstein does not.

A remarkable difference also occurs, of course, in the role of 'reality' in both views. In his *Reply to Criticisms*, Einstein (1949b, p. 680) puts the for him crucial message of Kantian philosophy thus:

> There is such a thing as a conceptual construction for the grasping of the inter-personal, the authority of which lies purely in its validation. This conceptual construction refers precisely

---

[23]E.g. Kant (1781, pp. 26 and 34–35): "Space does not represent any quality of objects by themselves, or objects in their relation to one another"; "time is no longer objective, if we remove the sensuous character of our intuitions, [...] and speak of things in general."

[24]Cf. also Beller (2000) for historical details on Einstein's life-long exposure to Kantian thought.

to the 'real' (by definition), and every further question concerning the 'nature of the real' appears empty.

Reference to a "conceptual construction for the grasping of the inter-personal" and the 'emptiness' of further questions makes Einstein appear as the subtle kind of realist that Fine (1986) depicts him as. But it has often been claimed that "mature Einstein was a realist [...] and that the 'prescriptive principles' [...] that he put forth [...] were quite unambiguously interpreted by him [...] as referring to mind-independent reality." (d'Espagnat 2011, p. 1705) This is not without reason; in Einstein (1950, p. 46 ff.), for instance, the latter describes a definitive scientifically realist stance as the framework of physical thinking up to the twentieth century, and expresses his affinities to it and to a belief in causal laws. And while young Einstein's thinking clearly had empiricist elements—as he acknowledged himself e.g. in conversations with Heisenberg (1969, cf. p. 80)—, as early as 1934 he held "it to be true that pure thought is competent to comprehend the real, as the ancients dreamed", and confessed to "still believe in the possibility of giving a model of reality, a theory, that is to say, which shall represent *events themselves* and not merely the probability of their occurrence." (Einstein 1934, pp. 167–169; my emphasis – FB)

Our original point of contention however was the synthetic a priori in Kant's thinking, and it seems useful for making sense of grounds for disagreement to take into account the notion of a *relativized* or *constitutive a priori*, as endorsed more recently by Friedman (1999, p. 62) and traced back by him to Reichenbach (1920, cf. in particular p. 47). The conviction underlying the constitutive a priori is that there are to any theory certain principles, 'axioms of coordination', which contrast with 'axioms of connection' (cf. Reichenbach 1920, p. 51; Friedman 1999, p. 61). The former "must be laid down *antecedently* to ensure [...] *empirical well-definedness* in the first place" (Friedman 1999, p. 61; my emphasis – FB), whereas the latter axioms are simply "empirical laws in the usual sense involving terms and concepts that are already sufficiently well defined." (ibid.) Don Howard (2010, p. 337) understands Friedman's version of the constitutive a priori "structurally and functionally as that without which the rest of a theory would lack content." In other words: constitutive a priori principles provide what is (objectively) *there*, according to the theory.

Notably, Bernard d'Espagnat (2011, p. 1704) suggests to distinguish two 'tenets' in Kantian philosophy, namely that (I) "science might just be a construction grounded on prescriptive principles *chosen* a priori" and (II) "the view that the principles in question are *given* a priori—that is, once and for all—by the very structure of our sensibility and understanding." (my emphasis—FB) Moreover, d'Espagnat (2011, p. 1705) notes that "the expression 'a priori' does not carry the same meaning in the two tenets", since in (I) a priori concepts may be allowed to vary in a way which (II) forbids. The constitutive a priori is freed from tenet (II).

Friedman, in fact, finds a distinction between the two inedependent uses of 'a priori' already in Reichenbach (1920, cf. p. 46): "according to Reichenbach, the a priori had two independent aspects in Kant: the first involves necessary and unrevisable validity, but the second involves only the [...] feature of 'constitutivity.'" (Friedman 1999, p. 7) This is just (I) and (II) in reverse.

Under the influence of Schlick (cf. Friedman 1999, p. 62 ff.), Reichenbach must have come to accept that his 'constitutive a priori' was merely *terminologically* distinguished from what Schlick referred to as *conventions*. But Friedman (1999, p. 64; my emphasis – FB) believes that "when Reichenbach [...] buys into Schlick's *underlying conception* [...] the most important element in his own earlier conception of the relativized a priori is actually lost." What is the difference? According to Friedman (1999, p. 67; emphasis in original), "Schlick's conception does not, in fact, yield a *distinction* between the constitutive and the empirical, between the conventional and factual parts of science at all—even relative to a particular given theory."

To give an example of how conventionalism eliminates this distinction, Friedman (1999, p. 67) refers back to Einstein (1921, p. 236) in his endorsement of Poincaré's (1905, p. 48 ff.) ideas on geometry and physics. Geometric laws (G), Einstein therein declares, only predicate anything about the behavior of real things *in concert with* physical laws (P). "Thus (G) may be chosen arbitrarily, and also parts of (P); *all these laws are conventions*." (Einstein 1921, p. 236; my emphasis – FB)

Such considerations of conventionalism were possibly what Einstein had in mind when he told Heisenberg that the theory decides what even *counts* as observable (cf. Sect. 2.1.2). But at the same time, declaring *all* laws conventions—and hence on the same footing in the theory—ultimately makes it possible to hold fast to certain principles *across theory change*, and to give them a much higher status on epistemological or metaphysical grounds than merely constitutive a priori principles (we shall lay this out in more detail below).

The way that Friedman (1999, p. 66) now suggests to single out what parts of a theory should be considered as having a constitutive a priori character, is by appeal to the theory's *invariance group(s)*: While in Newtonian physics, the group of Galilean transformations singles out a unique geometry of the spactime, in GR this is watered down to diffeomorphism-invariance, and "only the underlying topology and manifold structure remain constitutively a priori" (Friedman 1999, p. 66), *not* the geometry of the spacetime.

In modern physico-philosophical discourse, one usually encounters talk of *symmetries* rather than invariances, where "the symmetry of a 'something' (a figure, an equation,...) is defined in terms of its invariance with respect to a specified transformation group, its symmetry group." (Castellani 2003, p. 322) The basic intuition here is this: the elements of the group represent transformations of some 'object' (e.g. a spacetime manifold), and if that object has features which occur when 'viewed from all angles or perspectives', where switches between these 'angles and perspectives' are provided by the group's elements, then these features must *really* pertain to the object, not just as a 'perspectival effect'. This is quite similar to the 'eidetic variation' endorsed by phenomenologist philosophers in the philosophy of mind (cf. Gallagher and Zahavi 2008, p. 27).

But there is a 'trick' employed with the constitutive a priori here, which basically means turning the symmetry intuition upside down: instead of postulating specific sorts of objects *ad hoc* and then looking into their invariant properties, *we let the theory define what sorts of objects there are* by looking into what invariants *it prescribes*. Of course we find the same basic reasoning in Kant's 'transcendental deductions' already, e.g. in the deduction of transcendental consciousness from invariance under different observational situations[25]:

> it is the *one* consciousness which unites the manifold that has been perceived successively, and afterwards reproduced into one representation. This consciousness may often be very faint, and we may connect it in the effect only, and not in the act itself, with the production of a representation. But in spite of this, that consciousness, though deficient in pointed clearness, must always be there, and without it, concepts, and with them, knowledge of objects are perfectly impossible. (Kant 1781, pp. 103–104; emphasis in original)

But in Kant, there is no recursion to a specific theory. And a merely constitutive a priori allows that a given theory and its prescriptions may be found by appeal to all sorts of however pragmatic considerations, to encompass more (and new) evidence. The invariants (symmetries) implied are then a priori, *definitive* of objects, but only relative to that theory.

An interesting corollary that strongly connects to the previous discussion derives from this relativity to theory. Friedman (1999, p. 69) suggests "that Carnap's *L-rules* [...] can be profitably viewed as a precise explication of Reichenbach's notion of the constitutive or relativized a priori." (my emphasis—FB) These L-rules contrast with the aforementioned P-rules that Sellars and others identify with material inferences. L-rules are the "*logico-mathematical* transformation rules" of some language (Carnap 1937, p. 180; my emphasis – FB). The suggestion that these (or maybe only some of them) connect to, or even represent constitutive a priori principles is based on Friedman's (1999, ibid.) observation that Carnap (1937, e.g. pp. 178–179 and 327–328) considers formulae containing the metric tensor in geometries of constant curvature as L-rules, but in GR, where there is a mass-energy-dependent curvature, these become P-rules. Thus in Carnap's exposition, "geometry has itself undergone a transition from a nonempirical and constitutive status to an empirical and thus nonconstitutive status" (Friedman 1999, p. 69), much in the same way as suggested by Reichenbach (1920). P-rules (material inferences), to recall, may be seen as supplying at least *inferential* content to certain statements. So with theory change certain a priori principles may be demoted to merely factual statements that can become *false* under certain circumstances, and accordingly, the inferential content of one's belief-system may be altered or diminished quite drastically in the process.

On the one hand this insight can be used to explain how we can still *talk* of (material) 'objects' in QM, even if certain features are missing (certain NQMCs about them are 'disallowed') that we might have though *must* be there on any 'object'. We shall demonstrate this in more detail below, but the general reason is that what is constitutive of an object of a given sort may radically change with acceptance of a new theory.

---

[25]Cf. also Allison (2015, ch. 7) on this point.

There is, then, an impact on the semantic/referential gap here, that we claimed above was there in virtue of the loss of descriptivity of quantum states (cf. point (a) on p. 332 in this document): It seems that, once one is willing to accept *those* 'objects' that QM prescribes, there really is no such referential gap anymore. QM has components that make reference to objects possible just as much—*however much that may be*—as it is possible in classical theories and ordinary language, even if the quantum state is not a description of the system but a predictive tool that allows or disallows certain purported descriptions to greater or lesser extent. In particular, objects simply do not *have to* have the same sort of '*causal* etiology' in QM as they do in classical mechanics. The 'causal story' provided by the unitary evolution, which ensures that there *is* an object between creation and annihilation events to which we can (contextually) assign *some* properties, is enough.

But we also see, on the other hand, that inferential content may be altered or diminished with theory change *in far more cases* than just in the case of QM. This incidentally seems like a much 'deeper' connection between QM and GR than the similarities pointed out by Bohr (1935, pp. 701–702) at the end of his reply to EPR—or possibly, it is an elucidation of Bohr's underlying thoughts. This relativizes the often acclaimed 'weirdness' of QM.

To see the connections more clearly, take the following example. While it is constitutive of a triangle in *Euclidean* geometry to have angles that sum to π, in *non-Euclidean* spaces 'this is a triangle, so it has angles that sum to π' becomes a material inference that may or may not be supported by the underlying geometry. Now imagine, more concretely, a possible world in which physical space behaves geometrically like the surface of a solid sphere. Here the angles of some triangle would sum to $\pi + A/a^2$ instead of π, where A is the triangle's area and a the radius of the sphere (e.g. Hartle 2003, p. 18). For creatures living in such a physical space, the material inference just mentioned would be 'licensed' by that geometry *only for very small scales, compared to the size of the sphere*. And encountering only small scales could be the normal, everyday-life circumstances for these creatures, so the fact that their space is curved like a sphere's surface could go unnoticed for long periods of time, and recognition of it would require extraordinary evidence. This should be compared to the licensing of statements about definite properties of systems by their degree of immersion into an environment, which we can consider as quite high under normal, everyday-life circumstances, and to the fact that 'quantum features' went unnoticed for a long period of time in physics.

These are important but still frail connections to QM, and to make the connection more intimate, the procedure of determining what may count as constitutively a priori should be applied *to QM* just in the same way as suggested by Friedman for spacetime theories: Determine the (relevant) invariance group(s) of the theory and then find out the invariants.

For the a priori constraints on what it is to be an object in non-relativistic QM, this was already executed by Mittelstaedt (1995, 2009),[26] with the upshot (cf. Mittelstaedt 2009, p. 857) that the object is that which will be invariant under a unitary representation of the group G_{10} of Galilean transformations.[27] For any given situation, one can (ideally) single out a set of commuting projectors, each from a different PVM, in turn representing a different respective quantum observable. These projectors are then construed as *propositions*, i.e. property ascriptions in the sense of Sect. 2.1.5. However, it is a well-known fact that the *totality* of projectors so construed forms a *non-distributive* set or 'lattice',[28] such that not all properties that could be ascribed classically can be simultaneously ascribed *via* projectors. In any given situation one can hence single out only a subset (sublattice) of *commuting* projectors which then *is distributive* ('Boolean'). Since the *entire lattice* is invariant under a(n irreducible) unitary representation of G_{10}, and so also the *ascribability* of a sublattice from the total lattice to any observational situation, an object in QM can be constituted as "a *carrier* of properties" from the non-distributive lattice, "but not only in one contingent situation, which is given by the [measurement] apparatus and its space time coordinates, but also in all situations which can be obtained by Galilei-transformations." (Mittelstaedt 2009, p. 857; my emphasis – FB)

This is exactly an expression of the fact that QM licenses only a restricted set of NQMCs in conjunction: objects constituted in terms of these 'properties' do not feature sharp positions and momenta at the same time; NQMCs like 'system S is exactly this and that fast' and 'system S is here' are not both simultaneously licensed by QM, in any given reference frame. But the lattice itself remains.

Mittelstaedt's reasoning, however, only concerns non-relativistic QM, and did we not say that theorems by Malament (1996) and Halvorson and Clifton (2002a) put restrictions on *localizability* of particles in relativistic QM? True, but this does not affect the attribution of properties, even approximate localizations in a given situation, since the same procedure can be repeated using the *Poincaré group* instead of G_{10}.

[26] Mittelsteadt makes no mention of Friedman or Reichenbach or of the difficulties with the *synthetic a priori* and d'Espagnat's tenet (II). But his aim is to show how "objects can be constituted as new entities by means of invariance properties *of the theories in question*." (Mittelstaedt 2009, p. 847; my emphasis – FB) Since this is understood by him as an execution of the "Kantian way of reasoning" (p. 851), he seems to *unconsciously* appeal to a relativized a priori.

[27] 10 is the number of independent parameters needed to specify the elements of the group: 3 rotational, 1 time- and 3 space-translational, and 3 for velocity-shifts (e.g. Ballentine 2000, p. 68).

[28] *Lattices* are algebraic structures, certain of which ('Boolean' ones) can be understood as an abstract formulation of propositional logic. Lattices, interpreted thus, are basically sets of propositions which allow for combination by material implication, dis-, and conjunction. A complemented lattice has a negation and a Boolean one additionally allows for the usual distribution laws between disjunction and conjunction. Orthogonal projectors on a Hilbert space can be used to construct a non-Boolean *qua* non-distributive lattice in virtue of non-commutativity, and so define a non-classical 'quantum logic'. Cf. Redhead (1987, p. 176) for an elementary introduction and Piron (1976) or Bub (1997) for more detailed treatments.

In fact, using the same basic group theoretical methods with the Poincaré group singles out two properties as identifiers for *classes of particles*:

> The state space of a free elementary particle is the Hilbert space for an irreducible representation of the Poincare group.[...] The group-theoretic analysis shows there are two characteristics that are invariant under relativistic transformations. These characteristics are identified as the mass m and spin s. [P]ure relativistic considerations single out mass and spin as indices for the classification of various free elementary particles and put certain constraints on their values. (Auyang 1995, p. 37; emphasis omitted)

This is a consequence of a famous investigation by Wigner (1939), and we can see that the exact same general recipe is at play for constituting a particle as a 'carrier of properties' in the relativistic domain: finding out what is invariant under the relevant transformation group. Letting the theory constitute objects as carriers of properties in this way explains much of the talk of somewhat philosophically minded theoretical physicists, e.g. how Streater (1988, p. 144) can claim that

> Wigner [...] did not merely say that a particle is *well described* by [an irreducible projective representation of the Poincaré group]: this would leave the word particle still undefined. Thus a particle *is* a pair $(\mathcal{H}, \hat{U}_{m,s})$ where $\mathcal{H}$ is a Hilbert space, and $\hat{U}$ is a unitary continuous action of [the Poincaré group] on $\mathcal{H}$ [...]. (notation adapted; my emphasis—FB)

However, particles were supposed to be *pointlike* in interactions. Surely this fact is impaired by the aforementioned results in relativistic QM and QFT? Again the answer is 'no', in relevant cases, since once more considerations of invariance under transformations give rise to the constitution of objects which are 'pointlike' *in interactions*, in the sense of being 'structureless'. This was demonstrated as follows by Falkenburg (2007, p. 133 ff.), herself following Drell and Zachariasen (1961, p. 8).

Quite generally, scattering cross sections can be made 'dimensionless' by multiplying through by the appropriate terms, such as in $\left(\frac{\hbar c}{E}\right)^2 \frac{(ZZ'\alpha)^2}{16\sin^4(\theta/2)} \mapsto \frac{(ZZ'\alpha)^2}{16\sin^4(\theta/2)}$ for Rutherford scattering (cf. Falkenburg 2007, p. 134) If such a dimensionless quantity turns out to be *scale invariant*, i.e. "does *not* depend on any length, one concludes that the scattering center and the probe particles are structureless or *pointlike*." (Falkenburg 2007, p. 133; emphasis in original) This length scale is of course also coupled, through the de Broglie relations, to the energy scale used in the scattering.

In QFT, cross sections will include a matrix element which is then often transformed into so called 'form factors', generally dependent on the momentum transfer in the scattering. This is the case, for instance, in the cross section for Rosenbluth scattering as discussed by Drell and Zachariasen (1961, p. 8), where two form factors $F_{1/2}(q^2)$ are involved (q the momentum transfer between two scattering particles). Anticipating Falkenburg's analysis, Drell and Zachariasen (1961, ibid.) have it that "a particle has [...] structure—i.e. is *not* a point particle—if and only if the functions $F_1(q^2)$ and/or $F_2(q^2)$ are *not* constant." (my emphasis—FB) Hence it *is* 'pointlike' just in case they *are* constants, and the particle's scattering behavior does not depend on the momentum it receives. Or, using a pragmatic, pictorial lingo:

if two particles always scatter in the same way, no matter how 'hard we smash them together', they have no substructures that make them 'breakable into finer pieces' through 'harder smashing'. They are hence structureless or 'pointlike'.

But how can we ever talk of '*this* particle', when for particles independent of any measurement, it is 'inadmissible' (inferentially meaningless) to attribute *any* properties? I.e., how can we even *re-identify* a particular given particle over time? Here Mittelstaedt (1995, p. 1623) provides a treatment in terms of *repeated* measurements and answers some difficulties in terms of POVMs and weak requirements on probabilities and 'narrowness' of the measurement.

But of course for almost all situations, the much more satisfying specification comes from *decoherence theory*. Schlosshauer (2007, p. 67), for instance, describes decoherence in terms of "the environment" functioning "as a ubiquitous 'measuring device' which continuously performs effective measurements [...] on the system" and emphasizes "that this monitoring process does not require a human observer of any sort." In other words: when specific *isolation conditions* are *not* met, we can constitute persistent objects with properties specified by the theoretical mechanism (decoherence). The probabilistic part (the Born rule) just functions as a guide to (long run) expectations about the behavior of the objects thus constituted.

Decoherence, notably, *also involves an invariance* that one appeals to in order to specify the properties of the objects in question: A 'pointer observable' $\hat{O}$ is selected as an observable for which $[\hat{O}, \hat{H}_{\text{int}}] = 0$, i.e. which is invariant under the dynamics prescribed by the interaction Hamiltonian (cf. Zurek 1982, p. 1869). The invariance occurring in decoherence, in other words, determines which properties are constitutive of 'quantum objects' when the latter are immersed in an environment. And these properties will typically include 'unsharp' (approximate) positions *and* momenta. This ultimately allows the switch to a description in terms of NQMCs so favored, i.e. a return to a 'classical constitution' of objects, when these are sufficiently large and one can disregard the 'unsharpness'.

Above we noted that theory change can bring about the demotion of a priori principles to empirical contingencies on the relativized *a priori*-view. And one arguably encounters such a demotion when one moves from classical field theories to QFTs. Recall that renormalization methods form a corner stone of success in canonical QFTs, and that this seems 'odd' because certain quantities like charges become *scale-dependent* if (semi-)group theoretic methods are taken seriously. Moreover, the fact that 'bare quantities' may come out as infinite in renormalization schemes creates a *prima facie* problem. This is typically dealt with by claiming that it "is OK since *e* [the bare charge—FB] is not observable" (Schwartz 2014, p. 309), supplemented with a story about screening by vacuum polarization (e.g. Peskin and Schroeder 1995, p. 256). But the result is the specification of a 'running coupling constant' which depends on the energy (or momentum) scale and the immeasurable, formally infinite 'bare quantity'. As Auyang (1995, pp. 192–193) puts it:

> Renormalization is crucial for the triumph of quantum field theory and the standard model in elementary particle physics. It also shows the limitation of our conceptual ability to analyze the world. The theory is unable to specify the real parameters for the physically significant

entities; finally it has to appeal to experiments. Physicists regard this as a blemish of the theory.

From the present perspective, there is no such 'blemish'. What is and is not constitutive of some particle is relative to a given theory; what constitutes an electron in classical electrodynamics may be different from what constitutes an electron in QED. If coupling 'constants' such as (specifically valued) electric charges become scale dependent, i.e., are not invariants under a relevant transformation (semi-)group in an interacting QFT, they (the specific values) should not be viewed as properties constitutive of the respective objects anymore.[29] The measured, 'physical' quantities center around certain stable regions of the 'running constants'; and, put frankly, they are basically just the values of parameters occurring in a specific model of the respective set of experiments, since we can "*not* compute them from first principles." (Peskin and Schroeder 1995, p. 266; my emphasis – FB)

A little more specifically one finds that, as long as the corresponding Dyson series is valid, with "large energy-momentum transfer [...] the coupling constant gets larger and larger", which in turn provides "the physical picture that electric charge is largest close to a charged source (that is at small length scales), and dies away at large distances." (Lancaster and Blundell 2014, p. 307) This is perfectly acceptable on the view advocated here, and there is not even a need to appeal to infinite 'bare charges' as the 'true' replacements for constant ones, and 'virtual particles' or 'vacuum polarizations' as screenings for these to account for the measured *finite* values.[30]

Finally, we recognize that the precise correlations implied by, say, the singlet state also come out as a consequence of basically two invariances in the following way. The singlet state, first of all, implies $|\boldsymbol{S}| = 0$, meaning that the total spin is zero. This should be so, given an appropriate previous decay-situation; but in the quantum case, it holds *regardless of the axis* (cf. the discussion in Sect. 4.3.3 and e.g. Arntzenius 2012, p. 78 or Basdevant and Dalibard 2002, p. 276 for some calculations). The state is *rotation invariant*. So a *preferred axis* along which the particles are 'spinning' is not constitutive for conservation of angular momentum (spin) anymore. Moreover, the joint unitary evolution of the two has another important invariant: the moduli of the amplitudes. And this, we have urged in Sect. 6.5, should be seen as an implementation of *probability conservation*. This is one important sense, then, in which the correlations implied come out as *objective* features of such entangled pairs: They are a consequence of invanriances that are constitutive for the total particle-pair, and these invariances only secure perfect correlation for a joint axis of measurement, not a preferred axis.

---

[29] In principle the fact that the scaling transformations only form a *semi*group (do not all have inverses) introduces a subtlety; but that much of the literature suppresses the 'semi' can at least be seen as evidence for the fact that these scaling transformations are typically viewed as similarly important for the theory as are proper transformation groups.

[30] Cf. Teller (1988, p. 86) and Falkenburg (2007, p. 238), for some discussion on problems with these notions.

But with a new theory there may come a new a priori: Portmann and Wüthrich (2007, p. 849) have it that "there are theoretical grounds on which to expect a violation of the quantum mechanical prediction of perfect correlations. Some of the different approaches to quantum gravity, that is, suggest that tiny violations of Lorentz group invariance are to be expected." And we would not, upon acceptance of the aforementioned approaches to quantum gravity, expect an explanation of *all* violations of perfectly (anti-)correlated behavior *in terms of experimental error*. This is what it *means* to demote certain principles from a priori to empirical laws. Still, the correlations would be at least *approximately* invariant, similar to the pointer observable in decoherence. In other words: the *strongly* correlated character would remain objective in the flat space limit just as quasi-classical behavior is objective under suitable environmental circumstances.

With an eye on the earlier invariants/symmetries identification, the view of invariants as functioning constitutively in modern physical theories gains evidence e.g. from Weinberg's famous (1992, p. 142) observation that "there are symmetry principles that dictate the very existence of all the known forces of nature." Similarly, Weyl (1952, p. 126) thought that "we are [...] enabled to make predictions a priori on account of symmetry for special cases," and he even went so far as to suspect that "*all* a priori statements in physics have their origin in symmetry." (my emphasis—FB)

How do these considerations connect to the Bohr-Einstein-Kant problematic above, and what do they suggest for interpreting QM 'epistemologically'? Fine (1986, p. 97) holds, on the basis of textual evidence, that for Einstein causality—albeit in the quite strong sense of non-probabilistic laws—formed a crucial part of realism and physical thinking altogether, and in his *Physics and Reality*, Einstein (1936, p. 377) declared it "contrary" to his "scientific instinct", that "we shall never get any inside view of [...] alterations in the single systems, in their structure and their causal connections [...]." It seems that when Einstein declares such principles or foundational concepts as "the concept of causality [...] freely chosen conventions" (Einstein 1949a, p. 13), this ultimately allows him to give them a much *more* fundamental status than merely theory-relative axioms of coordination.

There is good evidence for an influence of Poincaré's philosophical writings on Einstein and Einstein's ultimately favoring Poincaré's conventions over the Kantian notion of an a priori altogether (cf. Howard 2010, p. 340 ff. and references therein). And Poincaré (1918, pp. 148–149) appeals to "a sure instinct" as a guide to the *choice* of conventions, or the 'grasp' ("vague consciousness") "of I know not what profounder and more hidden geometry" in the context of choosing geometrical axioms. It is at any rate conceivable that Einstein reserved a similar place for the choice of conventions in his thinking. He famously laid great emphasis, for instance, on the *simplicity* of the chosen conventions (cf. Einstein 1934, p. 167), and at the same time believed "that *in Nature* is actualized the ideal of mathematical simplicity" (ibid.; my emphasis—FB). This once more expresses his underlying (external and scientific) realism.

Einstein was also influenced by considerations of *theory holism* in the Duhemian, confirmational sense (cf. Fine 1986, p. 89; Howard 2010, p. 341), as evidenced by the fact that he believed the geometry/physics divide could be shifted quite arbitrarily. His only concerns regarding such holism were *contingent, time-dependent* limitations of incomplete theories; to him Poincaré was right *sub specie aeternis* (cf. Einstein 1921, p. 236). Thus Einstein, as we have reconstructed him, is at liberty to 'conventionally' accept such principles as (SEP) and (CEL) as suggested to him 'by a sure instinct', as indispensable, eventually, for science *as a whole*, and as ultimately referring to an external reality.

The alternative that we have advocated above now is to *systematically* shift the boundary between what counts as an axiom of coordination and what counts as an axiom of connection, i.e. what counts as a priori and what as empirical; and this requires relativization to a given theory—the theory that currently fares best regarding empirical success. This strong reference to success may be counted as a *pragmatic* element regarding theory choice, a pragmatic element of the kind that Folse (1994, p. 123) believes was also endorsed by Bohr:

> The original Kantian defense of the categories was based on the grounds of their pure necessity. For the pragmatized Kantians the defense of the categories lies in their utter practical contingency.

This "defense of the categories" certainly manifested in Bohr's writings in his claims that "however far the phenomena transcend the range of ordinary experience, the description of the experimental arrangement and the recording of observations must be based on common language", and that the "formalism, known as quantum mechanics, in which the elementary physical quantities are replaced by symbolic operators [...] can be regarded as a rational generalization of the conceptual framework of classical physics." (Bohr 1963, pp. 11–12)

But how does this pragmatism in the form of a 'rational generalization' of physical quantities relate to the constitutive a priori? Whether this is (once more) entirely fair to Bohr or not, a clear example for the sort of pragmatism we want to suggest here can be given by appeal to *local gauge invariance*.[31] Electromagnetism has, as is well known, a *gauge freedom* in the choice of the electromagnetic potentials. This means that in the 'gauge field' $A_\mu = (\phi, -\boldsymbol{A})$, a shift $A_\mu \mapsto A'_\mu = (\phi + \frac{\partial \chi}{\partial t}, -\boldsymbol{A} - \nabla\chi)$, short: $A_\mu \mapsto A_\mu + \partial_\mu \chi$, for some arbitrary (differentiable) function χ leaves the measurable quantities $\boldsymbol{E}$ and $\boldsymbol{B}$ invariant.

Now in the electromagnetic Hamiltonian, one will have to include $e\phi$ as a term for the potential energy contributed by $\boldsymbol{E}$ and replace $\boldsymbol{p}$ in the kinetic energy by $\boldsymbol{p} - e\boldsymbol{A}$ to ensure that the Lorentz-Coulomb force law $\boldsymbol{F} = e(\boldsymbol{E} + \boldsymbol{v} \times \boldsymbol{B})$ derives from it. In the Hamiltonian *operator* of the corresponding SE, however, this means that $\frac{(-i\hbar\nabla - e\boldsymbol{A})^2}{2m}$ now acts on a wavefunction ψ instead of just $\frac{(-i\hbar\nabla)^2}{2m}$ (similarly for the DE; e.g. Griffiths 2008, p. 358). A shift $A_\mu \mapsto A_\mu + \partial_\mu \chi$ now in turn implies the need for a simultaneous shift $\psi \mapsto \psi e^{-\frac{i}{\hbar}\chi}$, where χ, since it depends on $\boldsymbol{x}$ and t, is a *local phase*. In QED, A_μ is replaced by an operator-valued field $\hat{A}_\mu$, and it, being the creation operator of the photon, becomes the object of primary interest. Thus

[31] E.g. Healey (2007, p. 5 ff.) for the following.

the local gauge invariance of QED ultimately derives from *practical* considerations such as the conservation of the phenomenologically valid Lorentz-Coulomb force law.

Subsequent QFTs "share with electrodynamics the attractive feature[...] that the existence and some of the properties of the gauge fields follow from a principle of invariance under local gauge transformations." (Weinberg 1996, p. 1) An originally pragmatically endorsed principle thus dictates, if promoted to an a priori constraint, the very structure of the theory or the features that are objectively there according to it. This we take to be the essence of a 'pragmatized Kantianism', with an added (explicit) endorsement of constitutive a priori principles.

That this is at least somewhat in line with Bohr's thinking can be gathered from Faye's book-length investigation of Bohr's philosophical thinking. Faye (1991, p. 78), who has Bohr heavily influenced by the Danish philosopher Harald Høffding, holds that for Høffding

> categories of cognition emerge from an analysis of different kinds of judgments. [...] But opposite to what Kant thinks this does not make them a priori, since the logic of concepts is not fixed, according to Høffding. He holds the pragmatic view that the categories of cognition reflect the need to synthesize experience, and are thus relative both to our needs and to the experience which must be synthesized.

Separating *choice a priori* from *a prior givenness*, as suggested by Reichenbach (1920) and d'Espagnat (2011), we can say that it *still* makes them a priori, but only relative to the needs catered by a certain theory.

What is the philosophical lesson to be drawn from all this? Accepting a 'split' between the world as experienced and the world as it *is*, in concert with both, pragmatic values in theory choice and an a priori in the theory-relative, constitutive sense, one finds a path to a properly so called epistemological interpretation of QM: QM is just the *most startling expression* of how we construct a reality out of experience. And by trusting the theory in its guidance to expectations about future observations as well as its invariant properties as a guide to what is objectively there, we come to a world-construction that by far exceeds our ordinary modes of thinking.

This attitude allows one to *come to terms* with the apparent radically counter-intuitive character of QM as a science[32]: It is not like we *ever had* a privileged epistemic access in our pre-scientific intuitions anyway or that the categories of

[32]This 'coming to terms' is certainly reminiscent of Friederich's (2015) *therapeutic approach* to QM (therapeutic in the sense of Wittgenstein (1968)), whereby one is 'cured' from the OP by viewing QM (very roughly) as an *activity* that is *constituted* by certain rules (state assignment, decoherence mechanism, Born rule,...) and proceeds without problem. In Boge (2016a, p. 6), it was argued though, that this still raises the question why *exactly that activity* results, when one applies the rules; and for that reason we here pursue a quite different course. Friederich's own turn to the block universe to make sense of EPRB-correlations (his p. 143 ff.) and his defense of the in-principle possibility of sharp values for all observables (in the sense of hidden variables; his p. 161 ff.) may serve as evidence that his therapy has not really worked on himself—both considerations would otherwise be obsolete.

our 'ordinary' understanding were privileged in *that* sense. Why not let scientific progress guide us to entirely new 'categories'?

But now who do we agree with, Einstein or Bohr, the scientific realist conventionalist or the pragmatic Kantian? For did we not say, at some point, that *causal interpolability* is somewhat *constitutive* of an empirical reality, and that failure of such signals a 'gap' therein? And did we not deny that such causal interpolability is, in a relevant sense, secured for all physical situations? The answer is this: we agree with neither completely and with both in certain respects. Einstein is driven by certain *intuitions*, of which, we have argued, causality is a particularly important one. Thus when Einstein claims the 'right' to 'freely choose conventions' and lets his gut intuition guide him to principles like (CEL) or (SEP), he denies the possibility that scientific progress can spawn off theories that so radically change what is or is not constitutive of what is or is not *there* as does, arguably, QM.

We agree with Einstein to the extent that giving up on certain principles, such as 'causal closure', i.e. the in-principle possibility of finding a cause for any instance where we intuitively demand it (e.g. strong correlations between distant events), signals quite a radical departure from common sense intuitions, formative, to some extent, for *previous* scientific practice. But we disagree that sticking to these principles is a recommendable practice. Embracing merely constitutively a priori principles ('axioms of coordination'), rooted in a given theory, lifts the problematic aura from the break with common sense intuitions: It is our (empirically) best scientific theories that *should guide us* to an 'image', an 'empirical reality', formed by *immediate* sense experience (something like Carnap's (1928, p. 86) "Erlebnisstrom") *and* those theoretical concepts and principles that figure constitutively in those theories.

Since QM is understood here as having normative content (it 'licenses' only certain inferences, counterfactuals, NQMCs, etc., to recall), it seems perfectly capable of so guiding us to a constitution or construction of an empirical reality that strongly diverges from previous such constructions. And if the empirical reality thus constructed has *causal gaps* in the sense suggested here (no common cause for the outcomes in EPRB scenarios; no direct causation either), so be it. We thus agree with Bohr to the extent that scientific inquiry may undergo quite radical changes and require such things as "a final renunciation of the classical ideal of causality and a radical revision of our attitude towards the problem of physical reality." (Bohr 1935, p. 697)

To put things more clearly let us, with Putnam (1977, p. 487), refer to empirical reality as "the image"; and let us also acknowledge Sellars' (1963, p. 5) distinction between a *manifest* and a *scientific* image. Then we can, first of all, acknowledge that Kant may have been right about certain principles that are 'unconsciously operaitve' in experience and can only be made explicit by thorough epistemological investigation. And he may have also been right to assert that these are constitutive and a priori for low level, pre-scientific 'theories', leading to the formation of a *manifest image*. But he was wrong about the status of the science of his day as providing synthetic a priori judgements of general validity (tenet (II)). And he may, as a matter of fact, have even been wrong about the continuity between the principles that underlie (constitutively) the manifest image and those that underlie the scientific image of his day (e.g. also Gopnik and Meltzoff 1997, p. 212, for a critical appraisal along these lines). It is conceivable that Einstein was thus fooled, in his insistence on certain principles, by Kant into believing that there should be such a continuity.

Put in terms of the 'images', our above suggestion now is to *embrace* the *scientific* image (or patchwork of images; different sciences will generate different ones) as the best way to handle, in many respects, our difficult relation to mind-independent reality—regardless of how radical the break with our manifest image may be, and *especially* when there are suitable bridging principles (such as provided by decoherence) that specify under which circumstances we can resort to our everyday-life practice of entertaining the manifest image after all.

7.4.3 Issues of Truth and Our Access to the External World

While the relativization of certain a priori principles to a given theory eliminates the problematic status of the synthetic a priori (eliminates d'Espagnat's tenet (II)), and also somewhat lifts the problematic aura from the fact that we should find causal gaps in our scientific image, it does *not* thereby lift the 'radical spirit' off Kant's internally realist doctrine, relative to any metaphysically stronger realism. Putnam (1981, p. 63; my emphasis – FB), e.g., notices the following elements to the Kantian distinction between an empirical world and a noumenal one, which are independent of the status of the a priori:

> On Kant's view, any judgment about external[33] or internal objects (physical things or mental entities) says that the noumenal world *as a whole* is such that this is the description that a rational being (one with our rational nature) given the information available to a being with our sense organs (a being with our sensible nature) would construct. In *that* sense, the judgment ascribes a *Power*. But the Power is ascribed to the *whole noumenal world*; you must not think that because there are chairs and horses and sensations in our representation, that there are correspondingly noumenal chairs and noumenal horses and noumenal sensations. *There is not even a one-to-one correspondence between things-for-us and things in themselves.*

There are three crucial aspects to this quote: (i) that the noumenal world may have a degree of 'wholeness' not suggested to us by low-level theorizing, (ii) that the noumenal world is attributed a 'power' to affect our senses, and (iii) that a thorough internal realism might urge one to give up on the correspondence theory of truth. We will confront these in the order (ii), (i), (iii).

---

[33]NB: 'external' is here meant, as the quote makes explicit, in the 'sensuously external sense'. Kant (1781, p. 22) famously declared space and time "pure forms of intuition", and 'external objects' are meant here as objects *situated in* space and time. When we have talked about an 'external reality' above (and when we do so below), we thereby meant what Kant (1781, p. 372) means by "something [...] *transcendentally* [...] outside us[...]." (my emphasis—FB).

As a first thing, note that (ii) in part demarcates the fine line between internal realism and thorough idealism, according to which there simply *is no* external world. But why would we even suppose such a thing as the existence of a mind-independent world, given that we were eager to make the significance of QM depend on an agent's epistemic situation, i.e. that we let that which is 'objectively there' (in the scientific image) depend on the theory we use, and that we maintained that the very structure of the experienced world is at least in part a function of 'our' (every single user's, to borrow the QBists' phrase) cognitive interior and perceptive limitations?[34]

Strictly speaking, one cannot gain 'absolute certainty' as to the existence of such a mind-independent reality, as Kant was well aware (cf. his "Criticism of the Fourth Paralogism" in the first edition of the *Critique* (p. 367)). But we argued above that it would be puzzling if the correlations implied by entangled quantum states were not due to a 'rigidity' or 'recalcitrance' in nature; and this *experienced* recalcitrance is the strongest reason to infer something 'external' at all:

> We sometimes build up quite beautifully rational theories that experiments falsify. Something says no. This something cannot be 'us.' There must be something else than just 'us.' (d'Espagnat 1995, p. 314)

Schrödinger is accredited by Moore (2015, p. 348) with a quote that goes in the same direction: "Science is a game—but a game with reality, a game with sharpened knives..." With Schrödinger and d'Espagnat we take it, in other words, that besides our cognitive interior there is an external *source* of our sense-experience—a source which may be called a (mind-)independent or external reality, nature, the noumenon, the world in itself, or what have you, and that sometimes disappoints or 'punishes' certain of our convictions, formed (in part) on an a priori basis. If a theoretical expectation is disappointed by experiment, and we here include low-level 'theories' and 'experiments' formative of the manifest image, we will call this '*punishment by nature*'.[35] If an expectation is met, this may accordingly be called 'reward by nature'; if we do not form expectations, we are simply not playing.

---

[34]There are, in fact, also attempts more recent than Berkeley's (who in the end, to recall, merely replaced the external world by an all-perceiving god anyways; e.g. §66 of the *Principles*) of constructing *the impression* of a mind-independent world purely out of "interacting conscious agents" (Hoffman and Prakash 2014, p. 1), and with the ambitious aim to "show that [...] the quantum free [...] wave function [...] is identical in form to the harmonic functions that characterize the asymptotic dynamics of conscious agents [...]." (ibid.) While certainly interesting, these results have at best a preliminary character though, and rest, among other things, on questionable (formal) notions of what counts as a 'conscious agent'.

[35]We have used the term 'nature' here to emphasize the *dynamical* character (usually associated with the term) that this punishment has, *within experienced reality*. We have also used 'nature' above as a synonym for 'mind-independent reality' though, and d'Espagnat (1987, p. 527) equally identifies the use of this word as an indicator that his colleagues who otherwise decree versions of phenomenalism do endorse realist intuitions after all. But we have been careful to *separate* the dynamical content of 'nature' (which is *within* experience) from its strong metaphysical realist content (which expresses the conviction that the *source* of this punishment is *external*). We are *silent*, in other words, about the (non-)spatiotemporality of a mind-independent reality here, in contrast to Kant.

But besides the expression of a weak metaphysical realism, there is a second component to (ii), namely that the source is attributed a *power* to affect our senses in a particular way. This sounds as if *causation* was at play between this source and sense-impressions; and it was (again) Schrödinger (1964, p. 64) who commented that

> any effective causal relation between that 'existent' something and the ideal world constructed of simple data would be an entirely new relation and very much in need of explanation, having nothing whatever to do with the nexus of causality within the ideal world[...].[36]

We fully agree with Schrödinger here that the relation between the source and the experienced world cannot be causation proper, and that 'causation' would mean something entirely different if one would use the term in that context. But we disagree that the relation is in much need of explanation; for what could that explanation look like? We could only ever hope to accidentally 'grasp' the true nature of such a relation, e.g. by Poincaré's 'sure instinct', or by privileged metaphysical intuition.

The concept of an external source is introduced exactly for what we have used it: as a *tout court* explanation of the fact that certain correlations occur to us that we had never dreamed of, that some expectations are being disappointed in experiment and observation, that our actions and successful theorizing are in many ways limited. In other words: it acts as a *stabilizer* for the structure of experience; and the relation between source and empirical world is hence best described as that: a *stabilization* of the latter by the former.

Calling the source a cause of the correlations in EPRB-scenarios would incidentally run the risk of replacing the small λs by a capital Λ, and to give a causal account after all. This *need* not be construed so, since that Λ could be thought of merely as the cause of the experienced *correlations*, not of the correlated *values*.[37] Still, in the light of all that has been said it seems more sensible to say that the correlations, to the extent that they occur in actual experiments (or in experience more generally), are stabilized by an external source. Period. *That* is why we can endorse counterfactuals such as 'if Alice and Bob were to measure along the same axis, they would find opposing spin values'. We have been guided, by experience and formation, reformation, and reevaluation of relatively a priori principles, to a theory which has these implications. And we endorse it since it predicts successfully and robustly unlike almost any other theory. So we assume that it picks out *stable facts* in experience.

[36]Cf. also Allison (2004, chapter 3), for a detailed discussion of this problem.

[37]An interesting proposal along somewhat similar lines is defended by Gebharter A, Retzlaff N (*A new proposal how to handle counterexamples to Markov causation à la Cartwright, or: fixing the chemical factory*, unpublished manuscript), who introduce what they call "common cause triggered non-causal dependencies". The suggestion is to take, in the EPRB case, the quantum state as a common cause of the two subsequent detection events, but not of the correlated values. The correlated values are then rather explained in terms of a different, nomological dependency that does not allow for the kind of intervention that a causal connection would.

We can thus also make sense of the 'authoritative', normative character of QM that we accepted on account of the arguments by Healey, Stairs, and Friederich; and we can equally make sense of how it could be that a physical situation should *warrant* a state-assignment, even if one is not aware of this warrant (cf. Healey 2015, p. 4): Whenever one faces suitable backing conditions (*in experience*) for a particular QSA, one is *punished* by nature in the advice conditions in case one fails to assign that particular state so 'backed'; and if one makes that particular QSA, then one is *rewarded* by nature in the advice conditions instead. Backing and advice conditions are empirical, but reward and punishment are stabilized by an external source. In this way, we make up for the lack of fulfilling part (iii) of the DEC in Healey's account, the underspecification of the meaning of terms like 'physical reality', 'warrant', and so forth.

Turning, thus, to point (i) in the quote, the 'wholeness'-aspect Putnam attributes to Kant's noumenal world, we shall only say this much: Clearly, objects in experience are constituted or constructed in accord with contingent, theory-relative a priori principles that are subject to pragmatic revisions, on our account. So we can perfectly well allow that there are 'not *really* really' separate 'noumenal electrons', 'within' the source. And QM, of course, *has holistic* features that allow us to *reason about* experience, in which we *do* discern multiple objects,[38] in a way that these objects *need not* be 'mirrored' by multiple noumenal ones. But we must be careful as to how much we 'ontologize' these features, because we here understand QM precisely *not* in an ontological fashion.

Healey (2009, pp. 295–296), moreover, distinguishes three types of holism: (a) *explanatory*-, (b) *property*-, and (c) *ontological* holism; the convictions that (a) the *behavior* of a *system* cannot be *explained* in terms of its parts, that (b) the *properties* of a *whole* are not *determined* by the properties of its parts, and that (c) some *object* does not even *have* proper parts. Explanatory holism Healey finds in interference experiments of composite molecules like fullerenes, explanations of which are not provided in terms of the behavior of the fullerenes' parts, and property holism in e.g. the singlet state that specifies properties of the joint system (if any) which cannot be analyzed in terms of properties of the components. This is an important point: We claimed that, in virtue of an invariance (rotation), QM constitutes the particle-pair in a singlet state to have properties that do not reduce to or are not determined by the properties of the individual particles ($\hat{S}^2 |\chi\rangle = 0 |\chi\rangle$, whereas, say, $\hat{s}^2 \otimes \mathbb{1} |\chi\rangle \not\propto |\chi\rangle$). So the particle-pair has property-holistic features *in the image*; the image *violates* Einstein's (SEP).

But the temptation to also ascribe an *ontological holism*, in the sense that the noumenal world 'really is' indivisible in some respects, should be resisted. All that Putnam claims in the quote from which we have gathered (i) is that one is not at liberty to conclude, from the fact that we experience and theoretically constitute

[38] Cf. in particular the investigations of Muller and Saunders (2008), Muller and Seevinck (2009), and Caulton (2013), in this connection.

several objects, that for every object thus constituted there is a noumenal one. Reading into this that the noumenon or external source is just an 'indivisible whole' would fly in the face of the underlying internal realism. What QM allows, with its holistic features, is to renegotiate what *counts* as 'two separate objects' *in the scientific image* and what does not. This *also* allows for the *possibility* that the external source may stabilize our experience in whatever holistic ways. But nothing of that sort may be concluded about its 'mind-independent constitution'.

Another source of holism in QM was first proclaimed by Bohr (1963, p. 2). He therein has it that "Planck's discovery of the elementary quantum of action[...] revealed a feature of *wholeness* inherent in atomic processes" (emphasis in original) which he explicates (his p. 4) such that "interaction between object and apparatus [...] in quantum physics [...] forms an inseparable part of the phenomenon." In other words: what really counts as 'the phenomenon' cannot be strictly separated from what counts as 'the apparatus used to investigate it'; both form an 'indivisible whole'. Based on the modern treatment of measurement interactions, Zurek (2007, p. 3) similarly takes note of the somewhat arbitrary character of distinguishing between 'system', 'apparatus', 'environment', and so forth, in any entangled state. Building on this, he thinks that "*in absence of systems* measurement problem disappears [*sic*]: Schrödinger equation provides a deterministic description of evolution of such an indivisible Universe [*sic*], and questions about outcomes cannot be even posed." (my emphasis—FB) But Zurek's holism is thus just a holism of the ontological sort and hence not of interest to us here.

What about Bohr's? Bohr is aiming to introduce his notion of *complementary* here, which becomes clear when he later (Bohr 1963, p. 4) writes that "any attempt at a well-defined subdivision would demand a change in the experimental arrangement incompatible with the definition of the phenomena under investigation", and that "the notion of *complementarity* simply characterizes the answers we can receive by such inquiry, whenever the interaction between the measuring instruments and the objects forms an integral part of the phenomena." (ibid.; emphasis in original) In other words: To *fully* charatcterize the object (in terms of position, momentum,...), we would have to use a different aparatus; so one can never perfectly discern the object 'itself', but must resort to 'complementary' descriptions, based on mutually exclusive setups. In this sense, the phenomenon cannot strictly be separated from the respective setup; the 'momentum of the particle' is a feature of the entire setup designed to measure it.

To us, Bohr has parts of the story somewhat upside down. Relying on the role of theory-relative a priori principles, QM *in full, with* decoherence, provides a better guide to the—context sensitive—constitution of objects, since it also specifies under which conditions one *can* resort to a classical physical picture or rely on elements of our manifest image. Bohr's comments are mostly of interest for a partial restoration of classical *physics* in the *historical context*. Still, in practice one certainly often shifts between *different* scientific images, 'classical' and quantum, and appeals to everyday-life vocabulary in a quite arbitrary manner and without any considerations of decoherence. And we are, of course, more acquainted with the non-quantum images. To the extent that this is what Bohr (who could not have availed himself of decoherence yet) basically had in mind in stating such things as "the description of the experimental arrangement and the recording of observations must be given in plain language, suitably refined by the usual physical terminology" (Bohr 1963, p. 3), it seems completely agreeable.[39]

As we have seen, both these latter allusions to holism are of relatively low significance for the present investigation, and the freedom QM leaves to group together certain systems into an 'environment' only acquires relevance through backing conditions for quantum states, the decoherence mechanism, and the symmetries involved therein.

Finally, we turn to point (iii) in the Putnam-quote on Kant, the need to abandon the correspondence theory of truth on account of internal realism. While our discussion must fall short of an exhaustive treatment of the subject matter, we should nevertheless utter some comments.

We have clearly seen Healey advocate an abandonment of the correspondence theory of truth, which is why we had identified his position as closely related to internal realism. But we had cautioned against accepting his inferentialist pragmatism in full as a theory of meaning, and have offered ways to make sense of the uses of 'meaning' in physics which do not impair the fact that we can still understand those statements whose 'meaningfulness' is disputed. Here, however, *truth* is at stake. Note first that in interlude I, we argued, following Brown (1988), that Putnam is concerned with semantic considerations, whereas Kant is concerned with epistemic ones, in the formation of an internal realism. But now we have seen Putnam attribute to the *Kantian* position the consequence that one cannot entertain the correspondence theory of truth, which is so intimately connected with *reference*.

The crucial point is that Putnam (1981, p. 60 ff.) *reads Kant* as in need of disposing of the correspondence theory of truth in his defense against *Berkeley*,[40] who appeals to a "*similitude theory of reference*" (Putnam 1981, ibid.; my emphasis – FB) to dispute the existence of a *bare substratum* (cf. Berkeley 1710, §16), after having declared all sensible qualities *secondary*, i.e. dependent on the existence of minds (cf. Berkeley 1710, §9–10). In other words: while Putnam is *motivated* more directly by model-theoretic concerns and by considerations of *reference* (cf. in particular Putnam 1977) and Kant was *motivated* by epistemological ones (cf. in particular Kant 1781, p. 374 ff.) both ultimately end up, to a large extent, with the same conception (internal realism) and both feel the need to reject the correspondence theory of truth. This seems like a reasonable, reconciliatory view on these issues.

---

[39] In a quite similar vein, Camilleri and Schlosshauer (2015, p. 74) understand decoherence as "only the last step in a long line of attempts to undergird (or supplant) Bohr's doctrines by an explicit dynamical and physical account", noting that "[s]uch approaches were already pursued by a number of Bohr's followers [. . .]in the 1960s, who, far from seeing it as an invalidation of Bohr's basic insight, regarded it as providing a justification of his views."

[40] It has sometimes been argued that Kant was not even well-acquainted with Berkeley's writings when he wrote the first edition of the *Critique*, but there is also good historical evidence to the contrary (cf. Turbayne 1955).

Let us make things a little more precise though, and thereby reap some systematic insights about the palatability of the correspondence theory of truth, rather than just supplying Kant/Putnam exegesis. First of all note that the correspondence theory of truth, to recall, goes back at least to Aristotle's *Metaphysics* (1011b), where "truth" is identified as "the assertion that that which is is and that that which is not is not." Tarski (1944), trying to capture the intuition behind the Aristotelian notion, advanced his famous (T)-scheme, "*X is true if, and only if, p*" (his p. 344), where '*p*' may be replaced by any sentence of some language, capable of being true, and '*X*' by the 'name' of that sentence (cf. ibid.). To avoid such things as the liar-antinomy, Tarski (1944, p. 349) famously introduced a *meta-language* which would contain the names (X) of the sentences as well as the truth predicate.

Thus, investigating properties of *languages*, Tarski (1944, p. 345) called his conception of truth "semantic", where by 'semantics' he meant "a discipline which, speaking loosely, deals with certain relations between expressions of a language and the objects (or 'states of affairs') '*referred to*' by those expressions." (my emphasis—FB) We have expressed reservations about 'meaning' being exhausted by considerations of reference, so semantics, the study of (linguistic) meaning, may be concerned with more than Tarski says. The crucial point, though, is that on the correspondence theory of truth, truth and reference become inseparable. Put frankly, truth may be thought of as the successful reference of a declarative sentence of some language to an external states of affairs.

Herein lies the crux indeed. If correspondence requires reference to external states of affairs, then this means that the relations and properties expressed by the predicates of the language refer to relations and properties *in the outside world*. This cannot be an account of truth appropriate to an internally realist view, for then it would immediately collapse into strong metaphysical realism. But it is crucial to note that the correspondence-view of truth, stated in these terms, is already problematic *by itself*.

To wit, da Costa and French (2003, p. 10), referring back to Tarski's (1935) more detailed investigation of *formal* languages, have it that

> in order to talk rigorously of truth [...], we require not only a language $\mathbb{L}$ but also an *interpretation* $\mathcal{I}$ of $\mathbb{L}$ in a *structure* $\mathcal{A}$. This is what the metalanguage provides. A sentence of $\mathbb{L}$ is then true or false only with reference to $\mathcal{I}$; that is, truth and falsity are properties of sentences of a particular language $\mathbb{L}$, in accordance with an interpretation $\mathcal{I}$ for $\mathbb{L}$ in some structure $\mathcal{A}$. (my emphasis—FB)

A *structure*, $\mathcal{A}$, is (in the simplest case) formally defined as an n-tuple $\langle A, R_j \rangle_{j \in J}$, where A is a non-empty set of elements, representing a universe of discourse, and the R_j are k_j-ary relations defined thereover (cf. da Costa and French 2003, p. 38 for details). Note that k_j could also be 1, whence the 'relation' would really be a monadic property, and that any relation could be (e.g.) so constrained that it is really a *function*. The relations in such a structure are themselves constructed out of A by means of the cartesian product and power set operations, i.e. as sets of tuples of elements from A, or sets of tuples of tuples, or sets of tuples in which elements from A and tuples occur, or... and the interpretation $\mathcal{I}$ then assigns values in A to the constant symbols of $\mathbb{L}$ and values in $\langle R_j \rangle_{j \in J}$ to the predicates (cf. da Costa and French 2003, pp. 10 and 35 ff.). A structure thus specifies the properties of and relations among the referents of language $\mathbb{L}$ that we believe to obtain when we utter declarative sentences of $\mathbb{L}$. $\mathcal{A}$ may hence be said to provide a *model* of $\mathbb{L}$.

A formal language is obviously conceived of here as entirely *syntactic*. If that language is not interpreted, it is merely a concoction of symbols according to formation rules. These considerations may not transfer seamlessly to natural languages, but they can nonetheless be put in the service of natural-language semantics in multiple ways (e.g. Cresswell 2006, for an appraisal). Since there is, however, typically an *intended* interpretation for the sentences in a (formal or informal) language $\mathbb{L}$, "we may assert $\mathcal{S}$ of $\mathbb{L}$ as being true, without explicit mention of the interpretation, and may even forget that $\mathcal{S}$ is part of $\mathbb{L}$. If we are to be rigorous, however, we must consider the interpretation."

Given these prerequisites, a foundational problem for the correspondence theory of truth now arises as follows. According to da Costa and French (2003, p. 17), "[w]hen we say that some sentence $\mathcal{S}$ is 'true,' we may interpret it as strictly saying that $\mathcal{S}$ is true in a certain structure or model that represents a portion of reality" and that this means that "$\mathcal{S}$ can be said to '*point*' *to* the world *by means of a model* [...]." (my emphasis—FB) But the relation that we are really interested in, in the correspondence theory of truth, is not the relation between *model* and *sentence* (syntactically conceived), but between *interpreted* sentence and '*world*', i.e. the 'representation' or 'pointing to', as da Costa and French have it.

More precisely, da Costa and French (2003, p. 17) let a "domain of knowledge Δ" (which we can identify as the aforementioned "portion of reality") be represented, aspect wise, "by a 'data structure' $\mathcal{D}$", to which sentences of language $\mathbb{L}$ refer by means of some model $\mathcal{A}$: "aspects of $\mathcal{A}$ model $\mathcal{D}$[...]." (ibid.) *Here* we (typically) encounter *isomorphisms*, at least partial ones, between the model $\mathcal{A}$ that supplies meaning to the sentences of $\mathbb{L}$ and the model of the data $\mathcal{D}$. $\mathcal{A}$ then "effectively substitutes for Δ in our thought." (ibid.) In other words: we believe to 'grasp' reality, Δ, by means of some model $\mathcal{A}$ of language that isomorphically models aspects of (structured) data, $\mathcal{D}$.

Now a first difficulty with this is that "the manner in which the various elements of $\mathcal{D}$ are related to the 'objects' (for want of a better word) of Δ is [...] problematic", and da Costa and French even acknowledge that "it may be that, as Wittgenstein suggested, the nature of this relationship lies beyond linguistic expression." (da Costa and French 2003, ibid.) A second difficulty is that "strictly speaking, [...] an *isomorphism* cannot be said to hold between $\mathcal{A}$ and Δ, since this relation is rigorously defined as holding between *formal structures* only." (my emphasis—FB)

A correspondence view of truth clearly aims for an isomorphism, i.e. a (bijective) *structure preserving* map between an external reality and a(n interpreted) language. Our formal and informal languages, when interpreted, are supposed to (ideally) *match exactly*, or at least to a considerable extent,[41] the relational and non-relational *facts*, the states of affairs that are 'really there', by means of their predicates, function symbols, constants, etc. But an isomorphism cannot even be strictly *defined* between something that is *not* a formal structure and something that *is*; and suggesting that the relation between 'objects' of Δ—which one only talks of for want of a better word—and elements in $\mathcal{D}$ (the "data structure") is beyond linguistic expression does not imply any stronger commitments than does our 'stabilization'-condition above. In other words: A careful investigation such as that of da Costa and French already pretty much '*internalizes*' *truth*.

What, then, *are* truth and reference? Putnam's (1990, p. 41) appeal to "idealized rational acceptability" and "creatures with 'a rational and a sensible nature[...]'" seems to replace a problematic notion ('truth') by a host of other problematic notions ('rationality', 'ideal', 'sensible nature'...). Putnam (1981, e.g. pp. 14 ff. and 42 ff.) famously also advanced a *causal* theory of *reference*, according to which "a close causal connection" (his p. 14) between the use of a specific word of some language and its referent is required for (successful) reference. But this either makes reference a relation internal to the image as well (basically Putnam's choice) or else verges on questions of the possibility of causation between empirical and independent reality again; and, much worse, the *justification* for such causal theories of reference arguably relies far too much on specific intuitions, some of which have been put into doubt by cross-cultural studies with radical differences in linguistic intuition between westerners and far easterners (cf. Machery et al. 2004, 2013).

We shall refrain from advancing a theory of reference *or* truth here (or meaning for that matter), but only state the following comments, at least partly affirmative of Putnam's thinking on these issues: Firstly, however a given 'data structure' may be stabilized by an external source, as we here call it, all that a theory of *truth* really needs to provide in order to be satisfactory in *pre-philosophical* thinking is that there be sufficient *agreement* between different agents on the basis of any such 'data structure'—insofar as we can think of this structure as 'shared' among them—, such that these agents will know when to accept that someone is 'telling the truth'; e.g. in court, or under similar circumstances when assertions of 'truth' even matter. Such agreement could come about by watching a video tape from a surveillance camera or by consulting finger prints. That is *far less* than requiring (structural) correspondence to, or even 'truthmaking'[42] in virtue of, a 'transcendentally' external world or source, and it *only* has to do with the *experienced world*. As can be seen, we are here aiming at a theory of *empirical* truth.

And secondly, *regardless* of whether one embraces a causal, descriptivist, or what-have-you theory of reference (e.g. Schwartz 2006, for discussion), this theory

[41]Cf. also da Costa and French (2003, p. 18 ff.) for considerations on *partial* truth.

[42]Cf. Button (2013, pp. 9 and 18 ff.), for a brief discussion of the relation between correspondence theories and truthmaking.

need *only* secure that we be able to refer to 'that old chair in the attic', 'the car in the parking lot', 'grandma in the wheelchair', or 'the ions in that ion-trap', each of which items or states of affairs is, for the agent or user of language uttering these sentences, part of *her image*, scientific or manifest, and *not* part of the source.

Why can we even allow talk of other 'agents' or 'users of language', when we here take quite a methodologically solipsist stance? Because our abductive inferences to the existence of 'others' *are not punished by nature*, much in the same way as the general (abductive) inference to the existence of a source (the abandonment of thorough idealism) is not. On the contrary: many expectations are met precisely when we 'breathe life' into our image (in the sense of external stabilization) and 'breathe consciousness' into certain aspects thereof. The expectations formed on the basis of these inferences are *rewarded* by nature, in contrast to certain expectations formed on the basis of accepting classical physics, say.

Notably, we have been completely silent, for instance, about spatiotemporality in regard to the source, much in contrast to Kant. And we do not mean to disencourage educated speculations about the very nature of mind-independent reality altogether, since these have often proven fruitful in the past. But we do maintain that the structure of experience alone and the depth of past changes in relatively a priori concepts still allows for *doubt*—even about the existence of 'other minds' or a spacetime manifold.[43]

We do owe one final debt to make the general philosophical stance advertised here as a means of coming to terms with QM work. Namely, we have accepted a notion of 'experience' and even reserved a certain *immediacy* for it, despite the fact that we had outlined foundational difficulties for empiricist positions in the philosophy of science. These difficulties, to recall, arose from the fact that what counts as experience, and in particular what counts as a *possible* experience, is (a) notoriously hard to pin down, and (b) requires considerations of *modality* that go beyond any actual experience.

Our answer to both problems is, in part, immediate, and both answers are tightly connected. First off, we have not advocated any straightforward empiricism here, and have reserved a rather important place for a priori considerations (albeit in a theory-relative sense). This allows us to embrace theoretically informed constraints with a (relativized) a priorist character on what even *counts* as an 'experience'—which e.g. Nagel (2000), in fact, diagnoses van Fraassen to be in need of, or to entertain somewhat tacitly. Put frankly: to some extent what, (a), counts as an experience at one time, thereby shaping the face of science, may serve as a *normative corrective* for what *should* count as an experience at a later time, given e.g. progress in perceptual or cognitive psychology or in science more generally. This in turn implies, (b), that what is experienc*able* becomes a *theoretical matter*.

[43]Consider, in this connection, the understanding of time as "only an approximate concept" which "emerges from the separation into [...] different subsystems" (Kiefer 2009, p. 6) in some approaches to quantum gravity or the fact that treating GR as "an effective theory seriously involves rethinking physics without spactime" (Markopoulou 2009, p. 148).

Astrophysics, say, may well tell us what we 'could experience' on other planets if we had warp-drive or the like.

However, we maintain here, regarding (a), that there is still an 'impenetrable immediate core' to experience that is only so constrained by what is constitutive for the 'manifest image' in the narrowest sense—this much we had granted Kant.[44] And this immediacy and what these constraints are is, ultimately, not even a subject matter for intersubjective science but for *introspection*. This we invest as a premise not further justified.

We hence respond to both problems in a twofold, 'commanding' fashion: (i) let science, physical, social, biological, and psychological, be a guide to what counts as 'observable', and what as 'experience' in any more involved sense, and (ii) reflect on your own conscious content to find out what an experience is, in the *immediate*, pre-scientific sense.

7.4.4 A Brief Summary of the Proposal

We have helped ourselves to an epistemological basis for coming to terms with QM through a complicated analysis of different positions, intertwined with bits of history. Let us hence summarize our findings and sympathies in a brief, compact fashion.

We agree with the QBists that science should be viewed as concerned with experience, that there is a split between each 'user's' experienced world and an assumed mind-independent reality, and that the involvement of probabilities has something to do with the information or knowledge of users or agents. But we deny that probabilities or quantum states themselves are a representations of any actual agent's beliefs (doxastic states), that the 'weirdness' of QM *only* arises in virtue of the split between experience and that mind-independent reality, or that it arises in virtue of the effect of our interventions on the latter. A main reason for this denial are the strong, observed correlations predicted by QM, whose existence becomes completely elusive on such a view as QBism.

We agree, then, with Healey, Stairs, and Friederich that QM's probabilities have an objective character, that quantum states are objective relative to an agent's epistemic situation, and that QM does not support certain causal counterfactuals. With Healey and Friederich we agree, furthermore, that the probabilities supply a normative guide to future expectation, that decoherence provides a guide to situations in which the Born rule can be applied, and with Healey that this implies an impact on the pragmatic-inferential content of certain statements. We also endorse the notions of backing and advice conditions and that, within a given physical situation, a QSA can be warranted. Unlike Healey or the inferentialism he endorses,

[44]This view, of course, gains direct support from the debate on cognitive impenetrability of preception; Müller-Lyer illusions are an impressive, simple example (e.g. Pylyshyn 1999, p. 344).

however, we deny that there is an impact on *all* components of a statement's very meaning, i.e. that QM may render statements such as 'the electron took this and that path' *literally* meaningless. Based on the fact that we seem to *understand* such sentences very well, we here reserve a completely internal/mental component to meaning, much like Quine (1939, pp. 702–703).

Also unlike Healey, we hold the central impact of QM to be a limitation of the range of abductive inferences that we are entitled to draw based on observations, including and especially to hidden common causes when these seem to be dis-encouraged by QM (KS-theorem, Bell's theorem). A bridging principle between both views is provided by 'material inferences', expressing lawlike connections that can figure as a basis of the respective abduction (as 'known laws'), and which Healey claims not to be licensed by QM in some (surprising) cases. To make sense of statements like 'a physical situation warrants this and that QSA', we urge, in agreement with considerations of a 'split' between experienced and independent reality, to flesh out the conglomerate of the above considerations as promoting an internal realism, as endorsed by Kant and Putnam (at some stage). Unlike Kant, though, we do not endorse a synthetic a priori with unlimited, eternal validity, but merely a theory-relativized, constitutive one, like Reichenbach and Friedman. This enables us to let the theory dictate, to some extent, what 'is there'; objects such as particles can be constituted as 'bearers of properties', in the sense of being invariants under a relevant transformation group (e.g. an irreducible unitary representation of the Galilei or Poincaré group).

The endorsement of a relativized a priori allows for a shift in what is constitutive of objects or objective facts: In non-relativistic QM and special relativistic QFT, strict correlations are constitutive of a particle-pair in the singlet state *qua* the latter's rotation invariance; in quantum graivity this may not be the case anymore, and the strictness of the correlation becomes an empirical matter.

Much like empiricists such as van Frassen, we claim a certain immediacy for 'experience'; but in contrast to him (them), we allow for a priori principles being operative that determine what is or could be experienced, at an unconscious level in the manifest image and at a (more or less) conscious level in 'the' scientific image, or in a patchwork of such images. This allows to make sense of what counts as 'experiencable'.

Like Putnam, Healey, and Putnam's Kant, we express reservations about the correspondence theory of truth and our ability to successfully refer to a mind-independent reality altogether. But we refrain from advancing a positive, alternative account of these notions (truth and reference) at this point. We merely acknowledge that for everyday life and scientific conduct, much less is required of these notions than philosophical investigation may sometimes suggest.

7.5 Discussion (iii): Giving up on Science and Common Sense?

Certainly, our epistemological proposal is not the kind of interpretation everyone would want to accept, so we should confront at least a few possible objections.

Objection: This is nothing new! It is just Copenhagen all over again.

Reply: Granted, the interpretation suggested above stands in some continuity to the project undertaken by Bohr (in case that is what is meant by 'Copenhagen'). The same has been said about QBism, but there are also crucial differences (cf. Mermin 2014b). And Healey, in turn, admits influences from QBism (cf. Healey 2012d, p. 751), but also rejects many of its crucial ingredients. The interpretation advanced above bears connections to those of Healey, Bohr, and the QBists, but diverges from all of them in several respects. If one would like to classify it as 'neo-Copenhagen', then that is fine. All of this is hardly an objection: Bohr, as we noted, was himself influenced by Høffding who in turn was influenced by Kant (cf. Faye 1991, p. xix), and the Kantian elements in Bohr's thinking have long been acknowledged (e.g. Hooker 1972; Folse 1994; Bitbol and Osnaghi 2013). Multiple independent arguments and opinions have been developed and expressed above and these set the interpretation presented apart from all its influences. Should that not have been so, it would still hardly make for an objection.

Objection: It was argued before that logical empiricism was abandoned for good reasons and that constructive empiricism must endorse non-empirical modal notions. The interpretation proposed has it that the subject matter of science is experience; it hence falls prey to the same objections.

Reply: This objection was answered by the endorsement of constitutive a priori principles; a theory may dictate what *can* be experienced, and hence what may be expected to be experienced, should the appropriate conditions occur (e.g. interstellar flight).

Objection: All of this is too radical, too revisionary compared with the ways we normally think.

Reply: That is debatable, since many physicists talk quite nonchalantly and with pragmatic ease about the 'objects' that we have let QM constitute, and thereby (apparently) typically do not require much more of them than have we. In a sense, we are quite conservative w.r.t. scientific practice here—a feature we share with Friederich (2015)—, while acknowledging fundamental epistemological restrictions that many physicists, or scientists more generally, may not even have a definite opinion about. Moreover, there *is* no 'non-radical' option, as the discussion should have shown. ψ-epistemic models that preserve common sense intuitions can hardly be put to work; and otherwise one is faced with guiding fields on configurations spaces, myriads of worlds branching off, suddenly occurring and highly correlated flashes... common sense is lost!

Objection: Such an 'epistemological interpretation' means shying away from the 'real battle'. In a sense this is just giving up on science!

Reply: A point of departure for us were the actual difficulties exhibited by all the ontological interpretations. So none of them suggests itself as 'the real deal', and searching for alternatives seems indicated. But another important motivation for embracing internal rather than external realism are considerations of epistemic *humility*.[45] How could we ever claim to arrive at a 'god's eye view', how could we be *certain* to have sorted out the factors contributed *by us* from the factors 'imposed upon us'? By some means, our attitude is hence not anti-scientific, but rather as scientific as it gets. We allow to put basically *everything* under close scrutiny and let it be subject to revision in the face of evidence, instead of resorting to dogma.

Objection: The no-go theorems have shown that epistemic interpretations are impossible. Just not mentioning hidden variables does not make them go away. This knowledge has to be knowledge of something.

Reply: The no-go theorems do not show any such thing; they only impose restrictions on a certain brand of epistemic interpretation. Moreover, our interpretation here is radically different from hidden variables approaches. We do not, to recall, even assume that there is a 'real, outside world'-particle for every particle, i.e. every 'stream of impressions', if you will, that we reify into an object *via* constitutional principles. This obviates hidden variables λ that could subsume, say, definite particle momenta and positions. Still, with van Fraassen (1991, p. 243) we appreciate that, so long as there can be *different* interpretations of some theory, "an interpretation introduces factors not found in the theory originally—and what else does 'hidden variables' mean?" In this modest sense we can admit to also embrace 'hidden variables', but hidden variables very different from those occurring in ψ-epistemic models and not disallowed by the no-go theorems.

We still owe a classification of QBism, Healey's pragmatist approach (which we represent by PQM in the scheme) and our 'epistemological interpretation' (which we represent by EQM). QBism, to the extent that it relies on Quantum Bayesianism, modifies QM formally, since the latter introduces such things as 'quantum Bayes rules' and other adaptations from probability theory (cf. Fuchs and Schack 2011). It is obviously non-ontological and arguably conceptually revisionary, since quantum states are construed doxastically, not as pertaining to systems. It also accepts the PP (Lüders' rule) in the sense of a state update, so it should be classified as a collapse interpretation.

As for PQM, quantum states are not de- but *pre*scripitive, whence they do not describe anything at all. So the approach should count as non-ontological. And since Healey suggests a story in terms of "informational bridges" (cf. Healey 2015), this approach seems conceptually revisionary as well. Moreover, after decoherence, some NQMCs that correspond to projectors are preferred, so in that sense the interpretation is also 'collapse' (Fig. 7.1).

[45]Langton (1998) e.g. provides a book-length investigation of this aspect of Kant's internal realism.

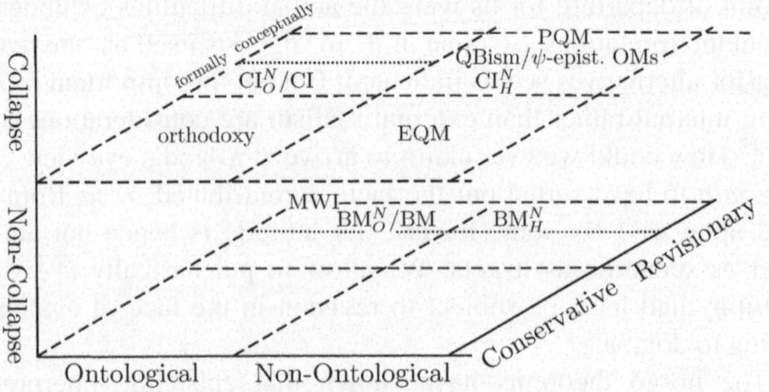

Fig. 7.1 Classification of all interpretations discusseds

How could our EQM differ? We suggest that, since the *physical world*, that which physics *treats of*, can only be the *empirical world* on our view, and since quantum states are assigned by physicists to whatever they *constitute* as 'objects' of the theory, a quantum state may well count as 'the physical state of the system', *insofar as any such thing exists*, even though by that no reference to a radically mind-independent world is implied. In other words: our interpretation is conceptually and formally *conservative*, as much as this is possible without repeating orthodoxy, and up to the fact that we construe everything in terms of an empirical, partially theoretically constructed reality only.

Much like that of Friederich (2015), our interpretation thus respects physical *practice*. But we have gone beyond Friederich's Wittgensteinian, therapeutic approach, in advancing our considerations of stabilization, punishment and reward by an external source, the scope of valid abductions, constitution of objects... and so forth.

The ability to remain conservative in this sense and with the help of these accommodations seems like a valuable achievement. In agreement with e.g. Maudlin (2014b, p. 796) and *pace* Wallace (2016), to recall, we had argued Chap. 2 that the 'textbook recipe' *is* operative—with suitable generalizations in terms of density matrices, POVMs, field operators and the like—in the actual *use* of QM, and that the theory harvests its success from this textbook recipe or orthodox approach. Any interpretation should respect that success, and as should have become obvious from the discussion in Chap. 6, all the ontological interpretations considered are ultimately aimed at harvesting it. Why else would one want to make sense of the *appearance* of definite outcomes distributed according to the Born rule in the MWI or include PVMs and POVMs into the description of *measurements* in BM or CI?

Certainly, many of the above claims are debatable, controversial, and counter-intuitive, depending on one's intuitions. But again, we must stress that there are no alternatives without these features. Lord Kelvin's (1901, p. 1) infamous "two clouds" have developed into raging thunderstorms from which there is no escape, and we have even highlighted some connections between both 'clouds' (QM and relativity) at a fundamental epistemological level (demotion of a priori principles of predecessor theories to empirical connections).

7.6 Conclusions (iii) and Coda

The treatment provided in this last chapter has lead to somewhat of a *reduction* of the *specific* problems associated with QM to general themes in the philosophy of science and epistemology. This, too, seems like a valuable step: if we can see QM and the problems raised by it by the same lights as those raised by other, previous theories, then the interpretation of QM does not pose a *special* problem at last. After all, Schrödinger (1931, p. 490) reminds us that there is a sense in which all of science treats of the subject-object relation.

A widespread objection to interpretations of this general sort is that 'the observer' figures *crucially* therein, which it should not do in a physical theory. But we have seen that there are important respects in which conscious observes do figure even in the MWI—being in a sense 'the most physical interpretation of them all', the one that tries to take the successful physical theory that is QM at face value—, namely in providing precise cut-offs for branches-as-experienced. And from the epistemological perspective taken on here, the inclusion of 'observes' seems like a *virtue*, not a vice.

Similar judgment has been advanced, of course, by the QBists. In particular, they (cf. Mermin 2014a, p. 421; Fuchs and Schack 2014, p. 2) refer back to Schrödinger (1954) who identifies an important pillar of modern science, as rooted in the ancient Greek tradition, in the fact that "[t]he scientist subconsciously, almost inadvertently, simplifies his problem of understanding Nature by disregarding or cutting out of the picture to be constructed, himself, his own personality, the subject of cognizance." (p. 92) This, Schrödinger (ibid.) thinks, "leaves gaps, enormous lacunae, leads to paradoxes and antinomies whenever, unaware of this initial renunciation, one tries to find oneself in the picture or to put oneself, one's own thinking and sensing mind, back into the picture." Mermin (2014b, p. 5) finds similar considerations in Freud (1961, p. 56) who regards "the problem of the nature of the world without regard to our percipient mental apparatus [. . .] an empty abstraction, devoid of practical interest."

Put less poetically, the upshot is this: A 'theory of everything' cannot be a theory of *everything* if it disregards the fact that it is being *applied by someone*, that it is relative to the epistemic conditions of an individual or an epistemic community. We do not *have* 'god's eye-view', and a neglect of this fact leaves out something scientifically important.

A final fly in the ointment we must admit though: How, if one allows that the correspondence theory of truth must be put in question, can the conception of internal realism itself even be accepted? I.e.: How could it be accepted as correctly depicting our epistemic situation, how could it be *true*? Stated differently, what would acceptance of that conception amount to? There seem to be two principled options: (a) accept the 'rational acceptability'-standard advanced by Putnam and apply it to internal realism itself, or (b) claim that the only TRUTH, if you will, in a stronger, structural correspondence-sense is the foundational conception of the subject-object relation at play in internal realism.

Now (a) seems 'a little weak' and too much contingent (to recall) on a problematic notion such as 'rationality'. But maybe feeling this way is just an over-endorsement of intuitions that should be disposed of anyways, and the suitable 'philosophical therapy' could cure one of these worries. (b), on the other hand seems quite *ad hoc*, and it seems to fall prey to the 'limit argument', as Moore (2012, p. 135) calls it. This argument states that "in order to be able to draw a limit to thought, we should have to find both sides of the limit thinkable (i.e. we should have to be able to think what cannot be thought)." (Wittgenstein 1961, p. 3) In other words: if we do recognize that we are so limited in our ability to conceptualize the external, 'nuomenal' world, how can we still entertain even a "merely limitative" (Kant 1781, p. 255; emphasis omitted) *concept* of such a world?

But think of it this way: Imagine a wall rising up at some place, and imagine yourself interested in what is behind that wall. Now if the wall extends sufficiently far and is sufficiently high, there may be no way around it. The only method of finding out what is behind it may consist in throwing rocks and listening to the resulting sounds. But imagine that there be a distortion-mechanism that radically modifies the sounds you hear, thereby creating distinctive patterns. Based on these patterns, you may draw an image, a structural diagram relating the elements suspected behind the wall to one another. But that diagram would only depict what you can gather based on your limited methods and the distorted information you obtain. That is, it will be informed *in some sense* by what is behind, but not *depict* the latter at all (not even structurally).

It seems to be no problem to *think* that wall, that something behind, and the distortion mechanism, even if all relevant information is missing about the realm behind the wall. All 'truths' *within* the diagram, that could be accepted, say, when one is able to *predict*, from the diagram, sounds that result when throwing rocks in a particular way, would be different from the TRUTH that one does sit in front of that wall and is merely throwing rocks. And note that one could never be *certain* that the sounds do originate from *behind* the wall, and that the rocks do not merely vanish *at the boundary*.

Is the metaphor compelling at all? Well... the dear reader should decide that for herself.

Appendix A
Required Mathematics (and a Little More)

In the following, we give a quick review of the central mathematical concepts. References to more detailed accounts will be given accordingly. In general, the expositions by Gustafson and Sigal (2011) and Heinosaari and Ziman (2012) may be used as comprehensive and somewhat complementary introductions to the mathematics of QM. Most textbooks on QM also provide sufficient introductions, presupposing however basic calculus and linear algebra (as shall we, to some extent).

A.1 Vector Spaces

In Schödinger's version of QM, the state of a system is mathematically represented by a *state function* ψ, in Heisenberg's version it is represented by a sort of 'matrix' (table of numbers). These collections of functions and matrices each form a *vector space* over the field of complex numbers. Very loosely speaking, a vector space is just a collection of mathematical objects that can be multiplied by numbers and added up. Imagine, for instance, a collection of little arrows that can be stretched or shortened, i.e. scaled (mathematically represented as multiplication by a number or 'scalar'), and glued together (mathematically represented as vector addition). Two arrows glued together tip to bottom count as a third one that points directly from the bottom of the first to the tip of the second.

Choosing three arrows as a 'basis' of some three dimensional vector space, joined together at their bottoms and not all lying in one plane, all other arrows can be described by a collection of three entries, loosely speaking specifying the respective steps one would have to go in the direction of each of the three 'basis arrows' to reach the point the arrow so represented would be pointing to if it were also glued to the basis at its bottom (without change in orientation). Vectors 'lying around' somewhere in space are identified with ones of same length and direction attached

F. J. Boge, *Quantum Mechanics Between Ontology and Epistemology*,
European Studies in Philosophy of Science 10,
https://doi.org/10.1007/978-3-319-95765-4

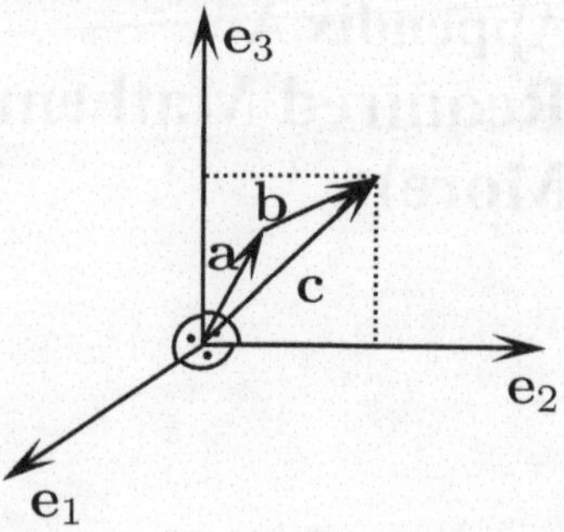

Fig. A.1 Arrows as a means of depicting vectors

to the origin of the basis. This makes perfect sense in their representation as triplets of numbers. It helps to familiarize oneself with these notions by adding columns of numbers (componentwise) and drawing arrows.

As a quick example hence take the six arrows depicted in Fig. A.1. We have conveniently chosen the basis arrows $\boldsymbol{e}_1, \boldsymbol{e}_2, \boldsymbol{e}_3$ to be at right angles, and we choose them to be of unit length. They have standard representations

$$\boldsymbol{e}_1 \doteq \begin{pmatrix} 1 \\ 0 \\ 0 \end{pmatrix}, \boldsymbol{e}_1 \doteq \begin{pmatrix} 0 \\ 1 \\ 0 \end{pmatrix}, \text{ and } \boldsymbol{e}_3 \doteq \begin{pmatrix} 0 \\ 0 \\ 1 \end{pmatrix},$$

where we use '$\doteq$' (as in the remainder of this book) to indicate the fact that this is actually a *choice* of representation. Then we can write

$$\boldsymbol{a} = 0 \cdot \boldsymbol{e}_1 + \frac{1}{6} \cdot \boldsymbol{e}_2 + \frac{1}{3} \cdot \boldsymbol{e}_3,$$

or equally

$$\boldsymbol{a} \doteq 0 \cdot \begin{pmatrix} 1 \\ 0 \\ 0 \end{pmatrix} + \frac{1}{6} \cdot \begin{pmatrix} 0 \\ 1 \\ 0 \end{pmatrix} + \frac{1}{3} \cdot \begin{pmatrix} 0 \\ 0 \\ 1 \end{pmatrix} = \begin{pmatrix} 0 \\ 1/6 \\ 1/3 \end{pmatrix}.$$

Since

$$\boldsymbol{a} + \boldsymbol{b} = \boldsymbol{c} \text{ and } \boldsymbol{c} \doteq \begin{pmatrix} 0 \\ 1/2 \\ 1/2 \end{pmatrix},$$

we must have

$$\boldsymbol{b} = \boldsymbol{c} - \boldsymbol{a} \doteq \begin{pmatrix} 0 \\ 1/2 - 1/6 \\ 1/2 - 1/3 \end{pmatrix} \doteq \begin{pmatrix} 0 \\ 1/3 \\ 1/6 \end{pmatrix}.$$

As '$\doteq$' indicates, there are different choices of columns of numbers that can represent the relations between $\boldsymbol{a}$, $\boldsymbol{b}$, and $\boldsymbol{c}$ just as well. We can now also see that $\boldsymbol{b}$ is described by a column of numbers that equally describes an arrow glued to the bottom of the basis, which justifies their identification.

Now a vector is formally speaking an element of some vector space. And what *precisely* is a vector space? To define this appropriately, one first needs the concept of a *field*.

Definition A.1 (Field) Let $\mathbb{F}$ be a nonempty set over which two operations, $+$ and $\cdot$ are defined, which map every pair[1] $(x, y) \in \mathbb{F} \times \mathbb{F}$ to some element $x + y \in \mathbb{F}$ or $x \cdot y \in \mathbb{F}$ respectively. If $\forall x, y, z \in \mathbb{F}$

(i) $(x+y)+z = x+(y+z)$ and $(x \cdot y) \cdot z = x \cdot (y \cdot z)$, i.e. $+$ and $\cdot$ are *associative*,
(ii) $x + y = y + x$ and $x \cdot y = y \cdot x$, i.e. $+$ and $\cdot$ are *commutative*,
(iii) $\exists 0, 1 \in \mathbb{F}$ such that $x + 0 = x$ and $y \cdot 1 = y$ (0 is called the *neutral element* w.r.t. $+$, and 1 w.r.t. $\cdot$),
(iv) $-x + x = 0$ and $x \cdot x^{-1} = 1$ in case $x \neq 0$, i.e. there exist a *negative element* $(-x)$ and an *inverse element* (x^{-1}),
(v) $x \cdot (y + z) = x \cdot y + x \cdot z$, i.e. $\cdot$ is distributive over $+$,

$(\mathbb{F}, (+, 0), (\,\cdot\,, 1))$ is called a *field*.

0 and 1 can be demonstrated to be unique (cf. Kerner and von Wahl 2013, p. 4). It is customary to abbreviate $(\mathbb{F}, (+, 0), (\,\cdot\,, 1))$ simply by $\mathbb{F}$. Notable fields are the real and complex numbers, $\mathbb{R}$ and $\mathbb{C}$. Any $c \in \mathbb{C}$ is of the form $c = \alpha + i\beta$ with $i = \sqrt{-1}$ the *imaginary unit* and $\alpha, \beta \in \mathbb{R}$. $c^* = \alpha - i\beta$ is called the *complex conjugate* of c. For $z \in \mathbb{R}$, i.e. $z = \alpha + i \cdot 0$, complex conjugation simply makes no difference.

$(\mathbb{F}, (+, 0))$ and $(\mathbb{F}, (\,\cdot\,, 1))$ each define a *group* in virtue of the properties (i), (iii), and (iv); in case (ii) is also satisfied, the group is called *Abelian* (cf. Kerner and von Wahl 2013, p. 113).

Given these notions, a vector space is defined as follows.

Definition A.2 (Vector space) A vector space, defined over some field $\mathbb{F}$ is a triple $(\mathcal{V}, +_{\mathcal{V}}, \cdot_{\mathcal{V}})$ consisting of a set $\mathcal{V}$, and two connections $+_{\mathcal{V}}$ (vector addition) and $\cdot_{\mathcal{V}}$ (scalar multiplication) such that, $\forall v, w \in \mathcal{V}, \forall \lambda, \mu \in \mathbb{F}$,

$$\mathcal{V} \times \mathcal{V} \xrightarrow{+_{\mathcal{V}}} \mathcal{V},\ (v, w) \longmapsto v + w, \qquad \mathbb{F} \times \mathcal{V} \xrightarrow{\cdot_{\mathcal{V}}} \mathcal{V},\ (\lambda, v) \longmapsto \lambda \cdot_{\mathcal{V}} v = \lambda v$$

and

(i) $(\mathcal{V}, +_{\mathcal{V}})$ is an Abelian group
(ii) $\lambda(\mu v) = (\lambda\mu)v$ (assoc. $\cdot_{\mathcal{V}}/\cdot$),

[1]We here use rounded brackets to denote k-tuples, to avoid confusion with the bras and kets introduced below. In the text the notation with angled brackets, $\langle x, y, z, \ldots \rangle$, which is more widespread in philosophy, is used when no ambiguity arises.

(iii) $(\lambda + \mu)v = \lambda v + \mu v$ (dist. $\cdot_{\mathcal{V}}$ over $+$)
(iv) $\lambda(v +_{\mathcal{V}} w) = \lambda v +_{\mathcal{V}} \lambda w$ (dist. $\cdot_{\mathcal{V}}$ over $+_{\mathcal{V}}$)
(v) $1v = v$

(cf. Kerner and von Wahl 2013, p. 118). Here we have used an index ($_{\mathcal{V}}$) to distinguish the field operations from the vector operations. It is, however, customary to omit indexation, and equally to refer to the triple $(\mathcal{V}, +_{\mathcal{V}}, \cdot_{\mathcal{V}})$ simply by $\mathcal{V}$. Note that (i) implies the existence of a zero element in $\mathcal{V}$, called a *null vector*, which we will denote by $0_{\mathcal{V}}$.

A.2 Hilbert Spaces

Hilbert spaces are strictly speaking just a special kind of vector space. They can be understood intuitively as a generalization of the underlying ideas in such a way that there may be arbitrarily large collections of basis-arrows that could still be joined together at their bottoms *all at right angles*, and where any of the arrows in the collection could be scaled ('stretched') by complex numbers, not just real ones. Of course, such a space of 'arrows' becomes unimaginable and the vectors in a Hilbert space would be representative of something more abstract.

In a Hilbert space of functions ψ, these functions, taking the place of the pictorial arrows, must satisfy the restriction that they be *square-integrable*, i.e.:

$$\int_{\mathbb{R}^n} \psi^*(\boldsymbol{\lambda})\psi(\boldsymbol{\lambda})\mathrm{d}^n\boldsymbol{\lambda} = \int_{\mathbb{R}^n} |\psi(\boldsymbol{\lambda})|^2\mathrm{d}^n\boldsymbol{\lambda} < \infty. \tag{A.1}$$

The set of these functions is often denoted by $L^2(\mathbb{R}^n)$, as their argument ($\boldsymbol{\lambda}$) is an n-tuple of real numbers. The tuples $\boldsymbol{\lambda}$ themselves constitute vectors of a different space, and for $n = 3$ they can easily be depicted as our neat little arrows from above. They could e.g. denote positions $\boldsymbol{x} \doteq (x, y, z)$ in the three dimensional space that we ordinarily consider ourselves to live in, or rather pointers (arrows) pointing to such points.

A function, to recall, maps elements from some set (or space) to numbers (the function's *values*). The values of the functions ψ may either be complex numbers or k-tuples thereof, where the latter is the case if one uses them to describe systems with *spin* (cf. Chap. 2). In that case one may write $\boldsymbol{\psi}(\boldsymbol{\lambda})$ for notational consistency, and speak of a 'vector valued function'.

Integrals, to recall, are usually introduced as computing the area under some curve between points a and b: One coarse grains that area into rectangles either directly above or below the curve and then lets the width of these shrink to zero. In case these two approximations from above and below converge to a common limit (the 'true' area under the curve), this defines the (*definite*) *Riemann integral* of the curve between a and b. A function F, unique up to an arbitrary constant, that can be used to compute the area between arbitrary points x and y on the curve by taking the difference $F(y) - F(x)$ is then the *indefinite integral* $F(x') = \int f(x')\,\mathrm{d}x'$. The fundamental theorem of calculus tells us that $f = \frac{\mathrm{d}F}{\mathrm{d}x}$, making differentiation and integration inverse operations of one another (cf. Kerner and von Wahl 2013, p. 81 ff.).

Of course one is typically not really interested in areas. Integrals are hence better understood as generalizations of sums, used where there is a continuum of items to be summed up. The items are then of the form $f(x)\,\mathrm{d}x$ or morel generally $f(\boldsymbol{\lambda})\,\mathrm{d}^n\boldsymbol{\lambda}$, and $\mathrm{d}^n\boldsymbol{\lambda}$ can then be understood as a (metaphorical) tiny 'volume' in the space defined by the $\boldsymbol{\lambda}$s, $f(\boldsymbol{\lambda})$ loosely as the (metaphorical) 'density' of something distributed across volumes of that space.

The 'L' in $L^2(\mathbb{R}^n)$ indicates, moreover, that the integrability is actually meant w.r.t. the *Lebesgue measure*, a different method of assigning a 'volume' or 'weight' to sets in the space of $\boldsymbol{\lambda}$s than in Riemann's theory. The details typically do not matter much in philosophical investigations of QM, and we will hence hardly concern ourselves with them here.[2] But due to their involvement in mathematical probability theory, we will say something more about measures in general below, and then also touch on the Lebesgue measure as an example.

A Hilbert space is generically denoted by the symbol $\mathcal{H}$ and it is common practice to use a special notation for its elements. This notation is due to Paul Dirac (1930), and called the *bra-ket notation* (from English: bracket). Thus, let $|\psi\rangle$ stand for a *ket-vector* (short: *ket*) and $\langle\psi|$ for a corresponding *bra*. This notation allows for creativity in labeling, as e.g. in |♣⟩ , |🐈⟩ , |↯⟩, or simply $|1\rangle$, $|2\rangle$, $|3\rangle$ and so forth. The space $\mathcal{H}$ only contains kets, and the bras form a so called *dual space* $\mathcal{H}^*$, isomorphic to $\mathcal{H}$ in virtue of the Riesz-Fréchet theorem (e.g. Tarasov 2008, p. 23). Formally, bras $\langle v|$ correspond to *linear maps* $v : \mathcal{H} \longrightarrow \mathbb{C}$, also referred to as *linear functionals*, which (linearly) map entire functions to numbers. Such linear functionals are usually denoted as $v[\cdot]$ ('$\cdot$' a slot to fill in some function), and an example are definite integrals $I[f] = \int_a^b f(x)\,\mathrm{d}x$.

All Hilbert spaces are (by definition) equipped with an *inner* or *scalar product*.

Definition A.3 (Inner product) An *inner* (or *scalar*) *product* on a complex vector space $\mathcal{V}$ is a map $\mathcal{V} \times \mathcal{V} \xrightarrow{\langle\cdot|\cdot\rangle} \mathbb{C}$, such that

(i) $\langle v|w\rangle = \langle w|v\rangle^*$ (*skew-symmetry* or *hermiticity*)
(ii) $\langle v|v\rangle \geq 0$ and $\langle v|v\rangle = 0$ *iff* $|v\rangle = 0_{\mathcal{V}}$ (*positive semidefiniteness*)
(iii) $\langle v|(\alpha\,|w\rangle + \beta\,|z\rangle) = \alpha\langle v|w\rangle + \beta\langle v|z\rangle$, $\alpha, \beta \in \mathbb{C}$ (*linearity in the ket*)

(cf. Goldhorn et al. 2009, p. 358; Heinosaari and Ziman 2012, pp. 1–2). (iii) together with (i) implies a form of linearity in the first argument, but with complex conjugated factors α^* and β^*. This property is referred to as *anti-linearity* in the bra. Note that the linearity in the ket coincides with the fact that the bras are linear functionals on $\mathcal{H}$, since in functional notation $\langle v|(\alpha\,|w\rangle + \beta\,|z\rangle) = v[\alpha w + \beta z] =$

[2] An overview is provided e.g. in Tetschl (2000, p. 259 ff.).

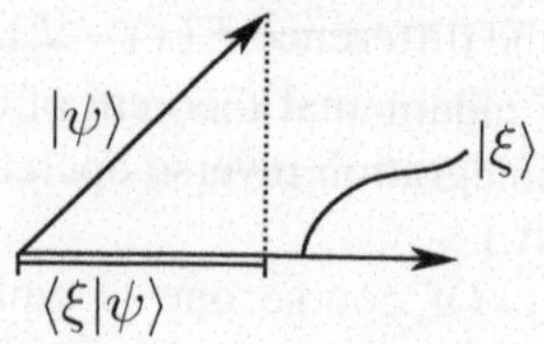

Fig. A.2 Orthogonal projection of $|\psi\rangle$ onto $|\xi\rangle$

$\alpha v[w] + \beta v[z] = \alpha\langle v|w\rangle + \beta\langle v|z\rangle$. One can easily check that (i) also ensures that $\langle v|v\rangle$ is always real.

So applying a bra to a ket is loosely speaking the same thing as computing the inner product of two vectors from $\mathcal{H}$, and one can intuitively think of the functionals (bras) as 'half filled' scalar products $\langle v|\cdot\rangle$.

In the arrow-image, the inner product can be quite literally understood as computing the (orthogonal) *projection* of one vector onto the other (cf. Fig. A.2). Think of one vector as an arrow lying on the ground and another one stuck in the ground at some angle, right at the foot of the first one. When light shines directly from above, the part of arrow 1 shadowed by arrow 2 corresponds to the orthogonal projection. In this image, the scalar product computes the length of the shadow. Even though this image breaks down again in $\mathcal{H}$, $\langle v|w\rangle$ is still sometimes called the *projection* of $|w\rangle$ onto $|v\rangle$.

The scalar product also induces a *norm*

$$\|v\| = \sqrt{\langle v|v\rangle} \tag{A.2}$$

on $\mathcal{H}$, for which two inequalities can be derived (stated here without proof): the *Schwartz inequality*

$$\begin{aligned} |\,\langle v|w\rangle\,| &\leq \|v\| \cdot \|w\| \\ \Leftrightarrow |\,\langle v|w\rangle\,|^2 &\leq \langle v|v\rangle\,\langle w|w\rangle\,, \end{aligned} \tag{A.3}$$

and *triangle inequality*

$$\|\,|v\rangle + |w\rangle\,\| \leq \|v\| + \|w\| \tag{A.4}$$

(cf. Heinosaari and Ziman 2012, pp. 2 and 4).

Moreover, a family (indexed set) of vectors, $\{|j\rangle\}_{j\in I}$ from $\mathcal{H}$ (with $I \subseteq \mathbb{N}$), which satisfies

$$\langle i|j\rangle = \delta_{ij} =: \begin{cases} 1, & \text{if } i = j \\ 0 & \text{else} \end{cases}\,, \tag{A.5}$$

is called *orthonormal*, as are the vectors that are its elements. The 'if' fixes their *normality*, i.e. their norm $\|j\|$ being finite (here: 1), the 'else' their *orthogonality*. δ_{ij} is called the *Kronecker-δ*(-function).

Using these relations, we can thoroughly introduce the notion of an *orthonormal basis*. Call a family of vectors $\mathcal{G} = \{|j\rangle\}_{j\in I}$ from $\mathcal{H}$ (or more generally: $\mathcal{V}$) a *generating set* of $\mathcal{H}$ ($\mathcal{V}$) if every vector $|v\rangle \in \mathcal{H}$ ($\mathcal{V}$) can be written as a *linear combination*

$$|v\rangle = \alpha_1 |1\rangle + \alpha_2 |2\rangle + \alpha_3 |3\rangle + \ldots = \sum_{j=1}^{n} \alpha_j |j\rangle, \tag{A.6}$$

with $|j\rangle \in \mathcal{G}$, $n = |I|$, $|\cdot|$ denoting the cardinality of the indexing set I, and $\alpha_j \in \mathbb{C}$ (more generally: $\mathbb{F}$), $\forall j \in I$. We can then say that the vectors in $\mathcal{G}$ *span* the space $\mathcal{H}$ ($\mathcal{V}$). In case $I = \mathbb{N}$ (i.e. '$n = \infty$'), this formula is only applicable for an important subclass of Hilbert spaces (cf. below). The sum is then actually a *series* of vectors, where a series is defined as a limit of partial sums, $\sum_{j=1}^{\infty} a_j := \lim_{n\to\infty} \sum_{j=1}^{n} a_j$, so long as that limit exists.

If no $|\ell\rangle$ in some family $\{|j\rangle\}_{j\in I}$ can itself be expanded as a linear combination $|\ell\rangle = \sum_j \lambda_j |j\rangle$ of $|j\rangle \in \{|j\rangle\}_{j\in I}$, or equally, if it holds that

$$\sum_j \lambda_j |j\rangle = 0 \textit{ iff } \lambda_j = 0, \forall j \in I,$$

then the $|j\rangle \in \{|j\rangle\}_{j\in I}$ are called *linearly independent*. A generating set of linearly independent vectors defines a *basis* $\mathcal{B}$ of $\mathcal{H}$. The *dimension* $\dim(\mathcal{H})$ of $\mathcal{H}$ is then given by the maximum number of linearly independent vectors in $\mathcal{H}$. A linear combination $|v\rangle = \sum_j \alpha_j |j\rangle$, where $|j\rangle \in \mathcal{B}, \forall j \in I$, is also called an *expansion* of $|v\rangle$ in $\mathcal{B}$, and the α_j are called the *expansion coefficients*. They are computed by $\langle j|v\rangle$, as becomes clear if one expands $|v\rangle$ in the respective basis and then computes the inner product with a given $|j\rangle$ (minding the Kronecker-δ).

Using the notion of an orthonormal basis, the norm induced by the Hermitian inner product can be understood more intuitively by comparison to the length of an arrow in our initial pictorial representation of vector spaces. It parallels, in this sense, the *Euclidean norm* $|\boldsymbol{a}| = \sqrt{\boldsymbol{a}\cdot\boldsymbol{a}} = \sqrt{a_1^2 + a_2^2 + a_3^2}$, with $\boldsymbol{a} = a_1\boldsymbol{e}_1 + a_2\boldsymbol{e}_2 + a_3\boldsymbol{e}_3$, where $\boldsymbol{a}\cdot\boldsymbol{b}$ is a real valued inner product, defined over an $\mathbb{R}$-vector space, which becomes symmetric ($\boldsymbol{a}\cdot\boldsymbol{b} = \boldsymbol{b}\cdot\boldsymbol{a}$) and linear in both arguments due to the exclusive involvement of real numbers. Given that the $\boldsymbol{e}_i$ are at right angles to one another, this length can be straightforwardly understood by appeal to the Pythagorean theorem in three dimensions. The norm introduced by the inner product on complex and high-dimensional vector spaces is an abstracting generalization of this intuitively accessible notion of a length.

Along with many a physicist, we will denote bases by sets in this book, even though they must strictly rather be represented as ordered n-tuples $\mathcal{B} = (|1\rangle, \ldots, |m\rangle)$, since sets are invariant under repetitions, i.e. $\{|\psi\rangle\} = \{|\psi\rangle, |\psi\rangle\}$, but bases are supposed to be collections of linearly independent vectors; and $|\psi\rangle$ is *not* linearly independent of itself ($|\psi\rangle = 1\cdot|\psi\rangle + \sum_j 0\cdot|\phi_j\rangle$).

Now while we have introduced most of the relevant features of Hilbert spaces somewhat informally already, a Hilbert space $\mathcal{H}$ may be more formally defined as a complex vector space with an inner product which is *complete* w.r.t. the norm induced by that product (cf. Goldhorn et al. 2009, pp. 200–201; Heinosaari and Ziman 2012, p. 6). Completeness w.r.t. the norm means that any sequence of vectors $|1\rangle, |2\rangle, |3\rangle, \ldots$ in $\mathcal{H}$ which eventually come closer and closer to one another in that norm with growing index—in the sense that $\forall\epsilon > 0 \exists N \forall n, m > N : \| \, |n\rangle - |m\rangle \, \| < \epsilon$—will converge to some limit $|v\rangle \in \mathcal{H}$. That $|v\rangle$ is *also an element of* $\mathcal{H}$ is the decisive point.

Not all Hilbert spaces will allow for basis expansions in terms of series $\sum_{j=1}^{\infty} \alpha_j \, |j\rangle$. Those that do, however, are usually called *separable* (e.g. Heinosaari and Ziman 2012, p. 6). This means that the space has a *countable* orthonormal basis. Finite-dimensional complex vector spaces with an inner product, i.e. spaces where the ∞ in the basis expansion can be replaced by some $n \in \mathbb{N}$, may also be called (finite) Hilbert spaces, as they are 'trivially' complete: the difference $|\phi\rangle - |\psi\rangle$ will always give back another vector of the space.

That there is always an *orthonormal* basis is actually secured by the following theorem (e.g. Shankar 1994, p. 14).

Theorem (Gram-Schmidt) *Given an arbitrary basis $\mathcal{B}$ of a space $\mathcal{H}$, one can always construct an orthonormal basis $\mathcal{B}'$ out of $\mathcal{B}$ by means of linear combination.*

It is also a general theorem of linear algebra that vector spaces over some field $\mathbb{F}$ of the same *finite* dimension are all isomorphic to one another, i.e. can be mapped onto one another linearly and bijectively (cf. Fischer 2014, p. 118) The same holds for all separable Hilbert spaces by another theorem (cf. Heinosaari and Ziman 2012, p. 6). This justifies occasional loose talk of 'Hilbert space' in general, rather than of some particular Hilbert space.

For finite Hilbert spaces, one may hence, in virtue of this isomorphism, bluntly appeal to the (respective) space $\mathbb{C}^n$ of complex column vectors to represent the elements of the space. Equipped with an inner product, these *are* Hilbert spaces. Addition of vectors then means adding up the entries from the same row, multiplication by a scalar $z \in \mathbb{C}$ means multiplying all entries by it. A bra is then obtained by forming the complex conjugated *transpose* of a ket, i.e.

$$\text{if } |v\rangle \doteq \begin{pmatrix} v_1 \\ v_2 \\ \vdots \end{pmatrix} \text{ then } \langle v| = (|v\rangle^T)^* \doteq (v_1^* \; v_2^* \ldots),$$

and the inner product can be computed as

$$\langle v|w\rangle \doteq (v_1^* \; v_2^* \cdots) \cdot \begin{pmatrix} w_1 \\ w_2 \\ \vdots \end{pmatrix} = v_1^* w_1 + v_2^* w_2 + \cdots$$

This in fact coincides with *matrix multiplication* for a (complex) 1×3 matrix and a (complex) 3×1 one, where matrix multiplication (denoted here simply by $\cdot$) can informally be summarized by the mantra 'row times column and then sum over'. More precisely, take two matrices A, B which can be described as collections $(a_{ij}), (b_{ij})$ of entries a_{ij}, b_{ij}, the first index referring to the row, the second one to the column. Then the first entry of the first row, m_{11}, of the resulting matrix $(m_{ij}) = M = A \cdot B$ is the sum $a_{11}b_{11} + a_{12}b_{21} + a_{13}b_{31} + \ldots$ and the second entry of the first row is $a_{21}b_{12} + a_{22}b_{22} + a_{23}b_{32} + \ldots$ and so forth. It may worth familiarizing oneself with these concepts, if not familiar with them yet.

The standard inner product for vectors ψ from $L^2(\mathbb{R}^n)$ is the integral

$$\langle \phi | \psi \rangle := \int \phi^*(\boldsymbol{\lambda})\psi(\boldsymbol{\lambda})\mathrm{d}^n\boldsymbol{\lambda}. \tag{A.7}$$

A function spits out values for a continuous infinity of points, so the intuition behind definition (A.7) is that as the number of arguments to sum up becomes infinite and at the same time denser and denser, the sum must become an integral, the continuous generalization of a sum.

$\boldsymbol{\lambda}$ can again be identified with a real column vector and usually either represents momenta $\boldsymbol{p}$ or positions in space $\boldsymbol{x}$. Setting e.g. $\boldsymbol{\lambda} = \boldsymbol{x}$ and allowing $\boldsymbol{x}$ to be defined in arbitrary, generalized coordinates q_i, we can write more explicitly:

$$\langle \phi | \psi \rangle = \int \cdots \int \phi^*(q_1, \cdots, q_n)\psi(q_1, \cdots, q_n)\mathrm{d}q_1 \cdots \mathrm{d}q_n = \int \phi^*(\boldsymbol{x})\psi(\boldsymbol{x})\mathrm{d}^n\boldsymbol{x}. \tag{A.8}$$

Setting $n = 3$, the q_i of which ϕ and ψ are functions could for instance be *spherical* coordinates, r, θ, φ, where r is the (variable) radius of some sphere and φ and θ are two angles that locate any point on that sphere. $\mathrm{d}^3\boldsymbol{x} = \mathrm{d}q_1\mathrm{d}q_2\mathrm{d}q_r = r^2 \sin(\theta)\, \mathrm{d}r\, \mathrm{d}\theta\, \mathrm{d}\varphi$ then is the corresponding 3-dimensional volume-element, which includes the so called *functional determinant* $r^2 \sin(\theta)$, a 'scaling factor' that takes track of the coordinate change (cf. Kerner and von Wahl 2013, p. 250 ff.). One must also not confuse $\mathrm{d}^n\boldsymbol{x}$ with $\mathrm{d}\boldsymbol{x}$ which should be read as an infinitesimal *line segment* for integrating along some (possibly curved) line.

The product thus defined fulfills all of the requirements (i)-(iii) from definition A.3. The integral is meant as a definite integral over the set of points over which the functions are defined. This may be some subset $\Omega \subset \mathbb{R}^n$, in which case the space is $L^2(\Omega)$ instead of $L^2(\mathbb{R}^n)$. The n comes from the fact that the coordinates (or momenta) of *multiple* (N) systems must be considered all at once in many cases. $\boldsymbol{\lambda}$ is thus often replaced by coordinates $\boldsymbol{X} = (\boldsymbol{x}_1, \ldots, \boldsymbol{x}_N)$ of points in an abstract ($3N$-dimensional) *configuration space*, or $\boldsymbol{P} = (\boldsymbol{p}_1, \ldots, \boldsymbol{p}_N)$ in a corresponding *momentum space* (we omit capitalization below). The space $\mathcal{H} = L^2(\mathbb{R}^n)$ with the norm introduced by this scalar product is a complete and separable Hilbert space, with a countable, orthonormal basis that can be provided in terms of Hermite functions (cf. Johnston 2014, for a nice proof).

A.3 Beyond the Hilbert Space

'Proper' Hilbert spaces, however, come with restrictions for application in physics. There are certain kinds of kets (e.g. $|x\rangle$ or $|\boldsymbol{x}\rangle$) often used in QM, which are neither countable nor properly normalizable ('are of infinite norm'), whence the convergence criterion cannot even be applied. Further reasons why Hilbert spaces may not always be the most convenient setting lie in facts about boundedness and unboundedness of operators (cf. de la Madrid 2005, p. 289, and below). One hence often appeals to an extension of $\mathcal{H}$, the *rigged Hilbert space* (e.g. Ballentine 2000, pp. 27–29; de la Madrid 2005). A rigged Hilbert space consist of a triple $(\Phi, \mathcal{H}, \Phi^{\times})$, where it holds that $\Phi \subset \mathcal{H} \subset \Phi^{\times}$. The space Φ is dense in $\mathcal{H}$, and $\mathcal{H}$ can be considered as the completion of Φ w.r.t. the norm induced by a scalar product (e.g. Tarasov 2008, p. 34). Φ is basically a space of *test functions* (see below). The space $\Phi^{\times}$ is called the *anti-dual* of Φ, and it contains all antilinear functionals over Φ, loosely speaking *all* the kets $|v\rangle$ (cf. de la Madrid 2005, pp. 300 ff. and 311).

An extensive discussion of the rigged Hilbert space can be found in Bohm and Gadella (1969) and de la Madrid (2005) provides a quite accessible and intuitive introduction. Some subtleties actually arise from the fact that Φ must strictly speaking satisfy a (topological) property called *nuclearity* (cf. Bohm and Gadella 1969, p. 11), which puts restrictions on the operators one may introduce (cf. de la Madrid 2005, p. 310 and references therein for details) and hence limits the generality of the rigged Hilbert space-formalism. For the purposes of this book, the details again do not matter much, and we will restrict ourselves to introducing a few basic ideas to convey an understanding of certain formulae.

For any $\boldsymbol{x} \in \mathbb{R}^3$, one can then, using kets from a rigged Hilbert space, understand $\langle \boldsymbol{x}|\psi\rangle$ as the projection of $|\psi\rangle$ onto $|\boldsymbol{x}\rangle$, and interpret this as the *value* of ψ *at* $\boldsymbol{x}$:

$$\psi(\boldsymbol{x}) = \langle \boldsymbol{x}|\psi\rangle \,. \tag{A.9}$$

But one may equally write $\langle \boldsymbol{p}|\psi\rangle = \tilde{\psi}(\boldsymbol{p})$, which strictly speaking defines a different function $\tilde{\psi}$. When given as a ket, a function is thus treated as an object somewhat independent of its domain (positions *or* momenta), and one also speaks of $\psi(\boldsymbol{x})$ and $\tilde{\psi}(\boldsymbol{p})$ as the *position- and momentum space representations* of $|\psi\rangle$ respectively.

It is important not to confuse position vectors $\boldsymbol{x} \in \mathbb{R}^n$ of which the $\psi \in \mathcal{H}$ are functions with the ket-vectors $|\boldsymbol{x}\rangle$. The position vectors $\boldsymbol{x}$ correspond to points a $n = 3N$-dimensional configuration space for N systems. The kets do not. Every $|\boldsymbol{x}\rangle$ corresponds to a (generalized) basis vector of some rigged Hilbert space, whence there are *uncountably many* basis vectors in that space. The three dimensional vector space $\mathbb{R}^3$ for a single system e.g. only has three basis vectors. It should hence be clear that these spaces are not identical (nor isomorphic).

The orthonormality-conditions for basis vectors in countable spaces were summarized, in (A.5), by appeal to the Kronecker-δ. For a rigged Hilbert space, this condition cannot be satisfied as a continuous number of basis vectors is needed.

Instead, a mathematical device called the *Dirac delta distribution* $\delta(x - x')$ (short: Dirac-δ) can be introduced, which leads to a similar condition. The Dirac-δ (inside an integral $I_\delta[\cdot] = \int_a^b \delta(x - x') \cdot \mathrm{d}x$) actually constitutes a functional, which (to recall) means that it maps a whole function to a single number. It is defined by the following two conditions:

(i) $\delta(x - x') = 0$ if $x \neq x'$

(ii) $\int_a^b \delta(x - x')\mathrm{d}x = 1$ for $a < x' < b$

Strictly speaking, the expression $\delta(x - x')$ is not even well-defined outside of an integral; a fact that physicists love to ignore, as shall we in this book. Functionals of this kind are also called *distributions*, and they are generally defined by appeal to so called *test functions*, i.e. suitably well-behaved functions such as *smooth functions* with *compact support*, where the former means that one can differentiate them arbitrarily often and will always get back a continuous function (with 'no jumps or gaps'), and the latter that they vanish somewhere on their domain (cf. Goldhorn et al. 2009, p. 350 ff.; Kerner and von Wahl 2013, p. 339 ff.).

One standardly writes $C^n(\Omega)$ to refer to the set of functions which are n-times differentiable on some domain Ω with the n-th derivative still continuous, and $C_0^n(\Omega)$ for the set of functions which also have compact support in Ω. Note that n may be ∞, and Ω may be $\mathbb{R}^k$ (for some $k \in \mathbb{N}$), in which case one would have the space $C_0^\infty(\mathbb{R}^k)$ as a space of test functions on $\mathbb{R}^k$. Another example for test functions would be a space of C^∞-functions that 'vanish sufficiently fast', like $x^k e^{-ax^2}$ on $\mathbb{R}$, where $0 < a \in \mathbb{R}$ (e.g. Tarasov 2008, p. 30 or Tetschl 2000, p. 139, for details).

The Dirac-δ is then defined as the joint limit of particular sequences of functions parameterized by some σ, e.g. of Gaussian bell-curves $\frac{1}{\sqrt{2\pi}\sigma}e^{-(x-x')^2/2\sigma^2}$, so that $\delta(x - x') = \lim_{\sigma \to 0} \frac{1}{\sqrt{2\pi}\sigma}e^{-(x-x')^2/2\sigma^2}$, but where the limit is actually taken within some integral over a test function $f \in C_0^\infty(\mathbb{R})$ (cf. Kerner and von Wahl 2013, p. 342 ff.). More intuitively, the δ-distribution can be thought of as a highly narrow and peaked curve which encloses a unit area and is centered around some value x'. The limit can then be understood as the width approaching zero while the height increases to infinity, notwithstanding the fact that the area enclosed is still 1. Figure A.3 depicts some functions that converge to $\delta(x - 0) = \delta(x)$ in the above sense.

When integrated over some suitable test function f between points a and b, $\delta(x - x')$ just gives back the value of f at x' so long as $x' \in (a, b)$:

$$\int_a^b f(x)\delta(x - x')\mathrm{d}x = f(x') \tag{A.10}$$

Using the Dirac-δ we can now define the promised sort of 'orthonormality' for basis vectors in spaces of uncountable dimension by

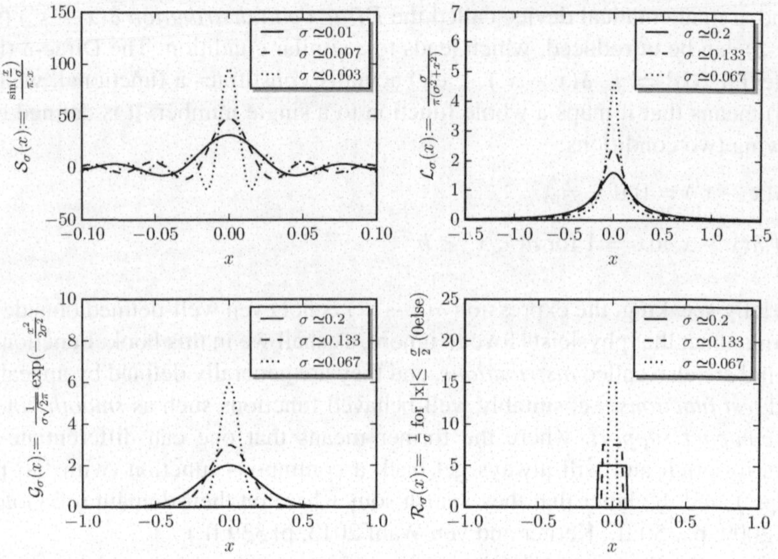

Fig. A.3 As the solid curves are narrowed down, they approximate a highly peaked and localized graph that can in turn be viewed as an approximate graph of the δ-distribution

$$\begin{aligned} \langle \boldsymbol{\lambda} | \boldsymbol{\lambda}' \rangle = \langle \lambda_1 | \lambda_1' \rangle \cdot \ldots \cdot \langle \lambda_n | \lambda_n' \rangle = \delta^n(\boldsymbol{\lambda} - \boldsymbol{\lambda}') := \\ := \delta(\lambda_1 - \lambda_1') \cdot \ldots \cdot \delta(\lambda_n - \lambda_n'), \text{ for } \boldsymbol{\lambda} \in \mathbb{R}^n. \end{aligned} \tag{A.11}$$

Note that this actually 'normalizes' the vectors $|x\rangle$ to *infinity*, as (informally) $\langle x | x' \rangle = \delta(x - x') = \infty$ for $x = x'$. The definition, however, provides what is needed in the given contexts. The Dirac-δ can again be given in arbitrary coordinates, which may imply the the need to divide it by the respective functional determinant (cf. Nolting 2013, p. 7).

$\langle \boldsymbol{x}' | \boldsymbol{x} \rangle = \delta^3(\boldsymbol{x}' - \boldsymbol{x})$ may, finally, also be thought of as a position space representation of the ket $|\boldsymbol{x}\rangle$. One may hence also think of $\langle \boldsymbol{x} | \psi \rangle$ as computing a scalar product $\langle \delta^3(\boldsymbol{x} - \boldsymbol{x}') | \psi(\boldsymbol{x}') \rangle$ as defined in (A.8) (e.g. Manoukian 2007, p. 38).

A.4 Rules for Combination

To combine the state vectors of multiple systems or of multiple independent (and compatible) degrees of freedom of the same system, one appeals to the *tensor* or *direct product* in QM. A little more mathematically, the tensor product is a multilinear map $\mathcal{H}_1 \times \mathcal{H}_2 \times \ldots \times \mathcal{H}_k \xrightarrow{\otimes} \mathcal{H}_1 \otimes \mathcal{H}_2 \otimes \ldots \otimes \mathcal{H}_k$, where the $\mathcal{H}_i$ $(i \in \{1, \ldots, k\})$ may be of different dimensionality, and $\mathcal{H}_1 \otimes \mathcal{H}_2 \otimes \ldots \otimes \mathcal{H}_k$

is called the *tensor product space*. If all dimensions are finite, i.e. $\dim(\mathcal{H}_1) = n, \dim(\mathcal{H}_2) = m, \dim(\mathcal{H}_3) = \ell \ldots$ and $n, m, \ell, \ldots \in \mathbb{N}$, then the tensor product space will be of dimension $n \cdot m \cdot \ell \cdot \ldots$. The resulting space will also be an inner product space (cf. Fischer 2014, p. 353; Heinosaari and Ziman 2012, p. 42 ff.).

When considered on vectors, '$\otimes$' is fully defined by the following properties:

(i) $c(|v_1\rangle \otimes |w_1\rangle) = c\,|v_1\rangle \otimes |w_1\rangle = |v_1\rangle \otimes c\,|w_1\rangle$,
(ii) $(|v_1\rangle + |v_2\rangle) \otimes |w_1\rangle = |v_1\rangle \otimes |w_1\rangle + |v_2\rangle \otimes |w_1\rangle$,
(iii) $|v_1\rangle \otimes (|w_1\rangle + |w_2\rangle) = |v_1\rangle \otimes |w_1\rangle + |v_1\rangle \otimes |w_2\rangle$,

with $|v_1\rangle, |v_2\rangle \in \mathcal{H}_1, |w_1\rangle, |w_2\rangle \in \mathcal{H}_2, c \in \mathbb{C}$, and where $\mathcal{H}_1, \mathcal{H}_2$ may already be a tensor product spaces. Loosely speaking, '$\otimes$' hence simply defines a *non-commutative* product for kets. For vectors $|v\rangle \otimes |w\rangle$ from a tensor product space $\mathcal{H}_1 \otimes \mathcal{H}_2$, we will usually just write $|v\rangle\,|w\rangle$ or even $|v, w\rangle$. The key feature is that either vector can be operated on individually by an appropriate operator (see below).

For kets from some $\mathcal{H}$ isomorphic to $\mathbb{C}^n$, we can also represent '$\otimes$' by the *Kronecker product* between two matrices (e.g. van Loan 2000), which we denote by the same symbol. For instance, let $|v\rangle \in \mathcal{H} \doteq \mathbb{C}^2, |w\rangle \in \mathcal{H}' \doteq \mathbb{C}^3$. Then we can write

$$|v\rangle\,|w\rangle \doteq \begin{pmatrix} v_1 \\ v_2 \end{pmatrix} \otimes \begin{pmatrix} w_1 \\ w_2 \\ w_3 \end{pmatrix} = \begin{pmatrix} v_1 w_1 \\ v_1 w_2 \\ v_1 w_3 \\ v_2 w_1 \\ v_2 w_2 \\ v_2 w_3 \end{pmatrix}.$$

The inner product of tensored vectors $|v_1\rangle \otimes \ldots \otimes |v_n\rangle, |w_1\rangle \otimes \ldots \otimes |w_n\rangle$ computes as

$$(\langle v_1|) \ldots \otimes \langle v_n|)\,|w_1\rangle \otimes \ldots \otimes |w_n\rangle = \langle v_1|w_1\rangle \cdot \ldots \cdot \langle v_n|w_n\rangle$$

A product between a ket $|w\rangle$ and a bra $\langle v|$ (in that order) is sometimes called an *outer* or *matrix product* $|w\rangle\langle v|$. The result formally corresponds to a linear map or operator (see below) $\mathcal{H} \xrightarrow{|w\rangle\langle v|} \mathcal{H}'$ that acts on vector space $\mathcal{H}$, since $(|w\rangle\langle v|)\,|u\rangle = |w\rangle\,\langle v|u\rangle = |w\rangle\,\alpha = \alpha\,|w\rangle$ gives another vector. A more natural understanding is that of an element from a tensor product space $\mathcal{H}' \otimes \mathcal{H}^*$, for two (possibly identical) spaces $\mathcal{H}, \mathcal{H}'$ (cf. Bongaarts 2014, p. 402), whence one may also write $|w\rangle \otimes \langle v|$. Using the Kronecker product again, we can give it a concrete representation in favorable cases. Letting e.g. $|v\rangle, |w\rangle \in \mathbb{C}^3$ (or rather: spaces isomorphic to it), we have

$$|w\rangle\langle v| = |w\rangle \otimes \langle v|) \begin{pmatrix} w_1 \\ w_2 \\ w_3 \end{pmatrix} \otimes \begin{pmatrix} v_1^* & v_2^* & v_3^* \end{pmatrix} = \begin{pmatrix} w_1 v_1^* & w_1 v_2^* & w_1 v_3^* \\ w_2 v_1^* & w_2 v_2^* & w_2 v_3^* \\ w_3 v_1^* & w_3 v_2^* & w_3 v_3^* \end{pmatrix},$$

which this time also coincides with a matrix multiplication for a 3×1 and a 1×3 matrix.

Finally, we introduce the (direct) *sum* of two (or more) vector spaces. Let $\{\mathfrak{h}_j\}_{j\in I}$ be a family of *subspaces* of some $\mathcal{H}$, i.e. nonempty sets of elements from $\mathcal{H}$ which satisfy the requirements of definition A.2 and are closed under the vector space addition and scalar multiplication. Then $\mathfrak{h}_1 + \mathfrak{h}_2 + \ldots + \mathfrak{h}_N$ is the set of vectors $\left\{\mathcal{H} \ni |v\rangle = \sum_j \alpha_j |w_j\rangle \mid \forall j \exists 1 \leq k \leq N : |w_j\rangle \in \mathfrak{h}_k, \alpha_j \in \mathbb{C}\right\}$. In case all the $\mathfrak{h}_k$ have only the null vector $0_{\mathcal{H}}$ in common, the sum is called *direct* and one writes $\mathfrak{h}_1 \oplus \mathfrak{h}_2 \oplus \ldots \oplus \mathfrak{h}_N$.

A.5 Linear Operators

Linear operators constitute the second key ingredient to QM, next to abstract vector spaces. A *linear operator* $\hat{O}$ on $\mathcal{H}$ is a linear map $\mathcal{H} \supseteq \mathrm{dom}(\hat{O}) \xrightarrow{\hat{O}} \mathcal{H}'$. The set $\mathrm{dom}(\hat{O})$ which the operator maps to $\mathcal{H}'$ is called its *domain*, the set of its values in $\mathcal{H}'$ is called its *range* $\mathrm{ran}(\hat{O}) \subseteq \mathcal{H}'$, just as with functions (e.g. Goldhorn et al. 2009, p. 225; Kerner and von Wahl 2013, p. 473). $\mathcal{H}$ and $\mathcal{H}'$ need not coincide, but for many operators of interest they do. Less formally one may think of operators as devices for transforming vectors, i.e.: $\hat{O}|v\rangle = |v'\rangle$. One may also sometimes encounter the notation $\hat{O}|v\rangle = |\hat{O}v\rangle = |v'\rangle$.

To give a concrete example for some general operator on a space of functions, consider that *differentiating* a function with respect to one of its variables, such as in $\frac{\partial \psi(x,t)}{\partial t}$ or $\frac{\partial \psi(x,t)}{\partial x}$, may also be understood in terms of a *differential operator*, $\frac{\partial}{\partial t}$ or $\frac{\partial}{\partial x}$, *acting on* ψ. To recall, *differentiation* of a function $f(x_1, \ldots, x_j, \ldots x_n)$ w.r.t. one of its arguments x_j means taking the limit

$$\lim_{\epsilon \to 0} \frac{f(x_1, \ldots, x_j + \epsilon, \ldots x_n) - f(x_1, \ldots, x_j, \ldots x_n)}{\epsilon} \left(=: \frac{\partial f(x_1, \ldots, x_j, \ldots x_n)}{\partial x_j}\right). \tag{A.12}$$

This corresponds to looking at how f changes over tiny length scales ϵ in the direction of x_j. But differentiation can equally be understood as a map $\frac{\partial}{\partial x} : f \mapsto f' = \frac{\partial f}{\partial x}$, whence $\frac{\partial}{\partial x}$ 'has a life of its own' as the partial differentiation operator w.r.t. x.

In finite spaces, linear operators can often times be expressed by matrices, whence we may write[3]

[3]In mathematics texts (e.g. Fischer 2014, p. 139) a reference to the particular bases $\mathcal{B}', \mathcal{B}$ w.r.t. which the operator has the given matrix form is sometimes included, which bases could be the same or could equally be bases of different spaces. For simplicity, we will make no such reference in this book.

$$\hat{O} \doteq \begin{pmatrix} O_{11} & \cdots & O_{1m} \\ \vdots & \ddots & \vdots \\ O_{n1} & \cdots & O_{nm} \end{pmatrix}$$

Operators may equally be combined by a tensor product to simultaneously operate on all the vectors from a tensor product space. This should be understood in the sense that

$$\hat{A} \otimes \hat{B} \, |v\rangle \otimes |w\rangle = \hat{A} \, |v\rangle \otimes \hat{B} \, |w\rangle = (\hat{A} \, |v\rangle)(\hat{B} \, |w\rangle) = |\hat{A}v\rangle \, |\hat{B}w\rangle \, , \tag{A.13}$$

using the simplification $|\phi\rangle \otimes |\psi\rangle = |\phi\rangle \, |\psi\rangle$. For two operators $\hat{A}$ and $\hat{B}$ with matrix representations A and B on some finite space(es), one can again appeal to the Kronecker product and compute

$$\hat{A} \otimes \hat{B} \doteq \begin{pmatrix} a_{11} & \cdots & a_{1n} \\ \vdots & \ddots & \vdots \\ a_{m1} & \cdots & a_{mn} \end{pmatrix} \otimes \begin{pmatrix} b_{11} & \cdots & b_{1\ell} \\ \vdots & \ddots & \vdots \\ b_{k1} & \cdots & b_{k\ell} \end{pmatrix} = \begin{pmatrix} a_{11}B & \cdots & a_{1n}B \\ \vdots & \ddots & \vdots \\ a_{m1}B & \cdots & a_{mn}B \end{pmatrix} \tag{A.14}$$

where B in the last part denotes the whole second matrix. More concretely, take

$$\begin{pmatrix} a & b \\ c & d \end{pmatrix} \otimes \begin{pmatrix} e & f & g \\ h & i & j \end{pmatrix} = \left(\begin{array}{ccc|ccc} ae & af & ag & be & bf & bg \\ ah & ai & aj & bh & bi & bj \\ \hline ce & cf & cg & de & df & dg \\ ch & ci & cj & dh & di & dj \end{array}\right),$$

with the horizontal and vertical lines merely visual aides.

The simplest operator one can think of (which incidentally turns out to be quite important) is the *identity* or *unit operator*, $\mathbb{1}$, that maps any given vector onto itself, $\mathbb{1} \, |v\rangle = |v\rangle$. In fact, there is not one unique unit operator but rather one for each space, so it would be more precise to write $\mathbb{1}_{\mathcal{H}}$ for a given $\mathcal{H}$, or $\mathbb{1}_n$ for the unit matrix on some $\mathbb{C}^n$. But we will omit the index (again, for notational simplicity) and take it as understood from context on which space a given $\mathbb{1}$ operates.

$\mathbb{1}$ can be expanded in some space $\mathcal{H}$ in terms of the basis vectors of $\mathcal{H}$, e.g. for $\mathcal{H} \doteq \mathbb{C}^2$ with basis vectors

$$|0\rangle \doteq \begin{pmatrix} 1 \\ 0 \end{pmatrix} \text{ and } |1\rangle \doteq \begin{pmatrix} 0 \\ 1 \end{pmatrix},$$

we obtain the matrix representation

$$\mathbb{1} = \sum_{j=1}^{2} |j\rangle\langle j| = |1\rangle\langle 1| + |2\rangle\langle 2| \doteq \begin{pmatrix} 1 & 0 \\ 0 & 0 \end{pmatrix} + \begin{pmatrix} 0 & 0 \\ 0 & 1 \end{pmatrix} = \begin{pmatrix} 1 & 0 \\ 0 & 1 \end{pmatrix}.$$

It can easily be verified that multiplying this matrix to some vector $|v\rangle \doteq \binom{\alpha}{\beta}$ from the left just gives back the original vector. The abstract formula $\mathbb{1} = \sum_j |j\rangle\langle j|$ is generally valid for finite and separable spaces and is usually called a *resolution of the identity*. This can be seen as

$$\mathbb{1}\,|\psi\rangle = \sum_j |j\rangle\langle j|\,|\psi\rangle = \sum_j \langle j|\psi\rangle\,|j\rangle = \sum_j \alpha_j\,|j\rangle\,, \tag{A.15}$$

which is just the expansion of $|\psi\rangle$ in the basis $\{\,|j\rangle\}_{j\in I}$ with expansion coefficients $\alpha_j = \langle j|\psi\rangle$.

Having familiarized ourselves with vector spaces of non-denumerable dimension, we can also generalize the resolution of $\mathbb{1}$ to

$$\mathbb{1} = \int\limits_{\mathbb{R}^n} |\boldsymbol{\lambda}\rangle\langle\boldsymbol{\lambda}|\,\mathrm{d}^n\boldsymbol{\lambda}\,, \quad \boldsymbol{\lambda} \in \mathbb{R}^n. \tag{A.16}$$

This allows for the following generalization of formula (A.15):

$$\mathbb{1}\,|\psi\rangle = \int\limits_{\mathbb{R}^n} |\boldsymbol{\lambda}\rangle\langle\boldsymbol{\lambda}|\,|\psi\rangle\,\mathrm{d}^n\boldsymbol{\lambda} = \int\limits_{\mathbb{R}^n} \langle\boldsymbol{\lambda}|\boldsymbol{\lambda}\rangle\,|\psi\rangle\,|\boldsymbol{\lambda}\rangle\,\mathrm{d}^n\boldsymbol{\lambda} = \int\limits_{\mathbb{R}^n} \psi(\boldsymbol{\lambda})\,|\boldsymbol{\lambda}\rangle\,\mathrm{d}^n\boldsymbol{\lambda}\,, \tag{A.17}$$

where $\psi(\boldsymbol{\lambda})\,\mathrm{d}^n\boldsymbol{\lambda}$ appears as a generalization of the expansion coefficient α_j.

As already noted, objects like $|0\rangle\langle 0|$ and $|1\rangle\langle 1|$ are themselves operators, which, when applied to a vector, only leave behind the component of the vector it has in common with $|0\rangle$ or $|1\rangle$ respectively: $|0\rangle\langle 0|\,|\psi\rangle = \langle 0|\psi\rangle\,|0\rangle$. Thus operators of this form are called *projection operators* or *projectors*, as they basically 'project' a vector $|\psi\rangle$ onto another vector $|\phi\rangle = \langle 0|\psi\rangle\,|0\rangle \propto |0\rangle$, parallel to the one that defines the projector (think back to our arrow-illustration with shadows and light shining from above; cf. Fig. A.2). Another common notation to denote projectors is $\hat{P}_0 = |0\rangle\langle 0|$, $\hat{P}_1 = |1\rangle\langle 1|$, or more generally $\hat{P}_v = |v\rangle\langle v|$.

As indicated above, the relevant operators in QM satisfy a *linearity* condition, which (as in the case of inner products or functionals) means that

$$\hat{O}(\alpha\,|v\rangle + \beta\,|w\rangle) = \alpha\hat{O}\,|v\rangle + \beta\hat{O}\,|w\rangle \quad (\alpha, \beta \in \mathbb{C}). \tag{A.18}$$

Of particular importance are so called *Hermitian* and *self-adjoint* operators. To understand these, first note that applying an operator $\hat{O}$ to a basis vector $|j\rangle \in \mathrm{dom}(\hat{O})$, and taking the inner product $\langle i|\hat{O}|j\rangle$ with another basis vector $|i\rangle \in \mathrm{dom}(\hat{O})$ yields a number called the i-j-th *matrix element*

$$\langle i|\hat{O}|j\rangle = O_{ij}, \tag{A.19}$$

which can be construed to define an actual entry in a matrix representation $O = (O_{ij})$ of $\hat{O}$ in the basis $\mathcal{B} = \{|j\rangle\}_{j\in I}$, whenever this representation exists. For spaces with a continuum of basis vectors, the notation

$$\langle\lambda|\hat{O}|\lambda'\rangle = O(\lambda, \lambda') \tag{A.20}$$

is quite common.

An operator $\hat{O}$ is now called *Hermitian* if for any two vectors $|i\rangle\,, |j\rangle \in \mathrm{dom}(\hat{O})$, it holds that

$$\langle i|\hat{O}|j\rangle = \langle j|\hat{O}|i\rangle^* \Leftrightarrow O_{ij} = O^*_{ji}. \tag{A.21}$$

Thus we equally call a *matrix* O Hermitian in case it holds that

$$O = (O^T)^*, \tag{A.22}$$

where O^T is the *transpose* of matrix O, computed by exchanging element O_{ij} with O_{ji}, i.e., 'flipping of entries over the main (left-to-right) diagonal'. The complex conjugation * then applies to the entries O_{ji} of O^T. Thus, for instance,

$$O = \begin{pmatrix} 3 & 5-2i \\ 5+2i & 7 \end{pmatrix}$$

is a Hermitian matrix, since transposing it and then complex conjugating the entries just gives back the very same matrix.

An important feature meets the eye here: the diagonal elements of this matrix *must* all be *real*, since the act of transposing the matrix leaves them in place and complex conjugation would then otherwise yield a different matrix. A unit matrix $\mathbb{1}$, e.g., is trivially Hermitian.

A *self-adjoint* operator is an operator which is equal to its *adjoint* $\hat{O}^\dagger$, where the adjoint $\hat{O}^\dagger$ of an operator $\hat{O}$ on $\mathcal{H}$ is formally defined by the property

$$\left\langle v \middle| \hat{O} w \right\rangle = \left\langle \hat{O}^\dagger v \middle| w \right\rangle, \tag{A.23}$$

for $|v\rangle\,, |w\rangle \in \mathcal{H}$. A widespread convention thus lets $\hat{O}^\dagger$ always operate on bras from the dual space $\mathcal{H}^*$, i.e.

$$\text{if } \hat{O}\,|v\rangle = \left|v'\right\rangle \text{ then } \langle v|\,\hat{O}^\dagger = \left\langle v'\right|.$$

This is kind of an ad hoc move, but captures the essence of a bunch of theorems.

If $\hat{O}$ is self adjoint, i.e. $\hat{O} = \hat{O}^\dagger$, we have

$$\langle v|\hat{O}|w\rangle = \langle v|\hat{O}^\dagger|w\rangle = \left\langle \hat{O}^\dagger v \middle| w\right\rangle \overset{\text{skew symmetry}}{=} \left\langle w \middle| \hat{O}^\dagger v\right\rangle^*$$
$$= \langle w|\hat{O}^\dagger|v\rangle^* = \langle w|\hat{O}|v\rangle^*,$$

so that every self-adjoint operator is Hermitian. A self-adjoint operator, in other words, is a Hermitian operator that satisfies the additional requirement that the *domains* of $\hat{O}$ and $\hat{O}^\dagger$ *coincide*.

The operators that represent *observables* in QM are *self-adjoint*. That this holds for the position operator(s) $\hat{x}_j$, the momentum operator(s) $-i\hbar\frac{\partial}{\partial x_j}$, or the free Hamiltonian $-\frac{\hbar^2}{2m}\Delta$ is not so trivial but provable.[4] This only holds on 'natural' subspaces of $\mathcal{H}$ however; for a space of square integrable functions, the domain of $-i\hbar\frac{\partial}{\partial x_j}$, say, is the subset of such functions, f, whose derivative exists and where $-i\hbar\frac{\partial}{\partial x_j} f$ is also square integrable (e.g. de la Madrid 2005, p. 296).

A self-adjoint operator $\hat{O}$ can be associated with a set $\mathcal{E}(\hat{O})$ of vectors called its *eigenvectors*. These satisfy the requirement that when $\hat{O}$ acts on them, they are simply multiplied by a (complex) number, i.e.:

$$\hat{O}\,|o_i\rangle = o_i\,|o_i\rangle\,,\ o_i \in \mathbb{C},\ \forall\,|o_i\rangle \in \mathcal{E}(\hat{O}). \tag{A.24}$$

The number o_i is then called the *eigenvalue* of $\hat{O}$ on $|o_i\rangle$, or equally, the eigenvalue of $|o_i\rangle$ for $\hat{O}$.

Denote the set of eigenvalues of an operator by $\sigma_e(\hat{O}) := \{\lambda \in \mathbb{C} \mid \exists\,|v\rangle : \hat{O}\,|v\rangle = \lambda\,|v\rangle\}$. This set may be a subset of what is called the operator's *spectrum* $\sigma(\hat{O}) \supseteq \sigma_e(\hat{O})$. To define the spectrum, note that since $\hat{O}\,|v\rangle = \lambda\,|v\rangle \Leftrightarrow \hat{O}\,|v\rangle = \lambda\mathbb{1}\,|v\rangle \Leftrightarrow (\hat{O} - \lambda\mathbb{1})\,|v\rangle = 0$ for some non-trivial $|v\rangle \in \mathcal{E}(\hat{O})$, we must assume that $(\hat{O} - \lambda\mathbb{1})$ now maps vectors onto 0. But this means that it cannot have an inverse[5] $(\hat{O} - \lambda\mathbb{1})^{-1}$ (where generally $\hat{A}^{-1}$ is defined by $\hat{A}^{-1}\hat{A} = \mathbb{1}$) since there is a whole set of vectors mapped to the same value, 0, not just the null vector $0_{\mathcal{H}}$. So $\sigma(\hat{O})$ is defined by the condition that $\lambda \in \sigma(\hat{O})$ *iff* $(\hat{O} - \lambda\mathbb{1})^{-1}$ does not exist (cf. Heinosaari and Ziman 2012, p. 16). $\sigma_e(\hat{O})$ may also be called the *eigenvalue spectrum* of $\hat{O}$.

Moreover, the *discrete spectrum* of some operator $\hat{O}$ is defined as the set $\sigma_d(\hat{O}) = \{o_j\}_{j\in J} \subseteq \sigma_e(\hat{O})$ of it's eigenvalues o_j for which there is a *finite* number of eigenvectors to each $o_j \in \sigma_d(\hat{O})$ and where it holds that for any $o_j \in \sigma_d(\hat{O})$ there is a 'neighborhood' $U_\epsilon(o_j) = \{\lambda \in \mathbb{C} \mid |o_j - \lambda| \leq \epsilon\}$ for some $\epsilon \in \mathbb{R}$,[6] such that $U_\epsilon(o_j) \cap (\sigma(\hat{O}) \setminus \{o_j\}) = \varnothing$; in words: the eigenvalues from the discrete spectrum

[4] E.g. Gustafson and Sigal (2011, p. 309) for a proof that $-i\hbar\frac{\partial}{\partial x_j}$ is self-adjoint.

[5] More precisely: no *bounded* (see below) inverse (cf. Gustafson and Sigal 2011, p. 47).

[6] $|c| = \sqrt{c^*c}$ here is the 'modulus' on complex numbers $c \in \mathbb{C}$.

'lie isolated' in $\sigma(\hat{O})$. The remainder $\sigma(\hat{O}) \setminus \sigma_d(\hat{O})$ of an operator's spectrum is called its *continuous spectrum*, and an operator may have a *purely discrete* spectrum, *no* discrete spectrum, or a mixed spectrum where both conditions apply to parts of the spectrum (cf. Gustafson and Sigal 2011, pp. 317–318; de la Madrid 2005, pp. 291–294). In the latter two cases, the operator's eigenvectors will obviously form a non-denumerable set.

The eigenvalues of self-adjoint operators are of special interest in QM, and we note the following important theorem (e.g. Ballentine 2000, p. 16).

Theorem *The eigenvalues of a Hermitian operator $\hat{O}$ are all real.*

Proof Let $\hat{O}\,|v\rangle = c\,|v\rangle$. Then taking the inner product with the same vector yields $\langle v|\hat{O}\,|v\rangle = \langle v|c\,|v\rangle \Leftrightarrow \langle v|\hat{O}\,|v\rangle = c\langle v|v\rangle$. Complex conjugating both sides gives $\langle v|\hat{O}|v\rangle^* = c^*\langle v|v\rangle$. By presupposition, $\hat{O}$ is Hermitian, so that $\langle v|\hat{O}|v\rangle^* = \langle v|\hat{O}|v\rangle$, whence we obtain $\langle v|\hat{O}\,|v\rangle = c^*\langle v|v\rangle$. By subtracting this from $\langle v|\hat{O}\,|v\rangle = c\langle v|v\rangle$, we get $0 = (c - c^*)\langle v|v\rangle \Leftrightarrow c = c^*$ which follows from positive semi-definiteness. □

The theorem of course implies that the same holds for self-adjoint operators. This is a crucial property for the interpretation of self-adjoint operators as physically measurable magnitudes or *observables* and their eigenvalues as measurable *values* of these observables.

We have not assumed here that there always exists a *unique* eigenvector for every eigenvalue of an operator, and in fact, this is not so. If there are $|v\rangle\,, |w\rangle$ with $|v\rangle \neq |w\rangle$ for which there is an $\hat{O}$ such that $\hat{O}\,|v\rangle = \alpha\,|v\rangle\,, \hat{O}\,|w\rangle = \alpha\,|w\rangle$, then $\sigma(\hat{O})$ is called *degenerate*.

For many purposes it is useful to consider only *bounded* operators. An operator $\hat{T}$ on $\mathcal{H}$ is called *bounded* if there is a $t \geq 0$ such that $\|\hat{T}\psi\| \leq t\|\psi\|, \forall\,|\psi\rangle \in \mathcal{H}$. The expression $\sup\limits_{\psi\in\mathcal{H}} \frac{\|\hat{T}\psi\|}{\|\psi\|}$ then defines a *norm* for such operators (cf. Heinosaari and Ziman 2012, pp. 11–12). The bounded operators $\mathfrak{B}(\mathcal{H})$ define a special kind of algebra (discussed below) which makes them particularly interesting. However, many important operators such as the position or the momentum operator are in fact *unbounded*, if no restrictions are imposed on their domains (cf. de la Madrid 2005, p. 292 ff.).

Call an operator $\hat{O}$ *diagonal* if $\forall i, j : O_{ij} = c_{ij}\delta_{ij}$. Then the following theorem holds (cf. Shankar 1994, p. 36 ff.; Tarasov 2008, p. 26):

Theorem *To every bounded self adjoint operator $\hat{O}$ on a separable Hilbert space $\mathcal{H}$, there exists a basis $\mathcal{B}$, consisting of orthonormal eigenvectors of $\hat{O}$, and $\hat{O}$ is diagonal in this basis, with the diagonal elements its eigenvalues.*

For *unbounded* operators the following more general theorem holds in a rigged Hilbert space (e.g. Tarasov 2008, p. 38):

Theorem *A self-adjoint operator on a rigged Hilbert space has a complete set of generalized eigenvectors corresponding to real eigenvalues.*

A complete set here means a family $\{|j\rangle\}_{j \in I}$ such that the condition $\langle w|j\rangle = 0, \forall j \in I$, with $|w\rangle$ in the rigged Hilbert space as well, implies that $|w\rangle$ is the null vector of that space. *Generalized eigenvectors* of some operator $\hat{O}$ are strictly speaking anti-linear functionals $\xi \in \Phi^{\times}$ such that for $\phi \in \Phi$ it holds that $\xi[\hat{O}\phi] = o\xi[\phi]$. Since these are anti-linear functionals, one can treat them directly as kets and write $\hat{O}|\xi\rangle = o|\xi\rangle$, or even $|o\rangle$ instead of $|\xi\rangle$ to indicate the eigenvalue for $\hat{O}$ (cf. Tarasov 2008, p. 39). We will call the set $\mathcal{E}(\hat{O})$ of (generalized) eigenvectors of an operator $\hat{O}$ its *eigenbasis*.

Since the mathematical steps for calculating an eigenvector $|o\rangle$ of operator $\hat{O}$ for some eigenvalue o never directly pick out a *unique* vector even in the non-degenerate case, but instead sets $\{\lambda|o\rangle \,|\, \lambda \in \mathbb{C}\}$, i.e. $|o\rangle$ up to a complex scaling λ, each eigenvector spans a one dimensional subspace of $\mathcal{H}$, also called a *ray*. Multiple eigenvectors $|v\rangle$, $|w\rangle$ with the same eigenvalue in $\sigma(\hat{O})$ that do not satisfy $|w\rangle = \lambda|v\rangle$ $(\lambda \in \mathbb{C})$ can be said to span a *degenerate* subspace of $\mathcal{H}$.

Notably, one can generally build new operators out of products (successive application) and/or linear combinations of other operators. A special kind of such a 'built' operator is the *commutator* $[\hat{A}, \hat{B}]$ of two operators $\hat{A}$ and $\hat{B}$, defined by

$$[\hat{A}, \hat{B}] = \hat{A}\hat{B} - \hat{B}\hat{A}. \tag{A.25}$$

In case $[\hat{A}, \hat{B}]|\psi\rangle = 0|\psi\rangle$ for any given $|\psi\rangle \in \mathcal{H}$, we say that $\hat{A}$ and $\hat{B}$ *commute*.

This allows to state the following theorem (e.g. Ballentine 2000, p. 24):

Theorem *If $\hat{A}$ and $\hat{B}$ are commuting self-adjoint operators, each of which possesses a complete set of eigenvectors, then there exists a complete set of vectors which are eigenvectors of both $\hat{A}$ and $\hat{B}$.*

For bounded operators with discrete spectra, this is unproblematic. More generally one can show that for a subclass of so called *essentially* self adjoint operators (i.e., Hermitian ones with a unique self adjoint extension), a joint eigenbasis in the generalized sense exists if they commute (cf. Bohm and Gadella 1969, p. 31 for further reference).

Expressions like $\hat{A}_1 \ldots \hat{A}_n\psi$ for any number n of operators are interpreted as the successive application of the $\hat{A}_i$ to ψ, starting with the *innermost* operator $\hat{A}_n$ and then going from right to left. This means, in other words, that for any i, $\hat{A}_i$ is applied to the result of applying the $\hat{A}_{i+1} \ldots \hat{A}_n$ to ψ. A notation such as $\hat{A}_1(\hat{A}_2(\hat{A}_3(\psi)))$ also makes this order more vivid, and it is reflected also in the rules of matrix multiplication for operators on finite spaces.

Furthermore define, depending on the dimension of the space, the *trace* of an operator $\mathrm{Tr}(\hat{O})$ as the sum or definite integral over the relevant domain Λ of its diagonal elements, i.e.

$$\mathrm{Tr}(\hat{O}) = \sum_i \langle i|\hat{O}|i\rangle \quad \text{or} \quad \mathrm{Tr}(\hat{O}) = \int_\Lambda \langle \lambda|\hat{O}|\lambda\rangle \, d\lambda \,. \tag{A.26}$$

Clearly the trace is not well-defined for *all* operators since it may be infinite when the space has an infinite number of basis vectors. Operators $\hat{O}$ for which the expression $(\hat{O}^\dagger \hat{O})^{1/2}$ has a finite trace are called *trace class*. These are linear as well (cf. Heinosaari and Ziman 2012, p. 32). The relevant example for this book are the so called *density operators*, $\hat{\rho}$ (cf. Heinosaari and Ziman 2012, p. 50), encountered from Chap. 2 on.

Another important class of linear operators besides self-adjoint and trace class ones are *unitary operators*, defined by the property that

$$\hat{U}\hat{U}^\dagger = \hat{U}^\dagger\hat{U} = \mathbb{1}. \tag{A.27}$$

A particular example is the *time evolution operator* $\hat{U}(t_1, t_0)$, which takes some initial vector $|v(t_0)\rangle$ as an input and gives back its value at some later time t_1:

$$\hat{U}(t_1, t_0)\,|v(t_0)\rangle = |v(t_1)\rangle\,. \tag{A.28}$$

It thus mathematically describes the time evolution of a system's quantum state.

Here, again, we note an important theorem (e.g. Shankar 1994, p. 28).

Theorem *Unitary operators preserve the inner product between the vectors they act on.*

Proof Let $\hat{U}\,|v\rangle = |v'\rangle$ and $\hat{U}\,|w\rangle = |w'\rangle$. Then $\langle w'|v'\rangle = \langle w|\hat{U}^\dagger\hat{U}\,|v\rangle = \langle w|\mathbb{1}\,|v\rangle = \langle w|v\rangle$. □

A.6 Probability and Measures

The interpretation of the word 'probability' is a complicated endeavor (cf. the first philosophical interlude in the text), but mathematically there is much less of a problem. Probability in the purely mathematical sense is usually introduced axiomatically as a function (or rather: map) p on a space of *events*, modeled by a collection $\mathcal{A}$ of sets called an *algebra* or *field of sets* (e.g. Heinosaari and Ziman 2012, p. 109; Williamson 2010, p. 11).

Definition A.4 (Algebra of sets) A collection $\mathcal{A}$ of subsets of a nonempty set Ω ($\mathcal{A} \subseteq \mathcal{P}(\Omega)$) such that

(i) $\varnothing \in \mathcal{A}$,
(ii) $\Omega \in \mathcal{A}$,
(iii) $\mathcal{A}$ is closed under finite unions,
(iv) $\mathcal{A}$ is closed under complements.

is called an *algebra of sets*.

We here use the qualifier '*of sets*' to distinguish this notion of an algebra from the one introduced below (*-algebras etc.). The elements ω of Ω may be considered as the *outcomes* of some experiment or more general observation, whence Ω is sometimes called an *outcome* or *sample space* (e.g. Williamson 2010, p. 11). The sets containing the $\omega \in \Omega$ are then understood as the *events* of their occurrence, whence $\mathcal{A}$ is also called an *event space* (cf. ibid.). (iii) and (iv) together imply that, by de Morgan, $\mathcal{A}$ is also closed under finite intersections. (i) is also actually implied by (ii) and (iv).

Philosophers may, in fact, be more inclined to think of probability as attaching to sentences α, β of some language $\mathbb{L}$. So long as logically equivalent sentences of $\mathbb{L}$ are attributed identical probabilities, one can indeed straightforwardly define everything equivalently by appeal to $\mathbb{L}$'s sentences. The events in $\mathcal{A}$ will then be interpreted as the *propositions* expressed by those sentences (e.g. Huber 2009, p. 3; Williamson 2010, p. 27). Using sentences rather than sets, a conjunction $\alpha \wedge \beta$ will replace the intersection $A \cap B$ of the respective propositions from $\mathcal{A}$, a disjunction $\alpha \vee \beta$ the union $A \cup B$, and a negation $\neg\alpha$ the complement $A^c = \Omega \setminus A$ (cf. also Schurz 2015, p. 9).

A probability function p on $\mathcal{A}$ is now required to satisfy the following (*Kolmogorov*) axioms (e.g. Roussas 2007, pp. 33–34):

(i) $p(A) \geq 0, \ \forall A \in \mathcal{A}$,
(ii) $p(\Omega) = 1$,
(iii) $p(\bigcup_{j=1}^{N} A_j) = \sum_{j=1}^{N} p(A_j), \ \forall A_j \in \mathcal{A}$, in case $A_j \cap A_k = \varnothing$ for $j \neq k$, and where $N \in \mathbb{N}$.

Note that (iii) implies that $p(\varnothing) = 0$, since for any $A \in \mathcal{A}$, $A \cup \varnothing = A$, so that $p(A) = p(A \cup \varnothing)$, and since $A \cap \varnothing = \varnothing$, $p(A) = p(A \cup \varnothing) = p(A) + p(\varnothing)$ by (iii) (cf. also Roussas 2007, p. 36). $\varnothing$ is sometimes also referred to as the *impossible event*, Ω as the *certain event* (cf. Roussas 2007, p. 7). They correspond to falsum $\bot$ and tautology $\top$ respectively; $A_j \cap A_k = \varnothing$ then e.g. means that $\alpha_j \wedge \alpha_k \leftrightarrow \bot$ ('$\leftrightarrow$' denoting logical equivalence).

Additionally, a *conditional* probability function for an event A, *given* that B is the case, is defined (or equally introduced axiomatically) as

$$p(A|B) = \frac{p(A \cap B)}{p(B)}. \tag{A.29}$$

In this book, we will write A, B for the joint occurrence of the events A and B, which set-theoretically corresponds to their intersection $A \cap B$.

For applications in mathematical probability theory, the function p is typically generalized to a *measure* μ. This requires the extension of the concept of an algebra of sets to that of a σ-*algebra*:

Definition A.5 (**σ-algebra**) An algebra of sets Σ defined over a nonempty set Ω which is also closed under countable unions is called a σ-*algebra*.

In case $\Omega = \mathbb{R}$, one can make use of the so called *Borel Algebra* $\mathcal{B}(\mathbb{R})$, which can be defined as the smallest σ-algebra containing all open sets on $\mathbb{R}$ (cf. Heinosaari and Ziman 2012, p. 115), or more intuitively as the σ-algebra generated by all intervals (open, closed, half open) on $\mathbb{R}$ (cf. Roussas 2014, p. 3). With the aid of a σ-algebra, we can now say precisely what a measure is (cf. Tetschl 2000, p. 210).

Definition A.6 (Measure) A map $\mu : \Sigma \longrightarrow [0, \infty]$ on a σ-algebra Σ is called a *measure*, in case

(i) $\mu(\varnothing) = 0$,
(ii) $\mu(\bigcup_{j=1}^{\infty} A_j) = \sum_{j=1}^{\infty} \mu(A_j)$, if $A_j \cap A_k = \varnothing, \forall j \neq k$ (*countable* or σ*-additivity*)

(Ω, Σ) may be called a *measurable space*, (Ω, Σ, μ) a *measure space* (cf. Roussas 2014, p. 19).

We initially hinted at the importance of the *Lebesgue measure* λ for the functions in spaces $L^2(\mathbb{R}^n)$. λ can be said to measure the 'volume' of n-dimensional 'rectangles', i.e. intervals $I \subset \mathbb{R}^n$ of the form $I = [a_1, b_1] \times \ldots \times [a_n, b_n]$ where the intervals $[a_i, b_i]$ could also be half open to the left or open. The Lebesgue measure then takes I and gives back the volume, i.e. the product of all the lengths in each of the n dimensions: $\lambda(I) = (b_1 - a_1) \cdot \ldots \cdot (b_n - a_n)$. Sets which are contained in any union of such rectangles with an arbitrarily small sum of measures are said to be *of zero measure* (w.r.t. λ). This goes, for instance, for singletons $\{p\}$ of points $p \in \mathbb{R}^n$, or countable unions thereof (cf. Kerner and von Wahl 2013, p. 275 ff.).

λ then also allows to define the *Lebesgue integral* which makes a larger range of functions integrable, e.g. functions which are only defined on the rational numbers $\mathbb{Q}$. In this book, the (cumbersome) details will not matter, so we refer to the literature for further reference (e.g. Capinski and Kopp 2004, p. 20 ff.).

Due to the probabilistic nature of QM, we are, however, interested in the notion of a *probability measure* (cf. Roussas 2014, p. 19), as a generalization of the probability function p.

Definition A.7 (Probability measure) A map $\mu : \Sigma \longrightarrow [0, 1]$ on a σ-algebra Σ over non-empty set Ω is called a *probability measure*, in case

(i) $\mu(A) \geq 0, \ \forall A \in \Sigma$,
(ii) $\mu(\Omega) = 1$,
(iii) μ is σ-additive.

(Ω, Σ, μ) and $(\Omega, \mathcal{A}, p)$, with p finitely additive and $\mathcal{A}$ only closed under finite unions, may then be called *probability spaces*; $(\Omega, \mathcal{A}, p)$ may be called a *finite* one.

The requirement of countable or σ-additivity cannot be endorsed in all interpretations of probability without problem though, so it should be treated with caution. The reason is that, if μ is construed as describing the limit of the relative frequency of some sequence of events, and if each of a countable set of mutually exclusive and exhaustive events E_j in fact occurs only finitely often, then each $\mu(E_j)$ will be 0, but $\mu(E_1 \cup E_2 \cup \ldots)$ must still be 1, since by presupposition one of the events is bound to happen. Similar concerns can be raised from the point of view of epistemic interpretations of probability (cf. Howson and Urbach 2006, pp. 27–28).

In mathematical probability theory, the notion of a *random variable* also plays quite an important role. It has a somewhat specific meaning therein, however, which is softened in the text (chapter 2). The specific meaning is that of a map $\underline{X} : \Omega \longrightarrow \mathbb{R}$ where the *inverse image* $\underline{X}^{-1}([a, \infty)) = \{\omega \in \Omega | X(\omega) \geq a\}$ is an element of some σ-algebra Σ or algebra $\mathcal{A}$ over Ω respectively (cf. Capinski and Kopp 2004, p. 66). The pair $(\mathbb{R}, \mathcal{B}(\mathbb{R}))$ will then define a measurable space for the (real) values of the random variable. In this sense, a random variable may be construed as a means of assigning numbers to the outcomes of some experiment or observation. If the variable maps to $\mathbb{R}^n$ instead and one has the measurable space $(\mathbb{R}^n, \mathcal{B}^n(\mathbb{R}))$, one may also talk of a *random vector* $\underline{\boldsymbol{X}}$ (e.g. Roussas 2014, p. 8).

If we have some probability space (Ω, Σ, P) and a random variable $\underline{X}$ defined on Ω, then the map $P_{\underline{X}} : \mathcal{B}(\mathbb{R}) \longrightarrow [0, 1]$, where for $B \in \mathcal{B}(\mathbb{R})$ it holds that $P_{\underline{X}}(B) = P(\underline{X}^{-1}(B))$, defines the *probability distribution* of $\underline{X}$ (cf. Roussas 2014, p. 66). $(\mathbb{R}, \mathcal{B}(\mathbb{R}), P_{\underline{X}})$ will constitute a probability space, and $P_{\underline{X}}$ measures the probability that $\underline{X}(\omega) \in B$, i.e. measures how probable it is that the value of $\underline{X}$ falls in some Borel set B (some union or join of intervals from $\mathbb{R}$).

With $P_{\underline{X}}$ one can now define the *expectation value* of random variable $\underline{X}$ or a measurable, real valued function g thereof by

$$\langle \underline{X} \rangle_P := \int_\Omega \underline{X} \, \mathrm{d}P \tag{A.30}$$

$$\langle g(\underline{X}) \rangle_P := \int_\Omega g(\underline{X}) \, \mathrm{d}P = \int_\mathbb{R} g(x) \, \mathrm{d}P_{\underline{X}} \,, \tag{A.31}$$

where it is a theorem that the last equality holds (cf. Roussas 2014, pp. 59 and 66). The case $g(x) = x$ provides the corresponding expression for (A.30). For a random vector $\underline{\boldsymbol{X}}$ one may replace $\mathbb{R}$ by $\mathbb{R}^n$ and $P_{\underline{X}}$ by $P_{\underline{\boldsymbol{X}}}$.

With these definitions, one can also state the following theorem (cf. Roussas 2014, p. 129):

Theorem *Let μ, ν be σ-finite measures on (Ω, Σ) and $\mu \ll \nu$, and let $\underline{X}$ be a random variable for which the integral $\int_\Omega \underline{X} \, \mathrm{d}\mu$ exists. Then*

$$\int_A \underline{X} \, \mathrm{d}\mu = \int_A \underline{X} \frac{\mathrm{d}\mu}{\mathrm{d}\nu} \, \mathrm{d}\nu \,, \forall A \in \Sigma.$$

The derivative here is the so called *Radon-Nikodym derivative* (cf. Capinski and Kopp 2004, p. 194; Roussas 2014, p. 129). A measure μ is called σ*-finite* in case $\mu(\Omega) = \infty$ but there is a partition $\{A_j\}_{j \in J \subseteq \mathbb{N}}$ of Ω, such that $\mu(A_j) < \infty, \forall j \in J$ (cf. Roussas 2014, p. 19). A measure ν is said, moreover, to *dominate* another measure μ, and one writes $\mu \ll \nu$, if it holds that $\nu(A) = 0$ implies $\mu(A) = 0$ for any set A in the σ-Algebra over which both μ and ν are defined (cf. Roussas 2014, p. 122).

The interesting part of this theorem is that it allows the often encountered use of *probability density functions* or *probability densities*. If one evaluates $\langle g(X) \rangle_P$ on some set A which is the inverse image $X^{-1}(B)$ for some $B \in \mathcal{B}(\mathbb{R})$ one can now write

$$\int_A g(\underline{X})\,\mathrm{d}P = \int_B g(x)\,\mathrm{d}P_{\underline{X}} = \int_B g(x)\varrho(x)\,\mathrm{d}\lambda \tag{A.32}$$

in case $P_{\underline{X}} \ll \lambda$, and where $\varrho(x) := \frac{\mathrm{d}P_{\underline{X}}}{\mathrm{d}\lambda}(x)$ is the probability density function (cf. Roussas 2014, p. 130). λ is usually the Lebesgue measure, and in practice one can often even write $\varrho(x)\,\mathrm{d}x$ instead.

Note that so long as ϱ exists, $P_{\underline{X}}((-\infty, x]) = \int_{-\infty}^{x} \varrho(x')\,\mathrm{d}x'$. $F_{\underline{X}}(x) = P_{\underline{X}}((-\infty, x])$ is also sometimes called the *cumulative distribution* or *distribution function* of $\underline{X}$, and $\varrho(x)$ can then be understood more directly as $\frac{\mathrm{d}F_{\underline{X}}(x)}{\mathrm{d}x}$ (cf. Capinski and Kopp 2004, pp. 109–110; Roussas 2014, pp. 66 and 129). In physics contexts, moreover, it is often simply *assumed* that a probability density exists and some such density is simply *defined* according to given needs. This approach is rather 'bottom up'.

To reconnect these concepts to QM, we turn to *positive operator valued measures* (POVMs). First note that a Hermitian operator $\hat{O}$ on $\mathcal{H}$ is called *positive* in case $\langle v|\hat{O}|v\rangle \geq 0, \forall |v\rangle \in \mathcal{H}$. Then the relation $\hat{Q} \leq \hat{P}$ on linear operators can be understood in the sense that $\hat{P} - \hat{Q}$ is a positive operator (cf. Heinosaari and Ziman 2012, pp. 18). With this, one can define the set of operators $\hat{E}$ such that $\mathbb{O} \leq \hat{E} \leq \mathbb{1}$ (with $\mathbb{O}$ a *zero operator* that gives zero when applied to any arbitrary vector in $\mathcal{H}$) and call it the set of *effects* $\mathcal{E}(\mathcal{H})$ on a given $\mathcal{H}$ (cf. Heinosaari and Ziman 2012, p. 70).[7]

A POVM is then defined as follows.

Definition A.8 (POVM) A map $\hat{E} : \Sigma \longrightarrow \mathcal{E}(\mathcal{H})$ for a measurable space (Ω, Σ) is called a *positive operator valued measure* (POVM), in case

(i) $\hat{E}(\varnothing) = \mathbb{O}$,
(ii) $\hat{E}(\Omega) = \mathbb{1}$
(iii) $\hat{E}(\bigcup_{j=1}^{\infty} A_j) = \sum_{j=1}^{\infty} \hat{E}(A_j)$, if $A_j \cap A_k = \varnothing, \forall j \neq k$

The convergence in the third condition is meant 'weakly', i.e. $\lim_{n\to\infty} \left| \langle\phi|\hat{T}|\psi\rangle - \langle\phi|\hat{T}_n|\psi\rangle \right| = 0$ for $\hat{T} = \hat{E}(\bigcup_{j=1}^{\infty} A_j)$ and $\hat{T}_n = \sum_{j=1}^{n} \hat{E}(A_j)$ (cf. Heinosaari and Ziman 2012, pp. 35 and 109). Details concerning the use and construction of POVMs can be found in, e.g., Busch et al. (1995, p. 25 ff.), Nielsen and Chuang (2010, p. 90 ff.), or Peres (2002, p. 282 ff.), and some details are also given in Chap. 4.

[7]In fact, the set of effects is defined over the set of *bounded* self-adjoint operators (cf. Heinosaari and Ziman 2012, p. 70). This is, however, implicit in the requirement that $\mathbb{O} \leq \hat{E} \leq \mathbb{1}, \forall \hat{E} \in \mathcal{E}(\mathcal{H})$, as $\sigma(\hat{E}) \subseteq [0, 1]$.

The simplest case of a POVM is a family of projectors $\{|j\rangle\langle j|\}_{j\in I}$. These also allow for the so-called *spectral decomposition* of a self-adjoint operator, which is ensured by the following theorem (cf. Tarasov 2008, p. 198).

Theorem (Spectral Theorem) *For each self-adjoint linear operator* $\hat{A}$ *on a separable Hilbert space* $\mathcal{H}$*, there exists a resolution of the identity* $\left\{\hat{E}_\lambda | \lambda \in \mathbb{R}\right\}$ *such that* $\hat{A}$ *can be presented by the* $\hat{E}_\lambda$ *via*

$$\langle v|\hat{A}|w\rangle = \int_{-\infty}^{\infty} \lambda \, \langle v|\mathrm{d}\hat{E}_\lambda|w\rangle \, , \; \textit{for } |w\rangle \in \mathrm{dom}(\hat{A}).$$

$|v\rangle \in \mathcal{H}$ *belongs to* $\mathrm{dom}(\hat{A})$ iff

$$\|\hat{A}v\| = \int_{-\infty}^{\infty} \lambda^2 \, \langle v|\mathrm{d}\hat{E}_\lambda|v\rangle < \infty.$$

That the $\hat{E}_\lambda$ resolve the identity means that $\int_{-\infty}^{\infty} \mathrm{d}\hat{E}_\lambda = \mathbb{1}$, and from the theorem it becomes apparent that we can use expressions of the form $\mathrm{d}\hat{E}_\lambda = |\lambda\rangle\langle\lambda| \, \mathrm{d}\lambda$. For $\lambda = x$, say, we would then have $\langle\psi|\hat{x}|\psi\rangle = \int_{-\infty}^{\infty} x \, \langle\psi|x\rangle \, \langle x|\psi\rangle \, \mathrm{d}x = \int_{-\infty}^{\infty} x\psi^*(x)\psi(x) \, \mathrm{d}x = \int_{-\infty}^{\infty} x|\psi(x)|^2 \, \mathrm{d}x$, which is the expectation value for position w.r.t. ψ, as $|\psi(x)|^2$ defines a probability density.

A self-adjoint operator $\hat{A}$ with a discrete spectrum can be expanded as a sum (or series)

$$\hat{A} = \sum_i a_i \hat{P}_{a_i}, \tag{A.33}$$

with a_i the eigenvalues of $\hat{A}$ and $\hat{P}_{a_i} = |a_i\rangle\langle a_i|$. This follows from the spectral theorem if the measure is chosen to be of the form $\mathrm{d}\hat{E}_\lambda = \sum_i \delta(\lambda - a_i)\hat{P}_{a_i} \, \mathrm{d}\lambda$ (cf. Tarasov 2008, p. 199). Examples of spectral decompositions are found in the text. Using the spectral decomposition, one can also write a *function* f of operator $\hat{A}$ as

$$f(\hat{A}) = \sum_j f(a_j)\hat{P}_{a_j}, \tag{A.34}$$

which proves extremely useful on occasion.

A.7 Abstract Algebras

Especially in the context of quantum field theory, notions from the abstract mathematical theory of *algebras* become important. An algebra in the present sense, however, is not the same as an algebra of sets as defined above. Here we mean by an algebra something that can be informally characterized as "a collection of elements along with a way of taking their products and linear combinations." (Ruetsche 2011, p. 73)

More precisely, an algebra $\mathcal{A}$ over a field $\mathbb{F}$ can be defined as follows.

Definition A.9 (Algebra) A set $\mathcal{A}$ of mathematical objects will be called an *algebra* over the field $\mathbb{F}$ if there are two operations $+, \cdot$ defined over $\mathcal{A}$ such that

(i) $+$ is *associative*, *commutative* and has a *null element* ($A + 0 = 0 + A = A, \forall A \in \mathcal{A}$),
(ii) $\cdot$ is *associative* and *distributive* w.r.t. addition ($A \cdot (B + C) = A \cdot B + B \cdot C, (A + B) \cdot C = A \cdot C + B \cdot C, \forall A, B, C \in \mathcal{A}$),
(iii) $\mathcal{A}$ is *closed* w.r.t. $+, \cdot$.

Associativity and commutativity are exactly defined as in the definition of a field, and we here equally use $\mathcal{A}$ to refer to the set *with* the operations defined on it. In the multiplication, we will generally omit the $\cdot$ (i.e. $A \cdot B = AB$). If $\mathcal{A}$ also has a *multiplicative identity* or *unit element* I, s.t. $IA = AI = A$, then it is called *unital.*

Moreover, for the QM context, there are special classes of algebras that are of interest. For the following, we only concern ourselves with the case $\mathbb{F} = \mathbb{C}$, since this is the case of interest in QM.[8]

Definition A.10 (*-algebra) An algebra $\mathcal{A}$ is called a **-algebra* if it is closed under an *involution* $^* : \mathcal{A} \to \mathcal{A}$ which satisfies for all $A, B \in \mathcal{A}, c \in \mathbb{C}$ that

(i) $(A^*)^* = A, (A + B)^* = A^* + B^*$
(ii) $(cA)^* = c^* A^*$,
(iii) $(AB)^* = B^* A^*$,

where c^* means the complex conjugate for $c \in \mathbb{C}$.

There are two interesting subclasses of *-algebras, namely:

Definition A.11 (C^*-algebra) A *-algebra $\mathfrak{A}$ is called a *C^*-algebra* if it is complete w.r.t. a norm $\| \cdot \|$ that satisfies

(i) $\|A^* A\| = \|A\|^2$,
(ii) $\|\alpha A\| = |\alpha| \|A\| (\alpha \in \mathbb{C})$,
(iii) $\|A + B\| \leq \|A\| + \|B\|$,
(iv) $\|AB\| \leq \|A\| \|B\|, \forall A, B \in \mathcal{A}$,

[8] The subsequent definitions are gathered from Ruetsche (cf. 2011, p. 75 ff.) and Haag (1996, p. 112–118).

and where only the null element 0 has zero norm.

The most important example of a C^*-algebra are the $\mathfrak{B}(\mathcal{H})$ on a Hilbert space $\mathcal{H}$. The involution $*$ is then the adjoint operation $\dagger$. The second interesting case is a special case of a C^*-algebra, the von Neumann algebra (cf. Ruetsche 2011, pp. 78 and 86–87).

Definition A.12 (von Neumann-algebra) A *von Neumann algebra* $\mathfrak{M}$ is the *closure* of a C^*-algebra of bounded operators $\mathfrak{B}(\mathcal{H})$ on some Hilbert space $\mathcal{H}$ w.r.t. the *strong* and *weak operator topologies*, i.e. the union of $\mathfrak{B}(\mathcal{H})$ with the set of operators $\hat{A}$ such that $|\langle\psi|(\hat{A}_n - \hat{A})|\phi\rangle| \to 0$ as $n \to \infty, \forall |\psi\rangle, |\phi\rangle \in \mathcal{H}$ and where $\hat{A}_n \in \mathfrak{B}(\mathcal{H}), \forall n \in \mathbb{N}$ (where in the strong case the same holds without $\langle\psi|$).

A von Neumann algebra $\mathfrak{M}$ is also its own *double commutant* $\mathfrak{M}''$, where $\mathfrak{M}' = \{B \in \mathfrak{B}(\mathcal{H}) | BA = AB, \forall A \in \mathfrak{M}\}$.

Appendix B
Mesoscopic Quantum Superposition

We here give a brief overview of the phenomenon of mesoscopic (quantum) superposition, i.e. superposition of states of comparatively large objects. To elaborate on this, we first need to introduce some of the basics of *superconductivity*. Superconductivity is a phenomenon in which an electric current can flow (almost) without any dissipation, i.e. without loss of (kinetic) energy into the surrounding medium. The surrounding medium in question will be a solid that can be modeled as a periodic lattice of evenly spaced nuclei (a 'crystal'). In such a solid, electrons can form so called *Cooper pairs*, whose occurrence can be plausibilized as follows (cf. Blundell 2009, p. 56 ff.).

In principle two electrons obviously repel each other since they are both equally negatively charged. But they also attract the positively charged nuclei of the lattice-like solid. The result is a distortion in the periodicity of the lattice which persists for some time after an electron has passed, and in turn attractively affects other electrons, since now some of the positive charge of the nuclei is accumulated more strongly. This leads to a pairwise *coupling* of electrons so brought into proximity of one another, mediated by the distortion of the lattice. This distortion, propagating with the electrons as a vibration of the lattice, will come in energetically discretized form. This leads to the notion of *phonons*, quanta of the lattice-vibration, not unlike the photons, construed as quanta of the electromagentic field. The electron-electron interaction which results in Cooper pairing is hence mediated by such a phonon. Notably, this only happens in materials in which the electrons interact more strongly with the vibrations (phonons), so in materials which are otherwise bad conductors.

The most successful *formal* model of superconductivity that describes this situation is known as the *BCS model*, after John Bardeen, Leon Cooper and Robert Schrieffer (1957). They found an effective Hamiltonian

$$\hat{H} = \sum_{p\sigma} \varepsilon_p \hat{c}^\dagger_{p\sigma} \hat{c}_{p\sigma} - \left(\frac{g}{2V}\right)^2 \sum_{pp'} \hat{c}^\dagger_{p'\uparrow} \hat{c}^\dagger_{-p'\downarrow} - \hat{c}_{p\uparrow} \hat{c}_{-p\downarrow}, \tag{B.1}$$

F. J. Boge, *Quantum Mechanics Between Ontology and Epistemology*, European Studies in Philosophy of Science 10,
https://doi.org/10.1007/978-3-319-95765-4

the first term describing the kinetic energy of all pairs involved, and the second one the attractive potential energy between pairs of electrons with opposite spins and momenta (cf. Lancaster and Blundell 2014, p. 401). $\varepsilon_{\boldsymbol{p}}$ is some fixed (kinetic) energy for given $\boldsymbol{p}$, g a momentum-independent potential, V a volume and $\sigma \in \{\uparrow, \downarrow\}$ a spin index. More importantly, Schrieffer 'invented' a trial many-particle state that can be written as

$$|\Psi_{BCS}\rangle = \prod_{\boldsymbol{p}} (u_{\boldsymbol{p}} \mathbb{1} + v_{\boldsymbol{p}} \hat{c}^{\dagger}_{\boldsymbol{p}\uparrow} \hat{c}_{-\boldsymbol{p}\downarrow}) \, |0\rangle \, , \tag{B.2}$$

where for any given $\boldsymbol{p}$ one finds that $|v_{\boldsymbol{p}}|^2$ gives the average pair-occupation number for states of momentum $\boldsymbol{p}$ and with a given fixed spin direction (so that $2 \sum_{\boldsymbol{p}} |v_{\boldsymbol{p}}|^2$ gives the total average number of occupied states), and $|u_{\boldsymbol{p}}|^2$ the number of unoccupied states with momentum $\boldsymbol{p}$. However, the number of electrons described by this state vector only has a fixed *average*, meaning that there is generally not a fixed number of electrons present in it (cf. Lancaster and Blundell 2014, pp. 402–403; Blundell 2009, p. 61).

Crucially, in many situations there is such a large (average) number of pairs involved, that the state is called 'macroscopically occupied'. Moreover, the state is an example of a *coherent state*, which formally means that it is an eigenstate of an annihilation operator, and implies that its position-representation $\Psi_{BCS}(\boldsymbol{x}_1\sigma_1, \ldots, \boldsymbol{x}_n\sigma_n)$ is a quite narrow wave-packet (in configuration space) for high occupation numbers that *maintains* a somewhat narrow spread over time (cf. Lancaster and Blundell 2014, p. 278).[1]

The paired up electrons in the BCS-state can be construed as forming *quasi particles*[2] called *bogolons*, which somewhat behave like bosons in virtue of the pair-creating operators $\hat{c}^{\dagger}_{\boldsymbol{p}\uparrow} \hat{c}_{-\boldsymbol{p}\downarrow}$ satisfying commutation-relations, not anti-commutation. However, the anti-commutation relations are not exactly those of ordinary bosons, as given in (2.50), whence the pairs should not literally be viewed as bosons (cf. Lancaster and Blundell 2014, p. 410).

Working out the energies of the BCS-Hamiltonian, one finds some value Δ that separates the ground state energy from that of the first excited state. More precisely, this holds for any individual one of the Cooper pairs, crucially implying that if all the pairs are in the low-lying ground state, they cannot be excited by small thermal vibrations of the lattice. In turn, this results in the resistance- and dissipationless current advertised above, sometimes also referred to as a *supercurrent* (cf. Schlosshauer 2007, p. 271).

---

[1]We here represent the spin degrees of freedom rather as 'coordinates' σ_i than by appeal to a column vector, which is a widespread notational convenience. For a single particle this can be read, for instance, as $\psi(\boldsymbol{x})\chi(\sigma)$, with $\chi(\uparrow) \doteq \binom{1}{0}$, and so forth (e.g. Annett 2004, p. 149). The fermionic BCS-wave function is of course overall antisymmetric.

[2]These are "excitations of a macroscopic many-particle system[…] [which] do not come on their own but belong to collective effects." (Falkenburg 2007, p. 238).

Such supercurrents can be used to bring about mesoscopic superpositions in a device called a *superconducting quantum interference device* (SQUID) (we focus on the simplest example; e.g. Schlosshauer 2007, p. 272 for some details). Such SQUIDs exploit what is called the *Josephson effect*, namely the (experimentally supported) fact that in ring-shaped superconducting materials with a thin insulating barrier inserted (a 'Josephson junction'), Cooper pairs will start to tunnel through the barrier, resulting in a supercurrent without any voltage applied (cf. Schlosshauer 2007, ibid.).

As we saw, the totality of all the Cooper pairs together is described by one total quantum state or wave function. But at least in the limit of a certain critical temperature, one can also describe each Cooper pair in terms of a joint wave function $\psi(\boldsymbol{x}_{\text{cm}}, t) = |\psi|e^{i\varphi(\boldsymbol{x}_{\text{cm}},t)}$ ($\boldsymbol{x}_{\text{cm}}$ being the center of mass coordinate) with local phase $\varphi(\boldsymbol{x}_{\text{cm}}, t)$ (cf. Leggett 2002, p. R439; Annett 2004, p. 127). This wave function appears in the BCS model as an 'order parameter' which depends on the energy gap Δ (cf. Annett 2004, pp. 127–128). One can use it to describe many properties of the total current of Cooper pairs and define, for instance, a kind of current density, similar to the probability current density (2.56), but multiplied by the charge $-2e$ of the Cooper pairs ($-e$ the charge of the electron), and with effective mass $2m_e$ (i.e. twice the electron mass).

If in addition an external magnetic field $\boldsymbol{B}_{\text{ext}}$ is applied to the SQUID, the Schrödinger equation has to be modified by replacing $\hat{p} \mapsto \hat{p} - 2e\boldsymbol{A}$, with $\boldsymbol{A}$ the vector potential of $\boldsymbol{B}_{\text{ext}}$ (i.e. $\boldsymbol{B}_{\text{ext}} = \nabla \times \boldsymbol{A}$). However, the corresponding modification of the Schrödinger equation together with a requirement that it remains form invariant in spite of the spacetime dependence of φ implies the need to perform a *gauge transformation* $\boldsymbol{A} \mapsto \boldsymbol{A} - \frac{\hbar}{2e}\nabla\varphi$ which leaves the measurable quantity $\boldsymbol{B}_{\text{ext}}$ unchanged (as $\nabla \times (\nabla\varphi) = 0$).

This gauge transformation in turn implies a change in the current expression, namely the appearance of a term $-\frac{2e^2\boldsymbol{A}}{m_e}\psi^*\psi$. Executing this equation on $\psi(\boldsymbol{x}_{\text{cm}}, t)$ from above, one can derive the condition

$$\hbar\nabla\varphi = \frac{m_e}{|\psi|^2 e}\boldsymbol{j}_s + 2e\boldsymbol{A} \tag{B.3}$$

($\boldsymbol{j}_s$ now being construed as the 'supercurrent density'). Taking the superconducting ring of the SQUID to be closed without the junction for the moment, and the wave function $\psi(\boldsymbol{x}_{\text{cm}}, t)$ to be confined to the ring, we would have to require that it has a unique value at any point inside the ring (in order for ψ to be single-valued). This is equivalent to requiring that

$$\oint_\gamma \nabla\varphi\,\mathrm{d}\boldsymbol{r} = n \cdot 2\pi, n \in \mathbb{N}, [3] \tag{B.4}$$

[3]The reason that we are only appealing to the natural numbers, not all integers, becomes clear below.

i.e. that along any closed curve γ inside the ring, the integral change of the phase φ is an integer multiple of 2π, so that $\psi(\boldsymbol{x}_{\text{cm}}, t)$ has a unique value at the coinciding initial and final points of the integral.

But using expression (B.3) in (B.4), we now have

$$\oint_\gamma \frac{m_e}{|\psi|^2 e} \boldsymbol{j}_s + 2e\boldsymbol{A}\,\mathrm{d}\boldsymbol{r} = n \cdot 2\pi\hbar = nh$$

$$\Leftrightarrow \oint_\gamma \frac{m_e}{|\psi|^2 e} \boldsymbol{j}_s\,\mathrm{d}\boldsymbol{r} + \underbrace{2e \int_{\mathscr{S}} \boldsymbol{B}_{\text{ext}}\,\mathrm{d}\boldsymbol{s}}_{=\Phi_{\text{ext}}} = nh$$

$$\Leftrightarrow \Phi = n \cdot \underbrace{(h/2e)}_{=:\Phi_0}.$$

Here the second line follows from Stokes' theorem and the definition of the magnetic flux Φ_{ext} as the integral amount of $\boldsymbol{B}_{\text{ext}}$ that passes through some surface $\mathscr{S}$. Φ is called a *fluxoid* rather than a flux, as it also encompasses the magnitude $\oint_\gamma \frac{m_e}{|\psi|^2 e} \boldsymbol{j}_s\,\mathrm{d}\boldsymbol{r}$ not present in the classical flux. And Φ_0 defines the so called *flux quantum*, meaning that the total fluxoid Φ is discretized, with Φ_0 the unit of discretization (cf. Annett 2004, p. 30 ff.; Caldeira 2014, p. 57).

However, the SQUID includes a junction (made out of a different material) and so there will be a phase shift $\Delta\varphi_j$ in addition to the $2\pi n = 2\pi\Phi/\Phi_0$. But since $\Delta\varphi_j$ is determined by the properties of the junction (and hence a fixed quantity) and Φ_0 is a constant, the behavior of all the Cooper pairs together is determined solely by the behavior of Φ, whence one can describe the whole collection in terms of a 'macroscopic wave function' $\Psi_n(\Phi) = \langle\Phi|n\rangle$, with n as above (cf. Schlosshauer 2007, p. 273).

Some 'ordinary' physical considerations (e.g. Caldeira 2014, p. 61) now lead to the definition of a Hamiltonian

$$\hat{H}_\Phi = \frac{\hat{p}_\Phi^2}{2C} + U(\Phi) = -\frac{\hbar^2}{2C}\frac{\mathrm{d}^2}{\mathrm{d}\Phi^2} + \left(\frac{(\Phi - \Phi_{\text{ext}})^2}{2L} - \frac{I_0\Phi_0}{2\pi}\cos(2\pi\Phi/\Phi_0)\right) \tag{B.5}$$

under which $\Psi_n(\Phi)$ evolves. Here L is the (self-) inductance of the loop, C is its capacitance, and I_0 is the peak of the current of tunneling Cooper pairs. Φ and $\hat{p}_\Phi := -i\hbar\frac{\mathrm{d}}{\mathrm{d}\Phi}$ play the role of position and momentum in the usual SE, C acts as a mass.

The most important ingredient of $\hat{H}_\Phi$, however, is the potential $U(\Phi)$. For the appropriate values of Φ_{ext} it has the shape of a double well, just as encountered in Sect. 2.1.1 (cf. Fig. 2.8). But the present double well potential $U(\Phi)$ is, first of all, not situated in position space, but rather in 'flux space'. Hence, the two wells correspond not to positions, but rather, "[b]roadly speaking, [...] to the two possible directions (clockwise and counterclockwise) of the supercurrent around the loop." (Schlosshauer 2007, p. 273) And secondly, under the influence of the external magnetic field $\boldsymbol{B}_{\text{ext}}$, the double well can be made unequally deep on both sides, as can be seen from the presence of Φ_{ext} in the first term of $U(\Phi)$ (cf. Caldeira 2014, p. 62).

Both wells contain a number of narrow states $|k\rangle$ which are eigenstates of $\hat{H}_{\Phi}$, meaning that we have a bunch of clockwise (left well) and counterclockwise (right well) supercurrent states, which represent approximate classical currents and each correspond to a narrow range of fluxoids Φ. However, due to the possibility of tunneling (which now occurs in the flux space), the states $|k\rangle$ have a non-vanishing possibility of states from one well ending up in the other well (e.g. by excitation and subsequent de-excitation into a state not located in the original well). These tunneling processes are accompanied by a change in flux (or fluxoid) which in turn can be accessed by measuring changes in the magnetic moment of the system (cf. Schlosshauer 2007, p. 274).

Most importantly, the SQUID (or the setup containing it) can be engineered in such a way that only two possible energies are relevant, since all the other states $|k\rangle$ become inaccessible. This means that we once more have a kind of qubit system, albeit this time with 'macroscopic' states of clockwise and counterclockwise supercurrent, $\left|\boldsymbol{j}_s^{\circlearrowright}\right\rangle, \left|\boldsymbol{j}_s^{\circlearrowleft}\right\rangle$. The Hamiltonian now takes the form $\hat{H} = -\epsilon\hat{\sigma}_z - \delta\hat{\sigma}_x$, where $\left|\boldsymbol{j}_s^{\circlearrowright}\right\rangle$ and $\left|\boldsymbol{j}_s^{\circlearrowleft}\right\rangle$ are eigenstates of $\hat{\sigma}_z$, $2\epsilon \propto \Phi_{\text{ext}} - \Phi_0/2$ is the 'difference in height' between the two wells, and δ is a matrix element for transition between $\left|\boldsymbol{j}_s^{\circlearrowright}\right\rangle$ and $\left|\boldsymbol{j}_s^{\circlearrowleft}\right\rangle$ due to tunneling (cf. Leggett 2002, p. R444; Schlosshauer 2007, pp. 274–275).[4]

The two relevant energy states can now be expanded as $|0\rangle = \cos\theta\left|\boldsymbol{j}_s^{\circlearrowright}\right\rangle + \sin\theta\left|\boldsymbol{j}_s^{\circlearrowleft}\right\rangle$ and $|1\rangle = \cos\theta\left|\boldsymbol{j}_s^{\circlearrowleft}\right\rangle - \sin\theta\left|\boldsymbol{j}_s^{\circlearrowright}\right\rangle$, where $\tan 2\theta = \delta/\epsilon$. Thus for the case that $\epsilon \gg \delta$, the two energy eigenstates approximate the clockwise/counterclockwise states, in virtue of the properties of the trigonometric functions in question. However, for $\Phi_{\text{ext}} = \Phi_0/2$,[5] ϵ vanishes (the potential becomes symmetric), and the states $|0\rangle, |1\rangle$ go over into the equal superpositions $|0\rangle = \frac{1}{\sqrt{2}}\left(\left|\boldsymbol{j}_s^{\circlearrowright}\right\rangle + \left|\boldsymbol{j}_s^{\circlearrowleft}\right\rangle\right), |1\rangle = \frac{1}{\sqrt{2}}\left(\left|\boldsymbol{j}_s^{\circlearrowleft}\right\rangle - \left|\boldsymbol{j}_s^{\circlearrowright}\right\rangle\right)$ (cf. Schlosshauer 2007, p. 276).

In other words, the totality of all the cooper pairs will be jointly in a quantum superposition of flowing clockwise and counterclockwise as a supercurrent. This is, on the face of it, almost as 'absurd' as a cat in a superposition of being dead and alive. But the situation of the two current directions is notably different in that it is experimentally accessible. In fact, Leggett's original intention was exactly that, when he first described the experimental situation elaborated on here: to find an experimental realization of a Schrödinger-cat-like situation (cf. Leggett 1980). The SQUIDs considered are of course still all much, much smaller.

---

[4]Choosing $\left|\boldsymbol{j}_s^{\circlearrowright}\right\rangle \doteq \binom{1}{0}$ and $\left|\boldsymbol{j}_s^{\circlearrowleft}\right\rangle \doteq \binom{0}{1}$, we can see that $\hat{\sigma}_x\left|\boldsymbol{j}_s^{\circlearrowright}\right\rangle \doteq \begin{pmatrix} 0 & 1 \\ 1 & 0 \end{pmatrix}\binom{1}{0} = \binom{0}{1} \doteq \left|j_s^{\circlearrowleft}\right\rangle$, which is why $\hat{\sigma}_x$ can be used to describe transitions.

[5]Strictly speaking, this must be construed as a limiting procedure, since letting $\Phi_{\text{ext}} \longrightarrow \Phi_0/2$ we have $\epsilon \longrightarrow 0$ and thus $\delta/\epsilon \longrightarrow \infty$. We can see that this is satisfied for $\theta \longrightarrow \pi/4$, which justifies the superposition states below.

How are measurements performed, and why should they count as evidence for the presence of a mesoscopic quantum superposition? First of all we note that in actual experiments currents were used which consisted of numbers of Cooper pairs in the order of 10^6 (cf. van der Wal et al. 2000, p. 773) or even 10^9 (Friedman et al. 2000, p. 45), meaning pretty large collections of microscopic systems. But the SQUIDs in question still only had sizes in the order of some μm, which is why we have preferred to call them 'meso-' rather than 'macroscopic'. Secondly, as regards methods of detecting the superposed states, one can exploit the fact that $|0\rangle$ and $|1\rangle$ are separated by some energy difference $\Delta E = 2\sqrt{\epsilon^2 + \delta^2}$ (which simply becomes 2δ for $\Phi_{\text{ext}} = \Phi_0/2$). Exciting, for various 'geometries' of the double-well for different values of epsilon, states in the one well by microwaves and then determining the probability of tunneling into the other well by de-excitation, one can map out whether energies must 'cross', i.e. become degenerate at some point (cf. Friedman et al. 2000). If they do not, this is taken to indicate the presence of the two well-seperated superposition states; a feature referred to as 'anticrossing'. Such an anticrossing was observed by Friedman et al. with the energy gap corresponding closely to the theoretical predictions. Other kinds of measurement can be performed in terms of Rabi oscillations (cf. Schlosshauer 2007, pp. 246 ff. and 276 for further reference).

The crucial point is that these measurements do not differ *in principle* from methods used to access the (superposed) quantum states of *atoms* or *molecules*, whence, if we do not doubt the existence of quantum superposition in the latter cases, we have no *special* reason to do so in the much larger SQUID-cases.

Appendix C
The GNS Construction

The GNS construction starts off from the realization that a *positive linear form*[1] $\omega : \mathfrak{A} \to \mathbb{C}$ over a C^*-algebra[2] $\mathfrak{A}$ will define a Hilbert space $\mathcal{H}_\omega$ and a representation π_ω of the elements of $\mathfrak{A}$ as linear operators on $\mathcal{H}_\omega$.

What is a *representation* in the present sense? Such a representation could be viewed as a 'representation of a representation' by empirically-minded, non-Platonist philosophers. The elements of some algebra $\mathfrak{A}$ and their interrelations are typically *already* used to *represent*, in a structurally-abstracting fashion, physical operations/properties/events that we encounter in experience and experiment. The representation in the present sense maps the more *abstract* objects from $\mathfrak{A}$ into (although not generally 'onto', i.e. not necessarily *surjectively*) a set of *mathematically* more concrete objects (in this case: the operators), in a structure-preserving way, i.e. such that the relations in $\pi(\mathfrak{A})$ mirror those in $\mathfrak{A}$.

In the case of C^*-algebras, the map π is *linear*, and it generally preserves the structure of the product operation and involution in $\mathfrak{A}$ (cf. Ruetsche 2011, pp. 77 and 83). Moreover, a representation is called *faithful iff* it maps only the zero-element of $\mathfrak{A}$ onto zero, and *irreducible iff* the only (closed) subspaces of $\mathcal{H}$ that are *invariant* under the action of the $\hat{A} \in \pi(\mathfrak{A})$ are $\mathcal{H}$ itself and $\{0_\mathcal{H}\}$ ($0_\mathcal{H}$ the null-vector of $\mathcal{H}$; cf. appendix A).

Because of its addition and scalar-multiplication properties, the algebra $\mathfrak{A}$ is also a $\mathbb{C}$-vector space, and the positive linear form can be understood as providing a (skew symmetric, positive semi-definite; cf. appendix A) scalar product $\langle A|B\rangle = \omega(A^*B)$ for the $A, B \in \mathfrak{A}$. There will be a subset $\mathcal{I} \subset \mathfrak{A}$ such that $\langle X|X\rangle = 0$

[1]Linearity, to recall, means $\omega(\alpha A + \beta B) = \alpha\omega(A) + \beta\omega(B)$, $A, B \in \mathfrak{A}$, $\alpha, \beta \in \mathbb{C}$, positivity here means $\omega(A^*A) \geq 0$ (cf. Haag 1996, p. 122).

[2]Strictly speaking, only a weaker Banach *-algebra is required for most of the construction, which need not satisfy condition (i) of def. A.11 (cf. Haag 1996, pp. 112, 118, and 122–123). But we here focus on C^*-algebras right away for simplicity.

F. J. Boge, *Quantum Mechanics Between Ontology and Epistemology*, European Studies in Philosophy of Science 10,
https://doi.org/10.1007/978-3-319-95765-4

and $AX \in \mathcal{I}$ for $X \in \mathcal{I}, A \in \mathfrak{A}$, and the set $\mathfrak{A}/\mathcal{I}$ of equivalence classes $[A]$ such that $A' \in [A]$ *iff* $A' = A + X$ for some $X \in \mathcal{I}$ will form a *pre-Hilbert space* (a Hilbert space up to completeness induced by the scalar product). A class $[A]$ hence defines a *vector* ψ. The completion of $\mathfrak{A}/\mathcal{I}$ w.r.t. the norm induced by $\omega(\psi^*\psi) = \langle\psi|\psi\rangle = \|\psi\|^2$ then leads to the desired Hilbert space $\mathcal{H}_\omega$.

If ω is *normed*, i.e. if $\omega(I) = 1$ (I the unit element),[3] it will be called an *algebraic state* (cf. Haag 1996, p. 122; Ruetsche 2011, p. 89). Intuitively, algebraic states provide *expectation values* of the operators that are the representation of the algebra. Moreover, if π_ω is *cyclic*, one regains, setting $\pi_\omega(\tilde{A}) = \hat{A}$, the familiar form $\langle A\rangle_\psi = \langle\psi|\hat{A}|\psi\rangle = \langle\psi|\pi_\omega(\tilde{A})|\psi\rangle =: \omega_\psi(\tilde{A})$, and the $\omega_\psi(\tilde{A})$ are then called *vector states* (cf. Haag 1996, p. 124).[4] What does 'cyclic' mean? It means that there is a ('cyclic') vector $\Omega \in \mathfrak{A}/\mathcal{I}$ such that $\pi_\omega(\mathfrak{A})\Omega$ is *dense* in $\mathcal{H}_\omega$, i.e. that any $\psi \in \mathcal{H}_\omega$ can be approximated arbitrarily well by applications of elements of the representation of the algebra $\mathfrak{A}$ to Ω. Intuitively, Ω is a vacuum state from which all other states are created. The *action* of the operators $\pi_\omega(A)$ on vectors $\psi = [B]$ here corresponds to the equivalence class $[AB]$ of respective products in $\mathfrak{A}$.

In such a cyclic representation one obtains $\omega(A) = \langle\Omega|\pi_\omega(A)|\Omega\rangle$, which justifies the expression $\omega_\psi(A) = \langle\psi|\pi_\omega(A)|\psi\rangle$, since some $B\Omega$ approximates ψ (where, to recall, B may be an arbitrary product of elements from $\mathfrak{A}$) and $\omega(B^*AB) = \langle\Omega|\pi_\omega(B^*)\pi_\omega(A)\pi_\omega(B)|\Omega\rangle$ approximates $\omega_\psi(A)$. One then says that *ω is represented by* Ω (cf. Haag 1996, p. 122–124). Moreover, the trace functional $\omega_{\hat{\rho}}(A) = \mathrm{Tr}\big(\hat{\rho}\pi_\omega(A)\big)$, where $\hat{\rho} \in \mathfrak{B}(\mathcal{H}_\omega)$ is positive and trace class, defines a set of states called the *folium* of π_ω, i.e. the set of all states ω that can be so constructed out of density operators (cf. Haag 1996, p. 124; Ruetsche 2011, p. 96).

The algebraic states ω on a C^*-algebra $\mathfrak{A}$ constitute a *convex set*, i.e. if ω_1, ω_2 are states, then so is $\omega = \lambda\omega_1 + (1 - \lambda)\omega_2, \forall\lambda \in [0, 1]$; this makes probabilistic weighting possible. Additionally, it is a theorem that the GNS-representation of *pure algebraic states*, i.e. those which cannot be non-trivially expressed as convex combinations are exactly those with an irreducible representation (cf. Ruetsche 2011, pp. 89 and 93).[5]

If one closes the representation $\pi_\omega(\mathfrak{A})$ in either the weak or strong operator topology, i.e. includes operators $\hat{A}$ such that sequences $|\langle\phi|\hat{A}_n|\psi\rangle - \langle\phi|\hat{A}|\psi\rangle|$ (with $\langle\phi|$ omitted in the strong topology) converge to 0 as $n \to \infty$ and with $\hat{A}_n \in \pi_\omega(\mathfrak{A}), \forall n \in \mathbb{N}$, this leads to a *von Neumann algebra* $\mathfrak{M}$. Again the algebra of bounded operators $\mathfrak{B}(\mathcal{H})$ is an example. But the algebra $\mathfrak{M}$ obtained by the

[3]This definition of 'normed' obviously only works in a *unital* algebra; otherwise normalization has to be defined in terms of the Hilbert space norm again (cf. Haag 1996, p. 122).

[4]We emphasize the distinction between A and $\tilde{A}$ again: $\tilde{A}$ is an element of $\mathfrak{A}$, an abstract mathematical symbolism. A is supposed to be the 'real world observable', i.e. a universal, a class of tropes, a set of operations, or whatever your favored metaphysics spits out.

[5]However, there is a subtlety involved since non-pure algebraic states will nonetheless be represented in terms of *state vectors* on a Hilbert space. That this is possible can be understood best when the 'behavior' of state vectors in the presence of superselection rules is considered (cf. footnote 24 and Ruetsche 2011, p. 93).

closure-procedure may be only a proper subalgebra of $\mathfrak{B}(\mathcal{H})$. $\mathfrak{M}$ is also the *double commutant* $\pi(\mathfrak{A})''$ of $\pi(\mathfrak{A})$, i.e. if we take all the operators in $\mathfrak{B}(\mathcal{H})$ that commute with all the operators in $\pi(\mathfrak{A})$ and then again take the operators that commute with those, they will constitute the von Neumann algebra $\mathfrak{M} = \pi(\mathfrak{A})''$ (cf. Ruetsche 2011, pp. 86–89).

Why is $\pi(\mathfrak{A})''$ of interest? Because in contrast to a C^*-algebra, a von Neumann algebra will be "rich with projections" (Ruetsche 2011, p. 88), whence algebraic states on $\pi(\mathfrak{A})''$ will allow for familiar probability-expressions such as $\omega_\psi(\hat{P}_j) = \langle\psi|\hat{P}_j|\psi\rangle = \langle\psi|j\rangle\langle j|\psi\rangle = |\langle j|\psi\rangle|^2 = \mathrm{Pr}^\psi(j)$.[6] However, not all algebraic states will be *countably additive* (will satisfy $\omega\left(\sum_{j=1}^\infty \hat{P}_j\right) = \sum_{j=1}^\infty \omega\left(\hat{P}_j\right)$ on a countable set $\left\{\hat{P}\right\}_{j\in\mathbb{N}}$ of projectors), and those that are are usually called *normal* (cf. Ruetsche 2011, p. 90).

Now the set of normal states on $\pi(\mathfrak{A})''$ and the folium of the state ω generating the representation π_ω *coincide* (cf. Haag 1996, p. 124; Ruetsche 2011, p. 95). In other words: all algebraic states ω' that produce a countably additive probability measure on the projectors $\hat{P}_j$ in the von Neumann algebra $\pi(\mathfrak{A})''$ that is the (weak/strong) closure of the representation π_ω generated by some ω will be of the form $\omega_{\hat{\rho}}(\hat{P}_j) = \mathrm{Tr}\left(\hat{\rho}\hat{P}_j\right)$, i.e. *can be provided by density operators*. This anticipates (or rather generalizes) the conclusion of *Gleason's* famous theorem that is briefly discussed in Chap. 7. Moreover, two *unitarily equivalent* representations $\pi_\omega, \pi_{\omega'}$ of a C^*-algebra $\mathfrak{A}$, generated from states ω, ω' respectively, will have coinciding folia (cf. Ruetsche 2011, p. 96). This demonstrates the importance of unitary equivalence also in this context.

[6]Note that the reconstruction of the the probability density $\psi^*\psi(x)$ from this algebraic basis will need some extra effort, since a 'point-projection operator' $|x\rangle\langle x|$ on $L^2(\Lambda)$ ($\Lambda \subseteq \mathbb{R}$) would correspond to a characteristic function $f(x') = \chi_{\{x\}}(x')$ when the 'expectation value' $\langle f(x')\rangle_\psi$ is supposed to generate a probability measure for finding the value x. But by its construction, a von Neumann-algebra can only provide such multiplication-by-$f(x)$-operators as equivalence classes of (Lebesgue-)measurable functions up to sets of measure zero, and a set $\{x\}$ is of Lebesgue-measure 0. The solution is a convergence-construction from characteristic functions for larger, measurable sets, discussed e.g. in Ruetsche (2011, p. 91).

Bibliography

Aaronson, S., A. Bouland, L. Chua, and G. Lowther. 2013. ψ-epistemic theories: The role of symmetry. *Physical Review A* 88(1–12): 032111.

Adams, S. 1997. *Relativity. An introduction to space-time physics*. London/Philadelphia: Taylor & Francis.

Aharonov, Y., and D. Rohrlich. 2005. *Quantum paradoxes. Quantum theory for the perplexed*. Weinheim: Wiley-VCH.

Albert, D. 2010. Probability in the Everett picture. In *Many worlds? Everett, quantum theory, & reality*, eds. S. Saunders, J. Barrett, A. Kent, and D. Wallace, 355–368. Oxford/New York: Oxford University Press.

Albert, D., and B. Loewer. 1988. Interpreting the many worlds interpretation. *Synthese* 77(2): 195–213.

Albert, D.Z. 1992. *Quantum mechanics and experience*. Cambridge/London: Harvard University Press.

Albert, D.Z. 1996. Elementary quantum metaphysics. In *Bohmian mechanics and quantum theory: An appraisal*, eds. J.T. Cushing, A. Fine, and S. Goldstein, 277–284. Berlin/Heidelberg: Springer.

Allison, H.E. 2004. *Kant's transcendental idealism*, rev. and enlarged ed. New Haven/London: Yale University Press.

Allison, H.E. 2015. *Kant's transcendental deduction: An analytical-historical commentary*. Oxford/New York: Oxford University Press.

Allori, V., S. Goldstein, R. Tumulka, and N. Zanghì. 2008. On the common structure of Bohmian mechanics and the Ghirardi–Rimini–Weber theory. *The British Journal for the Philosophy of Science* 59(3): 353–389.

Allori, V., S. Goldstein, R. Tumulka, and N. Zanghì. 2014. Predictions and primitive ontology in quantum foundations: A study of examples. *The British Journal for the Philosophy of Science* 65(2): 323–352.

Ampère, A. 1838. *Essai sur la philosophie des sciences: Ou, Exposition analytique d'une classification naturelle de toutes les connaissances humaines*. Paris: Bachelier.

Annett, J.F. 2004. *Superconductivity, superfluids, and condensates*. Oxford/New York: Oxford University Press.

Aristotle. 1998. *Metaphysics*. London/New York: Penguin Books. Translated with an introduction by Hugh Lawson-Tancred.

Armstrong, D.M. 1978. *Universals and scientific realism*. Cambridge/New York: Cambridge University Press. [2 volumes].

F. J. Boge, *Quantum Mechanics Between Ontology and Epistemology*,
European Studies in Philosophy of Science 10,
https://doi.org/10.1007/978-3-319-95765-4

Armstrong, D.M. 1989. *Universals: An opinionated introduction*. Boulder/San Francisco: Westview Press.

Arndt, M., O. Nairz, J. Petschinka, and A. Zeilinger. 2001. High contrast interference with C_{60} and C_{70}. *Comptes Rendus de l'Académie des Sciences-Series IV-Physics* 2(4): 581–585.

Arndt, M., O. Nairz, J. Vos-Andreae, C. Keller, G. van der Zouw, and A. Zeilinger. 1999. Wave-particle duality of C_{60} molecules. *Nature* 401: 680–682.

Arnold, E. 2010. Can the best-alternative justification solve Hume's problem? On the limits of a promising approach. *Philosophy of Science* 77(4): 584–593.

Arntzenius, F. 2012. *Space, time, and stuff*. Oxford/New York: Oxford University Press.

Ashcroft, N., and N. Mermin. 1976. *Solid state physics*. Fort Worth/Philadelphia: Harcourt College Publishers.

Aspect, A., J. Dalibard, and G. Roger. 1982. Experimental test of Bell's inequalities using time-varying analyzers. *Physical Review Letters* 49(25): 1804–1807.

Audretsch, J. 2007. *Entangled systems. New directions in quantum physics*. Weinheim: Wiley-VCH.

Auyang, S.Y. 1995. *How is quantum field theory possible?* Oxford/New York: Oxford University Press.

Ayer, A.J. 1971 [1936]. *Language, truth and logic*. London/New York: Penguin Books.

Baaquie, B.E. 2013. *The theoretical foundations of quantum mechanics*. New York/Heidelberg: Springer.

Bacchus, F., H.E. Kyburg, and M. Thalos. 1990. Against conditionalization. *Synthese* 85(3): 475–506.

Bacciagaluppi, G. 2010. Collapse theories as beable theories. *Manuscrito* 33(1): 19–54. http://philsci-archive.pitt.edu/8876/.

Bacciagaluppi, G., and A. Valentini. 2009. *Quantum theory at the crossroads: Reconsidering the 1927 Solvay conference*. Cambrdige/New York: Cambridge University Press.

Baker, D.J. 2007. Measurement outcomes and probability in Everettian quantum mechanics. *Studies in History and Philosophy of Science Part B: Studies in History and Philosophy of Modern Physics* 38(1): 153–169.

Ballentine, L. 2014. Ontological models in quantum mechanics: What do they tell us? arXiv preprint arXiv:1402.5689.

Ballentine, L.E. 1970. The statistical interpretation of quantum mechanics. *Reviews of Modern Physics* 42(4): 358–381.

Ballentine, L.E. 2000. *Quantum mechanics. A modern development*. Singapore/River Edge: World Scientific Publishing Co.

Ballentine, L.E., Y. Yang, and J.P. Zibin. 1994. Inadequacy of Ehrenfest's theorem to characterize the classical regime. *Physical Review A* 50(4): 2854–2859.

Bardeen, J., L.N. Cooper, and J.R. Schrieffer. 1957. Theory of superconductivity. *Physical Review* 108(5): 1175–1204.

Barnum, H. 2003. No-signalling-based version of Zurek's derivation of quantum probabilities: A note on 'Environment-assisted invariance, entanglement, and probabilities in quantum physics'. arXiv preprint quant-ph/0312150.

Barrett, J.A. 2002. On the nature of measurement records in relativistic quantum field theory. In *Ontological aspects of quantum field theory*, eds. M. Kuhlmann, H. Lyre, and A. Wayne, 165–180. New Jersey/London: World Scientific Publishing.

Bartlett, S.D., T. Rudolph, and R.W. Spekkens. 2012. Reconstruction of Gaussian quantum mechanics from Liouville mechanics with an epistemic restriction. *Physical Review A* 86(1): 012103(25 pp).

Basdevant, J.-L., and J. Dalibard. 2002. *Quantum mechanics*. Berlin/Heidelberg: Springer.

Bassi, A., and G. Ghirardi. 2003. Dynamical reduction models. *Physics Reports* 379(5): 257–426.

Bassi, A., K. Lochan, S. Satin, T.P. Singh, and H. Ulbricht. 2013. Models of wave-function collapse, underlying theories, and experimental tests. *Reviews of Modern Physics* 85(2): 471–527.

Bayes, T. 1763. An essay towards solving a problem in the doctrine of chances. *Philosophical Transactions (1683–1775)* 53: 370–418.

Bedingham, D., D. Dürr, G. Ghirardi, S. Goldstein, R. Tumulka, and N. Zanghì. 2014. Matter density and relativistic models of wave function collapse. *Journal of Statistical Physics* 154(1–2): 623–631.

Bedingham, D.J. 2009. Dynamical state reduction in an EPR experiment. *Journal of Physics A: Mathematical and Theoretical* 42(46): 465301.

Bedingham, D.J. 2011. Relativistic state reduction dynamics. *Foundations of Physics* 41(4): 686–704.

Belinfante, F.J. 1973. *A survey of hidden variables theories*. Oxford/New York: Pergamon Press.

Bell, J.S. 1987. Are there quantum jumps? In *Speakable and unspeakable in quantum mechanics*, ed. J.S. Bell, 201–212. Cambridge/New York: Cambridge University Press.

Bell, J.S. 1987[1964]. On the Einstein-Podolsky-Rosen paradox. In *Speakable and unspeakable in quantum mechanics*, ed. J.S. Bell, 14–21. Cambridge/New York: Cambridge University Press.

Bell, J.S. 1987[1966]. On the problem of hidden-variables in quantum mechanics. In *Speakable and unspeakable in quantum mechanics*, ed. J.S. Bell, 1–13. Cambridge/New York: Cambridge University Press.

Bell, J.S. 1987[1971]. Introduction to the hidden-variable question. In *Speakable and unspeakable in quantum mechanics*, ed. J.S. Bell, 29–39. Cambridge/New York: Cambridge University Press.

Bell, J.S. 1987[1976]. The theory of local beables. In *Speakable and unspeakable in quantum mechanics*, ed. J.S. Bell, 52–62. Cambridge/New York: Cambridge University Press.

Bell, J.S. 1987[1980]. de Broglie-Bohm, delayed-choice double-slit experiment, and density matrix. In *Speakable and unspeakable in quantum mechanics*, ed. J.S. Bell, 111–116. Cambridge/New York: Cambridge University Press.

Bell, J.S. 1987[1981]a. Bertlmann's socks and the nature of reality. In *Speakable and unspeakable in quantum mechanics*, ed. J.S. Bell, 139–158. Cambridge/New York: Cambridge University Press.

Bell, J.S. 1987[1981]b. Quantum mechanics for cosmologists. In *Speakable and unspeakable in quantum mechanics*, ed. J.S. Bell, 139–158. Cambridge/New York: Cambridge University Press.

Bell, J.S. 1987[1984]a. Beables for quantum field theory. In *Speakable and unspeakable in quantum mechanics*, ed. J.S. Bell, 173–180. Cambridge/New York: Cambridge University Press.

Bell, J.S. 1987[1984]b. Speakable and unspeakable in quantum mechanics. In *Speakable and unspeakable in quantum mechanics*, ed. J.S. Bell, 169–172. Cambridge/New York: Cambridge University Press.

Bell, J.S. 1990a. Against 'measurement'. In *Sixty-two years of uncertainty. Historical, philosophical, and physical inquiries into the foundations of quantum mechanics*, ed. A.I. Miller, 17–32. New York: Plenum Press.

Bell, J.S. 1990b. La Nouvelle cuisine. In *Between science and technology. Proceedings of the international conference between science and technology. Eindhoven University of Technology, The Netherlands, 29–30 June 1989*, eds. A. Sarlemijn, and P. Kroes, 97–115. Amsterdam/Oxford: North-Holland (Elsevier Science Publishers B.V.).

Beller, M. 2000. Kant's impact on Einstein's thought. In *Einstein: The formative years, 1879–1909*, ed. D. Howard, 83–106. Boston/Basel: Birkhäuser.

Beller, M., and A. Fine. 1994. Bohr's response to EPR. In *Niels Bohr and contemporary philosophy*, eds. J. Faye, and H.J. Folse, 1–31. Dordrecht: Springer Science + Business Media.

Berkeley, G. 2009 [1710-13]. *The principles of human knowledge and three dialogues*. Edited by Howard Robinson. Oxford/New York: Oxford University Press.

Bernstein, H.R. 1981. Passivity and inertia in Leibniz's Dynamics. *Studia Leibnitiana* 13(1): 97–113.

Binney, J., and D. Skinner. 2014. *The physics of quantum mechanics*. Oxford/New York: Oxford University Press.

Bird, A., and E. Tobin. 2015. Natural kinds. In *The Stanford encyclopedia of philosophy*, ed. E.N. Zalta. The Metaphysics Research Lab, Center for the Study of Language and Information (CSLI), Stanford University. http://plato.stanford.edu/entries/natural-kinds/.

Bitbol, M. 1996. *Schrödinger's philosophy of quantum mechanics*. Dordrecht/Boston: Kluwer.

Bitbol, M., and S. Osnaghi. 2013. Bohr's complementarity and Kant's epistemology. *Séminaire Poincaré* XVII: 145–166.

Blatt, J.M., and V.F. Weisskopf. 1979. *Theoretical nuclear physics*. New York/Heidelberg: Springer.

Bleck-Neuhaus, J. 2013. *Elementare Teilchen. Von den Atomen über das Standard-Modell bis zum Higgs-Boson*, 2nd ed. Berlin/Heidelberg: Springer.

Blundell, S. 2009. *Superconductivity. A very short introduction*. Oxford/New York: Oxford University Press.

Boge, F. 2016a. Book review: Simon Friederich: Interpreting quantum theory: A therapeutic approach. *Erkenntnis*. https://doi.org/s10670-016-9823-9.

Boge, F. 2016b. *On probabilities in the many worlds interpretation of quantum mechanics*. Bachelor thesis, Institue for Theoretical Physics, University of Cologne. http://kups.ub.uni-koeln.de/6889/.

Bogen, J., and J. Woodward. 1988. Saving the phenomena. *The Philosophical Review* 97(3): 303–352.

Bohm, A., and M. Gadella. 1969. *Dirac Kets, Gamow vectors and Gel'fand triplets. The Rigged Hilbert space formulation of quantum mechanics*. Berlin/Heidelberg: Springer.

Bohm, D. 1951. *Quantum theory*. New York: Dover Publications Inc.

Bohm, D. 1952a. A suggested interpretation of the quantum theory in terms of 'hidden' variables. I. *Physical Review* 85(2): 166–179.

Bohm, D. 1952b. A suggested interpretation of the quantum theory in terms of 'Hidden' variables. II. *Physical Review* 85(2): 180–193.

Bohm, D., and B. Hiley. 1993. *The undivided universe: An ontological interpretation of quantum theory*. London/New York: Routledge.

Bohr, N. 1934. *Atomic theory and the description of nature*. Cambrdige/New York: Cambridge University Press.

Bohr, N. 1935. Can quantum-mechanical description of physical reality be considered complete? *Physical Review* 48(8): 696–702.

Bohr, N. 1938. Natural philosophy and human cultures. *Nature* 143(3616): 268–272.

Bohr, N. 1948. On the notions of causality and complementarity. *Dialectica* 2(3–4): 312–319.

Bohr, N. 1963. *Atomic physics and human knowledge*. New York/London: Wiley.

Bohr, N. 1983 [1949]. Discussion with Einstein on epistemological problems in atomic physics. In *Quantum theory and measurement*, eds. J.A. Wheeler, and W.H. Zurek. Princeton: Princeton University Press.

Bokulich, P., and A. Bokulich. 2005. Niels Bohr's generalization of classical mechanics. *Foundations of Physics* 35(3): 347–371.

Bongaarts, P. 2014. *Quantum theory: A mathematical approach*. Cham/Heidelberg: Springer.

Born, M. 1926. Quantenmechanik der Stoßvorgänge. *Zeitschrift für Physik* 38(11–12): 803–827.

Born, M. 1927. Quantenmechanik und Statistik. *Die Naturwissenschaften* 15(10): 238–242.

Born, M., W. Heisenberg, and P. Jordan. 1926. Zur Quantenmechanik. II. *Zeitschrift für Physik* 35(8–9): 557–615.

Born, M., and P. Jordan. 1925. Zur Quantenmechanik. I. *Zeitschrift für Physik* 34(1): 858–888.

Borst, C. 2010. Solipsism. In *A companion to epistemology*, eds. J. Dancy, E. Sosa, and M. Steup, 747–749. Chichester/Malden: Wiley.

Branciard, C. 2011. Detection loophole in Bell experiments: How postselection modifies the requirements to observe nonlocality. *Physical Review A* 83(3): 032123.

Brandom, R.B. 1994. *Making it explicit. Reasoning, representing, and discursive commitment*. Cambridge: Harvard University Press.

Brandom, R.B. 2000. *Articulating reasons. An introduction to inferentialism*. Cambridge: Harvard University Press.

Brown, C. 1988. Internal realism: Transcendental idealism? *Midwest Studies in Philosophy* 12(1): 145–155.
Brune, M., E. Hagley, J. Dreyer, X. Maitre, A. Maali, C. Wunderlich, J. Raimond, and S. Haroche. 1996. Observing the progressive decoherence of the 'meter' in a quantum measurement. *Physical Review Letters* 77(24): 4887.
Bub, J. 1974. *The interpretation of quantum mechanics*. Dordrecht/Boston: D. Reidel Publishing Co.
Bub, J. 1997. *Interpretating the quantum world*. Cambridge/New York: Cambridge University Press.
Bub, J. 2016. *Bananaworld: Quantum mechanics for primates*. Oxford/New York: Oxford University Press.
Buchler, J., ed. 1955. *Philosophical writings of Peirce*. New York: Dover Publications Inc.
Burch, R. 2014. Charles sanders Peirce. In *The Stanford encyclopedia of philosophy*, ed. E.N. Zalta. The Metaphysics Research Lab, Center for the Study of Language and Information (CSLI), Stanford University. http://plato.stanford.edu/entries/peirce/.
Busch, P. 2003. Quantum states and generalized observables: A simple proof of Gleason's theorem. *Physical Review Letters* 91(12): 120403.
Busch, P., M. Grabowski, and P.J. Lahti. 1995. *Operational quantum physics*. Berlin/Heidelberg: Springer.
Busch, P., and G. Jaeger. 2010. Unsharp quantum reality. *Foundations of Physics* 40(9–10): 1341–1367.
Busch, P., and P. Lahti. 2009. Measurement theory. In *Compendium of quantum physics. Concepts, experiments, history and philosophy*, eds. D. Greenberger, K. Hentschel, and F. Weinert, 374–379. Berlin/Heidelberg: Springer.
Busch, P., P.J. Lahti, and P. Mittelstaedt. 1996. *The quantum theory of measurement*, 2nd rev. ed. Berlin/Heidelberg: Springer.
Butterfield, H. 1931. *The Whig interpretation of history*. Eliohs, Electronic Library of Historiography. Html edition by Guido Abbattista, 2002. http://www.eliohs.unifi.it/testi/900/butterfield.
Butterfield, J. 2007. Stochastic Einstein locality revisited. *British Journal for the Philosophy of Science* 58(4): 805–867.
Button, T. 2013. *The limits of realism*. Oxford/New York: Oxford University Press.
Byrne, P. 2010a. Everett and wheeler: The untold story. In *Many worlds? Everett, quantum theory, and reality*, eds. S. Saunders, J. Barrett, A. Kent, and D. Wallace, 521–542. Oxford/New York: Oxford University Press.
Byrne, P. 2010b. *The many worlds of Hugh Everett III: Multiple universes, mutual assured destruction, and the meltdown of a nuclear family*. Oxford/New York: Oxford University Press.
Cachro, J., and T. Placek. 2003. Cartwright's models are not adequate for EPR. In *Philosophical dimensions of logic and science. Selected contributed papers from the 11th international congress of logic, methodology, and philosophy of science, Kraków, 1999*. Dordrecht: Springer Science + Business Media.
Caldeira, A.O. 2014. *An introduction to macroscopic quantum phenomena and quantum dissipation*. Cambridge/New York: Cambridge University Press.
Callender, C. 2015. One world, one beable. *Synthese* 192(10): 3153–3177.
Camilleri, K., and M. Schlosshauer. 2015. Niels Bohr as philosopher of experiment: Does decoherence theory challenge Bohr's doctrine of classical concepts? *Studies in History and Philosophy of Science Part B: Studies in History and Philosophy of Modern Physics* 49: 73–83.
Cantor, G., J. Christie, M. Hodge, and R. Olby, eds. 1990. *Companion to the history of modern science*. London/New York: Routledge.
Cao, T.Y. 1997. *Conceptual developments of 20th century field theories*. Cambridge/New York: Cambridge University Press.
Capinski, M., and P. Kopp. 2004. *Measure, integral and probability*, 2nd ed. London: Springer.
Capri, A. 2002. *Nonrelativistic quantum mechanics*, 3rd ed. Singapore, River Edge: World Scientific Publishing Co.
Carnap, R. 1936. Testability and meaning. *Philosophy of Science* 3(4): 419–471.

Carnap, R. 1945. The two concepts of probability. *Philosophy and Phenomenological Research* 5(4): 513–532.

Carnap, R. 1950. *Logical foundations of probability*. Chicago: University of Chicago Press.

Carnap, R. 1955. Statistical and inductive probability. In *The structure of scientific thought*, ed. E.H. Madden, 269–279. Boston: Houghton Mifflin.

Carnap, R. 1956. The methodological character of theoretical concepts. In *Minnesota studies in the philosophy of science, vol. I: The foundations of science and the concepts of psychology and psychoanalysis*, eds. H. Feigl, and M. Scriven, 38–76. Minneapolis: University of Minnesota Press.

Carnap, R. 1958. Beobachtungssprache und theoretische Sprache. *Dialectica* 12(3–4): 236–248.

Carnap, R. 1974 [1928]. *Der logische Aufbau der Welt*, 3rd ed. Hamburg: Felix Meiner Verlag.

Carnap, R. 2001 [1937]. *Logical syntax of language*. London: Routledge.

Carroll, S. 2004. *Spacetime and geometry. An introduction to general relativity*. San Francisco/Boston: Addison Wesley.

Carroll, S. 2013. The most embarrassing graph in modern physics. Web log post: http://www.preposterousuniverse.com/blog/2013/01/17/the-most-embarrassing-graph-in-modern-physics/.

Carroll, S.M., and C.T. Sebens. 2014. Many worlds, the born rule, and self-locating uncertainty. In *Quantum theory: A two-time success story*, eds. D.C. Struppa, and J.M. Tollaksen, 157–169. Milan: Springer.

Cartwright, N. 1983. *How the laws of physics lie*. Oxford/New York: Clarendon Press.

Cartwright, N. 1988. How to tell a common cause: Generalization of the conjunctive fork criterion. In *Probability and causality. Essays in honor of Wesley C. Salmon*, ed. J.H. Fetzer. Dordrecht/Boston: D. Reidel Publishing Co.

Cartwright, N. 1989. *Nature's capacities and their measurement*. Oxford/New York: Oxford University Press. Reprinted 2002.

Cartwright, N. 1999. *The dappled world: A study of the boundaries of science*. Cambridge/New York: Cambridge University Press.

Case, W.B. 2007. Wigner functions and Weyl transforms for pedestrians. *American Journal of Physics* 76(10): 937–946.

Castellani, E. 2003. On the meaning of symmetry breaking. In *Symmetries in physics: Philosophical reflections*, K. Brading, and E. Castellani, 321–334. Cambridge/New York: Cambridge University Press.

Castelvecchi, D., and A. Witze. 2016. Einstein's gravitational waves found at last. *Nature news*. 10.1038/nature.2016.19361.

Caulton, A. 2013. Discerning 'indistinguishable' quantum systems. *Philosophy of Science* 80(1): 49–72.

Caves, C.M. 2004. Notes on Zurek's derivation of the quantum probability rule. http://info.phys.unm.edu/~caves/reports/ZurekBornderivation.ps. Last modified 29 July 2005.

Caves, C.M., C.A. Fuchs, and R. Schack. 2002a. Quantum probabilities as Bayesian probabilities. *Physical Review A* 65(2): 022305.

Caves, C.M., C.A. Fuchs, and R. Schack. 2002b. Unknown quantum states: The quantum de Finetti representation. *Journal of Mathematical Physics* 43(9): 4537–4559.

Caves, C.M., C.A. Fuchs, and R. Schack. 2007. Subjective probability and quantum certainty. *Studies in History and Philosophy of Science Part B: Studies in History and Philosophy of Modern Physics* 38(2): 255–274.

Chakravartty, A. 2001. The semantic or model-theoretic view of theories and scientific realism. *Synthese* 127(3): 325–345.

Chalmers, D. 2012. *Constructing the world*. Oxford/New York: Oxford University Press.

Chalmers, D.J. 1996. *The conscious mind. In search of a fundamental theory*. Oxford/New York: Oxford University Press.

Chen, G., D.A. Church, B.-G. Englert, C. Henkel, B. Rohwedder, M.O. Scully, and M.S. Zubairy. 2007. *Quantum computing devices. Principles, designs, and analysis*. Boca Raton/London: Chapman Hall/CRC.

Cheng, T. 2005. *Relativity, gravitation and cosmology: A basic introduction*. Oxford/New York: Oxford University Press.

Clauser, J.F., and M.A. Horne. 1974. Experimental consequences of objective local theories. *Physical Review D* 10(2): 526–535.

Clauser, J.F., M.A. Horne, A. Shimony, and R.A. Holt. 1969. Proposed experiment to test local hidden-variable theories. *Physical Review Letters* 23(15): 880–884.

Clauser, J.F., and A. Shimony. 1978. Bell's theorem. Experimental tests and implications. *Reports on Progress in Physics* 41(12): 1881–1927.

Cohen, J., and C. Callender. 2009. A better best system account of lawhood. *Philosophical Studies* 145(1): 1–34.

Colbeck, R., and R. Renner. 2012. Is a system's wave function in one-to-one correspondence with its elements of reality? *Physical Review Letters* 108(15): 150402(1–4).

Collett, B., and P. Pearle. 2003. Wavefunction collapse and random walk. *Foundations of Physics* 33(10): 1495–1541.

Conant, J. 1990. Introduction. In *Realism with a human face*, ed. J. Conant, xv–lxxiv. Cambridge: Harvard University Press.

Couder, Y., and E. Fort. 2006. Single-particle diffraction and interference at a macroscopic scale. *Physical Review Letters* 97(15): 154101.

Cramer, J.G. 1986. The transactional interpretation of quantum mechanics. *Reviews of Modern Physics* 58(3): 647–687.

Cresswell, M. 2006. Formal semantics. In *The blackwell guide to the philosophy of language*, eds. M. Devitt, and R. Hanley, 131–146. Malden/Oxford: Blackwell.

Crispino, L.C.B., A. Higuchi, and G.E.A. Matsas. 2008. The Unruh effect and its applications. *Reviews of Modern Physics* 80(3): 787–838.

da Costa, N., and S. French. 2003. *Science and partial truth: A unitary approach to models and scientific reasoning*. Oxford/New York: Oxford University Press.

Davies, M. 2006. Foundational issues in the philosophy of language. In *The blackwell guide to the philosophy of language*, eds. M. Devitt, and R. Hanley, 19–40. Malden/Oxford: Blackwell.

Davies, P.C.W. 2004. John Archibald Wheeler and the clash of ideas. In *Science and ultimate reality: Quantum theory, cosmology, and complexity*, eds. J.D. Barrow, P.C.W. Davies, and C.L. Jr. Harper, 3–23. Cambridge/New York: Cambridge University Press.

Davisson, C., and L.H. Germer. 1927. Diffraction of electrons by a crystal of nickel. *Physical Review* 30(6): 705–740.

Dawid, R., and K.P.Y. Thébault. 2015. Many worlds: Decoherent or incoherent? *Synthese* 192(5): 1559–1580.

de Broglie, L. 1927. La mécanique ondulatoire et la structure atomique de la matière et du rayonnement. *Journal de Physique et le Radium* 8(5): 225–241.

de Broglie, M.L. 1925. *Recherche sur la théorie des quanta*, vol. III. *Annales de physique*. Paris: Masson Et C. Éditeurs.

de Finetti, B. 1930. Funzione caratteristica di un fenomeno aleatorio. *Memorie della Reale Accademia dei Lincei* IV(5): 86–133.

De Finetti, B. 1937. La prévision: Ses lois logiques, ses sources subjectives. *Annales de l'institut Henri Poincaré* 7(1): 1–68.

de Finetti, B. 1990 [1970]. *Theory of probability: A critical introductory treatment*. Trans. A. Machi, and A. Smith. Chichester/New York: Wiley.

de la Madrid, R. 2005. The role of the rigged Hilbert space in quantum mechanics. *European Journal of Physics* 26(2): 287–312.

de Laplace, P.S. 1902 [1814]. *A philosophical essay on probabilities*. Trans. from the 6th French ed. by F.W. Truscott, and F.L. Emory. London: Chapman & Hall, Ltd.

De Zela, F. 2008. A non-local hidden-variable model that violates Leggett-type inequalities. *Journal of Physics A: Mathematical and Theoretical* 41: 505301(9pp).

Demtröder, W. 2009. *Experimentalphysik. Band 2. Elektrizität und Optik*, 5th ed. Berlin/Heidelberg: Springer.

Demtröder, W. 2010. *Experimentalphysik. Band 3. Atome, Moleküle und Festkörper*, 3rd ed. Berlin/Heidelberg: Springer.

Descartes, R. 1996 [1642]. *Meditations on first philosophy. With selections from the objections and replies*, rev. ed. Trans. and ed. J. Cottingham. Cambrdige/New York: Cambridge University Press.

d'Espagnat, B. 1979. The quantum theory and reality. *Scientific American* 241: 158–181.

d'Espagnat, B. 1987. Empirical reality, empirical causality, and the measurement problem. *Foundations of Physics* 17(5): 507–529.

d'Espagnat, B. 1990. Towards a separable 'empirical reality'? *Foundations of Physics* 20(10): 1147–1172.

d'Espagnat, B. 1995. *Veiled reality. An analysis of present day quantum mechanical concepts*. Reading: Addison-Wesley Publishing Co.

d'Espagnat, B. 1997. Aiming at describing empirical reality. In *Potentiality, entanglement and passion-at-a-distance. Quantum mechanical studies for Abner Shimony*, eds. R.S. Cohen, M. Horne, and J. Stachel. Dordrecht: Springer Science + Business Media.

d'Espagnat, B. 2011. Quantum physics and reality. *Foundations of Physics* 41(11): 1703–1716.

Deutsch, D. 1999. Quantum theory of probability and decisions. *Proceedings of the Royal Society of London A: Mathematical, Physical and Engineering Sciences* 455(1988): 3129–3137.

DeWitt, B.S. 1967. Quantum theory of gravity. I. The canonical theory. *Physical Review* 160(5): 1113.

DeWitt, B.S. 1973[1970]. Quantum mechanics and reality. In *The many worlds interpretation of quantum mechanics*, Princeton series in physics, eds. B. Dewitt, and N. Graham, 155–165. Princeton: Princeton University Press.

DeWitt, B.S. 1973[1971]. The many-universes interpretation of quantum mechanics. In *The many worlds interpretation of quantum mechanics*, eds. B. Dewitt, and N. Graham, 167–218. Princeton: Princeton University Press.

Dick, R. 2012. *Advanced quantum mechanics. Materials and photons*. Berlin/Heidelberg: Springer.

Dickson, M. 2007. Non-relativistic quantum mechanics. In *Philosophy of physics. Part A*, eds. J. Butterfield, and J. Earman, 275–416. Amsterdam/Boston: Elsevier.

Dieks, D. 2002. Events and covariance in the interpretation of quantum field theory. In *Ontological aspects of quantum field theory*, eds. M. Kuhlmann, H. Lyre, and A. Wayne. New Jersey/London: World Scientific Publishing.

Diósi, L. 1987. A universal master equation for the gravitational violation of quantum mechanics. *Physics Letters A* 120(8): 377–381.

Diósi, L. 1988. Continuous quantum measurement and Itô formalism. *Physics Letters A* 129(8): 419–423.

Diósi, L. 1989. Models for universal reduction of macroscopic quantum fluctuations. *Physical Review A* 40(3): 1165–1174.

Diósi, L., and B. Lukács. 1987. In favor of a Newtonian quantum gravity. *Annalen der Physik* 499(7): 488–492.

Dirac, P.A.M. 1958 [1930]. *The principles of quantum mechanics*, 4th (rev.) ed. Oxford: Clarendon Press.

Dizadji-Bahmani, F. 2013. The probability problem in Everettian quantum mechanics persists. *The British Journal for the Philosophy of Science* 66(2): 257–283.

Dorato, M. 2015. Laws of nature and the reality of the wave function. *Synthese* 192(10): 3179–3201.

Dowker, F., and I. Herbauts. 2005. The status of the wave function in dynamical collapse models. *Foundations of Physics Letters* 18(6): 499–518.

Dowling, J., and G. Milburn. 2003. Quantum technology: The second quantum revolution. *Philosophical Transactions of the Royal Society of London* 361: 1655–1674.

Drell, S., and F. Zachariasen. 1961. *Electromagnetic structure of nucleons*. London/Glasgow: Oxford University Press.

Dretske, F.I. 1977. Laws of nature. *Philosophy of Science* 44(2): 248–268.

Drezet, A. 2012. Should the wave-function be a part of the quantum ontological state? *Progress in Physics* 4: 14–17.
Duane, W. 1923. The transfer in quanta of radiation momentum to matter. *Proceedings of the National Academy of Sciences of the United States of America* 9(5): 158–164.
Dummett, M. 1982. Realism. *Synthese* 52: 55–112.
Dunningham, J., and V. Vedral. 2007. Nonlocality of a single particle. *Physical Review Letters* 99: 180404.
Dürr, D., S. Goldstein, T. Norsen, W. Struyve, and N. Zanghì. 2014. Can Bohmian mechanics be made relativistic? *Proceedings of the Royal Society A* 470(2162): 20130699.
Dürr, D., S. Goldstein, and N. Zanghì. 2012. *Quantum physics without quantum philosophy*. Berlin/Heidelberg: Springer.
Dürr, D., and S. Teufel. 2009. *Bohmian mechanics: The physics and mathematics of quantum theory*. Berlin/Heidelberg: Springer.
Earman, J. 1991. Kant, incongruous counterparts, and the nature of space and space-time. In *The philosophy of right and left. Incongruent counterparts and the nature of space*, eds. J. van Cleve, and R.E. Frederick, 131–149. Dordrecht/Boston: Kluwer.
Earman, J. 1992. *Bayes or bust? A critical examination of Bayesian confirmation theory*. Cambridge: MIT Press.
Egg, M., and M. Esfeld. 2014. Non-local common cause explanations for EPR. *European Journal for Philosophy of Science* 4(2): 181–196.
Egg, M., and M. Esfeld. 2015. Primitive ontology and quantum state in the GRW matter density theory. *Synthese* 192(10): 3229–3245.
Einstein, A. 1905. Über einen die Erzeugung und Verwandlung des Lichtes betreffenden heuristischen Gesichtspunkt. *Annalen der Physik* 17: 132–148.
Einstein, A. 1926. Letter to Sommerfeld. *Einstein Archive* 21–356.
Einstein, A. 1934. On the method of theoretical physics. *Philosophy of Science* 1(2): 163–169.
Einstein, A. 1935. Letter to Schrödinger. Archived at the duplicate Einstein Archive, Mudd Manuscript Library, Princeton University.
Einstein, A. 1936. Phyiscs and reality, Trans. J. Piccard. *Journal of the Franklin Institute* 221(3): 349–382.
Einstein, A. 1948. Quanten-Mechanik und Wirklichkeit. *Dialectica* 2(3–4): 320–324.
Einstein, A. 1949a. Atobiographical notes. In *Albert Einstein. Philosopher-Scientist*, ed. P.A. Schilpp, 1–94. New York: MJF Books.
Einstein, A. 1949b. Reply to criticism. In *Albert Einstein. Philosopher-Scientist*, ed. P.A. Schilpp, 665–688. New York: MJF Books.
Einstein, A. 1954 [1921]. Geometry and experience. In *Ideas and opinions*, ed. C. Seelig, 232–246. Lecture before the Prussian Academy of Sciences. New York: Crown Publishers, Inc.
Einstein, A. 1954 [1944]. Remarks on Bertrand Russell's theory of knowledge. In *Ideas and opinions*, ed. C. Seelig, 18–24. New York: Crown Publishers, Inc. Reprinted from *The philosophy of Bertrand Russell*, ed. P.A. Schilpp. Evanston: The Library of Living Philosophers.
Einstein, A. 2005 [1950]. Physics, philosophy and scientific progress. *Physics Today* 58(6): 46–48. Reprinted from *Journal of the International College of Surgeons* 14.
Einstein, A. 2011 [1939]. Letter to Schrödinger. In *Letters on wave mechanics: Correspondence with H. A. Lorentz, Max Planck, and Erwin Schrödinger*, ed. K. Przibram. New York: Open Road Media.
Einstein, A., B. Podolsky, and N. Rosen. 1935. Can quantum-mechanical description of physical reality be considered complete? *Physical Review* 47: 777–780.
Emerson, J., D. Serbin, C. Sutherland, and V. Veitch. 2013. The whole is greater than the sum of the parts: On the possibility of purely statistical interpretations of quantum theory. arXiv preprint arXiv:1312.1345.
Englert, B.-G., M.O. Scully, G. Süssmann, and H. Walther. 1992. Surrealistic Bohm trajectories. *Zeitschrift für Naturforschung A* 47(12): 1175–1186.
Esfeld, M. 2012. Das Messproblem in der Quantenmechanik heute: Übersicht und Bewertung. In *Philosophie der Physik*, ed. M. Esfeld. Berlin: Suhrkamp.

Esfeld, M., M. Hubert, D. Lazarovici, and D. Dürr. 2014. The ontology of Bohmian mechanics. *The British Journal for the Philosophy of Science* 65(4): 773–796.

Everett III, H. 1957. "Relative state" formulation of quantum mechanics. *Reviews of Modern Physics* 29(3): 454.

Everett III, H. 1973. The theory of the universal wavefunction. In *The many worlds interpretation of quantum mechanics*, eds. B. Dewitt, and N. Graham, 3–140. Princeton: Princeton University Press.

Falkenburg, B. 2007. *Particle metaphysics. A critical account of subatomic reality.* Berlin/Heidelberg: Springer.

Falkenburg, B. 2010. Wave-particle duality in quantum optics. In *EPSA philosophical issues in the sciences*, eds. M. Suárez, M. Dorato, and M. Rédei, 31–42. Dordrecht/Heidelberg: Springer Science+Business Media.

Faye, J. 1991. *Niels Bohr: His heritage and legacy. An anti-realist view of quantum mechanics.* Dordrecht: Springer Science + Business Media.

Faye, J. 2016. Darwinism in disguise? A comparison between Bohr's view on quantum mechanics and QBism. *Philosophical Transactions of the Royal Society A* 374(2068): 20150236.

Fayngold, M. 2002. *Special relativity and motions faster than light*, 1st ed. Weinheim: Wiley-VCH.

Feintzeig, B. 2014. Can the ontological models framework accommodate Bohmian mechanics? *Studies in History and Philosophy of Science Part B: Studies in History and Philosophy of Modern Physics* 48: 59–67.

Feynman, R. 1965. *The character of physical law*. Cambridge: The MIT Press.

Feynman, R., R.B. Leighton, and M. Sands. 2010 [1965a]. *The Feynman lectures on physics*, vol. III. New milennium ed. New York: Basic Books.

Feynman, R.P. 1982. Simulating physics with computers. *International Journal of Theoretical Physics* 21(6): 467–488.

Fine, A. 1982. Some local models for correlation experiments. *Synthese* 50(2): 279–294.

Fine, A. 1984. What is Einstein's statistical interpretation, or, is it Einstein for Whom Bell's theorem tolls? *Topoi* 3(1): 23–36.

Fine, A. 1986. *The Shaky game: Einstein, realism and the quantum theory*. Chicago/London: The University of Chicago Press.

Fine, A. 1989. Do correlations need to be explained? In *Philosophical consequences of quantum theory*, eds. J. Cushing, and E. McMullin, 175–194. Notre Dame: University of Notre Dame Press.

Fischer, G. 2014. *Lineare algebra*, 18th ed. Wiesbaden: Springer Spektrum.

Fleming, G. 2002. Comments on Paul Teller's book, 'An interpretive introduction to quantum field theory'. In *Ontological aspects of quantum field theory*, eds. M. Kuhlmann, H. Lyre, and A. Wayne, 135–144. New Jersey/London: World Scientific Publishing.

Fleming, G.N. 2000. Reeh-Schlieder meets Newton-Wigner. *Philosophy of Science* 67:S495–S515.

Fließbach, T. 2009. *Mechanik. Lehrbuch zur Theoretischen Physik I*, 6th ed. Heidlberg: Spektrum.

Fodor, J., and E. Lepore. 2001. Brandom's burdens: Compositionality and inferentialism. *Philosophy and Phenomenological Research* 63(2): 465–481.

Folse, H.J. 1994. Bohr's framework of complementarity and the realism debate. In *Niels Bohr and contemporary philosophy*, eds. J. Faye, and H.J. Folse, 119–139. Dordrecht: Springer Science + Business Media.

Foot, C. 2005. *Atomic physics*. Oxford/New York: Oxford University Press.

Fornasini, P. 2008. *The uncertainty in physical measurements. An introduction to data analysis in the physics laboratory*. New York: Springer Science + Business Media, LLC.

Frankel, T. 2004. *The geometry of physics: An introduction*, 2nd ed. Cambrdige/New York: Cambridge University Press.

French, S., and D. Krause. 2006. *Identity in physics. A historical, philosophical, and formal analysis*. Oxford/New York: Oxford University Press.

Freud, S. 1961. *The future of an illusion*. Trans. and ed. by J. Strachey. New York: W. W. Norton & Company Inc.

Friebe, C., M. Kuhlmann, H. Lyre, P. Näger, O. Passon, and M. Stöckler. 2015. *Philosophie der Quantenphysik*. Berlin/Heidelberg: Springer.
Friederich, S. 2015. *Interpreting quantum theory: A therapeutic approach*. Basingstoke: Palgrave Macmillan.
Friedman, J.R., V. Patel, W. Chen, S.K. Tolpygo, and J.E. Lukens. 2000. Quantum superposition of distinct macroscopic states. *Nature* 406(6791):43–6.
Friedman, M. 1983. *Foundations of space-time theories. Relativistic physics and philosophy of science*. Princeton: Princeton University Press.
Friedman, M. 1999. *Reconsidering logical positivism*. Cambridge/New York: Cambridge University Press.
Frigg, R. 2006. Scientific representation and the semantic view of theories. *Theoria* 55: 49–65.
Fuchs, C.A. 2002. Quantum mechanics as quantum information (and only a little more). arXiv preprint quant-ph/0205039.
Fuchs, C.A. 2010. QBism, the perimeter of quantum Bayesianism. arXiv preprint arXiv:1003.5209.
Fuchs, C.A. 2014. Introducing QBism. In *New directions in the philosophy of science*, eds. M.C. Galavotti, D. Dieks, W.J. Gonzalez, S. Hartmann, T. Uebel, and M. Weber, 385–402. Heidelberg/New York: Springer.
Fuchs, C.A. 2016. On participatory realism. arXiv preprint arXiv:1601.04360.
Fuchs, C.A., N.D. Mermin, R. Schack. 2014. An introduction to QBism with an application to the locality of quantum mechanics. *American Journal of Physics* 82(8): 749–754.
Fuchs, C.A., and A. Peres. 2000. Quantum theory needs no 'interpretation'. *Physics Today* 53(3): 70–71.
Fuchs, C.A., and R. Schack. 2011. A quantum-Bayesian route to quantum-state space. *Foundations of Physics* 41(3): 345–356.
Fuchs, C.A., and R. Schack. 2012. Bayesian conditioning, the reflection principle, and quantum decoherence. In *Probability in physics*, eds. Y. Ben-Menahem, and M. Hemmo, 233–248. Dordrecht: Springer.
Fuchs, C.A., and R. Schack. 2014. QBism and the Greeks: Why a quantum state does not represent an element of physical reality. *Physica Scripta* 90(1): 015104.
Gähler, R., and A. Zeilinger. 1991. Wave-optical experiments with very cold neutrons. *American Journal of Physics* 59(4): 316–324.
Gallagher, S., and D. Zahavi. 2008. *The phenomenological mind: An introduction to philosophy of mind and cognitive science*. London/New York: Routledge.
Galvan, B. 2015. Relativistic Bohmian mechanics without a preferred foliation. arXiv preprint arXiv:1509.03463.
Garner, A.J.P., O.C.O. Dahlsten, Y. Nakata, M. Murao, and V. Vedral. 2013. A framework for phase and interference in generalized probabilistic theories. *New Journal of Physics* 15: 093044(26pp).
Gearhart, C. 2009. Black-body radiation. In *Compendium of quantum physics. Concepts, experiments, history and philosophy*, eds. D. Greenberger, K. Hentschel, and F. Weinert, 39–42. Berlin/Heidelberg: Springer.
Gell-Mann, M., and J.B. Hartle. 1989. Quantum mechanics in the light of quantum cosmology. In *Proceedings of the 3rd international symposium on the foundations of quantum mechanics*, 321–343. Reading: Addison Wesley.
Gerlach, W., and O. Stern. 1922. Der experimentelle Nachweis der Richtungsquantelung im Magnetfeld. *Zeitschrift für Physik* 9(1): 349–352.
Ghirardi, G. 2016. Collapse theories. In *The Stanford encyclopedia of philosophy*, ed. E.N. Zalta. The Metaphysics Research Lab, Center for the Study of Language and Information (CSLI), Stanford University. http://plato.stanford.edu/entries/qm-collapse/.
Ghirardi, G., R. Grassi, and A. Rimini. 1990a. Continuous-spontaneous-reduction model involving gravity. *Physical Review A* 42(3): 1057–1064.
Ghirardi, G., and L. Marinatto. 2003. Entanglement and properties. *Fortschritte der Physik* 51(4–5): 379–387.

Ghirardi, G., L. Marinatto, and T. Weber. 2002. Entanglement and properties of composite quantum systems: A conceptual and mathematical analysis. *Journal of Statistical Physics* 108(1–2): 49–122.

Ghirardi, G., and P. Pearle. 1990a. Dynamical reduction theories: Changing quantum theory so the statevector represents reality. In *PSA: Proceedings of the Biennial meeting of the philosophy of science association*, 19–33.

Ghirardi, G., and P. Pearle. 1990b. Elements of physical reality, nonlocality and stochasticity in relativistic dynamical reduction models. In *PSA: Proceedings of the Biennial meeting of the philosophy of science association*, 35–47.

Ghirardi, G.C., R. Grassi, and F. Benatti. 1995. Describing the macroscopic world: Closing the circle within the dynamical reduction program. *Foundations of Physics* 25(1): 5–38.

Ghirardi, G.C., O. Nicrosini, A. Rimini, and T. Weber. 1988. Spontaneous localization of a system of identical particles. *Il Nuovo Cimento B* 102(4): 383–396.

Ghirardi, G.C., P. Pearle, and A. Rimini. 1990b. Markov processes in Hilbert space and continuous spontaneous localization of systems of identical particles. *Physical Review A* 42(1): 78–89.

Ghirardi, G.C., A. Rimini, and T. Weber. 1986. Unified dynamics for microscopic and macroscopic systems. *Physical Review D* 34(2): 470–491.

Ghirardi, G.C., A. Rimini, and T. Weber. 1987. Disentanglement of quantum wave functions: Answer to 'Comment on 'Unified dynamics for microscopic and macroscopic systems''. *Physical Review D* 36(10): 3287–3289.

Gillies, D. 2000. *Philosophical theories of probability*. London/New York: Routledge.

Giovannetti, V., S. Lloyd, and L. Maccone. 2004. Quantum-enhanced measurements: Beating the standard quantum limit. *Science* 306(5700): 1330–1336.

Giulini, D. 2009. Superselection rules. In *Compendium of quantum physics. Concepts, experiments, history and philosophy*, eds. D. Greenberger, K. Hentschel, and F. Weinert, 771–779. Berlin/Heidelberg: Springer.

Giustina, M., M.A. Versteegh, S. Wengerowsky, J. Handsteiner, A. Hochrainer, K. Phelan, F. Steinlechner, J. Kofler, J.-Å. Larsson, C. Abellán, et al. 2015. Significant-Loophole-Free test of Bell's theorem with entangled photons. *Physical Review Letters* 115(25): 250401.

Gleason, A.M. 1957. Measures on the closed subspaces of a Hilbert space. *Journal of Mathematics and Mechanics* 6(6): 885–893.

Glymour, C. 2006. Markov properties and quantum experiments. In *Physical theory and its interpretation*, eds. W. Demopoulos, and I. Pitowsky, 117–126. Berlin/Heidelberg: Springer.

Goldhorn, K.-H., H.-P. Heinz, and M. Kraus. 2009. *Moderne mathematische Methoden der Physik. Band 1*. Berlin/Heidelberg: Springer.

Goldstein, S., R. Tumulka, and N. Zanghì. 2012. The quantum formalism and the GRW formalism. *Journal of Statistical Physics* 149(1): 142–201.

Gomatam, R. 2006. Niels Bohr's interpretation and the copenhagen interpretation—are the two incompatible? *Philosophy of Science* 77(5): 736–748.

Good, I. 1983. *Good thinking: The foundations of probability and its applications*. Mineapolis: University of Minnesota Press.

Goodman, N. 1983 [1955]. The new riddle of induction. In *Fact, fiction, and forecast*, ed. N. Goodman, 59–83, 4th ed. Cambridge: Harvard University Press.

Gopnik, A., and A. Meltzoff. 1997. *Words, thoughts, and theories*. Cambridge: MIT Press.

Gottfired, K., and V.F. Weisskopf. 1986. *Concepts of particle physics*, vol. II. Oxford/New York: Oxford University Press.

Graham, N. 1973. The measurement of relative frequency. In *The many worlds interpretation of quantum mechanics*, eds. B. Dewitt, and N. Graham, 229–253. Princeton: Princeton University Press.

Grangier, P. 2001. Count them all. *Nature* 409(6822): 774–775.

Grangier, P., G. Roger, and A. Aspect. 1986. Experimental evidence for a photon anticorrelation effect on a beam splitter: A new light on single-photon interferences. *Europhysics Letters* 1(4): 173–179.

Graßhoff, G., S. Portmann, and A. Wüthrich. 2005. Minimal assumption derivation of a bell-type inequality. *The British Journal for the Philosophy of Science* 56(4): 663–680.

Greaves, H. 2004. Understanding Deutsch's probability in a deterministic multiverse. *Studies in History and Philosophy of Science Part B: Studies in History and Philosophy of Modern Physics* 35(3): 423–456.

Greaves, H. 2007. Probability in the Everett interpretation. *Philosophy Compass* 2(1): 109–128.

Greaves, H., and D. Wallace. 2013. Empirical consequences of symmetries. *The British Journal for the Philosophy of Science* 65(1): 59–89.

Greenstein, G., and A.G. Zajonc. 2006. *The quantum challenge. Modern research on the foundations of quantum mechanics*, 2nd ed. Sudbury/Mississauga: Jones & Bartlett.

Greiner, W., and J. Reinhardt. 1993. *Field quantization*. Berlin/Heidelberg: Springer.

Grice, H.P. 1975. Logic and conversation. In *Syntax and semantics, vol. 3: Speech acts*, eds. P. Cole, and J.L. Morgan, 41–58. New York: Academic Press.

Griffiths, D. 2008. *Introduction to elementary particles*, 2nd rev. ed. Weinheim: Wiley-VCH.

Griffiths, D.J. 1995. *Introduction to quantum mechanics*. New Jersey: Prentice Hall.

Griffiths, D.J. 1999. *Introduction to electrodynamics*. New Jersey: Prentice Hall.

Gross, R., and A. Marx. 2012. *Festkörperphysik*. München: Oldenbourg Verlag.

Gustafson, S.J., and I.M. Sigal. 2011. *Mathematical concepts of quantum mechanics*, 2nd ed. Berlin/Heidelberg: Springer.

Haag, R. 1996. *Local quantum physics: Fields, particles, algebras*, 2nd rev. and enlarged ed. Berlin/Heidelberg: Springer.

Hackermüller, L., K. Hornberger, B. Brezger, A. Zeilinger, and M. Arndt. 2004. Decoherence of matter waves by thermal emission of radiation. *Nature* 427(6976): 711–714.

Hájek, A. 2009a. Dutch book arguments. In *The handbook of rational and social choice*, eds. P. Anand, P. Pattanaik, and C. Puppe, 173–195. Oxford/New York: Oxford University Press.

Hájek, A. 2009b. Fifteen arguments against hypothetical frequentism. *Erkenntnis* 70(2): 211–235.

Haken, H., and H.C. Wolf. 1996. *Atom- und Quantenphysik. Einführung in die experimentellen und theoretischen Grundlagen*, 6th rev. and extended ed. Berlin/Heidelberg: Springer.

Halvorson, H. 2007. Algebraic quantum field theory. In *Handbook of the philosophy of science: Philosophy of physics, Part A*, eds. J. Butterfield, and J. Earman, 731–864. Boston: Elsevier.

Halvorson, H., and R. Clifton. 2002a. No place for particles in relativistic quantum theories? In *Ontological aspects of quantum field theory*, eds. M. Kuhlmann, H. Lyre, and A. Wayne, 181–214. New Jersey/London: World Scientific Publishing.

Halvorson, H., and R. Clifton. 2002b. Reconsidering Bohr's reply to EPR. In *Non-locality and modality*, eds. T. Placek, and J. Butterfield, 3–18. Dordrecht: Springer Science + Business Media.

Hardy, L. 1994. Nonlocality of a single photon revisited. *Physical Review Letters* 73: 2279–2283.

Hardy, L. 2013. Are quantum states real? *International Journal of Modern Physics B* 27(1–3): 1345012 (19pp).

Harman, P. 2003. Maxwell, James Clerk. In *The Oxford companion to the history of modern science*, ed. J.L. Heilbron. Oxford/New York: Oxford University Press.

Harrigan, N., and T. Rudolph. 2007. Ontological models and the interpretation of contextuality. arXiv preprint arXiv:0709.4266.

Harrigan, N., and R.W. Spekkens. 2010. Einstein, incompleteness, and the epistemic view of quantum states. *Foundations of Physics* 40: 125–157.

Hartle, J.B. 2003. *Gravity: An introduction to Einstein's general relativity*. San Francisco/Boston: Addison Wesley.

Hartmann, S. 1996. The world as a process. In *Modelling and simulation in the social sciences from the philosophy of science point of view*, eds. R. Hegelsmann, U. Mueller, and K.G. Troitzsch, 77–100. Dordrecht: Springer Science + Business Media.

Hatfield, B. 1992. *Quantum field theory of point particles and strings*. Cambridge: Perseus Books.

Hawley, K. 2009. Identity and indiscernibility. *Mind* 118(469): 101–119.

Hayashi, M. 2006. *Quantum information. An introduction*. Berlin/Heidelberg: Springer.

Healey, R. 2007. *Gauging what's real: The conceptual foundations of contemporary gauge theories*. Oxford/New York: Oxford University Press.

Healey, R. 2009. Holism in quantum mechanics. In *Compendium of quantum physics. Concepts, experiments, history and philosophy*, eds. D. Greenberger, K. Hentschel, and F. Weinert, 295–298. Berlin: Springer.

Healey, R. 2012a. How to use quantum theory locally to explain 'non-local' correlations. arXiv preprint arXiv:1207.7064.

Healey, R. 2012b. Quantum decoherence in a pragmatist view: Dispelling Feynman's mystery. *Foundations of Physics* 42(12): 1534–1555.

Healey, R. 2012c. Quantum decoherence in a pragmatist view: Resolving the measurement problem. arXiv preprint arXiv:1207.7105.

Healey, R. 2012d. Quantum theory: A pragmatist approach. *The British Journal for the Philosophy of Science* 63(4): 729–771.

Healey, R. 2015. Quantum states as objective informational bridges. *Foundations of Physics* 47(2): 1–13.

Healey, R.A. 1991. Holism and nonseparability. *The Journal of Philosophy* 88(8): 393–421.

Heil, J. 2003. *From an ontological point of view*. Oxford/New York: Oxford University Press.

Heinosaari, T., and M. Ziman. 2012. *The mathematical language of quantum theory. From uncertainty to entanglement*. Cambridge/New York: Cambridge University Press.

Heisenberg, W. 1925. Über quantentheoretische Umdeutung kinematischer und mechanischer Beziehungen. *Zeitschrift fur Physik* 33: 879–893.

Heisenberg, W. 1927. Über den anschaulichen Inhalt der quantentheoretischen Kinematik und Mechanik. *Zeitschrift für Physik* 43(3–4): 172–198.

Heisenberg, W. 1934. Wandlungen der Grundlagen der exakten Naturwissenschaft in jüngster Zeit. *Angewandte Chemie* 47(41): 697–708.

Heisenberg, W. 1949 [1930]. *The physical principles of the quantum theory*. Trans. by C. Eckart, and F.C. Hoyt. New York/Toronto: Dover Publications Inc.

Heisenberg, W. 1971 [1958]. *Physics and philosophy. The revolution in modern science*, 3rd ed. London: Allen & Unwin Ltd.

Heisenberg, W. 2011. *Quantentheorie und philosophie*. Stuttgart: Reclam.

Heisenberg, W. 2012 [1969]. *Der Teil und das Ganze*, 9th ed. München, Zürich: Piper.

Held, C. 2012. Die Struktur der Quantenmechanik. In *Philosophie der Physik*, ed. M. Esfeld, 71–87. Berlin: Suhrkamp.

Held, C. 2013. The Kochen-Specker theorem. In *The Stanford encyclopedia of philosophy*, ed. E.N. Zalta. The Metaphysics Research Lab, Center for the Study of Language and Information (CSLI), Stanford University. http://plato.stanford.edu/entries/bell-theorem/.

Hensen, B., H. Bernien, A. Dréau, A. Reiserer, N. Kalb, M. Blok, J. Ruitenberg, R. Vermeulen, R. Schouten, C. Abellán, et al. 2015. Loophole-free Bell inequality violation using electron spins separated by 1.3 kilometres. *Nature* 526(7575): 682–686.

Hentschel, K. 2009a. Light quantum. In *Compendium of quantum physics. Concepts, experiments, history and philosophy*, ed. D. Greenberger, K. Hentschel, and F. Weinert, 339–346. Berlin/Heidelberg: Springer.

Hentschel, K. 2009b. Spin. In *Compendium of quantum physics. Concepts, experiments, history and philosophy*, eds. D. Greenberger, K. Hentschel, and F. Weinert, 726–731. Berlin/Heidelberg: Springer.

Hessmo, B., P. Usachev, H. Heydari, and G. Björk. 2004. Experimental demonstration of single photon nonlocality. *Physical Review Letters* 92: 180401.

Hiley, B.J. 1999. Active information and teleportation. In *Epistemological and experimental perspectives on quantum physics*, eds. D. Greenberger, W.L. Reiter, and A. Zeilinger, 113–126. Dordrecht: Springer Science + Business Media.

Hintikka, J. 1962. *Knowledge and belief: An introduction to the logic of the two notions*. Ithaca/New York: Cornell University Press.

Hofer-Szabó, G., and P. Vecsernyés. 2014. Bell's local causality for philosophers. *philsci* eprint: http://philsci-archive.pitt.edu/10795/1/LocalizationofCC7.pdf.

Hofer-Szabó, G., M. Rédei, and L.E. Szabó. 1999. On Reichenbac's common cause principle and Reichenbach's notion of common cause. *The British Journal for the Philosophy of Science* 50(3): 377–399.

Hoffman, D.D., and C. Prakash. 2014. Objects of consciousness. *Frontiers in Psychology* 5: 577(22 pp).

Holland, P. 1995. *The quantum theory of motion: An account of the de Broglie-Bohm causal interpretation of quantum mechanics*. Cambridge/New York: Cambridge University Press.

Home, D., and F. Selleri. 1991. Bell's theorem and the EPR paradox. *Rivista del Nuovo Cimento* 14(9): 1–95.

Home, D., and M.A.B. Whitaker. 1992. Ensemble interpretation of quantum mechanics. A modern perspective. *Physics Reports* 210(4): 223–317.

Hooker, C.A. 1972. The nature of quantum mechanical reality: Einstein versus bohr. In *Paradigms and paradoxes: The philosophical challenge of the quantum domain*, ed. R.G. Colodny, 67–302. Pittsburgh: University of Pittsburgh Press.

Horodecki, R., P. Horodecki, M. Horodecki, and K. Horodecki. 2009. Quantum entanglement. *Reviews of Modern Physics* 81(2): 865–942.

Horwich, P. 2006. The nature of meaning. In *The blackwell guide to the philosophy of language*, eds. M. Devitt, and R. Hanley, 43–57. Malden/Oxford: Blackwell.

Howard, D. 1985. Einstein on locality and separability. *Studies in History and Philosophy of Science Part A* 16(3): 171–201.

Howard, D. 1989. Holism, separability, and the metaphysical implications of the Bell experiments. In *Philosophical consequences of quantum theory. Reflections on Bell's theorem*, eds. J.T. Cushing, and E. McMullin, 224–253. Notre Dame: University of Notre Dame Press.

Howard, D. 1990. 'Nicht sein kann was nicht sein darf,' or the Preshistory of EPR, 1909–1935: Einstein's early worries about the quantum mechanics of composite systems. In *Sixty-two years of uncertainty. Historical, philosophical, and physical inquiries into the foundations of quantum mechanics*, ed. A.I. Miller, 61–112. New York/London: Plenum Press.

Howard, D. 1994. What makes a classical concept classical? Toward a reconstruction of Niels Bohr's philosophy of physics. In *Niels Bohr and contemporary philosophy*, eds. J. Faye, and H.J. Folse. Dordrecht: Springer Science + Business Media.

Howard, D. 2002. Who invented the 'Copenhagen Interpretation'? A study in mythology. *Philosophy of Science* 71(5): 669–682.

Howard, D. 2010. 'Let me briefly indicate why I do not find this standpoint natural.' Einstein, general relativity, and the contingent a priori. In *Discourse on a new method, or a manifesto for a synthetic approach to history and philosophy of science*, eds. M. Domski, and M. Dickson, 333–356. Chicago/La Salle: Open Court.

Howson, C., and P. Urbach. 2006. *Scientific reasoning. The Bayesian approach*, 3rd ed. Chicago/La Salle: Open Court.

Huber, F. 2009. Belief and degrees of belief. In *Degrees of belief*, eds. F. Huber, and C. Schmidt-Petri. Dordrecht: Springer Science + Business Media.

Hughes, G.E., and M.J. Cresswell. 1996. *A new introduction to modal logic*. London/New York: Routledge.

Hughes, R.I.G. 1989. *The structure and interpretation of quantum mechanics*. Cambridge/London: Harvard University Press.

Hume, D. 1999 [1748]. *An enquiry concerning human understanding*. Edited by T.L. Beauchamp. Oxford/New York: Oxford University Press.

Hüttemann, A. 2013. *Ursachen*. Berlin/Boston: De Gruyter.

Inagaki, T., N. Matsuda, O. Tadanaga, M. Asobe, and H. Takesue. 2013. Entanglement distribution over 300 km of fiber. *Optics Express* 21(20): 23241–23249.

Isham, C.J., and J. Butterfield. 1998. Topos perspective on the Kochen-Specker theorem: I. Quantum states as generalized valuations. *International Journal of Theoretical Physics* 37(11): 2669–2733.

Ivanova, M. 2014. Is there a place for epistemic virtues in theory choice? In *Virtue epistemology naturalized*, ed. A. Fairweather, 207–226. Berlin/Heidelberg: Springer.

Jackson, J.D. 1990. *Classical electrodynamics*, 3rd ed. Hoboken: Wiley.

Jaeger, G. 2007. *Quantum information. An overview*. New York: Springer Science + Business Media, LLC.

Jaeger, G. 2009. *Entanglement, information, and the interpretation of quantum mechanics*. Berlin/Heidelberg: Springer.

Jammer, M. 1966. *The conceptual development of quantum mechanics*. New York/St. Louis: McGraw-Hill Book Company.

Jammer, M. 1974. *The philosophy of quantum mechanics: The interpretations of QM in historical perspective*. Hoboken: Wiley.

Jarrett, J.P. 1984. On the physical significance of the locality conditions in the bell arguments. *Noûs* 18(4): 569–589.

Jarrett, J.P. 1989. Bell's theorem: A guide to the implications. In *Philosophical consequences of quantum theory. Reflections on Bell's theorem*, eds. J.T. Cushing, and E. McMullin, 60–79. Notre Dame: University of Notre Dame Press.

Jaynes, E.T. 1980. Quantum beats. In *Foundations of radiation theory and quantum electrodynamics*, ed. A.O. Barut, 37–43. New York: Springer Science + Business Media, LLC.

Jaynes, E.T. 2003. *Probability theory. The logic of science*. Cambridge/New York: Cambridge University Press.

Jennings, D., and M. Leifer. 2015. No return to classical reality. *Contemporary Physics* 57(1): 60–82.

Johnston, W. 2014. The weighted hermite polynomials form a basis for $L^2(\mathbb{R})$. *The American Mathematical Monthly* 121(3): 249–253.

Joos, E., H. Zeh, C. Kiefer, D. Giulini, J. Kupsch, and I.-O. Stamatescu. 2003. *Decoherence and the appearance of a classical world in quantum theory*, 2nd ed. Berlin/Heidelberg: Springer.

Kadanoff, L.P. 2013. Theories of matter: Infinities and renormalization. In *The Oxford handbook of philosophy of physics*, ed. R. Batterman, 141–188. Oxford/New York: Oxford University Press.

Kant, I. 1881 [1781]. *Critique of pure reason*. Trans. by M. Müller, with an historical introduction by Ludwig Noiré. London: MacMillan and Co.

Kant, I. 1949 [1783]. *Prolegomena to any future metahphysics*. Edited in English by P. Carus. Chicago/London: Open Court Publishing.

Kelvin, L. 1901. I. Nineteenth century clouds over the dynamical theory of heat and light. *The London, Edinburgh, and Dublin Philosophical Magazine and Journal of Science* 2(7): 1–40.

Kemp, G. 2012. *Quine versus Davidson. Truth, reference, and meaning*. Oxford/New York: Oxford University Press.

Kenny, A. 2007. *Philosophy in the modern world. A new history of Western philosophy*, vol. IV. Oxford/New York: Oxford University Press.

Kent, A. 1990. Against many-worlds interpretations. *International Journal of Modern Physics A* 5(09): 1745–1762.

Kent, A. 2010. One world versus many: The inadequacy of everettian accounts of evolution, probability, and scientific confirmation. In *Many worlds? Everett, quantum theory, and reality*, eds. S. Saunders, J. Barrett, A. Kent, and D. Wallace, 307–368. Oxford/New York: Oxford University Press.

Kent, A. 2015. Does it make sense to speak of self-locating uncertainty in the universal wave function? Remarks on Sebens and Carroll. *Foundations of Physics* 45(2): 211–217.

Kerner, H., and W. von Wahl. 2013. *Mathematik für Physiker*, 3rd rev. ed. Berlin/Heidelberg: Springer.

Khrennikov, A. 2010. Two versions of the projection postulate: From EPR argument to one-way quantum computing and teleportation. *Advances in Mathematical Physics* 2010: 945460(11 pp).

Kiefer, C. 2003. *Quantentheorie*, 2nd ed. Frankfurt a. M.: Fischer Taschenbuch Verlag.

Kiefer, C. 2007. *Quantum gravity*, 2nd ed. Oxford/New York: Oxford University Press.

Kiefer, C. 2009. Does time exist in quantum gravity? arXiv preprint arXiv:0909.3767.

Kleppner, D., and R. Jackiw. 2000. One hundred years of quantum physics. *Science* 289(5481): 893–898.

Kochen, S., and E.P. Specker. 1975 [1967]. The problem of hidden variables in quantum mechanics. In *The logico-algebaic approach to quantum mechanics, vol. I. Historical evolution*, ed. C.A. Hooker. Dordrecht/Boston: D. Reidel Publishing Company.

Koetsier, T. 1994. Kinematics. In *Companion encyclopedia of the history and philosophy of the mathematical sciences, vol. 1 & 2*, ed. I. Grattan-Guiness, 994–1001. London/New York: Routledge.

Kraus, S., and D. Lehmann. 1988. Knowledge, belief and time. *Theoretical Computer Science* 58(1–3): 155–174.

Kripke, S. 1982. *Wittgenstein on rules and private language: An elementary exposition*. Cambridge: Harvard University Press.

Kuhlmann, M. 2010. *The ultimate constituents of the material world. In search of an ontology for fundamental physics*. Heusenstamm: Ontos Verlag.

Ladyman, J. 2000. What's really wrong with constructive empiricism? Van Fraassen and the metaphysics of modality. *The British Journal for the Philosophy of Science* 51(4): 837–856.

Ladyman, J. 2004. Constructive empiricism and modal metaphysics: A reply to Monton and van Fraassen. *The British Journal for the Philosophy of Science* 55(4): 755–765.

Lancaster, T., and S.J. Blundell. 2014. *Quantum field theory for the Gifted Amateur*. Oxford/New York: Oxford University Press.

Landé, A. 1965a. *New foundations of quantum mechanics*. Cambridge: Cambridge University Press.

Landé, A. 1965b. Why do quantum theorists ignore the quantum theory? *The British Journal for the Philosophy of Science* 15(60): 307–313.

Landau, L.D., and E.M. Lifshitz. 1965. *Quantum mechanics. Non-relativistic theory*, 2nd ed. Trans. by J.B. Sykes, and J.S. Bell. Oxford/London: Pergamon Press.

Landsman, N. 2009. Quasi-classical limit. In *Compendium of quantum physics. Concepts, experiments, history and philosophy*, eds. D. Greenberger, K. Hentschel, and F. Weinert, 626–629. Berlin/Heidelberg: Springer.

Langton, R. 1998. *Kantian humility. Our ignorance of things in themselves*. Oxford/New York: Clarendon Press.

Le Bellac, M. 2006. *Quantum physics*. Trans. by P. de Forcrand-Millard. Cambridge/New York: Cambridge University Press.

Leggett, A.J. 1980. Macroscopic quantum systems and the quantum theory of measurement. *Progress of theoretical physics supplement* 69: 80–100.

Leggett, A.J. 2002. Testing the limits of quantum mechanics: Motivation, state of play, prospects. *Journal of Physics: Condensed Matter* 14(15):R415–R451.

Leggett, A.J., and A. Garg. 1985. Quantum mechanics versus macroscopic realism: Is the flux there when nobody looks? *Physical Review Letters* 54(9): 857–860.

Leifer, M. 2014. Is the quantum state real? An extended review of ψ-ontology theorems. *Quanta* 3(1): 67–155.

Leifer, M.S. 2011. Can the quantum state be interpreted statistically? Web log post: http://mattleifer.info/2011/11/20/can-the-quantum-state-be-interpreted-statistically/.

Leifer, M.S., and O.J.E. Maroney. 2013. Maximally epistemic interpretations of the quantum state and contextuality. *Physical Review Letters* 110: 120401(1–5).

Leitgeb, H. 2011. New life for Carnap's *Aufbau*? *Synthese* 180(2): 265–299.

Lewis, D.K. 1973. *Conunterfactuals*. Oxford: Blackwell.

Lewis, D.K. 1974. Causation. *The Journal of Philosophy* 70(17): 556–567.

Lewis, D.K. 1983. New work for a theory of universals. *Australasian Journal of Philosophy* 61(4): 343–377.

Lewis, D.K. 1987. A subjectivist's guide to objective chance. In *Philosophical papers, vol. II*, ed. D.K. Lewis. Oxford/New York: Oxford University Press.

Lewis, D.K. 1994. Humean supervenience debugged. *Mind* 103(412): 473–490.

Lewis, P.G., D. Jennings, J. Barrett, and T. Rudolph. 2012. Distinct quantum states can be compatible with a single state of reality. *Physical Review Letters* 109(15): 150404(1–5).

Lewis, P.J. 2007. Quantum sleeping beauty. *Analysis* 67(1): 59–65.

Lockwood, M. 1996. 'Many Minds' interpretations of quantum mechanics. *The British Journal for the Philosophy of Science* 47(2): 159–188.

Loewer, B. 1996. Humean supervenience. *Philosophical Topics* 24(1): 101–127.

Lombardi, O., and D. Dieks. 2012. Modal interpretations of quantum mechanics. In *The Stanford encyclopedia of philosophy*, ed. E.N. Zalta. The Metaphysics Research Lab, Center for the Study of Language and Information (CSLI), Stanford University. http://plato.stanford.edu/entries/qm-modal/.

Lüders, G. 1951. Über die Zustandsänderung durch den Messprozess. *Annalen der Physik* 443(5–8): 322–328.

Lyre, H. 2005. Metaphysik im 'Handumdrehen': Kant und Earman, Parität und moderne Raumauffassung. *Philosophia Naturalis* 42(1): 49–76.

Mach, E. 1910. Die Leitgedanken meiner naturwissenschaftlichen Erkenntnislehre und ihre Aufnahme durch die Zeitgenossen. *Scientia* 4(7): 228–240.

Machery, E., R. Mallon, S. Nichols, and S.P. Stich. 2004. Semantics, cross-cultural style. *Cognition* 92(3):B1–B12.

Machery, E., R. Mallon, S. Nichols, and S.P. Stich. 2013. If folk intuitions vary, then what? *Philosophy and Phenomenological Research* 86(3): 618–635.

Mahler, D.H., L. Rozema, K. Fisher, L. Vermeyden, K.J. Resch, H.M. Wiseman, and A. Steinberg. 2016. Experimental nonlocal and surreal Bohmian trajectories. *Science Advances* 2(2):e1501466.

Malament, D. 1996. In defense of dogma: Why there cannot be a relativistic quantum mechanics of (localizable) particles. In *Perspectives on quantum reality*, ed. R. Clifton, 1–10. Dordrecht: Kluwer.

Mann, C., and R. Crease. 1988. John Bell, particle physicist (interview). *Omni* 10(8): 84–92, 121.

Manoukian, E.B. 2007. *Quantum theory: A wide spectrum*. Berlin/Heidelberg: Springer.

Markopoulou, F. 2009. New directions in background independent quantum gravity. In *Approaches to quantum gravity: Toward a new understanding of space, time and matter*, ed. D. Oriti, 129–150. Cambridge/New York: Cambridge University Press.

Maroney, O.J., and C.G. Timpson. 2014. Quantum- vs. macro-realism: What does the Leggett-Garg inequality actually test? arXiv preprint arXiv:1412.6139.

Maroney, O.J.E. 2012. How statistical are quantum states? arXiv preprint arXiv:1207.6906.

Massimi, M. 2005. *Pauli's exclusion principle. The origin and validation of a scientific principle*. Cambridge/New York: Cambridge University Press.

Maudlin, T. 1995. Three measurement problems. *Topoi* 14: 7–15.

Maudlin, T. 2007. *The metaphysics within physics*. Oxford/New York: Oxford University Press.

Maudlin, T. 2010a. Can the world be only wavefunction? In *Many worlds? Everett, quantum theory, and reality*, eds. S. Saunders, J. Barrett, A. Kent, and D. Wallace, 121–143. Oxford/New York: Oxford University Press.

Maudlin, T. 2010b. What Bell proved: A reply to blaylock. *American Journal of Physics* 78(1): 121–125.

Maudlin, T. 2011. *Quantum non-locality and relativity. Metaphysical intimations of modern physics*, 3rd ed. Malden/Oxford: Wiley-Blackwell.

Maudlin, T. 2013. The nature of the quantum state. In *The wave function. Essays on the metahphysics of quantum mechanics*, eds. A. Ney, and D. Albert, 126–153. Oxford/New York: Oxford University Press.

Maudlin, T. 2014a. Critical study: David Wallace, the emergent multiverse: Quantum theory according to the everett interpretation. *Noûs* 48(4): 794–808.

Maudlin, T. 2014b. What Bell did. *Journal of Physics A: Mathematical and Theoretical* 47(42): 424010(24pp).

Mayer-Kuckuk, T. 2002. *Kernphysik. Eine Einführung*, 7th rev. and extended ed. Stuttgart/Leipzig: Teubner.

Mączyński, M.J. 1971. Boolean properties of observables in axiomatic quantum mechanics. *Reports on Mathematical Physics* 2(2): 135–150.

McIntyre, D.H. 2012. *Quantum mechanics. A paradigms approach*. Boston/Columbus: Pearson.

McLaughlin, B., and K. Bennett. 2011. Supervenience. In *The Stanford encyclopedia of philosophy*, ed. E.N. Zalta. The Metaphysics Research Lab, Center for the Study of Language and Information (CSLI), Stanford University. http://plato.stanford.edu/entries/supervenience/.

Mehra, J., and H. Rechenberg. 1987. *The historical development of quantum theory, vol. 5: Erwin Schrödinger and the rise of wave mechanics. Part 2: The creation of wave mechanics: Early response and applications 1925–1926*. New York/Heidelberg: Springer.

Mellor, D.H. 2005. *Probability: A philosophical introduction*. London/New York: Routledge.

Mermin, N.D. 1989. What's wrong with this pillow? *Physics Today* 42(4): 9.

Mermin, N.D. 1993. Hidden variables and the two theorems of John Bell. *Reviews of Modern Physics* 65(3): 803–815.

Mermin, N.D. 2012. Commentary: Quantum mechanics: Fixing the shifty split. *Physics Today* 65(7): 8–10.

Mermin, N.D. 2013. QBism as CBism: Solving the problem of 'the Now'. arXiv preprint arXiv:1312.7825.

Mermin, N.D. 2014a. Physics: QBism puts the scientist back into science. *Nature* 507(7493): 421–423.

Mermin, N.D. 2014b. Why QBism is not the Copenhagen interpretation and what John Bell might have thought of it. arXiv preprint arXiv:1409.2454.

Miller, A. 2006. Meaning scepticism. In *The Blackwell guide to the philosophy of language*, eds. M. Devitt, and R. Hanley, 91–113. Malden/Oxford: Blackwell.

Milonni, P.W. 1994. *The quantum vacuum. An introduction to quantum electrodynamics*. Boston/San Diego: Academic Press Inc.

Mittelstaedt, P. 1995. Constitution of objects in classical mechanics and in quantum mechanics. *International Journal of Theoretical Physics* 34(8): 1615–1626.

Mittelstaedt, P. 1998. *The interpretation of quantum mechanics and the measurement process*. Cambridge/New York: Cambridge University Press.

Mittelstaedt, P. 2009. Cognition versus constitution of objects: From Kant to modern physics. *Foundations of Physics* 39(7): 847–859.

Mohrhoff, U. 2004. Probabilities from envariance? *International Journal of Quantum Information* 2(02): 221–229.

Möllenstedt, G., and C. Jönsson. 1959. Elektronen-Mehrfachinterferenzen an regelmäßig hergestellten Feinspalten. *Zeitschrift für Physik* 155(4): 472–474.

Monton, B. 2002. Wave function ontology. *Synthese* 130(2): 265–277.

Monton, B. 2004. The problem of ontology for spontaneous collapse theories. *Studies in History and Philosophy of Modern Physics* 35(3): 407–421.

Monton, B., and B.C. Van Fraassen. 2003. Constructive empiricism and modal nominalism. *The British Journal for the Philosophy of Science* 54(3): 405–422.

Moore, A.W. 2012. *The evolution of modern metaphysics: Making sense of things*. Cambridge/New York: Cambridge University Press.

Moore, W. 1994. *A life of Erwin Schrödinger*. Cambridge/New York: Cambridge University Press.

Moore, W. 2015. *Schrodinger: Life and thought*. Cambridge/New York: Cambridge University Press.

Mormann, T. 2007. The structure of scientific theories in logical empiricism. In *The Cambridge companion to logical empiricism*, eds. A. Richardson, and T. Uebel, 136–162. Cambridge/New York: Cambridge University Press.

Morrison, M. 2007. Spin: All is not what it seems. *Studies in History and Philosophy of Modern Physics* 38: 529–557.

Muller, F.A. 2014. Elementary particles and metaphysics. In *New directions in the philosophy of science*, eds. M.C. Galavotti, D. Dieks, W.J. Gonzalez, S. Hartmann, T. Uebel, and M. Weber, 417–431. Cham/Heidelberg: Springer.

Muller, F.A., and S. Saunders. 2008. Discerning fermions. *The British Journal for the Philosophy of Science* 59(3): 499–548.

Muller, F.A., and M.P. Seevinck. 2009. Discerning elementary particles. *Philosophy of Science* 76(2): 179–200.

Mumford, S. 1998. *Dispositions*. Oxford/New York: Oxford University Press.

Mumford, S. 2004. *Laws in nature*. London/New York: Routledge.

Nagel, J. 2000. The empiricist conception of experience. *Philosophy* 75(03): 345–376.

Näger, P. 2013a. Causal graphs for EPR experiments. *philsci* e-print: http://philsci-archive.pitt.edu/9915/3/Naeger_PM_2013_%2D_Causal_Graphs_for_EPR_Experiments.pdf.

Näger, P. 2015. The Causal Markov condition in the quantum realm. *Causality meets quantum mechanics*, Satellite Workshop at SOPhiA, University of Salzburg, Austria. Abstract available at https://www.phil-fak.uni-duesseldorf.de/philo/personal/thphil/workshops-und-weitere-wissenschaftliche-aktivitaeten/causality-meets-quantum-mechanics/.

Näger, P.M. 2013b. A stronger bell argument for quantum non-locality. arXiv preprint arXiv:1308.3455.

Näger, P.M. 2016. The causal problem of entanglement. *Synthese* 193(4): 1127–1155.

Nakahara, M. 2003. *Geometry, topology and physics*, 2nd ed. Bristol/Philadelphia: IOP Publishing.

Neander, K. 2006. Naturalistic theories of reference. In *The Blackwell guide to the philosophy of language*, eds. M. Devitt, and R. Hanley, 374–391. Malden/Oxford: Blackwell.

Newton, R.G. 2007. *From clockwork to crapshoot: A history of physics*. Cambridge: Harvard University Press.

Ney, A. 2013. Introduction. In *The wave function. Essays on the metahphysics of quantum mechanics*, eds. A. Ney, and D. Albert. Oxford/New York: Oxford University Press.

Ney, A., and D. Albert, eds. 2013. *The wave function. Essays on the metahphysics of quantum mechanics*. Oxford/New York: Oxford University Press.

Nicrosini, O., and A. Rimini. 1990. On the relationship between continuous and discontinuous stochastic processes in Hilbert space. *Foundations of Physics* 20(11): 1317–1327.

Nielsen, M., and I. Chuang. 2010. *Quantum computation and quantum information*, 10th anniversary ed. Cambridge/New York: Cambridge University Press.

Nigg, D., T. Monz, P. Schindler, E.A. Martinez, M. Chwalla, M. Hennrich, R. Blatt, M.F. Pusey, T. Rudolph, and J. Barrett. 2012. Can different quantum state vectors correspond to the same physical state? An experimental test. arXiv preprint arXiv:1211.0942.

Nolting, W. 2013. *Grundkurs Theoretische Physik 3. Elektrodynamik*, 10th ed. Berlin/Heidelberg: Springer.

Norsen, T. 2007. Against 'Realism'. *Foundations of Physics* 37(3): 311–340.

Norsen, T. 2009. Local causality and completeness: Bell vs. Jarrett. *Foundations of Physics* 39(3): 273–294.

Norsen, T. 2010. The theory of (exclusively) local beables. *Foundations of Physics* 40(12): 1858–1884.

Norsen, T. 2011. John S. Bell's concept of local causality. *American Journal of Physics* 79(12): 1261–1275.

Norsen, T. 2016. Quantum solipsism and nonlocality. In *Quantum nonlocality and reality: 50 years of Bell's theorem*, eds. M. Bell, and S. Gao, 204–237. Cambridge: Cambridge University Press.

Nortmann, U. 2008. *Unscharfe Welt? Was Philosophen über Quantenmechanik wissen möchten*. Darmstadt: WBG.

Parfit, D. 1984. *Reasons and persons*. Oxford: Clarendon Press.

Passon, O. 2004. Why isn't every physicist a Bohmian? arXiv preprint quant-ph/0412119.

Patra, M.K., S. Pironio, and S. Massar. 2013. No-Go theorems for ψ-epistemic models based on a continuity assumption. *Physical Review Letters* 111: 090402(1–4).

Paul, L.A., and N. Hall. 2013. *Causation. A user's guide*. Oxford: Oxford University Press.

Pauli, W. 1927. Zur Quantenmechanik des magnetischen Elektrons. *Zeitschrift für Physik* 43(9–10): 601–623.

Pauli, W. 1961. *Aufsätze und Vorträge über Physik und Erkenntnistheorie*. Wiesbaden: Springer Fachmedien.

Pawłowski, M., J. Kofler, T. Paterek, M. Seevinck, and C. Brukner. 2010. Non-local setting and outcome information for violation of Bell's inequality. *New Journal of Physics* 12: 083051(9pp).

Pearl, J. 2009. *Causality. Models, reasoning, and inference*, 2nd ed. Cambridge/New York: Cambridge University Press.

Pearle, P. 1968. Reply to Dr. Sach's letter. *American Journal of Physics* 36: 464–465.

Pearle, P. 1989. Combining stochastic dynamical state-vector reduction with spontaneous localization. *Physical Review A* 39(5): 2277.

Pearle, P., and E. Squires. 1994. Bound state excitation, nucleon decay experiments and models of wave function collapse. *Physical Review Letters* 73(1): 1–5.

Pearle, P., and E. Squires. 1996. Gravity, energy conservation, and parameter values in collapse models. *Foundations of Physics* 26(3): 291–305.

Peirce, C.S. [1878]. Deduction, induction, and hypothesis. In *The essential Peirce. Selected philosophical writings, vol. 1 (1867–1893)*, eds. N. Houser, and C. Kloesel, 186–199. Bloomington/Indianapolis: Indiana University Press.

Penrose, R. 1996. On gravity's role in quantum state reduction. *General Relativity and Gravitation* 28(5): 581–600.

Penrose, R. 2004. *The road to reality. A complete guide to the laws of the universe*. London: Jonathan Cape.

Peres, A. 1991. Two simple proofs of the Kochen-Specker theorem. *Journal of Physics A: Mathematical and General* 24(4):L175–L178.

Peres, A. 1999. All the Bell inequalities. *Foundations of Physics* 29(4): 589–614.

Peres, A. 2002. *Quantum theory: Concepts and methods*. New York/Boston: Kluwer.

Perschbacher, W.J., ed. 1990. *The new analytical greek Lexicon*. Peabody: Hendrickson Publishers, Inc.

Peskin, M.E., and D.V. Schroeder. 1995. *An introduction to quantum field theory*. Reading: Perseus Books.

Peters, D. 2014. What elements of successful scientific theories are the correct targets for 'selective' scientific realism? *Philosophy of Science* 81(3): 377–397.

Petersen, A. 1963. The philosophy of Niels Bohr. *Bulletin of the Atomic Scientists* 19(7): 8–14.

Piron, C. 1976. *Foundations of quantum physics*. London/Amsterdam: W. A. Benjamin, Inc.

Pitowsky, I. 2006. Quantum mechanics as a theory of probability. In *Physical theory and its interpretation. Essays in honor of Jeffrey Bub*, eds. W. Demopoulos, and I. Pitowsky, 213–240. Dordrecht: Springer.

Poincaré, H. 1905. *Science and hypothesis*. London/Newcastle-on-Tyne: The Walter Scott Publishing Co., Ltd.

Poincaré, H. 1918. *Science and method*. London/Edinburgh: Thomas Nelson and Sons.

Popper, K.R. 1957. The propensity interpretation of the calculus of probability, and the quantum theory. In *Observation and interpretation. A symposium of philosophers and physicists. Proceedings of the ninth symposium of the Colston research society*, ed. S. Körner. London: Butterworths Scientific Publications.

Popper, K.R. 1959. The propensity interpretation of probability. *The British Journal for the Philosophy of Science* 10(37): 25–42.

Popper, K.R. 1967. Quantum mechanics without 'the observer'. In *Quantum theory and reality*, ed. M. Bunge. Berlin/Heidelberg: Springer.

Popper, K.R. 1995[1990]. *A world of propensities*. Bristol: Thoemmes Press.

Portmann, S., and A. Wüthrich. 2007. Minimal assumption derivation of a weak Clauser-Horne inequality. *Studies in History and Philosophy of Modern Physics* 38: 844–862.

Povh, B., M. Lavelle, K. Rith, C. Scholz, and F. Zetsche. 2006. *Particles and nuclei: An introduction to the physical concepts*, 5th ed. Berlin/Heidelberg: Springer.

Praxmeyer, L., B.-G. Englert, and K. Wódkiewicz. 2005. Violation of Bell's inequality for continuous-variable EPR states. *The European Physical Journal D* 32(2): 227–231.

Psillos, S. 1999. *Scientific realism. How science tracks truth*. London/New York: Routledge.

Psillos, S. 2002. *Causation and explanation*. Stocksfield: Acumen.

Pusey, M.F., J. Barrett, and T. Rudolph. 2011. On the reality of the quantum state. arXiv preprint arXiv:1111.3328.

Pusey, M.F., J. Barrett, and T. Rudolph. 2012. On the reality of the quantum state. *Nature Physics* 8(6): 475–478.
Putnam, H. 1975a. The meaning of 'meaning'. In *Language mind and knowledge*, ed. K. Gunderson, 131–193. Minneapolis: University of Minnesota Press.
Putnam, H. 1975b. What is 'realism'? *Proceedings of the Aristotelian Society* 76: 177–194.
Putnam, H. 1977. Realism and reason. *Proceedings and Addresses of the American Philosophical Association* 50(6): 483–498.
Putnam, H. 1981. *Reason, truth and history*. Cambrdige/New York: Cambridge University Press.
Putnam, H. 1982. Why reason can't be naturalized. *Synthese* 52(1): 3–23.
Putnam, H. 1990. *Realism with a human face*. Edited by J. Conant. Cambridge: Harvard University Press.
Putnam, H. 1992. Replies. *Philosophical Topics* 20(1): 347–408.
Putnam, H. 2005. A philosopher looks at quantum mechanics (again). *British Journal for the Philosophy of Science* 56(4): 615–634.
Putnam, H. 2015. Intellectual autobiography. In *The philosophy of hilary putnam*, eds. R.E. Auxier, D. Anderson, and L.E. Hahn, 3–110. Chicago: Open Court.
Pylyshyn, Z. 1999. Is vision continuous with cognition? The case for cognitive impenetrability of visual perception. *Behavioral and Brain Sciences* 22(03): 341–365.
Quine, W.V. 1939. Designation and existence. *The Journal of Philosophy* 36(26): 701–709.
Quine, W.V. 1948. On what there is. *The Review of Metaphysics* 2(1): 21–38.
Rae, A.I. 2004. *Quantum physics: Illusion or reality?*, 2nd ed. Cambridge/New York: Cambridge University Press.
Ramsey, F.P. 1931[1926]. Truth and probability. In *The foundations of mathematics and other logical essays*, ed. R.B. Braithwaite, 156–198. London/New York: Routledge.
Redhead, M. 1987. *Incompleteness, nonlocality, and realism. A prolegomenon to the philosophy of quantum mechanics*. Oxford: Clarendon Press.
Reeh, H., and S. Schlieder. 1961. Bemerkungen zur Unitäräquivalenz von Lorentzinvarianten Feldern. *Nuovo Cimento* 22(5): 1051–1068.
Reichenbach, H. 1920. *Relativitätstheorie und Erkenntnis Apriori*. Berlin: Springer.
Reichenbach, H. 1944. *Philosophic foundations of quantum mechanics*. Mineloa/New York: Dover Publications, Inc.
Reichenbach, H. 1961. *Experience and prediction. An analysis of the foundations and structure of knowledge*. Chicago/London: The University of Chicago Press.
Reichenbach, H. 1965. *The direction of time*. Edited by Maria Reichenbach. Mineloa/New York: Dover Publications, Inc.
Reimer, M., and E. Michaelson. 2014. Reference. In *The Stanford encyclopedia of philosophy*, ed. E.N. Zalta. The Metaphysics Research Lab, Center for the Study of Language and Information (CSLI), Stanford University. http://plato.stanford.edu/entries/reference/.
Rindler, W. 2006. *Relativity. Special, general, and cosmological*, 2nd ed. Oxford/New York: Oxford University Press.
Rosenberg, J.F. 1974. *Linguistic representation*. Dordrecht/Boston: D. Reidel Publishing Co.
Rosenblum, B., and F. Kuttner. 2011. *Quantum enigma. Physics encountes consciousness*, 2nd ed. Oxford/New York: Oxford University Press.
Rosenfeld, L. 1983 [1967]. Bohr's reply. In *Quantum theory and measurement*, eds. J.A. Wheeler, and W.H. Zurek. Princeton: Princeton University Press.
Roussas, G. 2007. *Introduction to probability*. Amsterdam/Boston: Elsevier Academic Press.
Roussas, G.G. 2014. *An introduction to measure-theoretic probability*, 2nd ed. Amsterdam/Boston: Elsevier Academic Press.
Ruetsche, L. 2011. *Interpreting quantum theories*. Oxford/New York: Oxford University Press.
Russell, B. 1914. *Our knowledge of the external world*. London: Allen & Unwin Ltd.
Sakurai, J.J. 1994. *Modern quantum mechanics*, rev. ed. Reading/New York: Addison-Wesley.
Salmon, W. 1984. *Scientific explanation and the causal structure of the world*. Princeton: Princeton University Press.
Salmon, W.C. 1998. *Causality and explanation*. Oxford/New York: Oxford University Press.

Sankey, H. 2008. *Scientific realism and the rationality of science*. Aldershot: Ashgate.
Saunders, S. 1998. Time, quantum mechanics, and probability. *Synthese* 114(3): 373–404.
Saunders, S. 2010. Many words? An introduction. In *Many worlds? Everett, quantum theory, and reality*, eds. S. Saunders, J. Barrett, A. Kent, and D. Wallace, 1–49. Oxford/New York: Oxford University Press.
Schilpp, P.A., ed. 1949. *Albert Einstein. Philosopher-scientist*. New York: MJF Books.
Schlosshauer, M. 2004. Decoherence, the measurement problem, and interpretations of quantum mechanics. *Reviews of Modern Physics* 76(4): 1267–1305.
Schlosshauer, M. 2007. *Decoherence and the quantum to classical transition*, 2nd ed. Berlin/Heidelberg: Springer.
Schlosshauer, M., and A. Fine. 2005. On Zurek's derivation of the Born rule. *Foundations of Physics* 35(2): 197–213.
Schlosshauer, M., and A. Fine. 2012. Implications of the Pusey-Barrett-Rudolph quantum no-go theorem. *Physical Review Letters* 108: 260404(1–4).
Schlosshauer, M., and A. Fine. 2014. No-go theorem for the composition of quantum systems. *Physical Review Letters* 112: 070407(1–4).
Schlosshauer, M., J. Kofler, and A. Zeilinger. 2013. A snapshot of foundational attitudes toward quantum mechanics. *Studies in History and Philosophy of Modern Physics* 44(3): 222–230.
Schrenk, M. 2014. Better best systems and the issue of CP-laws. *Erkenntnis* 79(10): 1787–1799.
Schrödinger, E. 1926a. Quantisierung als Eigenwertproblem. (Vierte Mitteilung.). *Annalen der Physik* 385(13): 437–490.
Schrödinger, E. 1926b. Quantisierung als Eigenwertproblem. (Zweite Mitteilung.). *Annalen der Physik* 384(6): 489–527.
Schrödinger, E. 1926. Über das Verhältnis der Heisenberg-Born-Jordanschen Quantenmechanik zu der meinen. *Annalen der Physik* 384(8): 734–756.
Schrödinger, E. 1935a. Discussion of probability relations between separated systems. *Mathematical Proceedings of the Cambridge Philosophical Society* 31(4): 555–563.
Schrödinger, E. 1964. *My view of the world*. Trans. by C. Hastings. Cambridge/New York: Cambridge University Press.
Schrödinger, E. 1983 [1935]b. The present situation in quantum mechanics. A translation of Schrödinger's 'Cat Paradox' paper. In *Quantum theory and measurement*, eds. J.A. Wheeler, and W.H. Zurek. Trans. by J.D. Trimmer. Princeton: Princeton University Press.
Schrödinger, E. 2011 [1931]. Letter to sommerfeld. In *Eine Entdeckung von ganz außerordentlicher Tragweite*, ed. K. von Meyenn, 489–492. Berlin/Heidelberg: Springer.
Schrödinger, E. 2014 [1951–1954]. *Nature and the greeks and science and humanism*. Cambridge: Cambridge University Press. With a foreword by R. Penrose.
Schurz, G. 2008. Patterns of abduction. *Synthese* 164(2): 201–234.
Schurz, G. 2009. Meta-induction and social epistemology: Computer simulations of prediction games. *Episteme* 6(02): 200–220.
Schurz, G. 2014. *Philosophy of science. A unified approach*. New York/London: Routledge.
Schurz, G. 2015. *Wahrscheinlichkeit*. Berlin/Boston: De Gruyter.
Schurz, G., and A. Gebharter. 2016. Causality as a theoretical concept: Explanatory warrant and empirical content of the theory of causal nets. *Synthese* 193(4): 1073–1103.
Schwabl, F. 2006. *Statistical mechanics*, 2nd ed. Berlin/Heidelberg: Springer.
Schwabl, F. 2007. *Quantenmechanik (QM I). Eine Einführung*, 7th ed. Berlin/Heidelberg: Springer.
Schwabl, F. 2008. *Advanced quantum mechanics*, 4th ed. Trans. by R. Hilton, and A. Lahee. Berlin/Heidelberg: Springer.
Schwartz, M. 2014. *Quantum field theory and the standard model*. Cambridge/New York: Cambridge University Press.
Schwartz, S.P. 2006. General terms and mass terms. In *The Blackwell guide to the philosophy of language*, eds. M. Devitt, and R. Hanley, 274–287. Malden/Oxford: Blackwell.
Schwindt, J.-M. 2013. *Tutorium quantenmechanik*. Berlin/Heidelberg: Springer.
Sellars, W. 1953. Inference and meaning. *Mind* 62(247): 313–338.

Sellars, W. 1963. Philosophy and the scientific image of man. In *Science, perception and reality*, eds. W. Sellars, 1–40. Atascadero: Ridgeview Publishing Co.

Sexl, R.U., and H.K. Urbantke. 1992. *Relativity, groups, particles*. Wien/New York: Springer.

Sextus Empiricus. 2000. *Outlines of scepticism*. Edited by J. Annas, and J. Barnes. Cambrdige/New York: Cambridge University Press.

Shalm, L.K., E. Meyer-Scott, B.G. Christensen, P. Bierhorst, M.A. Wayne, M.J. Stevens, T. Gerrits, S. Glancy, D.R. Hamel, M.S. Allman, et al. 2015. Strong loophole-free test of local realism. *Physical Review Letters* 115(25): 250402.

Shankar, R. 1994. *Principles of quantum mechanics*, 2nd ed. New York: Plenum Publishers.

Shankar, R. 2010. Quantum notes. In *Physics 201b. Open Yale course* (lecture notes). http://oyc.yale.edu/sites/default/files/notes_quantum_7.pdf.

Shimony, A. 1984. Contextual hidden variables theories and Bell's inequalities. *The British Journal for the Philosophy of Science* 35(1): 25–45.

Shimony, A. 1990. An exposition of Bell's theorem. In *Sixty-two years of uncertainty. Historical, philosophical, and physical inquiries into the foundations of quantum mechanics*, ed. A.I. Miller, 33–44. New York/London: Plenum Press.

Shimony, A. 2009. Bell's theorem. In *The Stanford encyclopedia of philosophy*, ed. E.N. Zalta. The Metaphysics Research Lab, Center for the Study of Language and Information (CSLI), Stanford University. http://plato.stanford.edu/entries/bell-theorem/.

Smart, J.J.C. 1993. Laws of nature as a species of regularities. In *Ontology, causality and mind: Essays in honour of D. M. Armstrong*, eds. J. Bacon, K. Campbell, and L. Reinhardt, 152–168. Cambrdige/New York: Cambridge University Press.

Smeenk, C., and C. Wüthrich. 2010. Time travel and time machines. In *The Oxford handbook of time*, ed. C. Callender, 577–630. Oxford: Oxford University Press.

Soames, S. 2014. *Analytic philosophy in America and other historical and contemporary essays*. Princeton/Oxford: Princeton University Press.

Spekkens, R. 2014. Quasi-quantization: Classical statistical theories with an epistemic restriction. arXiv preprint arXiv:1409.5041.

Spekkens, R.W. 2005. Contextuality for preparations, transformations, and unsharp measurements. *Physical Review A* 75: 052108.

Spekkens, R.W. 2007. Evidence for the epistemic view of quantum states: A toy theory. *Physical Review A* 75: 032110.

Spekkens, R.W. 2008. Why the quantum? Insights from classical theories with a statistical restriction. PIRSA: 08020051.

Spekkens, R.W. 2012. Why I am not a ψ-ontologist. PIRSA: 12050021.

Spirtes, P., C. Glymour, and R. Scheines. 2000. *Causation, prediction, and search*. Cambridge: MIT Press.

Squires, E.J. 1992. Explicit collapse and superluminal signals. *Physics Letters A* 163(5–6): 356–358.

Srinivas, M.D. 1980. Collapse postulate for observables with continuous spectra. *Communications in Mathematical Physics* 71: 131–158.

Stairs, A. 2011. A loose and separate certainty: Caves, Fuchs and Schack on quantum probability one. *Studies in History and Philosophy of Science Part B: Studies in History and Philosophy of Modern Physics* 42(3): 158–166.

Stalnaker, R.C. 1968. A theory of conditionals. In *Ifs*, eds. W.L. Harper, R. Stalnaker, and G.A. Pearce, 41–55. Dordrecht: Springer.

Stapp, H. 2009. Orthodox interpretation of quantum mechanics. In *Compendium of quantum physics. Concepts, experiments, history and philosophy*, eds. D. Greenberger, K. Hentschel, and F. Weinert, 444–447. Berlin/Heidelberg: Springer.

Strapp, H.P. 1972. The copenhagen interpretation. *American Journal of Physics* 40: 1098–1116.

Strapp, H.P. 1975. Bell's theorem and world process. *Il Nuovo Cimento* 29(2): 270–276.

Straumann, N. 2004. *General relativity. With applications to astrophysics*. Berlin/Heidelberg: Springer.

Streater, R.F. 1988. Why should anyone want to axiomatize quantum field theory? In *Philosophical foundations of quantum field theory*, eds. H.R. Brown, and R. Harré, 137–148. Oxford: Clarendon Press.
Struyve, W. 2010. Pilot-wave theory and quantum fields. *Reports on Progress in Physics* 73(10): 106001.
Struyve, W., and H. Westman. 2007. A minimalist pilot-wave model for quantum electrodynamics. *Proceedings of the Royal Society of London A: Mathematical, Physical and Engineering Sciences* 463(2088): 3115–3129.
Suárez, M. 2004. Causal processes and propensities in quantum mechanics. *Theoria* 51(3): 271–300.
Summers, S.J., and R. Werner. 1985. The vacuum violates Bell's inequalities. *Physics Letters* 110A(5): 257–259.
Suppe, F. 2000. Understanding scientific theories: An assessment of developments, 1969–1998. *Philosophy of Science* 67: 102–115.
Suppes, P. 1998. Pragmatism in physics. In *The role of pragmatics in contemporary philosophy*, eds. P. Weingartner, G. Schurz, and G. Dorn, 236–252. Vienna: Hölder-Pichler-Tempsky.
Suppes, P. 2002. *Representation and invariance of scientific structures*. Stanford: CSLI Publications.
Sushkov, A.O., W.J. Kim, D.A.R. Dalvit, and S.K. Lamoreaux. 2011. Observation of the thermal Casimir force. *Nature* 7: 230–233.
Susskind, L., and A. Friedman. 2014. *Quantum mechanics: The theoretical minimum*. New York: Basic Books.
Susskind, L., and G. Hrabovsky. 2013. *The theoretical minimum. What you need to know to start doing physics*. New York: Basic Books.
Takeda, S., M. Zwierz, H.M. Wiseman, and A. Furusawa. 2015. Experimental proof of nonlocal wavefunction collapse for a single particle using homodyne measurements. *Nature Communications* 6: 6665.
Tan, S.M., D.F. Walls, and M.J. Collett. 1991. Nonlocality of a single photon. *Physical Review Letters* 66: 252–255.
Tarasov, V.E. 2008. *Quantum mechanics of non-hamiltonian and dissipative systems*. Amsterdam/Boston: Elsevier.
Tarski, A. 1944. The semantic conception of truth: and the foundations of semantics. *Philosophy and Phenomenological Research* 4(3): 341–376.
Tarski, A. 1956 [1935]. The concept of truth in formalized languages. In *Logic, semantics, mathematics*, ed. A. Tarski, 152–278. Trans. J.H. Woodger, and reprinted from *Studia philosophica*, vol. 1, 261–405. Oxford: Clarendon Press.
Taylor, E.F., and J.A. Wheeler. 1963. *Spacetime physics*. San Francisco: W. H. Freeman and Company.
Tegmark, M. 1993. Apparent wave function collapse caused by scattering. *Foundations of Physics Letters* 6(6): 571–590.
Tegmark, M. 1998. The interpretation of quantum mechanics: Many worlds or many words? *Fortschritte der Physik* 46(6–8): 855–862.
Teller, P. 1986. Relational holism and quantum mechanics. *British Journal for the Philosophy of Science* 37(1): 71–81.
Teller, P. 1988. Three problems of renormalization. In *Philosophical foundations of quantum field theory*, eds. H.R. Brown, and R. Harré, 73–89. Oxford: Clarendon Press.
Teller, P. 1990. Prolegomenon to a proper interpretation of quantum field theory. *Philosophy of Science* 57(4): 594–618.
Teller, P. 1995. *An interpretive introduction to quantum field theory*. Princeton: Princeton University Press.
Tetschl, G. 2000. *Mathematical methods in quantum mechanics. With applictaions to Schrödinger operators*. Providence: Amercian Mathematical Society.
Thaller, B. 1992. *The dirac equation*. Berlin/Heidelberg: Springer.
Thaller, B. 2005. *Advanced visual quantum mechanics*. Berlin/Heidelberg: Springer.

Timpson, C.G. 2013. *Quantum information theory and the foundations of quantum mechanics*. Oxford: Oxford University Press.

Tomonaga, S. 1974. *The story of spin*. Trans. by T. Oka. Chicago/London: The University of Chicago Press.

Tonomura, A., J. Endo, T. Matsuda, T. Kawasaki, and H. Ezawa. 1989. Demonstration of single-electron buildup of an interference pattern. *American Journal of Physics* 57(2): 117.

Tooley, M. 1977. The nature of laws. *Canadian Journal of Philosophy* 7(4): 667–698.

Tumulka, R. 2006a. A relativistic version of the Ghirardi-Rimini-Weber model. *Journal of Statistical Physics* 125(4): 821–840.

Tumulka, R. 2006b. On spontaneous wave function collapse and quantum field theory. *Proceedings of the Royal Society of London A: Mathematical, Physical and Engineering Sciences* 462(2070): 1897–1908.

Turbayne, C.M. 1955. Kant's refutation of dogmatic idealism. *The Philosophical Quarterly (1950-)* 5(20): 225–244.

Vaidman, L. 2012. Probability in the many-worlds interpretation of quantum mechanics. In *Probability in physics*, eds. Y. Ben-Menahem, and M. Hemmo, 299–311. Dordrecht: Springer.

Valentini, A. 1996. Pilot-wave theory of fields, gravitation and cosmology. In *Bohmian mechanics and quantum theory: An appraisal*, eds. J.T. Cushing, A. Fine, and S. Goldstein, 45–66. Berlin/Heidelberg: Springer.

van der Wal, C.H., A. Ter Haar, F. Wilhelm, R. Schouten, C. Harmans, T. Orlando, S. Lloyd, and J. Mooij. 2000. Quantum superposition of macroscopic persistent-current states. *Science* 290(5492): 773–777.

van Enk, S. 2007. A toy model for quantum mechanics. *Foundations of Physics* 37(10): 1447–1460.

van Fraassen, B. 1989. *Laws and symmetry*. Oxford/New York: Oxford University Press.

van Fraassen, B. 2002. *The empirical stance*. New Haven/London: Yale University Press.

van Fraassen, B.C. 1980. *The scientific image*. Oxford: Clarendon Press.

van Fraassen, B.C. 1982a. Rational belief and the common cause principle. In *What? Where? When? Why?*, ed. R. McLaughlin, 193–209. Dordrecht/Boston: D. Reidel Publishing Co.

van Fraassen, B.C. 1982b. The Charybdis of realism: Epistemological implications of Bell's inequality. *Synthese* 52(1): 25–38.

van Fraassen, B.C. 1991. *Quantum mechanics: An empiricist view*. Oxford/New York: Oxford University Press.

van Fraassen, C. 1984. Belief and the will. *The Journal of Philosophy* 81(5): 235–256.

van Loan, C.F. 2000. The ubiquitous Kronecker product. *Journal of Computational and Applied Mathematics* 123(1): 85–100.

Vaziri, A., G. Weihs, and A. Zeilinger. 2002. Experimental two-photon, three-dimensional entanglement for quantum communication. *Physical Review Letters* 89(24): 240401.

Vedral, V. 2005. *Modern foundations of quantum optics*. London: Imperial College Press.

Vervoort, L. 2013. Bell's theorem: Two neglected solutions. *Foundations of Physics* 43(6): 769–791.

von Meyenn, K. 2011. *Eine Entdeckung von ganz außerordentlicher Tragweite*. Berlin/Heidelberg: Springer.

von Mises, R. 1981 [1957]. *Mathematical theory of probability and statistics*, 2nd rev. ed. New York: Dover Publications Inc.

von Neumann, J. 1955 [1932]. *Mathematical foundations of quantum mechanics*. Trans. by R.T. Beyer. Princeton: Princeton University Press.

Vértesi, T., and N. Brunner. 2014. Disproving the Peres conjecture by showing Bell nonlocality from bound entanglement. *Nature Communications* 5: 5297(1–5).

Walker, J., D. Halliday, and R. Resnick. 2012. *Fundamentals of physics*, 10th ed. Hoboken: Wiley.

Wallace, D. 2002. Quantum probability and decision theory, revisited. arXiv preprint quant-ph/0211104.

Wallace, D. 2003. Everettian rationality: Defending Deutsch's approach to probability in the Everett interpretation. *Studies in History and Philosophy of Science Part B: Studies in History and Philosophy of Modern Physics* 34(3): 415–439.

Wallace, D. 2006. In defence of naiveté: The conceptual status of Lagrangian quantum field theory. *Synthese* 151(1): 33–80.
Wallace, D. 2008. Philosophy of quantum mechanics. In *The Ashgate companion to contemporary philosophy of phyiscs*, ed. D. Rickles, 16–98. Hants/Burlington: Ashgate.
Wallace, D. 2011. Taking particle physics seriously: A critique of the algebraic approach to quantum field theory. *Studies in History and Philosophy of Science Part B: Studies in History and Philosophy of Modern Physics* 42(2): 116–125.
Wallace, D. 2012. *The emergent multiverse. Quantum theory according to the Everett interpretation*. Oxford: Oxford University Press.
Wallace, D. 2014. Life and death in the tails of the GRW wave function. arXiv preprint arXiv:1407.4746.
Wallace, D. 2016. What is orthodox quantum mechanics? arXiv preprint arXiv:1604.05973.
Wallace, D., and C.G. Timpson. 2010. Quantum mechanics on spacetime I: Spacetime state realism. *The British Journal for the Philosophy of Science* 61(4): 697–727.
Wayne, A. 2002. A naive view of the quantum field. In *Ontological aspects of quantum field theory*, eds. M. Kuhlmann, H. Lyre, and A. Wayne, 127–134. New Jersey/London: World Scientific Publishing.
Weinberg, S. 1992. *Dreams of a final theory*. New York: Vintage Books.
Weinberg, S. 1995. *The quantum theory of fields*, vol. 1. Cambridge/New York: Cambridge University Press.
Weinberg, S. 1996. *The quantum theory of fields*, vol. 2. Cambridge/New York: Cambridge University Press.
Weinberg, S. 1997. What is quantum field theory, and what did we think it is? arXiv preprint hep-th/9702027.
Weinberg, S. 2013. *Lectures on quantum mechanics*. Cambridge/New York: Cambridge University Press.
Werner, R.F. 1989. Quantum states with Einstein-Podolsky-Rosen correlations admitting a hidden-variable model. *Physical Review A* 40(8): 4277–4281.
Werner, S.A., R. Colella, A.W. Overhauser, and C. Eagen. 1975. Observation of the phase shift of a neutron due to precession in a magnetic field. *Physical Review Letters* 35(16): 1053–1055.
Weyl, H. 1950. *The theory of groups and quantum mechanics*. Trans. from the 2nd (rev.) ed. by H.P. Robertson. New York: Dover Publications Inc.
Weyl, H. 1952. *Symmetry*. Princeton: Princeton University Press.
Whitaker, A. 1996. *Einstein, Bohr and the quantum dilemma*. Cambridge/New York: Cambridge University Press.
Whitaker, M. 2004. The EPR paper and Bohr's response: A re-assessment. *Foundations of Physics* 34(9): 1305–1340.
Wightman, A.S. 1956. Quantum field theory in terms of vacuum expectation values. *Physical Review* 101(2): 860–866.
Wigner, E. 1939. On unitary representations of the inhomogeneous Lorentz group. *Annals of Mathematics* 40(1): 149–204.
Wigner, E. 1983 [1961]. Remarks on the mind-body question. In *Quantum theory and measurement*, eds. J.A. Wheeler, and W.H. Zurek, 168–181. Oxford/New York: Oxford University Press. Reprinted from *The scientist speculates*, ed. I.J. Good, 284–302. London: Heinemann.
Wigner, E.P. 1970. On hidden variables and quantum mechanical probabilities. *American Journal of Physics* 38(8): 1005–1009.
Williamson, J. 2010. *In defence of objective Bayesianism*. Oxford/New York: Oxford University Press.
Williamson, T. 2002. *Knowledge and its limits*. Oxford/New York: Oxford University Press.
Wilson, K.G. 1971a. Renormalization group and critical phenomena. I. Renormalization group and the Kadanoff scaling picture. *Physical Review B* 4(9): 3174–3183.
Wilson, K.G. 1971b. Renormalization group and critical phenomena. II. Phase-space cell analysis of critical behavior. *Physical Review B* 4(9): 3184–3205.

Wiseman, H.M. 2014. The two Bell's theorems of John Bell. *Journal of Physics A: Mathematical and Theoretical* 47(42): 424001(31pp).

Wiseman, H.M., S.J. Jones, and A.C. Doherty. 2007. Steering, entanglement, nonlocality, and the Einstein-Podolsky-Rosen paradox. *Physical Review Letters* 98(14): 140402.

Wiseman, H.M., and G.J. Milburn. 2010. *Quantum measurement and control*. Cambridge/New York: Cambridge University Press.

Wittgenstein, L. 1968. *Philosophical investigations*. Trans. by G.E.M. Anscombe. Oxford: Blackwell.

Wittgenstein, L. 2002 [1961]. *Tractatus logico-philosophicus*. Trans. by D.F. Pears, and B.F. McGuinness. London/New York: Routledge.

Wittmann, B., S. Ramelow, F. Steinlechner, N.K. Langford, N. Brunner, H.M. Wiseman, R. Ursin, and A. Zeilinger. 2012. Loophole-free Einstein-Podolsky-Rosen experiment via quantum steering. *New Journal of Physics* 14: 053030(12pp).

Wong, C. 2013. *Introduction to mathematical physics: Methods & concepts*, 2nd ed. Oxford/New York: Oxford University Press.

Wood, C.J., and R.W. Spekkens. 2015. The lesson of causal discovery algorithms for quantum correlations: Causal explanations of Bell-inequality violations require fine-tuning. *New Journal of Physics* 17(3): 033002.

Woodward, J. 2003. *Making things happen: A theory of causal explanation*. Oxford/New York: Oxford University Press.

Wootters, W.K. 2006. Distinguishing unentangled states with an unentangled measurement. *International Journal of Quantum Information* 04(01): 219.

Wroński, L. 2014. *Reichenbach's paradise. Constructing the realm of probabilistic common 'causes'*. Warsaw/Berlin: De Gruyter Open.

Wüthrich, A. 2014. Local acausality. *Foundations of Physics* 44: 594–609.

Zeh, H. 2000. The problem of conscious observation in quantum mechanical description. *Foundations of Physics Letters* 13(3): 221–233.

Zeh, H. 2012. *Physik ohne Realität: Tiefsinn oder Wahnsinn?* Berlin/Heidelberg: Springer.

Zeh, H.D. 1970. On the interpretation of measurement in quantum theory. *Foundations of Physics* 1(1): 69–76.

Zurek, W.H. 1982. Environment-induced superselection rules. *Physical Review D* 26(8): 1862.

Zurek, W.H. 2003. Environment-assisted invariance, entanglement, and probabilities in quantum physics. *Physical Review Letters* 90(12): 120404.

Zurek, W.H. 2005. Probabilities from entanglement, Born's rule $p_k = |\psi_k|^2$ from envariance. *Physical Review A* 71(5): 052105.

Zurek, W.H. 2007. Relative states and the environment: Einselection, envariance, quantum darwinism, and the existential interpretation. arXiv preprint arXiv:0707.2832.

Zurek, W.H. 2009. Quantum darwinism. *Nature Physics* 5(3): 181–188.

Index

F. J. Boge, *Quantum Mechanics Between Ontology and Epistemology*, European Studies in Philosophy of Science 10,
https://doi.org/10.1007/978-3-319-95765-4

The manufacturer's authorised representative in the EU is Springer Nature Customer Service Centre GmbH, Europaplatz 3, 69115 Heidelberg, Germany. If you have any concerns regarding our products, please contact ProductSafety@springernature.com

Printed and bound by CPI Group (UK) Ltd, Croydon, CR0 4YY
05/10/2026
02228135-0016